Organic Farming

The Editors

Dr K. A. Gopinath holds a PhD in Agronomy from Indian Agricultural Research Institute, New Delhi, India. He joined Agricultural Research Service (ARS) in 2003 and worked for 5 years at ICAR-Vivekananda Institute of Hill Agriculture, Almora. In 2008, he joined ICAR-Central Research Institute for Dryland Agriculture, Hyderabad as Senior Scientist and is presently working as Principal Scientist (Agronomy). He is working on organic farming since 15 years. His other research interests include integrated farming systems and weed management particularly in rainfed production systems. His research contributions have been recognized at the national level and he is the recipient of Indian Council of Agricultural Research (ICAR) Team Award for Outstanding Multidisciplinary Research (2005-06), Lal Bahadur Shastri Young Scientist Award (2007-08) of ICAR and Young Agronomist Award (2007) of Indian Society of Agronomy. He has published 70 papers in reputed journals and is serving as an editorial board member of 'International Journal of Bio-resource and Stress Management' and was awarded 'The Best Editor Award' for outstanding contribution in the year 2013.

Dr A. V. Ramanjaneyulu has completed his B.Sc. (Ag.) and M.Sc. (Ag.) Agronomy from ANGRAU in 1998 and 2000, respectively and PhD in Agronomy from Indian Agricultural Research Institute (IARI), New Delhi, India in 2006. Presently, he is working as Senior Scientist (Agronomy) at RARS, Palem, Professor Jayashankar Telangana State Agricultural University, Telangana. He is a recipient of Young Agronomist Award of Indian Society of Agronomy. His specialization includes castor agronomy, dryland agriculture with main emphasis on farm pond technology, micro irrigation, and organic farming. He has reviewed several articles for many national and international journals. He has 40 research papers, two edited textbooks, 13 book chapters, seven technical bulletins and one practical manual to his credit.

Organic Farming

– Editors –

K. A. Gopinath

A. V. Ramanjaneyulu

2019

Daya Publishing House®

A Division of

Astral International Pvt. Ltd.

New Delhi – 110 002

ISBN: 9789389569155 (Int. Edition)

Published by : **Daya Publishing House®**
A Division of
Astral International Pvt. Ltd.
– ISO 9001:2015 Certified Company –
4736/23, Ansari Road, Darya Ganj
New Delhi-110 002
Ph. 011-43549197, 23278134
E-mail: info@astralint.com
Website: www.astralint.com

Digitally Printed at : **Replika Press Pvt. Ltd.**

Preface

Organic farming is native to this country. Indian farmers are known to have evolved nature-friendly farming systems and practices such as mixed farming, mixed cropping, resource recycling and crop rotation. Organic farming is a holistic system which promotes and enhances agro-ecosystem health, including biodiversity, biological cycles, and soil biological activity. The practice of farming described as 'organic' is being promoted under many names including natural farming, biological farming, *Agnihotra* farming, ecological farming, homa farming *etc.*

The Government of India is promoting organic farming through various schemes like National Project on Organic Farming (NPOF), National Horticulture Mission (NHM), Horticulture Mission for North East and Himalayan States (HMNEH), National Project on Management of Soil Health and Fertility (NPMSH and F), Paramparagat Krishi Vikas Yojana (PKVY), and Mission Organic Value Chain Development for North Eastern Region (MOVCD-NER). In addition, government agencies of many States are involved in either promotion of organic programmes or formulation of organic policies. The interest of several States in promoting organic farming indicates that organic agriculture is being viewed as a precursor to dynamic change for an otherwise stagnant agricultural sector.

There has been significant increase in the area under certified organic farming during the last 13 years. With less than 42,000 ha under certified organic farming during 2003-04, the area under organic farming grew by almost 25 fold, during the next 5 years, to 1.2 million ha during 2008-09. Presently (2016-17), about 1.44 million ha area is under certified organic cultivation and India ranks 9th in terms of total land under organic cultivation. Further, India has about 4.2 million ha under organic wild collection and non-agricultural areas. During 2016-17, India had the largest number of organic producers of about 0.84 million and accounted for 1.18 million tons of certified organic produce.

Despite the initiatives and rapid progress, apprehension about the economic viability and environmental and human health benefits of organic farming continue to bother agricultural researchers and policy makers. It is evident from limited short-term research findings that many crops respond better to organic management particularly after an initial conversion period of 2-3 years. Organic farming can significantly contribute to improving the livelihoods of smallholders particularly in rainfed, hill and mountain areas as it generates higher incomes and involves less risk. Organic farming also has the potential of increasing natural capital, such as improved water retention in the soil, reduced soil erosion, improved organic matter in soils, increased biodiversity and carbon sequestration.

This book contains 30 chapters from distinguished experts, practitioners and dedicated researchers. It provides a comprehensive coverage of diverse topics of organic farming including history, concept and principles, nutrient management, pest management, nutritional quality of organic produce, organic livestock production management, organic processing of meat, other forms of organic farming, economics, case studies, certification, market for organic produce and policy issues.

We hope that this book will be of immense use to researchers, teachers, students, policy makers, extension personnel and others interested in organic farming. We thank all the contributors of this book and are grateful for their valuable contributions.

K. A. Gopinath

A. V. Ramanjaneyulu

Contents

List of Contributors

Arun K. Sharma *ICAR-Central Arid Zone Research Institute, Jodhpur – 342003, Rajasthan*
A. Srinivas *Regional Agricultural Research Station, Professor Jayashankar Telangana State Agricultural University, Palem – 509215, Telangana*
A. Velmurugan *ICAR-Central Island Agricultural Research Institute, Port Blair – 744101, Andaman and Nicobar Islands*
A.K. Yadav *Former Director, National Centre of Organic Farming, Ghaziabad – 201002, Uttar Pradesh*
A.V. Ramanjaneyulu *Regional Agricultural Research Station, Professor Jayashankar Telangana State Agricultural University, Palem – 509215, Telangana*
Aryakrishak. Mohan Shankar Deshpande *Shri Samarth Agriculture Research Centre, Ajara, Kolhapur – 416505, Maharashtra*
B. Gangaiah *ICAR-Central Island Agricultural Research Institute, Port Blair – 744101, Andaman and Nicobar Islands*

B. Shivanna *College of Agriculture, University of Agricultural Sciences, Bengaluru – 500065, Karnataka*
B. L. Manjunath *ICAR-Indian Institute of Horticultural Research, Hessaraghatta, Bangalore – 560089, Karnataka*
Baswa Reddy *ICAR-NRC on Meat, Hyderabad – 500092, Telangana*
Bharat Mansata *C/o Earthcare Books, 10 Middleton Street, Kolkata – 700017, West Bengal*
C. S. Vaidya *H.P. University, Shimla – 171005, Himachal Pradesh*
Ch. Srinivasa Rao *ICAR-National Academy of Agricultural Research Management, Hyderabad – 500030, Telangana*
D. B. V. Ramana *ICAR-Central Research Institute for Dryland Agriculture, Hyderabad – 500059, Telangana*
Deeksha Joshi *ICAR-Indian Institute of Sugarcane Research, Lucknow – 226002, Uttar Pradesh*
Dibakar Mahanta *ICAR-Vivekananda Parvatiya Krishi Anusandhan Sansthan, Almora – 263601, Uttarakhand*
G. Chandra Sekhar *Centre for Sustainable Agriculture (CSA), Hyderabad – 500017, Telangana*
G. Rajashekar *Centre for Sustainable Agriculture (CSA), Hyderabad – 500017, Telangana*
G. Venkatesh *ICAR-Central Research Institute for Dryland Agriculture, Hyderabad – 500059, Telangana*

G. V. Ramanjaneyulu
Centre for Sustainable Agriculture (CSA), Hyderabad – 500017, Telangana

J. C. Bhatt
ICAR-Vivekananda Parvatiya Krishi Anusandhan Sansthan, Almora – 263601, Uttarakhand

J. S. Mishra
ICAR Research Complex for Eastern Region, Patna – 800 014, Bihar

K. Adilakshmi
ICAR-Central Research Institute for Dryland Agriculture, Hyderabad – 500059, Telangana

K. A. Gopinath
ICAR-Central Research Institute for Dryland Agriculture, Hyderabad – 500059, Telangana

K. C. Nataraj
Agricultural Research Station, ANGRAU, Anantapur – 515001, Andhra Pradesh

K. R. Kiran
ICAR-Central Island Agricultural Research Institute, Port Blair, Andaman and Nicobar Islands – 744101

Kewalanand
Department of Agronomy, College of Agriculture, G. B. Pant University of Agriculture and Technology, Pantnagar – 263145, Uttarakhand

Krishan Chandra
National Centre of Organic Farming, Ghaziabad – 201002, Uttar Pradesh

Lata Nain
ICAR-Indian Agricultural Research Institute, New Delhi – 110012

M. Jayalakshmi
KVK-Banavasi, Acharya NG Ranga Agricultural University, Banavasi – 518360, Andhra Pradesh

M. Muthukumar
ICAR-NRC on Meat, Hyderabad – 500092, Telangana

M. Srinivasa Rao *ICAR-Central Research Institute for Dryland Agriculture, Hyderabad – 500059, Telangana*
M. Venkata Ramana *Regional Agricultural Research Station, Professor Jayashankar Telangana State Agricultural University, Palem – 509215, Telangana*
Mahesh Chander *ICAR-Indian Veterinary Research Institute, Izatnagar – 243122, Uttar Pradesh*
Malleswari Sadhineni *Agricultural Research station, ANGRAU, Anantapur-515 001, Andhra Pradesh*
Minakshi Grover *ICAR-Indian Agricultural Research Institute, New Delhi – 110012*
Nina Osswald *Independent Researcher and Consultant, Hyderabad*
P. Spandana Bhatt *Krishi Vigyan Kendra, Professor Jayashankar Telangana State Agricultural University, Palem – 509215, Telangana*
P. S. Prabhamani *ICAR-Central Research Institute for Dryland Agriculture, Hyderabad – 500059, Telangana*
Prabhat Kumar Pankaj *ICAR-Central Research Institute for Dryland Agriculture, Hyderabad – 500059, Telangana*
R. A. Ram *ICAR-Central Institute for Subtropical Horticulture, Lucknow – 226101, Uttar Pradesh*
R. K. Pathak *ICAR-Central Institute for Subtropical Horticulture, Lucknow – 226101, Uttar Pradesh*
Ravi Koushik *Bhaikaka Krishi Kendra, Ravipura, Anand – 388440, Gujarat*

S. Girish Patil

ICAR-NRC on Meat, Hyderabad – 500092, Telangana

S. R. Kumar

ICAR-Indian Institute of Millets Research, Rajendranagar, Hyderabad – 500030, Telangana

Sachidananda Swain

ICAR-Central Island Agricultural Research Institute, Port Blair, Andaman and Nicobar Islands – 744101

Shilpanjali Deshpande Sarma

The Energy and Resources Institute, Indian Habitat Centre, Lodhi Road, New Delhi – 110003

Sreedevi Shankar

ICAR-Central Research Institute for Dryland Agriculture, Hyderabad – 500059, Telangana

Sunita T. Pandey

Department of Agronomy, College of Agriculture, G. B. Pant University of Agriculture and Technology, Pantnagar – 263145, Uttarakhand

T. Subramani

ICAR-Central Island Agricultural Research Institute, Port Blair, Andaman and Nicobar Islands – 744101

T. L. Neelima

Agril. Polytechnic, Regional Agricultural Research Station, Palem – 509215, Telangana

T. P. Swarnam

ICAR-Central Island Agricultural Research Institute, Port Blair, Andaman and Nicobar Islands – 744101

T. V. Prasad

ICAR-Central Research Institute for Dryland Agriculture, Hyderabad – 500059, Telangana

V. Vasudeva Rao

All India Network Project on Vertebrate Pest Management, PJTSAU, Rajendranagar, Hyderabad – 500030, Telangana

V. Visha Kumari

ICAR-Central Research Institute for Dryland Agriculture, Hyderabad – 500059, Telangana

V. P. Singh

ICAR-Indian Institute of Sugarcane Research, Lucknow – 226002, Uttar Pradesh

Zakir Hussain

Centre for Sustainable Agriculture (CSA), Hyderabad – 500017, Telangana

Chapter 1

Introduction

K.A. Gopinath

ICAR-Central Research Institute for Dryland Agriculture,
Hyderabad – 500 059, Telangana
E-mail: gopinath.icar@gmail.com

1. INTRODUCTION

Indian farmers are known to have evolved nature-friendly farming systems and practices such as mixed farming, mixed cropping, resource recycling and crop rotation. The first "scientific" approach to organic farming can be quoted back to the Vedas of the "Later Vedic Period", 1000 BC to 600 BC (Randhawa, 1986; Pereira, 1993). The essence is to live in partnership with, rather than exploit, nature. In this regard, the "*Vrikshayurveda*" (Science of plants), the "*Krishisastra*" (Science of agriculture) and the "*Mrugayurveda*" (Animal Science) are the main works (Mahale and Soree, 1999). However, organic movement owes its origin primarily to the work of Sir Albert Howard, often referred to as the father of modern organic agriculture, who believed that a shift from nature's methods of crop production to adoption of newer methods leads to the loss of soil fertility (Howard, 1943). From 1905 to 1924, he worked as an agricultural adviser in India, where he documented traditional Indian farming practices, and came to regard them as superior to his conventional agriculture science. His research and further development of these methods is recorded in his writings, notably in his book, *An Agricultural Testament*. It is this pioneering work which sowed the seeds of organic movement in India, placing greater emphasis on the use of compost and other organic sources of plant nutrients to the total exclusion of chemical fertilizers.

Outside India, the organic movement grew from the influential publications of workers such as Howard (1943), Balfour (1943) and Rodale (1946) which resulted from concern in the inter-war years over problems such as soil erosion and health (Scofield, 1986). The true roots of modern organic farming, however, lie earlier in the agriculture of the anthroposophic followers of Rudolf Steiner (1924). This movement, known as biodynamic agriculture, was developed from a series of eight lectures given by Steiner at the request of a group of German farmers concerned about the increasing degeneration they had noticed in seed-strains and in many cultivated plants. The origin of the term 'organic farming' also lay here for Lord Northbourne, who first used the term 'organic' in his forgotten classic 'Look to the Land' (1940) (Harwood, 1983), was a practitioner of biodynamic farming, and this provided him with the inspiration of his vision of the farm as a sustainable, ecologically stable, self-contained unit, biologically complete and balanced-a dynamic living organic whole. Therefore, the original use of the term 'organic' is clear. It refers first and foremost to a balanced, yet dynamic, living whole. It is also important to distinguish this meaning of 'organic' as it applies to a system of farming from the common misunderstanding that 'organic' specifically refers to the carbon based chemistry of the fertilizers that are often used in organic farming. In 1942, Rodale began publishing *'Organic Farming and Gardening'* magazine, which taught people how to grow better food by cultivating a healthier soil using natural techniques.

The practice of farming described as 'organic' is promoted under many names (Merrill, 1983). Some of the descriptive names which are, or have been, used in reference to it are:

- ✰ Fertility farming or agriculture (Howard, 1943; Turner, 1951)
- ✰ Humus farming or agriculture (Howard, 1943)
- ✰ Natural farming or agriculture (Cocannouer, 1958)
- ✰ Organic farming or agriculture (Rodale, 1942)
- ✰ Bio-Dynamic farming or agriculture (Pfeiffer, 1947)
- ✰ Biological farming or agriculture (Aubert, 1970)
- ✰ Ecological farming or agriculture (Walters, 1975)
- ✰ Holistic farming or agriculture (Hill, 1982)
- ✰ Alternative farming or agriculture (Boeringa, 1980)
- ✰ Sustainable farming or agriculture (Fisher, 1978)
- ✰ Scientific ecological farming or agriculture (Hyams, 1976)

In general, these names are used interchangeably, the choice being determined by personal preference as much as by the audience for whom it is used. The only real exception to the general synonymity of these names is Bio-Dynamic which is used by and in reference to the methods developed by the agricultural followers of Rudolf Steiner. It is perhaps of interest to note that organic, which is used fairly widely, continues to carry the heaviest load of negative connotations.

2. Definition of Organic Agriculture

The term "organic agriculture" refers to a process that uses methods respectful of the environment from the production stages through handling and processing. Organic production is not merely concerned with a product, but also with the whole system used to produce and deliver the product to the ultimate consumer.

Two main sources of general principles and requirements apply to organic agriculture at the international level. One is the Codex Alimentarius Guidelines for the Production, Processing, Labelling and Marketing of Organically Produced Foods (FAO, 2001). According to Codex, "Organic agriculture is a holistic production management system which promotes and enhances ecosystem health, including biological cycles and soil biological activity. Organic agriculture is based on minimising the use of external inputs, avoiding the use of synthetic fertilizers and pesticides. Organic agriculture practices cannot ensure that products are completely free of residues, due to general environmental pollution. However, methods are used to minimize pollution of air, soil and water. Organic food handlers, processors and retailers adhere to standards to maintain the integrity of organic agriculture products. The primary goal of organic agriculture is to optimize the health and productivity of interdependent communities of soil life, plants, animals and people."

The other is the International Federation of Organic Agriculture Movements (IFOAM) (IFOAM, 2002), a private-sector international body, with some 750 member organizations in over 100 countries. IFOAM defines and regularly reviews, in consultation with its members, the Basic Standards that shape the "organic" term. According to the IFOAM 2002 Basic Standards, "organic agriculture is a whole system approach based upon a set of processes resulting in a sustainable ecosystem, safe food, good nutrition, animal welfare and social justice. Organic production therefore is more than a system of production that includes or excludes certain inputs."

As per India's National Standards for Organic Production (NSOP), organic agriculture is a system of farm design and management to create an eco system, which can achieve sustainable productivity without the use of artificial external inputs such as chemicals, fertilizers and pesticides.

3. Concept of Organic Agriculture

The essential concept of organic farming is "Give back to nature", where the philosophy is to feed the soil rather them the crop to maintain the soil health. The concept of organic farming has been perceived differently by different people. To most of them, it implies the use of organic manures and natural methods of plant protection instead of using synthetic fertilizers and pesticides. Organic farming involves giving back to the nature what we take from it. Organic farming is a system of farming based on integral relationship of processes, inputs, farming and; animal and human community in harmony with nature.

In its most developed form, organic farming is both a philosophy and a system of agriculture. The objectives of environmental, social, and economic sustainability lie at the heart of organic farming. The term "organic" is not directly related to the type of inputs used, but refers to the concept of the farm as an organism, first proposed by Steiner (1924), in which all the components of farming – soil minerals, organic matter, microorganisms, insects, plants, animals, and humans – interact to create a coherent whole. Attention to the uniqueness of every operation is considered in relation to ecological, economic and ethical imperatives, with an awareness of local and global implications.

The basic concepts behind organic farming (Behera *et al.*, 2012) are:

- ☆ It concentrates on building up the biological fertility of the soil so that the crops take the nutrients they need from steady turnover within the soil nutrients produced in this way and are released in harmony with the need of the plants.
- ☆ Control of pests, diseases and weeds is achieved largely by the development of an ecological balance within the system and by the use of bio-pesticides and various cultural techniques such as crop rotation, mixed cropping and cultivation.
- ☆ Organic farmers recycle all wastes and manures within a farm, but the export of the products from the farm results in a steady drain of nutrients.
- ☆ Enhancement of the environment in such a way that wild life flourishes.

4. Principles of Organic Agriculture

The principles of organic agriculture are the roots from which organic agriculture grows and develops. They express the contribution that organic agriculture can make to the world, and a vision to improve all agriculture in a global context. Composed as inter-connected ethical principles to inspire the organic movement in its full diversity (Source: https://www.ifoam.bio/sites/default/files/poa_english_web.pdf).

4.1 Principle of Health

Organic agriculture should sustain and enhance the health of soil, plant, animal, human and planet as one and indivisible. This principle points out that the health of individuals and communities cannot be separated from the health of ecosystems - healthy soils produce healthy crops that foster the health of animals and people. Health is the wholeness and integrity of living systems. It is not simply the absence of illness, but the maintenance of physical, mental, social and ecological well-being. Immunity, resilience and regeneration are key characteristics of health. The role of organic agriculture, whether in farming, processing, distribution, or consumption, is to sustain and enhance the health of ecosystems and organisms from the smallest in the soil to human beings. In particular, organic agriculture is intended to produce high quality, nutritious food that contributes to preventive health care and well-being. In view of this it should avoid the use of fertilizers, pesticides, animal drugs and food additives that may have adverse health effects.

4.2 Principle of Ecology

Organic agriculture should be based on living ecological systems and cycles, work with them, emulate them and help sustain them. This principle roots organic agriculture within living ecological systems. It states that production is to be based on ecological processes, and recycling. Nourishment and well-being are achieved through the ecology of the specific production environment. For example, in the case of crops this is the living soil; for animals it is the farm ecosystem; for fish and marine organisms, the aquatic environment. Organic farming, pastoral and wild harvest systems should fit the cycles and ecological balances in nature. These cycles are universal but their operation is site-specific. Organic management must be adapted to local conditions, ecology, culture and scale. Inputs should be reduced by reuse, recycling and efficient management of materials and energy in order to maintain and improve environmental quality and conserve resources.

Organic agriculture should attain ecological balance through the design of farming systems, establishment of habitats and maintenance of genetic and agricultural diversity. Those who produce, process, trade, or consume organic products should protect and benefit the common environment including landscapes, climate, habitats, biodiversity, air and water.

4.3 Principle of Fairness

Organic agriculture should build on relationships that ensure fairness with regard to the common environment and life opportunities. Fairness is characterized by equity, respect, justice and stewardship of the shared world, both among people and in their relations to other living beings. This principle emphasizes that those involved in organic agriculture should conduct human relationships in a manner that ensures fairness at all levels and to all parties – farmers, workers, processors, distributors, traders and consumers. Organic agriculture should provide everyone involved with a good quality of life, and contribute to food sovereignty and reduction of poverty. It aims to produce a sufficient supply of good quality food and other products. This principle insists that animals should be provided with the conditions and opportunities of life that accord with their physiology, natural behavior and well-being.

Natural and environmental resources that are used for production and consumption should be managed in a way that is socially and ecologically just and should be held in trust for future generations. Fairness requires systems of production, distribution and trade that are open and equitable and account for real environmental and social costs.

4.4 Principle of Care

Organic agriculture should be managed in a precautionary and responsible manner to protect the health and well-being of current and future generations and the environment. Organic agriculture is a living and dynamic system that responds to internal and external demands and conditions. Practitioners of organic agriculture can enhance efficiency and increase productivity, but this should not be at the risk of jeopardizing health and well-being. Consequently, new technologies need to be assessed and existing methods reviewed. Given the incomplete understanding of ecosystems and agriculture, care must be taken. This principle states that precaution and responsibility are the key concerns in management, development and technology choices in organic agriculture. Science is necessary to ensure that organic agriculture is healthy, safe and ecologically sound. However, scientific knowledge alone is not sufficient. Practical experience, accumulated wisdom and traditional and indigenous knowledge offer valid solutions, tested by time.

Organic agriculture should prevent significant risks by adopting appropriate technologies and rejecting unpredictable ones, such as genetic engineering. Decisions should reflect the values and needs of all who might be affected, through transparent and participatory processes.

5. Need for Promotion of Organic Farming

As per the Agriculture Census 2010-11, the total number of operational holdings in the country has almost doubled from 71.01 million in 1970-71 to 138.35 million 2010-11. During the same period, the average size of operational holding has declined to 1.15 ha from 2.28 ha in 1970-71 (GOI, 2014). The small and marginal holdings taken together (below 2.0 ha) constituted 85.01 per cent with operated area of 44.58 per cent in 2010-11 against 83.29 per cent with 41.14 per cent of operated area in 2005-06.

Vast majority of rural households in developing countries lack the ecological resources or the financial means to shift to intensive modern agricultural practices. For many (especially small) farmers the purchase of manufactured fertilizers and pesticides is and will continue to be constrained by their high costs relative to output prices and risks or simply by unavailability (FAO, 2003). The approach of resource poor farmers' farming without the use of agrochemicals is often driven by poverty and lack of resources and characterized by low inputs and unsustainable practices, rather than by conscious adoption of organic farming techniques or acceptable levels of productivity (FAO, 2003). As productivity of traditional systems is often very low, organic agriculture could provide a solution to the food needs of poor farmers while relying on natural and human resources (Scialabba, 2000).

Thus many small farmers may not use agrochemicals because they cannot access them. This is often termed organic by default, but is often far from sustainable, as it uses few, if any, organic methods and is not based on an organic philosophy. Lack of access to agrochemicals may constrain conventional intensification, but lack of knowledge seems to be as much of a constraint on their ecological improvement. Systems that depend upon sustainable use of locally available natural resources and farmers' knowledge and labour are far more likely to meet the needs and aspirations of resource-poor farmers than those which require costly or scarce external inputs. In this respect organic farming is a technology that, to quote Mahatma Gandhiji, 'puts the last man first'.

Organic farming has the potential of increasing natural capital, such as improved water retention in the soil, improved water tables, reduced soil erosion, improved organic matter in soils, increased biodiversity and carbon sequestration (Scialabba and Hattam, 2002; Rasul and Thapa, 2004). Organic farming generally tends to reduce exposure to toxic chemicals, and risk of crop failure reduces outgoings and exposure to debt, builds environmental resilience and can build on local knowledge and socio-cultural practices. It is assumed that organic agriculture in developing countries facilitates women's participation, as it does not rely on purchased inputs and thus reduces the need for credit (Scialabba and Hattam, 2002; FAO, 2003).

Thus, for the organic movement there are both advantages and disadvantages of becoming pre-occupied with debates about yields, yield potential and potentials for feeding the world. On the one hand such a debate (and research to inform that debate) is necessary to legitimate organic approaches as a strategy for meeting food security. For this perspective, improving understanding of the effects of organic farming on yields (and the mechanisms employed) can be seen as a priority.

6. Development and State of Organic Farming in the World

According to the latest FiBL survey on certified organic agriculture worldwide, there were 57.8 million hectares of organic agricultural land in 2016, including in-conversion areas (Willer and Julia, 2018). The regions with the largest areas of organic agricultural land are Oceania (27.3 million ha, which is almost half the world's organic agricultural land) and Europe (13.5 million ha, 23 per cent). Latin America has 7.1 million ha (12 per cent) followed by Asia (4.9 million ha, 9 per cent), North America (3.1 million ha, 6 per cent), and Africa (1.8 million ha, 3 per cent). The countries with the most organic agricultural land are Australia (27.4 million ha), Argentina (3 million ha), and China (2.3 million ha).

Organic farmland increased by 7.5 million hectares or 15 percent in 2016. This is mainly because 5 million additional hectares were reported from Australia. However, many other countries reported an important increase and thus contributed to the global growth, such as China (42 per cent increase; over 0.67 million ha more) Uruguay (27 per cent increase; more than 0.3 million ha more), and India and Italy, both with an additional 0.3 million hectares. There has been an increase in organic agricultural land in all regions.

Apart from the organic agriculture land, there is organic land dedicated to other activities, most of which area for wild collection and beekeeping. Other areas include aquaculture, forests, and grazing areas on non-agricultural land. The areas of non- agricultural land constitute more than 39.7 million hectares.

There were at least 2.7 million organic producers in 2016.1 Forty percent of the world's organic producers are in Asia, followed by Africa (27 per cent) and Latin America (17 per cent). The countries with the most producers are India (835000), Uganda (210352), and Mexico (210000). There has been an increase in the number of producers of over 300000, or over 13 per cent, compared to 2015. A quarter of the world's organic agricultural land (14.3 million hectares) and more than 87 per cent (2.4 million) of the producers were in developing countries and emerging markets in 2016.

Organic food and drink sales have increased from less than 15 billion US dollars to almost 90 billion US dollars over two decades according to Ecovia Intelligence (Willer and Julia, 2018). Although the positive trend is likely to continue, there remain challenges. These include demand concentration (about 90 per cent of sales are in North America and Europe), proliferating standards, and the fact that the farmland growth is slowing in parts of Europe and North America. In 2016, the countries with the largest organic markets were the United States (38.9 billion euros), Germany (9.7 billion euros), and France (6.7 billion euros). The largest single market was the United States (47 per cent of the global market), followed by the European Union (30.7 billion euros, 37 per cent), and China (5.9 billion euros, 6 per cent). The highest per-capita consumption with more than 200 euros was found in Switzerland and Denmark. The highest organic market shares were reached in Denmark (9.7 per cent), Luxembourg (8.6 per cent), and Switzerland (8.4 per cent).

7. Development and State of Organic Farming in India

There has been significant increase in the area under certified organic farming during the last 13 years. With less than 42,000 ha under certified organic farming during 2003-04, the area under organic farming grew by almost 25 fold, during the next 5 years, to 1.2 million ha during 2008-09. Later, however, the area under certified organic farming fluctuated between 0.78-1.1 million ha. Presently (2016-17), about 1.44 million ha area is under certified organic cultivation (Table 1.1) and India ranks 9th in terms of total land under organic cultivation. Further, India has about 4.2 million ha under organic wild collection and non-agricultural areas. During 2016-17, India had the largest number of organic producers of about 0.84 million and accounted for 1.18 million tons of certified organic produce (Tables 1.2 and 1.3).

Table 1.1: State-wise Farm Area (including in conversion) under NPOP (ha) during 2016-17

State	*Area*	*State*	*Area*
Madhya Pradesh	464859	Telangana	9688
Maharashtra	224008	Meghalaya	9630
Rajasthan	151610	Tamil Nadu	5713
Odisha	92190	West Bengal	5176
Karnataka	81089	Haryana	5012
Sikkim	75218	Nagaland	4700
Gujarat	64241	Arunachal Pradesh	4011
Uttar Pradesh	56249	Punjab	1033
Uttarakhand	30907	Lakshadweep	896
Jharkhand	26814	Manipur	241
Kerala	24813	Mizoram	210
Assam	23870	Tripura	204
Jammu and Kashmir	22608	New Delhi	9
Andhra Pradesh	17684	Pondicherry	3
Goa	15762	Bihar	1
Chhattisgarh	12712	Andaman and Nicobar Islands	0
Himachal Pradesh	12377	**Total**	**1443538**

Source: apeda.gov.in.

The total volume of export during 2014-15 was 285608 tons (Table 1.4). The organic food export realization was around 298 million USD. Organic products are exported to European Union, US, Canada, Switzerland, Korea, Australia, New Zealand, South East Asian countries, Middle East, South Africa *etc.* Oilseeds (50 per cent) lead among the products exported followed by processed food products (25 per cent), cereals and millets (17 per cent), tea (2 per cent), pulses (2 per cent), spices (1 per cent), dry fruits (1 per cent), and others.

Table 1.2: State-wise Farm Production (including in conversion) under NPOP (tons) during 2016-17

State	*Production*	*State*	*Production*
Madhya Pradesh	391598	Andhra Pradesh	8373
Maharashtra	255935	Haryana	8193
Karnataka	164165	Goa	4759
Uttar Pradesh	88183	Chhattisgarh	3108
Rajasthan	64245	Himachal Pradesh	1837
Gujarat	41611	Meghalaya	1111
Assam	32396	Nagaland	878
Odisha	30661	Punjab	711
Uttaranchal	28161	Telangana	521
West Bengal	17977	Tripura	339
Kerala	14625	Sikkim	185
Tamil Nadu	11085	Arunachal Pradesh	38
Jammu and Kashmir	9408	Pondicherry	3
Total			**1180106**

Source: apeda.gov.in.

Table 1.3: Category-wise Production of Organic Products under NPOP (2014-15)

Category	*Production (tons)*
Sugar crops	338193
Oilseed crops	228414
Fibre crops	208931
Cereals	159500
Pulses	34717
Plantation crops	33930
Medicinal/herbal and aromatic plants	32663
Fruits	20219
Spices and condiments	18176
Vegetables	10824
Dry fruits	7348
Ornamental plants and flowers	2358
Others	269
Tuber crops	166
Fodder crops	44

Source: apeda.gov.in.

Table 1.4: Category-wise Export of Organic Products under NPOP (2014-15)

Category	*Quantity (tons)*
Oilseeds	160559
Cereals and millets	63622
Processed foods	23626
Sugar crops	19450
Tea	5488
Pulses	2547
Dry fruits	2417
Spices and condiments	2403
Medicinal, aromatic and herbal products	1223
Coffee	1214
Others	1165
Essential and aromatic oils	866
Fibre crops	397
Tuber crops	139
Edible oils	130
Fruits	124
Vegetables	109
Ornamental plants and flowers	78
Plantation crops other than tea and coffee	32
Honey	18
Total	**285608**

Source: apeda.gov.in.

8. Organic Farm Management

Organic farm management is focused on the whole farm system and its interactions with climate, environment, social, and economic conditions, rather than considering the farm as comprised of individual enterprises. Crop production in organic systems is characterized by an increased diversity of cropping patterns in time and in space compared to intensive conventional crop production systems. The major objectives of such diversity are to operate a closed system for nutrients and organic matter and maintain crop health.

The National Programme for Organic Production (NPOP) provides for Standards for organic production, systems, criteria and procedure for accreditation of Certification Bodies, the National (India Organic) Logo and the regulations governing its use. The standards and procedures have been formulated in harmony with other International Standards regulating import and export of organic products.

8.1 Conversion Period

Among the requirements for organic agricultural production, the conversion period from conventional to organic farming is of particular significance. The conversion period is basically the time between the start of organic management and the certification of crops or animal husbandry as organic. It is the time taken to neutralise chemical residues, if any, left behind in the soil by practised agricultural techniques. This intermediate phase also allows farmers a period of time to meet requirements for organic farming in maintaining and increasing the fertility of the organic production areas.

Plant products produced can be certified when the national standards requirements have been met during a conversion period of at least two years before sowing for annual crops or at least three years in the case of perennial crops other than grassland. The certification program may allow plant products to be sold as "produce of organic agriculture in process of conversion" or a similar description during the conversion period of the farm. However, the certification authority has the discretion of extending or reducing the duration of the conversion period depending on past use of the land and the ecological situation.

The conversion period may turn out to be a difficult phase for the farmer owing to several direct and indirect costs involved in the process and associated yield reductions. The yield behavior of crops during conversion period largely depend on the type of crops grown, initial soil fertility levels and nutrient requirement of crops, insect-pests and disease pressure, and on the agricultural practices followed before conversion. The crop yields will be generally lower compared to those under conventional practices, although they may turn out to be equal or even higher after the conversion period. This is because when the application of chemical fertilizers and pesticides stopped suddenly, the soil may not have healthy diversity of soil microorganisms, natural enemies and other helpful living organisms. It takes time for the soil microorganisms to re-establish the equilibrium that had earlier been disrupted by conventional agriculture. Hence, pest problems are also expected to be higher during the conversion period. Furthermore, it is difficult to achieve a good level of balanced soil fertility with the available organic manures.

8.2 Suitable Crop Varieties

Crop improvement efforts during the second half of 20th century has been focused almost entirely on breeding for conventional farming systems. Conventional varieties have been developed with the aim of combining high productivity and standardized product quality under high-input conditions. The consequence is that organic farmers have to depend on varieties bred for cultivation with high external inputs. The varieties often perform differently in different environments due to genotype-environment interactions. The amount of stress on the crop is expected to be more under organic management than under input-intensive conventional farming. The stress may be in the form of available nutrients, weed pressure, insect-pests and diseases. Varieties are needed that can respond to the sometimes sub-optimum conditions (typical of organic farming conditions). Therefore, choice of variety is more critical in organic situations than for conventional crops where problems can be solved at a later stage by application of pesticides or mineral fertilizers.

8.3 Nutrient Management

The aim of nutrient management within organic farming systems is to work, as far as possible within a closed system. Organic farming practices aim to maximize the efficiency of nutrient cycling within the farm ecosystem (*e.g.* avoidance of losses from manure heaps, optimizing mineralization of soil organic N), and maximizing the fixation of atmospheric N by legumes. Organic standards minimize or eliminate use of synthetic or manufactured inputs and encourage maximum use of local natural resources. The inputs allowed as fertilizers in organic production are generally lower and more variable in nutrient content and plant-availability than commercial fertilizers. Therefore, they have to be applied at high rates to meet all the crop needs. Furthermore, there is greater likelihood of supplying some nutrients at excess rates, which may lead to increased risk of loss and negative environmental impact. There are a number of organic sources of nutrition and among them green manuring, composting, biofertilizers, vermi-compost and biodynamics are important.

As per the NSOP, biodegradable material of microbial, plant or animal origin produced on organic farms should form the basis of the fertilisation programme. Non-synthetic mineral fertilisers and brought in fertilisers of biological origin should be regarded as supplementary and not a replacement for nutrient recycling.

8.4 Pest Management

Pest control strategies under organic farming are largely preventive rather than reactive. The balance of cropped and uncropped areas, crop species, variety, temporal and spatial pattern of the crop rotation seek to maintain a diverse population of pests and their natural enemies and disrupt the life cycle of pest species. A combination of appropriate cultural techniques such as cultivation, crop rotation, smother crops, trap crops, irrigation, solarization *etc.* and the use of biological control agents are used to manage insect-pests, weeds and diseases.

8.5 Certification Bodies

Certification bodies are the agencies accredited by the National Accreditation Body under NPOP for certifying organic products. The bodies shall certify organic products as per the scope of accreditation approved by the NAB. It is the body responsible for inspection and certification of the operators as per NPOP standards. Certification shall refer to the procedure by which the accredited certification body by way of a scope certificate assures that the production or processing system of the operator has been methodically assessed and conforms to the specified requirements as envisaged in the NPOP. In India, there are 28 accredited certification bodies across different states (Table 1.5).

9. Research on Organic Farming

The Government of India is promoting organic farming through various schemes like National Project on Organic Farming (NPOF), National Horticulture Mission (NHM), Horticulture Mission for North East and Himalayan States (HMNEH), National Project on Management of Soil Health and Fertility (NPMSH and F), Paramparagat Krishi Vikas Yojana (PKVY) and Mission Organic Value Chain

Development for North Eastern Region (MOVCD-NER). In addition, government agencies of many States are involved in either promotion of organic programmes or formulation of organic policies. The interest of several States in promoting organic farming indicates that organic agriculture is being viewed as a precursor to dynamic change for an otherwise stagnant agricultural sector.

Table 1.5: List of Accredited Certification Bodies

1	Aditi Organic Certifications Pvt. Ltd.
2	APOF Organic Certification Agency (AOCA)
3	Bureau Veritas Certification India Pvt. Ltd.
4	Chhattisgarh Certification Society, India (CGCERT)
5	Control Union Certifications
6	Ecocert India Pvt. Ltd.
7	Fair Cert Certification Services Pvt. Ltd.
8	Food Cert India Pvt. Ltd.
9	Global Certification Society
10	GreenCert Biosolutions Pvt. Ltd
11	Gujarat Organic Products Certification Agency (GOPCA)
12	IMO Control Pvt. Ltd.
13	Indian Organic Certification Agency (INDOCERT)
14	Indian Society for Certification of Organic Products (ISCOP)
15	Intertek India Pvt. Ltd.
16	Karnataka State Organic Certification Agency
17	LACON Quality Certification Pvt. Ltd.
18	Madhya Pradesh State Organic Certification Agency
19	Natural Organic Certification Agro Pvt. Ltd.
20	Odisha State Organic Certification Agency (OSOCA)
21	One Cert Asia Agri. Certification Pvt. Ltd.
22	Rajasthan Organic Certification Agency (ROCA)
23	SGS India (Pvt.) Ltd.
24	Sikkim State Organic Certification Agency (SSOCA)
25	Tamil Nadu Organic Certification Department (TNOCD)
26	Uttar Pradesh State Organic Certification Agency
27	Uttarakhand State Organic Certification Agency (USOCA)
28	Vedic Organic Certification Agency (VOCA)

Source: apeda.gov.in.

Despite the initiatives and rapid progress, apprehension about the economic viability and environmental and human health benefits of organic farming continue to bother agricultural researchers and policy makers. Indian Council of Agricultural Research (ICAR) initiated a network project on organic farming in 2003-04 under the Project Directorate for Farming Systems Research (PDFSR),

Modipuram to study productivity, profitability, sustainability, quality and input-use-efficiencies of different crops and cropping systems under organic farming in different agro-ecological regions. Thirteen research stations from all over the country are participating in the project. Similarly, The All India Network Project on Biofertilizers (AINPB) under the Indian Institute of Soil Science, Bhopal is mandated to formulate and testing of mixed biofertilizers for diverse cropping systems and to improve biofertilizer technology with particular reference to quality, carriers, consortia and delivery systems. The University of Agricultural Sciences, Dharwad (Karnataka) established the Institute of Organic Farming in 2006 with an objective to cater to the needs of organic farmers and other stakeholders in the State. Similarly, in 2006, the University of Agricultural Sciences, Bangalore (Karnataka) established the Organic Farming Research Center at Shimoga, and Organic Farming Research Station at Naganahalli (near Mysore) to scientifically validate and analyze the claims of organic agricultural produce and help popularize it among farmers. The Department of Organic Agriculture was established by the CSKHPKV, Palampur (Himachal Pradesh) in 2009 with a specific mandate to promote organic farming in the State. In addition, many ICAR Institutes and State Agricultural Universities have initiated a number of research projects/centers to develop location-specific organic farming modules.

It is evident from limited short-term research findings that many crops respond better to organic management particularly after an initial conversion period of 2-3 years. Organic farming can significantly contribute to improving the livelihoods of smallholders as it generates higher incomes and involves less risk. Although many ICAR institutes and state agricultural universities have initiated research on organic farming, the literature is dominated by comparisons of organic and conventional agriculture. There are several researchable issues and more are likely to emerge as researchers begin to explore it. Some of these include:

- ✰ Delineation of the potential areas/zones including hill and tribal areas for organic farming by identifying contiguous blocks of areas with little or no chemical input use and where productivity can be enhanced by using permitted inputs to enable group certification to farmers.
- ✰ Carry out a country wide survey/inventorisation of areas in arid, semi-arid and dry sub humid regions about the level of chemical input use, productivity in selected commodities which have potential to fetch price premiums in international markets.
- ✰ Survey, documentation and critical evaluation of indigenous technological knowledge on organic farming.
- ✰ Inter-disciplinary and location-specific research has to be taken up for development of package of practices for organic farming. Organic production packages will be more location-specific than inorganic package of practices as the input use depends largely on locally available resources.
- ✰ Identification of suitable varieties from existing pool for optimum productivity, quality and pest resistance

- ☆ Understand the nutrient release patterns of different organic sources in combination and alone
- ☆ Development of cost effective technologies for on-farm organic manure production as well as large-scale production of compost from domestic, agricultural, and industrial wastes.
- ☆ Development of appropriate machines, tools and machine/bullock driven devices for organic farming operations such as manure spreader, mechanical weeding machines, seed drills for multi-crop sowing and planting *etc.*
- ☆ Generation of adequate scientific information on the yield, quality, economics and post-harvest aspects of various crops under different management levels and agro-climatic conditions.
- ☆ Study the role organic agriculture in mitigating the climate change and the potential of organic farming to adapt to climate change
- ☆ Developing methods which link production systems to product quality and onwards into both livestock and human health and well-being.

10. Constraints in Scaling Up

Besides the well known limitation of the availability of FYM and other organic forms of nutrients in desired quantities as highlighted by Chhonkar (2003), water availability also is an important constraint for adoption of organic farming, particularly in arid and dry semi-arid tropics. Absence of surplus rainwater for harvesting and long periods of low soil moisture can limit the overall biomass production for recycling, green leaf manuring and on-farm composting. Application of 5-10 t FYM/ha is required in most crops to produce on par yields with recommended chemical fertilizers. Such level of inputs use can only be possible in limited areas for specific crops. However, biomass production during the off season (without competition with the kharif crop) through a legume cover cropping and its incorporation in the soil can be another strategy to overcome the limitation of organic matter availability (Venkateswarlu *et al.*, 2007). Since the overall biomass production is linked to rainfall, using crop biomass either by composting or through recycling should be a major strategy in relatively high rainfall receiving areas in moist semi-arid and dry subhumid regions (750 - 1200 mm) while the dry semi-arid and arid areas (300 - 750 mm) may depend on use of FYM as the principal source, since livestock is a strong component in these regions.

Considering the low organic matter and fertility status of Indian soils, the yield decline during conversion period could be sharp in the absence of external inputs. In view of the limited biomass and organic resources available for use in rainfed areas, organic production either for domestic or export markets should be encouraged in highly selected areas and commodities. This strategy alone can sustain the production and marketing of organic food on a long term basis.

11. Focus on Niche Areas and Commodities

The inherent advantages of organic farming should be capitalized by encouraging organic farming in highly selected areas and commodities with edapho-climatic and price advantages. The primary focus should be on commodities which have export potential with price premiums. Having selected the commodities, a two pronged strategy need to be followed for popularizing organic farming. Firstly, areas where relatively low or no inputs are used and which are climatically well endowed with reasonable productivity levels may be identified. Farmers in contiguous areas can be encouraged to adopt farm management practices that are required in organic production. Yield levels in such areas may be further enhanced by using permitted inputs. A commodity and area oriented group certification system may be possible with the support of the Government agencies and service providers. As a second strategy, areas where farmers are already realizing higher yields but using chemical inputs need to be identified and a systematic conversion protocols need to be introduced based on research data. Besides training and capacity building of farmers on production of inputs required for organic farming at farm level, the availability of other bio inputs like biofertilisers and bio pesticides need to be increased in selected areas by encouraging the setting up of bio resource centers. Forward linkages with certifying agencies and markets will be essential to sustain the initiative (Venkateswarlu, 2008).

REFERENCES

Aubert C. 1970. *L'Agriculture Biologique; une Agriculture pour la Sante et l'Epanouissement de l'Homme*. Le Courrier du Livre, Paris.

Balfour EB. 1943. *The Living Soil*. Faber and Faber, London.

Behera KK, Alam A, Vats S, Sharma HP and Sharma V. 2012. Organic farming history and techniques. In: *Agroecology and Strategies for Climate Change* (Lichtfouse E. Ed), Sustainable Agriculture Reviews, vol 8. Springer, Dordrecht. pp. 287-328.

Boeringa R. 1980. *Alternative Methods of Agriculture*. Elsevier Scientific Publishing Co., Amsterdam.

Chhonkar, P.K. 2003. Organic Farming: Science and Belief, Journal of the Indian Society of Soil Science, 51(4): 365-377.

Cocannouer JA. 1958. *Water and the Cycle of Life*. Devin-Adair, New York.

FAO. 2003. Strengthening Coherence in FAO's Initiatives to Fight Hunger Conference. Thirty second Session, Rome, 29 November - 10 December 2003.

Fisher C. 1978. Introduction to the conference theme "Towards a Sustainable Agriculture". In: *Towards a Sustainable Agriculture*. 1st International Conference, Sissach, Switzerland. (Besson JM and Vogtmann H, Eds.), Verlag Wirz AG, Aarau. pp. 11-17.

GOI. 2014. Agriculture Census 2010-11. All India report on number and area of operational holdings, Ministry of Agriculture, Government of India. 87 p.

Harwood RR. 1983. International overview of regenerative agriculture. In: *Proceedings of Workshop on Resource-Efficient Farming Methods for Tanzania*, May 16-20, 1983 (Semoka JMR, Shao FM, Harwood RR and Liebhardt WC, Eds.). Rodale Press, Emmaus, PA.

Hill SB. 1982. Steps to a Holistic Ecological Food System. In: *Basic Techniques in Ecological Farming*. 2nd International Conference held by the IFOAM (Hill SB, Ed.) Montreal. October 1-5, 1978. Birkhauser Verlag, Basel. pp. 15-21.

Howard A. 1943. *An Agricultural Testament*. The Oxford University Press, Oxford.

Hyams E. 1976. *Soil and Civilization*. Harper Colophon Books, New York.

IFOAM. 2002. Basic Standards for Organic Production and Processing, International Federation of Organic Agricultural Movements, Bonn, Germany.

Mahale P and Soree H. 1999. Cosmovisions in health and agriculture in India In: *Food for Thought: Ancient visions and new experiments of rural people* (Haverkort B and Hiemstra W, Eds.), Books for Change (India), Bangalore. pp. 33-42.

Merrill MC. 1983. Eco-agriculture: A review of its history and philosophy. Biological Agriculture and Horticulture, 1: 181-210.

Pereira W. 1993. *Tending the Earth*. Earthcare Books, Mumbai. 315p.

Pfeiffer E. 1947. *Soil Fertility, Renewal and Preservation, Bio-dynamic Farming and Gardening*, Lanthorn Press, East Grinstead, UK.

Randhawa MS. 1986. *A History of Agriculture in India 1980-1986*, Vol. I-IV, Indian Council of Agricultural Research, New Delhi.

Rasul G and Thapa GB. 2004. Sustainability of ecological and conventional agricultural systems in Bangladesh: An assessment based on environmental, economic and social perspectives. Agricultural Systems, 79: 327-351.

Rodale JI. 1942. An introduction to organic farming. *Organic Farming and Gardening*. May 1942.

Scialabba N and Hattam H. 2002. Organic agriculture, Environment, and Food Security. Environment and Natural Resources Series No. 4. FAO, Rome.

Scialabba N. 2000. Factors influencing organic agriculture policies with a focus on developing countries. Proceeding from the 13th IFOAM Scientific Conference, Basel, Switzerland.

Scofield AM. 1986. Organic farming: The origin of the name. Biological Agriculture and Horticulture, 4: 1-5.

Steiner R. 1924. *In* "Agriculture: A Course of Eight Lectures". Rudolph Steiner Press/Biodynamic Agriculture Association, London.

Turner N. 1951. *Fertility farming*. Faber and Faber, 24 Russell Square, London. 264p.

Venkateswarlu, B. 2008. Organic Farming in Rainfed Agriculture: Prospects and Limitations. In: *Organic Farming in Rainfed Agriculture: Opportunities and Constraints*. (B. Ventateswarlu, S.S. Balloli and Y.S. Ramakrishna, eds.). Central Research Institute for Dryland Agriculture, Hyderabad. pp. 7-11.

Venkateswarlu B, Srinivasa Rao Ch, Ramesh G, Venkateswarlu S, and Katyal JC. 2007. Effect of long-term incorporation of legume biomass on soil organic carbon, microbial biomass, nutrient build-up and grain yields of sorghum/ sunflower under rainfed conditions. Soil Use and Management, 23: 100-107.

Walters C. 1975. *The case for Eco-Agriculture.* Acres, USA, Raytown, Mo.

Chapter 2

History of Organic Farming: Return to Right

Arun K. Sharma

ICAR-Central Arid Zone Research Institute,
Jodhpur – 342 003, Rajasthan
E-mail: arun.k_sharma@yahoo.co.in

1. INTRODUCTION

History of agriculture is as old as human civilization. It passed millenniums to reach this stage of human development and same is true with agriculture. As agriculture is basically a nature's process that gradually understand by the humans and they mimicked it for satisfying the needs. Till 15th century, agriculture was the main source of income and given top priority all over the world and in India it was considered as the sacred profession due to lifeline of society. However, the story of agriculture started to change with the advent of industrial development in Europe and later on other parts of the world that changed the attitude towards agriculture production thinking that the production is possible with the use of synthetic chemicals and improved seeds rather with the harmony of nature. The system was very well fitted in the industrialization process as so many chemicals have been produced, marketed and used. Therefore, it was given a good name green revolution in tune of industrial revolution. The system gave miraculous results for a short time till the buffer capacity of soil and ecosystem as a whole exhausted in later part of twentieth century. Also, the industry became as main source of income, and therefore, slowly agriculture started getting lesser attention. The declining trend in productivity of soil and agriculture system as a whole in second half of 20th century as a result of this approach is visible around the globe. Now it is being realized all over the world to revisit the agriculture production system that was based on harmony with nature and prevailed for millenniums and fed the world without any negative consequences. This nature friendly system was studied by some of the agriculturists in middle of twentieth century and given the name "Organic farming". Rather than describing chronology that is monotonous and to make a logical analysis of the past, the history of this system can be divided into three phases namely Revisit, Recognize and Reshape the present form of organic farming.

2. Revisiting Phase

2.1 Organic Farming in the World

As mentioned earlier, all farming before the advent of chemicals was organic only and it is really worth to study the shifting from organic to chemical and then after sometimes a general thinking all over the world to return on the right methods of farming *i.e.* the nature's friendly organic farming. For this analysis the period was divided into two parts *viz.*, Development of concept and technology to use chemicals in farming and Consequences that compelled to revisit traditional system.

2.1.1 Development of Concept and Technology to Use Chemicals in Farming

In 1840, Justus von Liebig developed a theory on mineral plant nutrition. He believed that mineral salts were the only nutrients that plants needed and they could completely replace manure. He was a German chemist who made major contributions to agricultural and biological chemistry, and worked on the organization of organic chemistry. He is known as the "father of the fertilizer industry" for his discovery of nitrogen as an essential plant nutrient, and his formulation of the Law of the Minimum which described the effect of individual nutrients on crops. In 1910, chemists Fritz Haber and Carl Bosch developed an ammonia synthesis process, making use of nitrogen from the atmosphere. This form of ammonia already had been used to manufacture explosives, and it was made available for fertilizer in farming. With these two major breakthrough inventions the use of chemicals started in Farming.

2.1.2 Consequences that Compel to Revisit Ancient System

Within 20 years of use of chemicals in farming, the ill effects were seen by many of the visionary farmers and environmentalists all over the world and they started thinking to go back or revisit the traditional nature's friendly farming system that was later named as organic farming/agriculture. The chronology of development of organic philosophy and movement is detailed below:

- ☆ **1905 to 1924 -** Organic farming began more or less simultaneously in Central Europe and some parts of India. The British botanist Sir Albert Howard, often referred to as the father of modern organic farming, while working as an agricultural adviser in Pusa, Bengal, documented traditional Indian farming practices, and considered them as superior to his conventional farming science.
- ☆ **1909 -** American agronomist F.H. King toured China, Korea, and Japan, studying traditional fertilization, tillage, and general farming practices. He published his findings in "Permanent Agriculture: Farmers of Forty Centuries" (1911, Courier Dover Publications, ISBN 0-486-43609-8). King visualized a "world movement for the introduction of new and improved

methods" of farming and in later years his book became an important organic reference.

- ☆ **1924** - In Germany, Rudolf Steiner's publishes his "Spiritual Foundations for the Renewal of Farming" which might be the first comprehensive organic farming system and led to the development of "biodynamic agriculture". His book began with a series of lectures. Steiner presented at a farm in Koberwitz (now in Poland) during 1924. Steiner emphasized the farmer's role in guiding and balancing the interaction of the animals, plants and soil. Healthy animals depend upon healthy plants (for their food), healthy plants depend upon healthy soil, and healthy soil depends upon healthy animals (for the manure).
- ☆ **1939** - The first use of the term "organic farming" was by Lord Northbourne. The term derived from his concept of "the farm as organism", which he expounded in his book, "Look to the Land" (1940). He described a holistic, ecologically balanced approach to farming. Northbourne had written article "chemical farming versus organic farming".
- ☆ **1939** - Influenced by Sir Albert Howard's work, Lady Eve Balfour launched the Haughley Experiment on farmland in England. It was the first scientific, side-by-side comparison of organic and conventional farming.
- ☆ **1940** - Sir Albert Howard's book, "An Agricultural Testament", was influential in promoting organic techniques, and his 1947 book "The Soil and Health, A Study of Organic Agriculture" adopted Northbourne's terminology and was the first book to include "organic" agriculture or farming in its title (Howard, 1947).
- ☆ **1940** - In Japan, Masanobu Fukuoka, a microbiologist working in soil science and plant pathology, begun to doubt the modern agricultural movement. In the early 1940s, he quit his job as a research scientist, returned to his family's farm, and devoted the next 30 years to developing a radical no-till organic method for growing grain, now known as "Fukuoka farming".
- ☆ **1943** - Lady Eve Balfour published "The Living Soil", based on the initial findings of the Haughley Experiment. It led to the formation of a key international organic advocacy group, the Soil Association.

2.2 Organic Farming in India

The story of Organic farming in India can be divided into three parts (i) Before the advent of Europeans (ii) During the Europeans' ruling (iii) After independence

2.2.1 Before the Advent of Europeans

The Indus civilization had great economy that was mainly based on agriculture which is evident through trade and transportation. During the period of Mahajanpada and Maurya Empire (600-300 BC), trade flourished and economy was very good. The next 1500 years India was known as the largest economy of the world that dates back to the 1st and the 17th centuries AD. Before the Europeans,

India was ruled for over 700 years by the Mughal emperors and had some of the most powerful rulers who had good knowledge of agriculture, trade and commerce. India was better economically during these times and saw some of the most golden years in terms of economy.

Our ancient literature between 6,000 BCE and 1,000 ACE contains a lot of information on good agriculture. This includes the four Vedas, the nine Brahmanas, the Aranyakas, Sutra literature, the Sushruta Samhita, the Charaka Samhita, the Upanishads, the epics Ramayana and Mahabharata, the 18 Puranas, and texts such as the Krishi-Parasharas, Kautilya's Arthashastra, the Manusmriti, Varahamihira's Brhat Samhita, the Amarkosha, the Kashyapiya-Krishisuktiand Surapala's Vrikshayurveda. Kautilya's Arthashastra deals with the agriculture of his time; Vrikshayurveda provides information on how to combat plant problems through various traditional practices and utilizing available resources (Sadhle, 1996); Even in the poems of Ghagh one comes across descriptions of agro-management, timing and forecasting of weather, and crop yields.

Traditional farming systems appear to be complex and advanced as they exhibit important elements of sustainability: for instance, they are well adapted to the particular environment, rely on local resources, are decentralized, and, overall, tend to conserve the natural resource base. The ancient texts referred to contain information on farm implements to be used, types of land, monsoon forecasts, manure, irrigation, seeds and sowing, pests and their management, horticulture, *etc.* The fertile status of the soil in most parts of our country is a result of the wisdom of our forefathers (Katyal and Rattan, 2006).

Till the 17th century, India was not totally under the rule of the Europeans and hence it was not yet plundered by these colonial rulers. However, it still experienced some unprecedented downs in economy. Studies have revealed that the gross domestic surplus goods contributed the world's 25.1 per cent economy. It was estimated to be the second largest in the world, more than the treasury of Great Britain.

2.2.2 During Europeans Rulings

With the advent of Europeans, everything changed forever. They were responsible for ruining the Indian economy to a large scale through their reforms. It was a two way depletion of resources. The most prominent reason for the decline is the way British exploited India for 200 years. They bought raw material at a cheaper rates in India and sold the finished good at a very higher rate as compared to the rates in the Indian market. Thus, this changed the whole scenario of the most powerful economy of the world to decline from its position from 22.3 per cent in 1700 AC to a drastic dip of 3.8 per cent in 1952. In India, the production of rice has been recorded up to 9.0 t/ha during 17th century (Alvares, 2009); Later on productivity has gone down during British period and recorded only 700 kg/ha in 1947.

The colonial era had a tremendous impact on the economy due to the changes in the process of taxation and trade resulted in the breakdown of the economy. Indian industries were shut down and people were forced to buy English goods only, due to which the produce suffered losses and people lost their money. We may start with looking at eighteenth-century Mughal India, before the British had entrenched themselves as an invincible territorial power. The view that eighteenth century Mughal India was undergoing a deep economic crisis and decline has been pervasive among historians. It has been seen as the decisive broader context within which we may locate the decline of the Mughal Empire. But some later historians have refuted this view, and have instead drawn attention to the rise of new rebellious groups into power, to account for the fall of the empire. They have argued that the Mughal period was in fact a period of overall well-being and economic growth rather than stagnation or crisis. Within the political structure, there was sufficient space and autonomy in the hands of local landed elites and urban guilds to generate and accumulate surplus. Moradabad-Bareilly, Awadh. Banaras and Bengal were some such 'surplus areas'. Forests were being cleared to expand cultivation. Consequent rise in agricultural yield and the establishment of a cash nexus made surplus accumulation possible in the hands of erstwhile landlords and zamindars, who challenged Mughal paramountcy to emerge as the new regional power elite.

After the diwani of Bengal, Bihar and Orissa was granted to the East India Company in 1765, the maximization of revenue from the colony became the primary objective of the British administration. Agricultural taxation was the main source of income for the company, which had to pay dividends to its investors in Britain. Therefore, the British administration tried out various land revenue experiments to this aim. These experiments also partly determined the relationship that the colonial state would share with the people it governed.

In 1772, the Governor of Bengal, Warren Hastings, introduced a system of revenue farming in the province of Bengal. In this system, European District Collectors would farm' out the right to collect revenue to the highest bidder. This system was a total failure and ruined the cultivators because of the arbitrarily high revenue demands.

To undo this disaster, Cornwallis introduced the system of Permanent Settlement in 1793. Under this system, 'zamindars', who earlier only had the right to collect revenue, were established as the proprietors or owners of land. The state's demand for land revenue was permanently fixed but if the zamindars were unable to pay the full tax on time, their lands would be taken away and auctioned by the state. Through this system, the state tried to create an enterprising class of landowners, who would try to improve crop production in their fields to earn profits. Besides, it would be simpler for the state to deal with a limited number of zamindars than with every peasant, and a powerful section of society would become loyal to the British administration.

But this system led to greater impoverishment of the tenant-cultivator because of the burden of high revenue assessment. It also caused great difficulty for zamindars, many of whom were unable to pay the revenue on time and lost their lands. A large number of traditional zamindar houses collapsed. The system also encouraged many layers of intermediaries between the zamindar and cultivator, adding to the woes of the peasantry.

As a result of the revenue policies of the British, agriculture stagnated and peasants almost became tenants at will. They also increased the number of landed intermediaries, and strongly entrenched the figure of the moneylender in the countryside. Landlords and zamindars became an important class and collaborators of British colonial rule. It is often believed that the colonial administration encouraged the commercialization of agriculture that improved the position of peasants in many areas of the Indian colony. From the 1860s onwards, the nature of agricultural production was determined by the demands of the overseas markets for Indian primary products. The items exported in the first half of the nineteenth century included cash crops like indigo, opium, cotton and silk. Gradually raw jute, food grains, oil seeds and tea replaced indigo and opium. Raw cotton remained the most in demand item. This expansion in cash crop production was accompanied by the building of railways, after 1850, to improve trade networks.

But commercialization seems to have been a forced artificial process that led to very limited growth in the agricultural sector. It led to differentiation within the agricultural sector, but did not create the figure of the 'capitalist landowner' as in Britain. The lack of any simultaneous large scale industrial development meant that accumulated agrarian capital had no viable channels of investment, for it to be converted into industrial capital. Initiatives to expand the productive capacity and organization of agriculture was also a risky proposition, as the sector catered to a distant foreign market with wildly fluctuating prices, while the colonial state provided no protection to agriculturists. Commercialization thus, increased the level of sub infatuation in the countryside and money was channelized into trade and usury.

The larger part of the profits generated by the export trade went to British business houses, which controlled shipping and insurance industries, besides commission agents, traders and bankers. Those who benefited in the colony were big farmers, some Indian traders and moneylenders. Commercialization further intensified the feudal structure of landlord-moneylender exploitation in rural areas.

The so called process of commercialization, which was supposed to lead to capitalist agriculture, was often carried out through very exploitative and almost unfree forms of labour. Tea was grown in plantations in Assam, owned by whites, and they used indentured labour, which was almost like slavery. White planters had to force farmers to grow indigo because it yielded low profits and upset the harvesting cycle. This involved inhuman levels of coercion, which eventually led to the indigo-rebellion in 1859-60. Commercialization did lead to limited phases of success in the cotton producing areas of western India in 1860s and in jute production in eastern India, but they were because of increases in demand rather than capitalist innovation in production and organization.

Farmers were forced to grow cash crops also because they had to pay the high revenue, rents and debts in cash. The shift away from food crops like sorghum, pearlmillet and pulses to cash crops often created disaster in famine years. A decline in world demand for Indian cotton led to heavy indebtedness, famine and agrarian riots in the Deccan cotton belt in the 1870s. The jute industry collapsed in the 1930s, which was followed by a devastating famine in 1943 in Bengal. Although, causes of these famines have been widely debated by historians, it is undeniable that the aggregate production of food crops remained far behind population growth, and millions of people died of starvation and epidemics. This whole agriculture history of India during colonial period has been analyzed in detail by Shiva and Bhar, 2001); in their book 'an ecological history of food and farming in India'.

This whole history reveals that-

- ☆ Indian economy that was based on agriculture was at the top in the world up to 17th century
- ☆ After the advent of European rulers, the fertile lands of rice were shifted to cash crops *e.g.* cotton, sugarcane, jute, indigo *etc.*
- ☆ Several policies and taxes were imposed and incentives were stopped to farming that not only made farmer poor but also trapped in debt and ultimately they lost interest and avoided investment in farming before independence period.

With this whole analysis, it is proved that the productivity of ancient Indian agriculture was sufficiently high that was based on organic inputs and this has been endorsed by A. Haward. However, when the productivity gone down drastically due to policies of British rulers, the whole blame imposed on traditional farming practices to justify the use of chemicals that invented in early 20th century.

2.2.3 After the Independence

Grow More Food campaign that started in 1950s and later on continued with new term green revolution, had the base of technologies for targeted or cure and control approach *e.g.* use the synthetic form of nutrients that found deficient in soil or kill the pest and weed with poisonous pesticides and herbicides. In other words, these chemicals were used to cure the problem of food scarcity in Indian agriculture. Although at that time (1950-60), food was on the top agenda in the nation's program, so nothing is to be criticized as it was the need of the hour. In late 1970s, India achieved self sufficiency in food grain production. However, even after, the use of chemicals was continued. In other words, the medicine was continued even after the disease is cured because this has been found as an easy way to sustain the production and the basic nature of biological system has been continuously ignored that converted remedy into malady. In 1980s, the consequences of disturbance in agricultural production with chemicals were becoming visible and the production growth become almost stagnated or in some of the cases following negative trend with several socio-economical and environmental problems. The major reason of this negative trend is the agriculture has been treated like machine using the chemicals and Liebig theory. This system is named as conventional system (CS) in this text.

Productivity of CS that mainly dependent on external inputs.e.g chemicals, irrigation, seed,exotic animals *etc.* gets a stagnation in productivity or in many cases it started following negative trend (Timsina and Connor, 2001; Ladha *et al.*, 2003;). This also affected not only the productivity of agriculture *per se* but also the regenerative capacity of natural resources like soil, sharp decrees in population of beneficial flora and fauna (particularly pollinators), deficiency of micronutrients *e.g.*

Zinc, Iron, Boron *etc.*, resistance in pest to pesticides, secondary salinization (Pingali and Shah, 2001), decline of ground water table (Ambast *et al.*, 2006) *etc.*, or decrease in soil organic carbon content (Lal, 2004); *etc.* are being observed. This led to decline in factor productivity and unsustainablity. Now, soil health deterioration is the more serious thing in CS, Since 1990s, increased incidence of farmer suicides in India has been the most dramatic outcome of the hopelessness faced by many farmers, due to a combination of factors like high input prices, crop failure, indebtedness, *etc.* (Mishra, 2007); An estimated 27 per cent of farmers did not like farming because it was not profitable. In all, 40 per cent felt that, given a choice, they would take up some other career (NSO, 2005). There are several reports and references on the decreasing productivity of CS and its adverse effects on environments and society and, lower marginal returns with continuous intensification (Gupta and Seth, 2007). Dr M.S.Swamiainathan who was the pioneer of Green revolution in India has also accepted the need of evergreen revolution described it as the increasing productivity in perpetuity without ecological harm. He laid stress on the 'organic agriculture' which meant cultivation without use of chemical pesticides and 'green agriculture' which meant conservation agriculture with the help of integrated pest management, integrated nutrient supply and integrated natural resource management. Agro forestry system involving fertilizer trees was another component of evergreen revolution (Swaminathan, 2011).

3. Recognizing Phase

3.1 Recognition in the World

Just after realization of ill effects of CS, the efforts were started in various parts of the world to recognize ecofriendly/organic farming. The movement started after Second World War in the following chronology:

- ☆ **1946** - Creation of the Soil Association in 1946 by a group of farmers, scientists and nutritionists who observed a direct connection between farming practice and plant, animal, human and environmental health. Today, it is the UK's leading organic organization, with over 180 staff based in Bristol headquarters.
- ☆ **1947** - In France, principals of organic farming were introduced because doctors and consumers blamed agricultural chemicals for causing the development of cancer and mental disorders.
- ☆ **1950** - During 1950s, sustainable agriculture was a topic of scientific interest, but research seems to concentrate on developing the new chemical approaches. In the United States, J.I. Rodale began to popularize the term and methods of organic growing, particularly to consumers through promotion of organic gardening.
- ☆ **1959** - Creation of Groupement d'agriculteurs biologiques de l'Ouest in France. (Association of organic farmers from the west)
- ☆ **1962** - Rachel Louise Carson, (1907-1964) a prominent scientist and naturalist, published "Silent Spring", describing the effects of DDT and other pesticides on the environment.

- **1970s** - Global movements had concerns with pollution and the health of the Earth's environment. There is an increase in focus toward organic farming. As the distinction between organic and conventional food became clearer, one goal of the organic movement is to encourage consumption of locally grown food based on the slogan "Know Your Farmer, Know Your Food".
- **1972** - In Versailles, France, there was the creation of The International Federation of Organic Agriculture Movements (IFOAM). Its goal is to communicate and exchange throughout the world information relating to the principles and practices of organic agriculture of all schools and across national and linguistic boundaries.
- **1975** - Fukuoka released his first book, "One Straw Revolution", with a strong impact in certain areas of the agricultural world. His approach to small-scale grain production emphasized a meticulous balance of the local farming ecosystem, and a minimum of human interference and labor.
- **1980s** - Various farming and consumer groups throughout the world seriously began pressurizing for government regulation of organic production. This leads to legislation and certification standards being enacted through the 1990s and to date. Currently, most aspects of organic food production are government-regulated in the United States and the European Union.
- 1989 - Organic food production as a necessity in Cuba–. Cuba has the world's only state-supported infrastructure to support urban food production. It is called organopónicos.

Interestingly, many a times OF is projected as a new system with several apprehensions, however, before the synthetic chemicals were invented, it was all OF based. As mentioned earlier, in India, the production of rice was very good during 17th century (Alvares, 2009) but later on productivity gone down during British period and recorded only 700 kg/ha in 1947.Therefore, food production technologies (obviously organic only) were capable to maintain sufficiency level during ancient time and these technologies can be revalidated and improved with the integration of modern ecofriendly technologies. Modern science has not developed only the synthetic chemicals (fertilizers, pesticides, herbiicides) but also the several ecotechnologies *i.e.* use of enriched compost, biofertilisers, biopesticides, rainwater conservation, crop rotation, mulching, agroforestry *etc.* OF is just integration of them (Table 2.1) considering the local conditions and traditions with a set ideology (Sharma, 2014). Therefore, this is a highly scientific system with all possibility of need based improvements.

3.2 Recognition of Organic Farming in India

As mentioned earlier, the ill effects of chemical farming were being realized by producers and consumers. In June 2001, the Government of India announced the National Programme for Organic Production (NPOP) with an aim to promote sustainable production, environmental conservation, reduction in the use and import of agrochemicals, the promotion of export and rural development.

Table 2.1: Ideological Differences between Organic Agriculture and Conventional Agriculture (Sharma, 2014)

Organic Agriculture	*Conventional (Chemical) Agriculture*
Holistic approach: Any technology applied considering the system as a whole- No imbalance	**Reductionist approach**: Targeted approach for one commodity or one pest or deficiency of nutrient- creates imbalance in system
Decentralize production: Most of the inputs *e.g.* seed, manure, biopesticides *etc.* produced at farm/village level- suitable to local environment+ generate employment+ low cost of production	**Centralize production**: Produced in factories/ farms, away from the place of use - no proper use of local resources+ least employment+ increase cost of production
Harmony with Nature: Harness the benefit of natural resources, flora and fauna by using or giving favorable environment to them- sustained productivity of natural resources	**Domination on Nature**: Agriculture system is forced to produce more- Regenerative capacity of natural resources decreased+ decrease productivity in long term.
Diversity: Includes all possible organisms complimentary in a system. Work as mutual service providers for nutrient and pest management. Least cost and time required of system owner.	**Specialization:** Only one crop or tree or animal. All cost and time of nutrient and pest management has to be borne by system owner/farmer.
Input optimization: best use/recycling of available resources. System regenerative capacity and owners' economic capacity maintained/enhanced.	**Output maximization**: Over use of resources disturbs system and resources productivity in long term- increasing cost.
Knowledge intensive: Only few resources but need how timely and best integrated. Least dependency on experts/imported technologies, once farmer trained-possible in remotest area.	**Input intensive**: Comprehensive list of chemicals with time and method. Needs experts for timely updating. Only possible in resources sufficient areas.
Preventive, protective and proactive approach: All the actions/applications are done in anticipation of system requirement-least use of inputs.	**Cause and control approach**: Most of the actions/applications are done to control the damage to system-heavy use of inputs.
Decreasing input use: As the system reaching at perfection it conserve/generate its own resources *e.g.* for nutrition and protection- decreasing requirement of inputs	**Increasing input use**: Target and action approach that rather, deteriorate systems regenerative capacity- increasing requirement of inputs.

4. Reshaping Phase

4.1 Reshaping in the World

After getting recognition the major challenge was to revalidate, revive and improve traditional system with incorporation of modern eco-friendly technologies at the production end and how to make sure the availability of genuine organic products to the consumers at marketing level. Some of the landmark developments in this direction are detaied below:

- ☆ **1990s** - The retail market for organic farming in developed economies is growing by about by about 20 per cent annually due to increasing demand by consumers. Concern for the quality and safety of food, and the potential for environmental damage from conventional agriculture, are apparently responsible for this trend.
- ☆ **1991** - European Union provided a legal framework for the organic agriculture designation.

- ☆ **2002** - The United States of America adopts the National Organic Program (NOP), providing a development framework for organic agriculture.
- ☆ **2006**- IFOAM - International Federation of Organic Agriculture Movements World Board the following definition for Organic Agriculture:

"Organic agriculture is a production system that sustains the health of soils, ecosystems and people. It relies on ecological processes, biodiversity and cycles adapted to local conditions, rather than the use of inputs with adverse effects. Organic agriculture combines tradition, innovation and science to benefit the shared environment and promote fair relationships and a good quality of life for all involved" (IFOAM, 2006).

The last two decades of 20th century witnessed an overwhelming popularity and scientific acceptance of organic farming in the western world, esp. USA, Germany and the Scandinavian countries. In-depth research has gone into the different aspects, stages and shades of organic agriculture. Many universities such as The Institute of Ecological Agriculture, Bonn University, Germany offer courses on ecological agriculture and long-term research, The FiBL, organic agriculture research institute in Switzerland, international federations like the IFOAM with HQ in Germany and scores of other institutions in Europe do exclusive research in organic farming and support organic farmers. Organically grown food produce is already being exported to the West from Latin America and Asia under the supervision and certification of inspectors from the West. Organic agriculture is now practiced in almost all countries of the world, and its share of agricultural land and farms is growing. The total organically managed area is more than 22 million hectares worldwide. In addition, the area of certified "wild harvested plants" is at least a further 10.7million hectare, according to various certification bodies. The market for organic products is growing, not only in Europe and North America (which are the major markets) but also in many other countries, including many developing countries. Official interest in organic agriculture is emerging in many countries (Yussefi and Mitscke, 2003). At present, a price premium of about 20-30 per cent over conventional products can be received (FAO, 2002).

4.2 Reshaping in India

India produces primary organic products and less of processed foods. Organic products grown in various agro-climatic zones are coffee, tea, spices, fruits, vegetables and cereals as well as honey and cotton. Organic animal husbandry, poultry and fisheries came later. Domestic organic markets and consumer awareness were underdeveloped till 2010, but, later on, it started increasing. In the domestic market, organic food is usually sold directly by the farmer or through specialized shops and restaurants.

4.3 Organic Marketing

Government of India started development of organic agriculture production and making rules and regulation after getting huge demand of certified organic from developed countries and from within the country. Firstly the initiatives was taken by ministry of commerce and later on other ministries and NGOs started work.

A major development was in 2001 India Started National Programme on Organic Production (NPOP) and National standards for organic production have been formulated. APEDA Agriculture Processed food Export Development Authority, Ministry of Commerce, Govt. of India) was made nodal regulatory agency for NPOP. The marketing, export and making rules and regulations part is being taken care by APEDA but the second major aspect is research and developmental work on various aspects of increasing production of organic agriculture to meet the demand of market, that is being done by various agencies.

4.4 Research

Research related to organic agriculture started in 1950's and continued till date on the name of eco-friendly farming technologies/conservation of natural resources. The issue of technology fatigue in agriculture is well known now. There is a need to shift away from individual crop-oriented research focused essentially on irrigated areas towards research on crops and ecofriendly cropping systems in the dry lands, hills, tribal and other marginal areas (Swaminathan, 2007); Some landmarks in research are:

- ☆ **2002:** An all India society of organic farmers was registered as society with a name Organic farming Association of India (OFAI) at Goa, for documentation of traditional knowledge and farmers experimentations.
- ☆ **2004:** National Centre on Organic Farming was established at Ghaziabad with six regional centers all over India
- ☆ **2004:** ICAR started research on organic farming through a network project coordinated from PDFSR, Modipuram. At present it is running at 20 centre's all over the country.
- ☆ **2011:** National Initiative on Climate Resilient Agriculture (NICRA) is a network project of the Indian Council of Agricultural Research (ICAR) was launched. The project aims to enhance resilience of Indian agriculture to climate change and climate vulnerability through strategic research and technology demonstration. Organic farming is one of the components of mitigating climate change and this technology is being disseminated partly or fully through 100 KVKs in eight zones, 25 co-operating centers of AICRP on Dryland Agriculture and 7 technology transfer divisions of core institutes of ICAR.

The All India Network Project on Biofertilizers (AINPB) under the Indian Institute of Soil Science, Bhopal is mandated to formulate and testing of mixed biofertilizers for diverse cropping systems and to improve biofertilizer technology with particular reference to quality, carriers, consortia and delivery systems. The University of Agricultural Sciences, Dharwad (Karnataka) established the Institute of Organic Farming in 2006 with an objective to cater to the needs of organic farmers and other stakeholders in the State. Similarly, in 2006, the University of Agricultural Sciences, Bangalore (Karnataka) established the Organic Farming Research Center at Shimoga, and Organic Farming Research Station at Naganahalli (near Mysore) to scientifically validate and analyze the claims of organic agricultural produce and help popularize it among farmers. The Department of Organic Agriculture was established by the CSKHPKV, Palampur (Himachal Pradesh) in 2009 with a specific mandate to promote organic farming in the State. In addition, many ICAR Institutes and State Agricultural Universities have initiated a number of research projects/centers to develop location-specific organic farming modules.

4.4.1 Revalidation and Standardize of Traditional Technologies/System

Under this group all the traditional knowledge/technologies developed in the millenniums are being revalidated and documented at various research organizations. Tamil Nadu Agriculture University, University of Agriculture Sciences, Bengaluru are the leading institutes working on this aspect. They standardize technique of Panchgavya, a product of cows (indigenous breed) five products *i.e.* dung, urine, milk, curd and butter. Panchgavya is an elixir or promoter of soil health and plant growth. At ICRISAT, Hyderabad (a CGIAR institute) experiment was conducted for 8 years with panchgavya. They observed significant increase in the population of beneficial soil fauna (Rupela *et al.*, 2006). Several other preparation from botanicals as plant growth promoter or as biopesticides have been revalidated and standardized. The beauty of these traditional technologies is the cost effective, locally available and socially acceptable.

4.4.2 Development of Ecofreindly Organic Inputs

This does not require gestation period for system development has some basic science/biotechnology, easy funding availability and can be commercialized. Several such products have been developed by Universities as well as private entrepreneurs. Some of the examples are enriched compost (with natural minerals and microbes), neem/botanical based biopesticides, isolation of local effective fauna for biofertilizers/biopesticides *etc.*

4.4.3 Organic System Research

This is the most difficult and time consuming aspect therefore only at few locations this type of research is going on. One interesting work worth mentioning is survey of productivity, soil health, economics of selected organic farms done by Ramesh *et al.* (2010). Some of the NGOs like OFAI, CSA, Green foundation, Navdhanya *etc.* have documented the organic systems available at various places (Alvares, 2009). Organized research after development of organic system is progressing at limited places. One such system has been developed for the low rainfall areas at Central Arid Zone Research Institute, Jodhpur (Sharma, 2013). Crop based organic protocol has been developed for basmati rice, cotton, tea, and spices and continue on some other high value crops.

4.5 Developmental Programmes

Govt. of India started several programme for direct/indirect development support to organic farming in India. Major programmes that were initiated beyond 2004 are National Horticulture Mission and Rashtriya Krish Vikas Yojna. Recently in 2015, a new scheme on Paramparagat Krish Vikas Yojna (PKVY) has also been initiated. Under these schemes, programmes like organic clusters development and capacity building, markets linking *etc.* are running at various levels. Several states have implemented organic farming policies, particularly, the north eastern states, AP, HP, Kerala, Tamil Nadu, Karnataka, M.P. *etc.* Sikkim has become first 100 per cent organic state state in India is followed by Kerala.

5. Conclusion

The history of agriculture or farming is as old as the history of human beings. Before 19th century, in no document, there was a reference of use of synthetic chemicals in agriculture, rather, different preparations with the ingredients of biological origin were in use as mentioned in several ancient documents in India. This was purely eco-friendly farming prevailed all over the world till 16th century. However, the situation changed after 16th century, when Europeans started making their colonies all over the world and started exploiting agriculture that decreased the productivity. After the industrial revolution in Europe, several chemicals were invented and scientist like Liebig and others advocated the use of chemicals in agriculture in the beginning of 20th century. But soon the ill effect of this approach was realized, thus 'organic farming' means no chemical use has come to the surface. It is based on the fact that agriculture is primarily a nature's business and can only be sustainable with the use of eco-friendly technologies.

REFERENCES

Alvares C. 2009. *Organic Farming Sourcebook*. Other India Press, Goa, India. pp 458.

Ambast S, Tyagi, N and Raul S. 2006. Management of declining groundwater in the Trans Indo-Gangetic Plain (India): some options. Agricultural Water Management, 82: 279–296.

FAO. 2002. *Organic Agriculture, environment and food security*. FAO, Rome, Itly. pp 252.

Gupta R and Seth A. 2007. A review of resource conserving technologies for sustainable management of the rice-wheat cropping systems of the Indo-Gangetic Plains (IGP). Crop Protection, 26: 436–447.

Howard A. 1947. *The Soil and Health: A Study of Organic Agriculture*. Reprint by Schocken Books, 1972 USA, pp. 307.

IFOAM. 2006. http: //www.ifoam.org/growing_organic/definitions/doa/index.html

Katyal JC and Rattan RK. 2005. Soil management in ancient, medieval and premodern India and its relevance in modern day sustainable agriculture. In: Agriculture heritage of Asia (Ed. Y.L. Nene), Asian Agri history foundation, Secundrabad, pp. 61-77.

Ladha J Dawe D, Pathak H, Padre A, Yadav R, Singh B and Singh Y. 2003. How extensive are yield declines in long-term rice-wheat experiments in Asia?. Field Crops Research, 81: 159–180.

Lal R. 2004. Soil carbon sequestration in India. Climatic Change, 65: 277-296.

Mishra S. 2007. Risks, farmers' suicides and the agrarian crisis in India: is there a way out? Indira Gandhi Institute of Development Research, Mumbai. 2007. WP-2007-014. 42p.

NSSO. 2005. Situation Assessment Survey of Farmers: Some Aspects of Farming, NSS 59th Round (January–December 2003) Report No. 496(59/33/3) National Sample Survey Organization, Ministry of Statistics and Programme Implementation Government of India. 281p.

Pingali P. and Shah M. 2001. Policy re-directions for sustainable resource use: the rice-wheat cropping system of the indo-gangetic plains. Journal of Crop Production, 3: 103–118.

Ramesh P, Singh AB, Ramana S and Subharao A. 2010. Status of organic farming in India. Current Science, 98(9): 1190-1194.

Rupela O P, Humanyun P, Venkaleswarlu B. and Yadav AK. 2006. Comparing conventional and organic farming crop production systems: Inputs, minimal treatments and data needs. Organic farming newsletter, 2(2): 3-17

Sadhale N. 1996. Surpala's Vrikshayurveda.Agri-histry bulletin No.1, Agrihistry foundation, Secundrabad, India. pp. 72.

Sharma A.K. 2013. Organic agriculture programming for sustainability in primary sector of India: action and adoption. Productivity, 55(1): 1-17.

Sharma AK. 2013. Organic system management. Indian Farming, 63(4): 13-16.

Shiva V and Bhar RH. 2001. An ecological history of food and farming in India - Vol.2. RFSTE/NAVDANYA, New Delhi. pp. 108.

Swaminathan MS. 2007. Agriculture cannot wait: New horizons in Indian agriculture, Academic Foundation, New Delhi, India. pp. 550.

Swaminathan M.S. 2011. Evergreen revolution. http: //www.icar.org.in/en/node/2826

Yussefi M and Willer H. (Eds.). 2003. The World of Organic Agriculture Statistics and Future Prospects. IFOAM,Germany. http: //orgprints.org/13883/1/willer-yussefi-2003-world-of-organic.pdf.

Chapter 3

Components of Organic Farming

S.R. Kumar[1], P. Spandana Bhatt[2],
A.V. Ramanjaneyulu[3] and A. Srinivas[3]

[1]ICAR-Indian Institute of Millets Research,
Rajendranagar, Hyderabad – 500 030, Telangana
[2]Krishi Vigyan Kendra, [3]Regional Agricultural Research Station,
Professor Jayashankar Telangana State Agricultural University,
Palem – 509 215, Telangana
E-mail: avr_agron@rediffmail.com

1. INTRODUCTION

Agriculture continues to be a key sector for the economic development of most developing countries. It is critically important for ensuring food security and alleviating poverty (Stockdale *et al.*, 2001). In early times (industrial age), food security of the growing population was a priority and hence crop sciences had the goal of making a quantum jump in food production. In most of the crops, this was achieved by bringing in traits and technology that helped increase the harvest index. In Figure 3.1 one can clearly see at various levels of total dry matter production (M), the harvest index could be increased, which was aided by external input of nitrogen fertilizer, especially in cereals. High input agriculture had positive effect in terms of increased production (accumulating food stocks) but also had implications in terms of environmental pollution and contaminations due to pesticide residues in the food chain.

Further, in recent times, there is a widespread problem of imbalanced use of inorganic fertilizers. In general, too much N and P fertilizers are being applied and too little potassium and micronutrients (Sofa *et al.*, 2006). In the process of attaining higher levels of food production for matching the demand of growing population during the past four decades, emphasis was laid on intensive agricultural practices. Modern Agriculture helped to enhance the productivity; however, it also led to soil

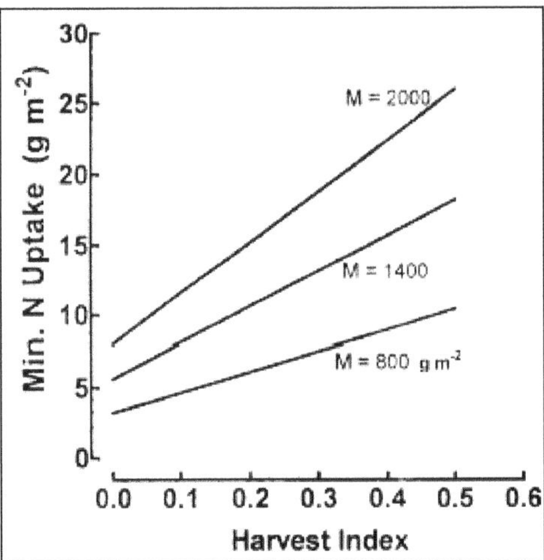

Figure 3.1: The Minimum Amount of Total Nitrogen Uptake Required by a Crop is Plotted as a Function of Harvest Index. Three values of constant total crop mass (*M*, g/m²) are plotted in the figure (*Source*: Sinclair, 1998).

erosion, environmental degradation, declining soil fertility and factor productivity, depletion of natural resources, pollution of soil, water and air. Further, soil and food quality is declining over the years. Due to indiscriminate use of pesticides, pests are developing resistance to pesticides and the pesticide residues are being detected in agricultural and dairy products (Dhaliwal and Pathak, 1993). Panda and Behera (2003) have emphasized that despite several decades of research on management practices in agriculture to minimize water pollution, agriculture remains a major source of water pollutants and water quality degradation throughout the world. Surface and groundwater quality degradation due to agricultural practices can be categorized as follows: (a) increased erosion and soil loss due to agricultural practices; (b) chemical pollution by fertilizer and pesticides. Antil *et al.* (2004) while studying the land degradation pattern in the state of Haryana have reasoned that imbalanced use of fertilizers in different areas have shown more depletion of secondary and micro nutrients.

The soil, food products and underground waters have been found to be contaminated with pesticide residues. The physico-chemical properties and toxic metal ions in soil varied according to the composition of the sewer water and duration of irrigation. Further, Gupta and Abrol (2000) reported serious problems of groundwater contamination due to dissolution of arsenic-bearing minerals under altered soil moisture regimes conducive to oxidation of pyritic sediments.

Centre for Science and Environment (CSE) used European Union (EU) norms to evaluate the Indian beverages and found 11-70 times more levels than the stipulated level of pesticides in various cold drinks as detailed in Table 3.1.

Table 3.1: CSE Report on Cold Drinks and Level of Pesticides

Cold Drinks	*Level of Pesticide above EU Norms*	*Cold Drinks*	*Level of Pesticide above EU Norms*
Mirinda Lemon	70 times	Limca	30 times
Coca-Cola	45 times	Blue Pepsi	29 times
Fanta	43 times	Mountain Dew	28 times
Mirinda Orange	39 times	Thums Up	22 times
Pepsi	37 times	Diet Pepsi	14 times
7 UP	33 times	Sprite	11 times

Source: www.myenjoyzone.com

Further inorganic fertilizers and other agro-chemicals do show negative effects on soils and plants. When water soluble fertilizers are applied to soil, a good portion of the added nutrients are lost to the atmosphere due to facilitation that of the autotrophic nitrifying organisms, thereby hindering the immobilization of nutrients. This results in rapid rate of nutrients loss in different forms and increases the soil acidity with nitrification. Emission of ammonia, methane, nitrous oxide and elemental nitrogen from the soil system is a result of denitrification (Signor and Cerri, 2013). Depletion of secondary and micronutrients especially Sulphur and Zinc occurs. Deficiency of micronutrients limits productivity of many field crops especially in rice. Adding high doses of N-fertilizers in modern agriculture without the use of organic manures, leads to humus depletion and fall in crop production (Handa, 1995). When high levels of N-fertilizers especially nitrate forms are applied to soil, nitrate pollution of drinking water is a serious health hazard found in extensively irrigated coarse textured highly percolating soils of central Punjab, where 40-50 per cent of applied nitrogen is lost in leaching and the mean concentrations of nitrate nitrogen was 3.88 ppm during 1982 (rainy season) and 1.02 ppm in 1975. In 10 per cent of the ground water samples nitrate concentration was 10 ppm which was the upper tolerance limit in drinking water against nil in 1975 (Signor and Cerri, 2013).

Alarming issue to human health is regular use of phosphatic fertilizer in large quantities often causes the buildup of trace metal contamination such as arsenic, fluoride, cadmium *etc.* in soil and plants. Cadmium in single super phosphate is available to plants as the Cd in cadmium chloride (Dwivedi *et al.*, 2004). Similarly, chloride contained in MOP and NH_4Cl creates toxicity to many crops like beans, citrus, grapes lettuce, potatoes *etc.* (Chadha *et al.*, 1997). These trace metal toxic contaminants reach the human body, through food chain and cause health problems. The water soluble nutrients when carried to lakes and stream through leaching and surface run off cause eutrophication as manifested by the luxuriant growth of algae and other water weeds on the water surface leading to oxygen deficient condition. This situation is not conducive to healthy aquatic life (Signor and Cerri, 2013).

Demand for products with quality attributes has given rise to a movement called the organic farming system (Gopalan, 1997). Unlike traditional farming where standards or indices had lower importance, these ecologically sound organic farming approaches calls for various intermediaries who certify both inputs and outputs of a system so that the consumer is guaranteed for a products quality. Thus, organic farming has been considered to be a sound and viable option in most of the countries to address the above problems (Bhattacharya and Chakraborty 2005).

Organic farming is a production system which favors maximum use of organic materials (crop residue, animal residue, legumes, on and off farm wastages, growth regulators, bio-pesticides) and discourages use of synthetically produced agro-inputs, for maintaining soil productivity and fertility and pest management under conditions of sustainable natural resources and healthy environment (Gaur *et al.*, 2002). According to the National Organic Standards Board of the US Department of Agriculture (USDA), the word 'Organic' has the following official definition: *"An ecological production management system that promotes and enhances biodiversity, biological cycles and soil biological activity. It is based on the minimal use of off-farm inputs and management practices that restore maintain and enhance ecological harmony".* According to Codex Alimentarius (FAO, 2001) *"Organic agriculture is a holistic production management system which promotes and enhances agro-ecosystem health, including biodiversity, biological cycles and soil biological activity. The primary goal of organic agriculture is to optimize the health and productivity of interdependent communities of soil life, plants, animals and people". "Organic agriculture is a production system that sustains the health of soils, ecosystems and people. It relies on ecological processes, biodiversity and cycles adapted to local conditions, rather than the use of inputs with adverse effects. Organic agriculture combines tradition, innovation and science to benefit the shared environment and promote fair relationships and a good quality of life for all involved."*

2. Components of Organic farming

Organic farming is a multidisciplinary subject involving Agronomists, Soil scientists, Plant protectionists, Microbiologists, Quality controllers and Environmentalists. While selecting the inputs or components for practicing organic farming, one should keep in mind, consumer preferences and ethics too. More or less, the following components can be included in organic farming (Figure 3.2).

2.1. Soil Fertility Enhancing and Erosion Controlling Methods

2.1.1. Manures and Concentrates

During conversion period, soil fertility can be improved and maintained initially through use of organic inputs like well decomposed organic manure/vermicompost, green manure and biofertilizers in appropriate quantity (Chhonkar, 2003). These organic inputs are used for feeding the soil. Well fed healthy soil rich in microflora and microfauna takes care of the crop nutrient requirement. Plant biomass, FYM, Cattle dung manure, enriched compost, biodynamic compost, Cow-pat-pit compost and vermicompost are key sources of on-farm inputs. Among off-farm inputs, important components are non-edible oil cakes, poultry manure, biofertilizers, mineral grade rock phosphate and lime *etc.* Loppings from Glyricidia and other plants grown on bunds, on-farm produced compost and vermicompost, animal dung and urine and crop residue should form the major source of nutrient and

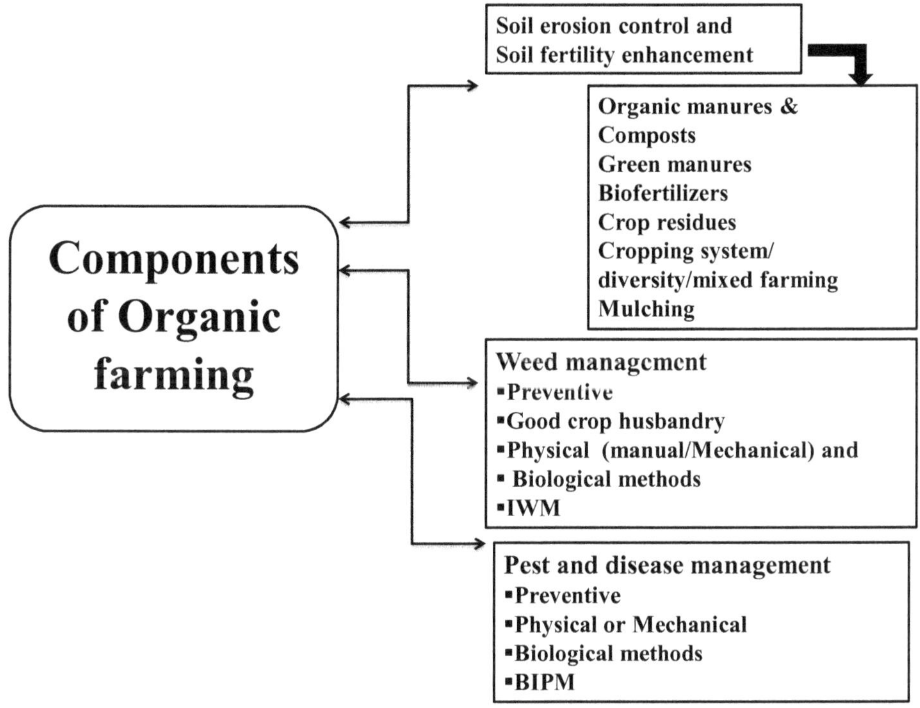

Figure 3.2: Various Components of Organic Farming.

concentrated manures such as crushed oil cakes, poultry manure, vegetable market waste compost and other novel preparations such as biodynamic formulations *etc.* can be used in appropriate quantity. Use of high quantities of manures should be avoided (Amir and Fouzia, 2011). Changing crop rotations and multiple crops ensure better utilization of resources. Depending upon the type of crop and requirement of nutrients for different crops, the quantity of externally produced inputs is determined. Application of liquid manure (for soil enrichment) is essential to maintain the activity of microorganisms and other life forms in the soil. 3-4 applications of liquid manure is essential for all types of crops (Alvares, 1996). Vermiwash, compost tea, cow urine, excellent growth promoters when used as foliar spray, 3-5 sprays after 25-30 days of sowing ensure good productivity. Use of Biodynamic preparations, such as BD-500 and BD-501 as foliar spray has also been found to be effective in growth promotion (Amir and Fouzia, 2011).

Manures can be grouped into bulky organic manures and concentrated organic manures based on concentration of the nutrients.

Bulky Organic Manures

Bulky organic manures will have to be applied in large quantity as they contain all essential elements in small percentage. Farmyard manure (FYM), compost and green manure are the most important and widely used bulky organic manures.

Advantages of these manures include supply of all plant nutrients including micronutrients, improve soil physical properties like structure, water holding capacity *etc.*, besides, increase the microorganism load that help in the decomposition process, thus increasing the availability of nutrients.

Farmyard Manure (FYM)

FYM is the most common organic manure in India. It contains partially decomposed dung, urine and straw. It contains approximately 5-6 kg nitrogen, 1.2-2.0 kg phosphorus and 5-6 kg potash/tonne. The composting pits should be free from inorganic material, which is not generic to the manure like plastics, metals and dry cells. Care should be taken to avoid any contamination from heavy metals and plastics. Agro-economic study of practices of growing maize with compost and liquid manure top dressing in low-potential areas showed significantly better performance than those of current conventional farmer practices of a combined application of manure and mineral fertilizers. Maize grain yields were 11-17 per cent higher than those obtained with conventional practices (Onduru *et al.*, 2002).

Sheep and Goat Manure

The droppings of sheep and goats contain higher nutrients than farmyard manure and compost. On an average, the manure contains 3 per cent N, 1 per cent P_2O_5 and 2 per cent K_2O. It is applied to the field in two ways. The sweeping of sheep or goat sheds are placed in pits for decomposition and it is applied later to the field. The nutrients present in the urine are *wasted* in this method. The second method is sheep penning, wherein sheep and goats are kept overnight in the field and urine and fecal matter added to the soil is incorporated to a shallow depth by working blade harrow or cultivator or cultivator. Alaa *et al.* (2014) reported that the application of extract of sheep manure applied at 40 per cent concentration and as a foliar spray produced superior results on both vegetative growth and flower parameters of marigold.

Poultry Manure

The excreta of birds ferment very quickly. If left exposed, 50 per cent of its nitrogen is lost within 30 days. It contains higher nitrogen and phosphorus compared to other bulky organic manures. The average nutrient content is 3.03 per cent N, 2.63 per cent P_2O_5 and 1.4 per cent K_2O.

2.1.2 Concentrated Organic Manures

Concentrated organic manures have higher nutrient content than bulky organic manures. These include oilcakes, blood meal, fish manure *etc.* Many oil cakes such as the castor, neem, madhuca, karanja, linseed, rape seed and cotton seed which are non-edible may serve as useful organic manure as these contain high amounts of plant nutrients (Table 3.2). They are valued much for their alkaloid contents which inhibit the nitrification process in soils. Neem cake contains the alkaloids such as nimbin and nimbicidine which effectively inhibit the nitrification process. Similarly, Karanjin (*Pongamia pinnata*) is a potent nitrification inhibitor equal in efficiency to nitropyrin in retarding the nitrification process of ammoniacal nitrogen and increasing the yield, nitrogen uptake and grain protein content of rice. Madhuca cake has been successfully used in coastal saline soils for cultivation of rice.

Table 3.2: Nutrient Content in Various Oilseed Cakes

Oil Cakes	*Nutrient Content (per cent)*		
	N	*P_2O_5*	*K_2O*
Non edible oil-cakes			
Castor cake	4.3	1.8	1.3
Cotton seed cake (undecorticated)	3.9	1.8	1.6
Karanj cake	3.9	0.9	1.2
Mahua cake	2.5	0.8	1.2
Safflower cake (undecorticated)	4.9	1.4	1.2
Edible oil-cakes			
Coconut cake	3.0	1.9	1.8
Cotton seed cake (decorticated)	6.4	2.9	2.2
Groundnut cake	7.3	1.5	1.3
Linseed cake	4.9	1.4	1.3
Niger cake	4.7	1.8	1.3
Rape seed cake	5.2	1.8	1.2
Safflower cake (decorticated)	7.9	2.2	1.9
Sesamum cake	6.2	2.0	1.2

2.1.3. Composts

Vermicompost

Vermicompost is prepared by decomposing the various non-toxic organic solids and liquid wastes available from cities, dairies, sugar and distillery units, pulp and paper mills, tanneries, fermentation industries and food processing units with the help of earthworms. Non-burrowing earthworms such as *Eisenia foetida* or *Eudrilus euginiae* are used for composting. Vermicastings consisting of excreta of earthworms and the cocoons released by them are rich in organic matter and plant nutrients. In the presence of vermicasts, the decay of organic refuses and the formation of compost is accelerated. The process could be continued uninterruptedly by the addition of dung and other organic wastes at regular intervals to serve as food source for the earthworms. Vermibeds are to be watered to maintain the optimum moisture of 50-55 per cent. The process of composting takes 60-90 days. Gopinath *et al.* (2008) reported that vermicompost amended soils do get benefitted due to buildup of organic matter (C-sequestration). Bhaskaran *et al.* (2009) noticed perceptible increase in cationic exchange capacity, C/N ratio and available N, P and K in vermicompost amended soils. Well prepared vermicompost can be applied @ 5 t/ha to millets and pulses, 7.5-12.5 t/ha to oilseed crops, 10 t/ha for vegetable and flower crops, 12.5 t/ha for commercial crops, 2-3 kg/tree in fruit crops and 5 kg/tree in plantation crops (Ramanjaneyulu *et al.*, 2012). Further they also stated that vermicompost can also be applied for top dressing just like chemical fertilizers for better growth and development. Reddy *et al.* (1998) recorded maximum plant height at harvest, days to first flowering, and branches/plant with the application of vermicompost (10 t/ha). Similarly, Tomar *et al.* (1998) reported that the application of vermicompost significantly increased leaf area in carrot (*Daucus carota* L.) plants. Addition of organic amendments and casting of earthworms to soil also proved effective in controlling diseases in pea (*Pisum sativum* L.), mustard (*Brassica juncea* L. Coss.) and chickpea (*Cicer arietinum* L.) during winter season. Nitrogen, phosphorus, potassium, calcium and magnesium accumulation also increased with increasing doses of vermicompost as well as with fertilizers (Yadav *et al.*, 2013). Samawat *et al.* (2001) reported that vermicompost application of 15 kg/m^2 gave the highest yield in tomato crop.

Other Composts

Good quality compost if applied to agricultural fields, they improve the nutrient cycling, soil and crop health through reduction in chemical usage. However, inferior quality compost may have a detrimental effect on crops or soils. Good quality compost should have pH in the range of 7 - 7.5 range, low C/N ratio (17:1), high degree of biological activity, cation exchange capacity and add humus to total organic matter. It should also not show any toxic anaerobic compounds that are harmful to plant growth (www.organicfarming.com.au). The compost can be prepared by Indore, Bangalore and NADEP methods. Municipal solid waste can be a potential source, however, its safe use should be ensured by using compost that is mature and sufficiently low in metals and salt content. Sewage sludge should not be added to the compost at any point since it will raise the metal content of the compost (Richard and Woodbury, 1992) as it increases the concentration of Ni, Pb, Se, and Zn (He *et al.*, 1995).The application of mineral enriched compost indirectly satisfies the N and P needs of plants in nutrient deficient soils. Apart from being a source of macro and micro nutrients for plants, compost is also believed to suppress soil-borne diseases in plants (Mamo *et al.*, 1999). Bioremediation by composting is an economically attractive method for cleaning organic pollutants like xenobiotic, petroleum products, polycyclic aromatic hydrocarbons (Moldes *et al.*, 2007). The vigorous biological activity during composting can be used to accelerate the decomposition of xenobiotics in soil (*Khalil et al.*, 2001).

2.1.4. Green Manuring and Green Leaf Manuring

Complete dependence on chemical fertilizers makes the soil infertile and less productive in absence of organic materials. Further, chemical fertilizers are expensive and non-renewable. Hence, renewable biological base nutrients are the very good alternative. Among the different renewable sources of nutrition, green manuring and green leaf manuring is the most important one. Green manuring is growing of crops and incorporating in-situ the fresh or green mass at 50 per cent flowering stage (45-60 DAS). While green leaf manuring is incorporation of branches/twigs/ leaves/lopping's of trees/bushes/shrubs of trees grown on bunds, wastelands and nearby forest areas. Green leaf manuring is generally followed in the central and eastern India. Mostly legumes are preferred for the purpose of green manuring they fix the atmospheric nitrogen in the soil in available form, improves the soil health, physical structure, exercises protective action against erosion, prevents leaching and conserve more soil moisture thus improve soil fertility. It increases soil organic matter, concentration of nutrients near the soil surface in available form especially the available nitrogen and reduces N losses through leaching and soil erosion. The soil aggregation due to increased organic matter in soil improves the soil physical properties. Green manure crops ensure ecological sustainability by maintaining the productivity of the soil over a long period by protecting soil from erosion. Depending upon the species and locations, green manure crops supply 40 to 120 kg N/ha. This amount would be equal to the application of three to ten tons of FYM on the basis of organic matter and its N contribution. Green manuring is useful in minimizing the ill effects of intensive agriculture particularly on natural resources. Plant nutrients are provided in a better form and over a longer period for the crops grown after green manuring. However, the choice of green manuring crops has to be made in relation to soil, climate and time available to raise the green manure crop and the facilities for irrigation. It has received a new impetus in recent years with an urgent need for increased food production in the country (Barzegar *et al.*, 2003). List of green leaf manuring crops/trees and green manuring are furnished in Tables 3.3 and 3.4.

Table 3.3: List of Green Leaf Manure Crops

Green Manure Crop	*Scientific Name*	*Nutrient Composition (per cent) on Dry Matter Basis*
Gliricidia	*Glyricidia sepium/ maculate*	Each plant gives 5-10 kg of green leaves annually 2.76 per cent N, 0.28 per cent P_2O_5, 4.60 per cent K_2O
Subabul	*Leucaena leucocephala*	Leaves contain 3-4 per cent N Fixes 500-600 kg N/ha/year
Peltophorum	*Peltoforum ferrugenum*	2.63 per cent N, 0.37 per cent P_2O_5, 0.50 per cent K_2O
Pongania	*Pongamia glabra*	3.31 per cent N, 0.44 per cent P_2O_5, 2.39 per cent K_2O
Neem	*Azadirachta indica*	2.83 per cent N, 0.28 per cent P_2O_5, 0.35 per cent K_2O
Gulmohur	*Delonix regia*	2.76 per cent N, 0.46 per cent P_2O_5, 0.50 per cent K_2O

Other commonly used green leaf manuring crops/tress are *Pongamia, Calotropis, Thespesia, Peltophorum, Morianga oleflorea, Agave sisalana, Thesphesia populanea, Cassia siamea, Vitex negundo, Anacardium spp., Bauhina vahii, Bretusa spp.*, Lentil, *Pisum sativum*, Chrysopogon, *Musa sapietum*, (ubiquitous in nature), *Quercus leucotrichophora, Quercus semicarpifolia, Daphiniphyllum himalayaense, Bridelia retusa, Nyctanthes arbortristis (lesser Himalayalan regiona), Melia azidirach, Cocos nucifera, Pongamia pinnata* (Deccan plateau), *Mangifera indica, Artocarpus spp., Adina cardifolia,Terminalia paniculata, Pterocarpus marsupium, Gmelina arborea, Tectona grandis* (Western ghats) and *Syzgium cumini, Terminalia bellerica, Cleistanthes collinus, Glyricidia spp.*(West Bengal). Life fences and hedge rows include *vitex negundo, Nerrium thevitifolia, Adathoda vasica, Datura stramonium, Artemisia nilagirica.*

Hemalatha *et al.* (2000) reported that amongst the organic treatments, *in-situ* incorporation of dhaincha @ 12.0 t/ha recorded taller plants, higher number of tillers per hill, leaf area index and dry matter production, grain yield and straw yield of

Table 3.3: List of Green Manures

Green Manure Crop	*Scientific Name*	*Special Characters*	*Fresh Matter (t/ha)*	*Fixation of N (kg/ha)*
Dhaincha	*Sesbania aculeate*	Commonly cultivated for green manure, fodder, non perennial temporary shade in crop field and as wind breaks; Highly resistant to drought and withstands water logging and salinity (Kalidurai and Kannaiyan, 1988). Root nodulating legume Seed rate: 50 kg/ha for green manure 3.50 per cent N, 0.60 per cent P_2O_5, 1.20 per cent K_2O	15-20	75-80
Sesbania	*Sesbania speciosa*	2.71 per cent N, 0.53 per cent P_2O_5, 2.21 per cent K_2O	-	-
Sunhemp	*Crotalaria juncea*	Quick growing green manure cum fibre crop and can't withstand heavy irrigation or waterlogging; 25-35 kg seed/ha for green manure 2.30 per cent N, 0.50 per cent P_2O_5, 1.80 per cent K_2O	13-15	-
Sesbania	*Sesbania rostrata*	Stem and root nodulating green manure crop; Thrives well under waterlogged condition; Seed rate: 30 to 40 kg/ha	15- 20	150-180
Wild indigo or kolingi	*Tephrosia purpurea*	Slow growing green manure crop not useful for fodder; Suitable for light soils Resists drought but does not withstand water stagnation; Seed rate: 20-25 kg/ha	8 to 10	-
Cowpea	*Vigna unguiculata*	Most productive heat adapted annual legume Used as grain/green manure crop, for animal fodder and also as a vegetable; 40 kg seed/ha for green manure; 1.4-1.5 per cent N	9-10	140-150
Pillipesara	*Phaseolus trilobus*	Dual propose crop which can produce green fodder and manure; 10-15 kg seed/ha is required for green manure purpose	6-7	-
Clusterben	*Cyamposis tetrago-naloba*	Drought tolerant annual legume	-	-
Azolla	*Azolla pinnata*	50-90 kg seed/ha for green manure crop Incorporation at 35 DAS	8-10	52

rice over no manuring. Singh *et al.* (2004) observed that application of 50 per cent per cent of the recommended dose of nitrogen (RDN) through inorganic fertilizer (IF) + dhaincha (*Sesbania aculeata*) at 2.5 tonnes/ha to rice gave significantly higher mean grain yield of rice (31.8 q/ha and wheat (28.4 q/ha) over the rest of the treatments, except for 50 per cent per cent RDN through chemical fertilizers + pressmud at 5 tonnes/ha in rice. The yield of succeeding wheat crop increased by 43.1-48.9 per cent because of the integrated nutrient management over the control. Rice crop gave significantly higher grain yield (4214 kg/ha) when either total plant (root+shoot) or only shoot portion of sunhemp was incorporated rather than only root portion and no incorporation. With shoot or total incorporation, 60 kg N appeared to enough as the grain yields with this dose were comparable with that of 120 and 180 kg N/ha. When root portion was alone incorporated, N requirement was 120 kg/ha while, it was 180 kg/ha without green manuring (Neelima *et al.*, 2007).

2.1.5. Biofertlizers

Biofertilizers are ready to use live formulates of such beneficial microorganisms which on application to seed, root or soil mobilize the availability of nutrients by their biological activity in particular, and help build up the micro-flora and in turn the soil health in general. They may be Nitrogen fixing, Phosphorus solubilising or mobilizing and plant growth promoting as detailed below in Table 3.5. Bio-fertilizers help to increase the economic yield by 20-30 per cent besides reduces N and P consumption by 25 per cent across various crops. Further, they stimulate

Table 3.5: Classification of Biofertilizers

Sl.No.	*Groups*	*Examples*
N fixing biofertilizers		
1.	Free-living	*Azotobacter, Beijerinkia, Clostridium, Klebsiella, Anabaena, Nostoc,*
2.	Symbiotic	*Rhizobium, Frankia, Anabaena azollae*
3.	Associative Symbiotic	*Azospirillum*
Phosphate solubilising		
1.	Bacteria	*Bacillus megaterium* var. *phosphaticum, Bacillus subtilis Bacillus circulans, Pseudomonas striata*
2.	Fungi	*Penicillium* sp., *Aspergillus awamori*
Phosphate mobilizing		
1.	Arbuscular mycorrhiza	*Glomus* sp., *Gigaspora* sp., *Acaulospora* sp., *Scutellospora* sp. and *Sclerocystis* sp.
2.	Ectomycorrhiza	*Laccaria* sp., *Pisolithus* sp., *Boletus* sp., *Amanita* sp.
3.	Ericoid mycorrhizae	*Pezizellaericae*
4.	Orchid mycorrhiza	*Rhizoctonia solani*
Biofertilizers for micronutrients		
1.	Silicate and Zinc solubilizers	*Bacillus* sp.
Plant growth promoting rhizobacteria		
1.	*Pseudomonas*	*Pseudomonas fluorescens*

plant growth and maintains/improves soil fertility. They are also cost effective. However, care should be taken for safe storage under tropical conditions. Farmers should be educated bout biofertilizers regarding their specificity to the crop, method and time of application and safe storage methods.

Major biofertilisers and target crops include *Rhizobium* for Leguminous crops among pulses, oilseeds and fodder), *Azatobacter* for wheat, rice, vegetables, *Azospirillum* for rice, and sugarcane, Blue Green Algae (BGA) and Azolla for rice (Ghayur Alam, 2000). Total biomass production was significantly higher with the application of 40 kg of N/ha, as compared to the check treatment (0 kg of N/ha), but the differences due to seed treatment with either of the *Azospirillum* strains were not significant (Table 3.6). Biological activity in the rhizoshpere is influenced by soil moisture and as well the organic carbon content. Alvey *et al.* (2003) analyzed the changes in rhizosphere and found that cropping system had a highly significant effect on eubacterial and ammonia-oxidizing community structure ($p < 0.005$), irrespective of plant species or sampling time. Ideal rhizosphere conditions are thus essential for the growth and consequent beneficial effects of *Azospirillum* on sorghum since it has a loose association with the plant system Vikas *et al.* (2013).

Table 3.6: *Azospirillum* and Nitrogen Effect on Sorghum Total Biomass Yield during *Kharif* Season (mean values of Parbhani, Indore and Udaipur)

Azospirillim Strain	*Plant Stand ('000/ha)*		*Total Biomass (t/ha)*	
	No N	*40 kg N/ha*	*No N*	*40 kg N/ha*
Dharwar strain	149	176	11.67	14.58
Coimbatore strain	155	176	11.76	14.73
No seed treatment	158	182	11.93	15.40
CD (5 per cent)	8.7		0.72	

Jat and Maitry (2006) reported that seed inoculation with PSB had significant positive effect on wheat yields, net returns (Rs. 31336/ha) and B:C ratio (3.18) as compared to no inoculation. According to Ramanjaneyulu *et al.* (2007), half RDF+biofertilizers (*Azotobacter*+PSB) performed more or less equally with that of RDF in respect of seed, oil and protein yields and total productivity of fodder sorghum-mustard sequence. Besides, saving half the recommended dose of N and P, this treatment also gave higher BC ratio. Further, biofertilizers bring about overall improvement in soil, physical, chemic and biological properties in the long run.

2.1.6. Crop Residues

Substantial quantities of crop residues are produced in India every year (Table 3.7). Major crops like rice, wheat sorghum, pearl millet and maize alone yield approximately 236 MT straw per year. The nutrient potential of cereal straw/residue from five crops comes to 1.13 MT N, 1.41 MT P_2O_5 and 3.54 MT K_2O Crop residues can be recycled either by compositing or by way of mulch or direct incorporation in the soil (Anthony *et al.*, 2003).

Table 3.7: Potential of Farm Residues and Plant Nutrients

Crop Residue	*Crop Residue Production (MT)*	*Per cent Oven Dry Basis*		
		N	*P_2O_5*	*K_2O*
Rice	106.01	0.58	0.23	1.66
Wheat	80.99	0.49	0.25	1.28
Sorghum	21.04	0.40	0.23	2.17
Pearlmillet	15.58	0.65	0.75	2.50
Maize	12.50	0.59	0.31	1.31
Total pulses	13.70	1.60	0.15	2.00
Pigeonpea	6.65	1.10	0.58	1.28
Chickpea	5.05	1.19	NA	1.25
Sugarcane	40.92	0.35	0.04	0.50
Oilseeds	35.78	NA	NA	NA

Annual applications of 1500 kg/ha of leaf litter from different locally grown shrubs for five seasons resulted in increases in rice grain yield in 1997 of between 20 and 26 per cent above the no-leaf litter control. Nutrient balances, determined by the difference between the inputs (fertilizer and added leaf litters) and outputs (grain and straw), indicated net positive balances of up to 457 kg N/ha and 60 kg P/ha, after five seasons of leaf litter applications. Soil carbon (C) concentrations increased significantly only where higher fertilizer rate and rice stubble retention were combined. According to Collins *et al.* (1992), burning residues reduced microbial biomass to 57 per cent of that in plots receiving farmyard manure. Microbial C represented 4.3, 2.8, and 2.2 per cent and microbial N 5.3, 4.9, and 3.3 per cent of total soil C and N under grass pasture, annual cropping, and wheat-fallow, respectively (Chadha *et al.*, 2004).

2.1.7. Cropping System/Diversity/Mixed Cropping

Crop Rotations

The practice of growing crops sequentially on a unit of land is called crop rotation. Crop rotation has many agronomic, economic and environmental benefits compared to monoculture cropping. Appropriate crop rotation increases organic matter in the soil, improves soil structure, reduces soil degradation, and can result in higher yields and greater farm profitability in the long-term. Increased levels of soil organic matter enhances water and nutrient retention, and decreases synthetic fertilizer requirements. Better soil structure in turn improves drainage, reduces risks of water-logging during floods, and boosts the supply of soil water during droughts. Moreover crop rotation effectively delivers on climate change mitigation. Leguminous crops in the rotation fix atmospheric nitrogen and bind it in the soil, increasing fertility and reducing the need for synthetic fertilisers. The objective is to maximize profit with least investment without impairing soil fertility. The selection of crops should be based on demand, problem, pest, disease and weed menace. Besides, resources available and needs of the farmers or market will also be considered.

Kumar *et al.* (2003) while analyzing the All India Coordinated Sorghum Improvement project (AICSIP) data found that across the two locations in Maharahstra, the depressing effect on grain yield was clearly seen among all the nutrient management practices in the sorghum after sorghum annual rotation (Table 3.8). The reduction ranged from (377 kg) in the check (crop residue incorporation) treatment to (1341 kg) in inorganic recommended dose of fertilizer treatment across the cropping systems. Inclusion of Pigeon pea and Cotton in rotation with the sorghum crop has the advantage of a nitrogen fixing legume on one hand and exploitation of soil layers at different depths with reference to Cotton crop.

Table 3.8: Cropping System and Nutrient Management Interaction Effect on Sorghum Grain Yield

Rotation	*Grain Yield (kg/ha)*			
	Nutrient Management			
	Crop Residue	*FYM*	*RDF*	*FYM + ½ RDF*
Cotton - sorghum	1936	2490	3794	3258
Pigeonpea – sorghum	1993	2566	4258	3764
Sorghum - sorghum	1559	1939	2917	2647
CD (5 per cent)	165			

Mean values of Parbhani and Akola locations for 4 years.

Across the Mauranipur and Udaipur locations the positive effect on grain yield was clearly seen among all the nutrient management practices in the sorghum after soybean annual rotation (Table 3.9). The increase ranged from 16 kg to 329 kg/ha across fertilizer treatments when compared to groundnut-sorghum and sorghum-sorghum cropping systems. Supplementing the inorganic fertilizers with organic fertilizers to an extant of 50 per cent resulted in a decline of 85 to 396 kg grain across all the cropping systems.

Table 3.9: Cropping System and Nutrient Management Interaction Effect on Sorghum Grain Yield

Rotation	*Grain Yield (kg/ha)*			
	Nutrient Management			
	Crop Residue	*FYM*	*RDF*	*FYM + ½ RDF*
Groundnut - Sorghum	1747	1979	2350	1954
Soyabean – Sorghum	1763	2053	2444	2283
Sorghum - Sorghum	1703	1996	2302	2217
CD (5 per cent)	121			

Mean values of Mauranipur and Udaipur locations for 4 years.

Mixed Cropping

The practice of cultivation of a number of crops in a unit of land is called mixed cropping. Mixed cropping may be random mixed (seeds of several crops are mixed together and sown), line sown mixed (sowing in separate lines) or strip cropping (different crops in different strips). Navdhanya agro-biodiversity farm is a living example of mixed farming. In the farm, one can observe a combination of crops in a unit of land. Mixed cropping is a strategy for crop protection. Likewise, many farmers practice mixed cropping wherein, they grow ginger or turmeric as cash crops, maintain castor plants along the borders of their fields in order to check the spread of diseases. In the wheat chickpea association, chickpea serves to stop rats from entering wheat crop. Mixed cropping and rotation of crop has been traditional practice in Indian farming. For eg; while drilling a crop such as jowar (*Sorghum vulgare*) or some other millet, arhar (*Cajanus cajan*), a legume also will be drilled.

The International Institute for Tropical Agriculture (IITA, Ibadan, Nigeria) has shown that soil erosion and run– off losses are proportionately lower from mixed as compared to monocultures. Mixed cropping also conserves moisture by reducing evaporation and improving water use efficiency. Chickpea alone used 12.51 cum of water and gave 10.68 Q/ha with a water use efficiency of 0.85 and barley alone used 14.91 cum of water and gave 16.41Q/ha with a water use efficiency of 1.85, a mixture of barley and grain used 15.89 cum of water, yielded 17.92 Q/ha grain used and increased water use efficiency to 1.91.

In Garhwal Himalayas, the practice of Baranaja a mixture of 12 grains is still prevalent. The grains are Phapra (*Fagopyrum tataricum*), Mandua (*Eleusine coracana*), Marsha (*Amaranthuys frumentaceous*), Bhat (*Glycine soja*), Lobia (*Vigna catiang*), Gahath (*Dolichos biflorus*), Rajma (*Phaselous vulgaris*), Jakhia (*Cleome viscosa*), Navrangi (*Vigna umbellata*), Jowar (*Sorghum vulgare*), Urad (*Phaselous mungo*). In the hill system, amaranth/mandua based cropping system is prevailing (60 per cent). There are also sesamum based cropping systems (30 per cent), buckwheat based farming systems and fruits (5 per cent). This agro biodiversity is yet one of the important aspects of the sustainable food system that has been practiced in the Himalayan region of the country. In the West arid zone of Rajasthan, the practice of mixed cropping is still a prevalent practice wherein the farmers grow a multitude of crops in the same patch of field. It is a survival imperative in which the farmers manage to harvest crop diversity during the *kharif* season, which is the only season they can grow crop due to paucity of rains. The crops grown primarily are millets-bajra and then jowar are the most important grain crops and guar-the main pulse-often with mung, moth, lobia, arhar and urad also being added as mixture. The grain varieties that provide fodder are a very important consideration with these farmers, as a large percentage of the population is pastoral. In the Southern part India, the farmers' totally relied on the indigenous varieties of ragi, pulses and groundnut (Scialabba and Müller-Lindenlauf, 2010). In Karnataka *e.g.* ragi or finger millet (*Eleusine coracona*) is the staple crop of the area. This crop is extremely drought resistant. As there are varieties of ragi available, the farmers chose the variety they wish to sow keeping its gastronomic, nutritional quality in mind. It has been observed that mixed agro-forestry is also practiced in the Western Ghats of Karnataka and Kerala. The spice gardens nowadays are grown in a 3 tier fashion in which in the first year Banana saplings are planted, during the second year, areca saplings are planted at a distance of 16 feet in a triangular shape. After a span of 15-20 years they again grow the third canopy of young sapling. The saplings of the next crop replace the old tree. Thus, a system has been evolved that ensures a systematic and sustained yield for the farmer. The spice garden has banana, cardamom and black pepper as an intercrop. Each one of these crops has its own advantage and they contribute to healthy growth of spice gardens. Likewise Tamil Nadu, the southernmost state of India, has a rich agro-biodiversity of millets, oilseeds (sesame and sometimes cardamom) and pulses such as red gram and lablab seeds dropped separately by hand in the furrows and covered with the plough (Salvador, 2003).

Cover Crops

Cover crops are planted to cover the surface of soil during fallow period of the cropping cycle. The benefits include soil and moisture conservation, improvement in soil structure and fertility, suppression of weed growth and reduction in chemical fertilizer requirements. Plants selected for this purpose should be fast growing, pest and disease resistant and suitable for wide range of soil types. *Dolichus lablab, Crotalaris sp., Canavalia sp.,Vigna sp., Tephrosia sp., Dioscroea sp., and pomea batatas* are commonly used cover crops (Roy Rabindra *et al.,* 2002).

2.1.8. Mulching

It is method of covering the surface of the soil with any decomposable material (grass, hay, paper, kitchen wastes, leaves, twigs, and plant residues) so that the soil is not exposed to the drying action of the sun or the desiccating action of the wind (Jacks *et al.,* 1955). Benefits of mulching include addition of organic matter and humus on decomposition and improves organic carbon content there by improves beneficial micro flora which plays an important role in recycling of nutrients and nitrogen fixation, phosphate solubilization and photosynthesis activity, cellulolytic activity; improves infiltration of rain water, moderation of soil temperature, minimises evaporation thus retain/conserve soil moisture thus relieve the plants from stress during dry spell. Surface mulch has been reported to conserve soil moisture and improve water use efficiency (Hameeda *et al.,* 2006). In the long term experiment at ICRISAT, it has been reported that mulch applied in this manner on the hottest day of summer (April 30) in 2002 the soil temperature at 5 and 10 cm depth in the mulch applied plots was 6.5 to 7.3° C lower planting different types of trees like neem, amla, tamarind, gular, zizipus bushes, gliricidia on bunds.

2.2. Weed Management

The main aim of weed management is not eradication but to arrest the weed growth and keep it under check. Wide ranging options are available for this purpose as furnished below (Table 3.10).

Different crops have different competitive abilities. In annual crops, cereals are considered to have the strongest competitive ability against weeds followed by oilseed rape, peas and potatoes/vegetables (Håkansson, 2003). Lundkvist *et al.* (2008) showed that peas, a weak competitor, had significantly higher weed biomass at harvest compared with oats and winter wheat. Autumn-sown cereals and oilseed rape also seem to have a stronger weed suppressing ability compared with corresponding spring-sown crops. Well established perennial leys or pastures are usually very competitive against weeds, while first year leys or pastures may be rather susceptible to weed competition. The relative competitive power is also affected by seed rate (relative plant density) and also time of sowing (relative emergence time) which affects the emergence of the crop in comparison with the weeds. The choice of crop rotation strongly affects the abundance and diversity of the weed flora (Bicksler and Masiunas, 2009). Since different crops favour different types of weed species, it is important to change between annual and perennial crops in the crop rotation. Autumn and spring sown annual crops also favour different types of weed species, which makes it important to rotate between such crops within a crop rotation.

Table 3.10: A Summary of Non-chemical Weed Management Tools

Weed Management Method	*Practices*
Preventive	Use of clean seed, avoid soil and sand transport from weed infested areas, avoid feeding cattle with weed seed containing fodders, use well rotten and decomposed organic manure, keep irrigation channels weed free, clean the farm machinery and tools before and after use
Good crop husbandry	Selective crop stimulation, proper planting method/time, crop rotation, stale seed bed, smother cropping, summer ploughing, minimum or zero tillage, flooding and drainage and lowering area under bunds, competitive crops and varieties, allelopathic crops, use of fertilizers and manures
Physical (manual/ Mechanical)	Hand pulling/weeding, hoeing, tillage, mowing, chaining, cutting, dredging, burning, mulching, soil solarisation, cropping system, cover crops, flame weeders
Biological	Parasites, predators and pathogens
IWM (Integrated weed management)	Combination of all above methods depending on the resources availability, crop, cost effectiveness, problem and demand

Integrated weed management combines various preventative, cultural, genetic and mechanical, biological weed control practices into a single program. While no single control measure is likely to provide complete weed control, the systematic implementation of the various components of integrated weed management can make significant contributions to weed control efforts (Knezevic *et al.*, 2012). Non-chemical weed control relies primarily on tillage and hand weeding, practices which are labor intensive and expensive Lack of available labor, and high wage rates prohibit use of these techniques for agricultural production. (Gianessi and Reigner, 2007). At the Wageningen University in the Netherlands, researchers documented the efficacy of multiple preventive measures (stale seedbed, including weed-competitive crops in the rotation, timely post-harvest tillage) in reducing weed pressure. A review of five studies comparing organic and conventional production of field corn and soybeans showed that organic fields had many more weeds (29 to 2000 per cent increase), but showed only a 0 to 18 per cent decrease in yield compared to conventional. In other words, the weeds in the organic system did not compete as severely against the crop as expected based on their abundance. The organic systems were more diverse, with more plant species in the crop rotation, and a greater variety of nutrient sources (Johnson *et al.*, 2010). St. Johnswort (*Hypericum perforatum* L.) on rangelands in the USA and Canada is being controlled by means of the leaf beetle *Chrysolina hyperici* Forster showed that part of this success may be attributed to the fungus *Colletotrichum gloeosporioides* Penz., which is transferred by the leaf beetle (Sheppard *et al.*, 2006).

2.3. Insect-Pest and Disease Management

Insect-pest and disease management presents a challenge to organic farmers due to non-availability of green pesticides and host plant resistance for the major pests and disease across different crops **(Lewis, 1997)**. However, cultural, physical/ mechanical and biological methods if employed either alone or in combinations depending on the severity and target pest or disease, economic loss can be minimized (Table 3.11).

Table 3.11: A Summary of Non-chemical Pest and Disease Management Tools

Management Method	*Practices*
Preventive	Use of clean seed, effective weed control use well rotten and decomposed organic manure
Physical/Mechanical	Hand picking, summer deep tillage soil solarisation, crop rotation, cropping system, trap crops, light and pheromone traps, resistant cultivars, field sanitation, neem oil and other preparations, Propalis, plant based repellants, silicates, soft soap, planting time, removal of affected plants and plant parts, stem application, seed treatment, collection and destruction of egg masses and larvae, installation of bird perches
Biological	Parasites, predators and pathogens
Biointensive Pest Management (BIPM)	Combination of all above methods depending on the resources availability, crop, cost effectiveness, problem and demand

In organic management, protection measures are used only in the case of problematic situations. Use of disease free seed stock and resistant varieties is the best option. For *e.g.* In case of redgram, PRG-158 (light soils) and Asha (heavy soils) which is resistant to wilt has to be used in wilt endemic areas. Likewise, in rice, RNR 15048 has to be used in stem borer endemic areas. In certain cases, hot water treatment at 53°C for 20-30 minutes for sugarcane, cow urine or cow urine-termite mound soil paste and *Trichoderma viride* (10gm/kg seed) or *Pseudomonas fluorscens* (10gm/kg seed) and Biofertilizers (*Rhizobium*/*Azotobacter* +PSB) can be used for seed treatment of different crops.

Kumar and Subbarayudu (2004) while discussing the measures for shoot fly management in sorghum emphasized the importance of sowing time and integrated pest management modules. Planting time is an important component of the management strategy to minimize the pest incidence levels and also gain initial advantage of the high soil temperature, that drives crop growth and development processes. Multi-location trials in AICSIP have indicated an advantage of 10 per cent grain yield increase due to dry sowing ahead of the onset of monsoon. The percent difference in shoot fly eggs and its dead hearts in module I over module II was 29.97 per cent and 41.25 per cent, respectively. Integrating eco-friendly components *viz.*, early sowing, seed treatment, spraying of botanicals like Neem Seed Kernel Extract *etc.*, helped develop a cost effective biotic stress management strategy for a pest like shoot fly in sorghum. Krishna Moorthy, (2005) has suggested tillage operations, soil solarization, adjusting the time of sowing, wider spacing, growing pest resistant cultivars, trap or barrier cropping, field sanitation, crop rotation under cultural methods, physical removal of pests or use of light and pheromone traps under mechanical methods, along with permitted inputs to protect the crop from insects/disease in organic crop production.

Use of pest predators and pathogens has also proved to be effective method of keeping pest problem below ETL (Singh *et al.*, 2012). Inundative release of *Trichogramma sp.* @ 40,000 to 50,000 eggs/ha, *Chelonus blackburni* @15,000 to 20, 000/ha, *Apanteles* sp.@15,000 to 20,000/ha and *Chrysoperla sp.*@ 5,000/ha after 15 days of sowing and other parasites and predators after 30 days after sowing can also effectively control pest problem in organic farming (Srijita, 2015). Biopesticide approach in which a biocontrol agent is applied as and when required (often repeatedly), in the same way as a chemical control agent is used. Examples of this include the use of *Bacillus thuringensis, Phoebiosis gigantean*, and *Agrobacterium radiobacter* (Suman Gupta, 2010). Use of Biopesticides like *Trichoderma viride* or *T. harazianum* or *Pseudomonas fluorescence* formulation @ 4gm/kg seed either alone or in combination, manage most of the seed borne and soil borne diseases. There are other formulations *viz. Beauvaria bassiana, Metarizium anisopliae, Numeria rileyi, Verticillium* sp, which are available in the market and can manage their specific host pest (Rupela *et al.*, 2005). Temperature control can also be achieved by viral biopesticides of baculovirus group *viz.* granulosis viruses (GV) and nuclear polyhedrosis viruses provided a great scope in plant protection field. Spray of nuclear polyhedrosis viruses (NPV) of *Helicoverpa armigera* (H) or *Spodoptera litura* (S) @ 250 larval equivalents are very effective tools to manage the *Helicoverpa* sp.or *Spodoptera* sp., respectively (Thakore, 2006).

Neem (*Azadirachta indica*) – Neem has been found to be effective in the management of approximately 200 insects, pests and nematodes. Neem is very effective against grasshoppers, leaf hoppers, plant hoppers, aphids, jassids, and moth caterpillars. Neem extracts, are also very effective against beetle larvae, butterfly, moth and caterpillars such as Mexican bean beetle, Colorado potato beetle and diamondback moth. Neem is very effective against grasshoppers, leaf minor and leaf hoppers such as variegated grasshoppers, green rice leaf hopper and cotton jassids. Neem is fairly good in managing beetles, aphids and white flies, mealy bug, scale insects, adult bugs, fruit maggots and spider mites (Subbalakshmi *et al.*, 2012).Many organic farmers and NGOs have developed large number of innovative formulations which are effectively used for control of various pests. Some of the popular formulations are listed below:

Cow urine (diluted with water in ratio of 1:20) as foliar spray is not only effective in the management of pathogens and insects, but also acts as effective growth promoter for the crop. Neem-Cow urine extract – (Crush 5 kg neem leaves in water, add 5 litres cow urine and 2 kg cow dung, ferment for 24 hrs with intermittent stirring, filter squeeze the extract and dilute to 100 lit), can be used as foliar spray over one acre against sucking pests and mealy bugs. Likewise, Mixed leaves extract (Crush 3 kg neem leaves in 10 lit cow urine. Crush 2 kg custard apple leaf, 2 kg papaya leaf, 2kg pomegranate leaves, 2 kg guava leaves in water. Mix the two and boil 5 times at some interval till it becomes half. Keep for 24 hrs, then filter squeeze the extract) Dilute 2-2.5 lit of this extract to 100 lit for 1 acre) is useful against sucking pests, pod/fruit borers.

Chilli-garlic extracts (Crush 1 kg Ipomea (besharam) leaves, 500 gm hot chilli, 500 gm garlic and 5 kg neem leaves in 10 lit cow urine. Boil the suspension 5 times till it becomes half. Filter squeezes the extract. Store in glass or plastic bottles. It can be used as 2-3 lit extract diluted to 100 lit for one acre and is useful against leaf roller, stem/fruit/pod borer.

2.4. Chemicals Permitted in Organic farming

According to USDA, the following most important 10 chemicals which do not contain heavy metals and exert adverse effect, are permitted for use in organic farming for various purposes.

2.4.1. Alcohols

Alcohols that include ethanol and isopropanol are permitted in organic farming for use as a disinfectant, algicide and sanitizer. These alcohols are approved for use during irrigation system cleaning.

2.4.2. Hydrogen Peroxide

It can be used for pest control, and as a disinfectant, algicide and sanitizer for irrigation systems

2.4.3. Soap

The use of soaps is allowed for large animal control, provided that no soap comes in contact with the food product. Further, it is also used as herbicides for roadways, ornamental crops and insecticides provided no soap comes in contact with soil

2.4.4. Vitamin D_3

Vitamin D3 can be used for slug and snail bait and also for plant disease control.

2.4.5. Boric Acid

Boric acid is allowed as an insecticide for structural buildings provided there is no contact with soil or crops.

2.4.6. Lignin Sulfonate

Lignin sulfonate is used as a soil amendment, chelating agent, dust suppressant and a floatation agent (post-harvest handling).

2.4.7. Magnesium Sulfate

Magnesium sulfate is allowed as a supplement to correct the deficiency only when the deficiency has been documented by a third party.

2.4.8. Streptomycin

Streptomycin is allowed on apples and pears only. It is used to fight fire blight, a bacterial infestation. Several other chemicals may be used in organic farming, provided they are only used for their stated purpose. These include:

- Tetracycline: For fire blight control
- Peracetic acid: For fire blight control
- Hydrated lime: For plant disease control
- Oils: For plant disease control
- Potassium bicarbonate: For plant disease control
- Humic acids: As a soil amendment
- Micronutrients: As a soil amendment once a deficiency has been documented
- Sulfates: As a soil amendment
- Ethylene gas: For regulation of pineapple flowering
- Sodium silicate: For tree fruit and fiber processing in post-harvesting
- Liquid fish products: As a soil amendment
- Pheromones: As insect management
- Ammonium carbonate: As bait in insect traps
- Ozone gas: For irrigation system cleaning
- Chlorine materials: For irrigation system cleaning
- Calcium hypochlorite: For irrigation system cleaning
- Sodium hypochlorite: For irrigation system cleaning
- Rock phosphate from mined source for P nutrition
- Spinosad (from sea algae) for lepidopteran pest control

3. Conclusions

The central government has initiated Paramparagat Krishi Vikas yojana (PKVY) to promote organic farming in the country besides establishing/promoting institutions of certification, channelizing the produce for marketing to get a premium price. Due to increase in prices of fertilizers and chemicals, the cost of cultivation has increased without corresponding increase in price of the end product. Conventional farmers have begun their search for ways to decrease input costs and as well are showing interest towards organic cultivation. Growing awareness across consumers for quality/nutritious food, has led to the growth of organic farming which is economically viable due to reduction in the use of external inputs and increased use of farm organic inputs. This has greatest potential to benefit the soil health, farmer profits and consumer preferences. However, farmers should be trained in production and use bio-fertilizers, bio-pesticides, botanical pesticides and other organic inputs. Green manure seeds, bio- fertilizers and bio-pesticides should be made available to the farmers at an affordable price. Urban wastes should be properly collected, composted and made available to the farmers. The location specific and profitable organic packages/modules developed in different crops by the Indian Council of Agricultural Research (ICAR), State Agricultural Universities (SAU's) and other organizations across the country, should be digitally made available. Government may create introduce minimum support price (MSP) for organically produced crops. Organic food can be introduced as a part of mid-day meal scheme for children across the country so as to address child health issues as well create demand for these nutritious products.

REFERENCES

Alaa Hasan E, Karim Bhiah M and Mushtaq TH. 2014. The impact of peat moss and sheep manure compost extracts on marigold (Calendula officinalis L.) growth and flowering. Journal of Organic Systems, 9(2): 25-29.

Alvares C. 1996. The organic farming source book. The Other India Press, Mapusa, Goa, India.

Alvey S, Yang CH, Buerkert A and Crowley DE. (2003). Cereal/legume rotation effects on rhizosphere bacterial community structure in west african soils. Biology and Fertility of Soils, 37: 73-82.

Amir K and Fouzia, 2011. Chemical nutrient analysis of different composts (Vermicompost and Pitcompost) and their effect on the growth of a vegetative crop Pisum sativum. Asian Journal of Plant Science and Research, 1(1): 116–130.

Anthony W, Graeme B, Yothin K, Rod L and Kunnika N 2003. Managing crop residues, fertilizers and leaf litters to improve soil C, nutrient balances, and the grain yield of rice and wheat cropping systems in Thailand and Australia. Agriculture, Ecosystems and Environment, 100 (2–3): 251–263.

Antil RS, Vinod K, Kathpal TS, Narwal RP, Sharma SK, Mittal SB, Joginder Singh and Kuhad, M.S. 2004. Extent of land degradation in different agro-climatic zones of Haryana. Fertiliser News, 49(3): 47-54.

Barzegar AR, Asoodar MA, Khadish A, Hashemi AM and Herbert, SJ. 2003. Soil physical characteristics and chickpea yield responses to tillage treatment. Soil Tillage Research, 71: 48-57.

Bhaskaran, Usha P and Devi K. 2009. Effect of organic farming on soil fertility, yield and quality of crops in the tropics. In: *The proceedings of the International Plant Nutrition Colloquim XVI*, UC Davis. Available online: http: //www. escholarship. org/uc/item/7k12w04m.

Bhattacharya P. and Chakraborty G. 2005 "Current status of organic farming in India and other countries," Indian Journal of Fertilizers, 1(9): 111–123.

Bicksler AJ and Masiunas J B. 2009. Canada thistle (Cirsium arvense) suppression with buckwheat or sudangrass and mowing. Weed Technology, 23 (4): 556-563.

Chadha GK, Sen S and Sharma HR. 2004. State of the Indian farmer: A millennium study, Land Resources. Department of Agriculture and Cooperation, Ministry of Agriculture, Government of India, New Delhi.

Chadha SL, Gopinath N. and Shekawat S. 1997. Urban–rural differences in the prevalence of coronary heart disease and its risk factors in Delhi. Bull. World Health Organisation, 75: 31–8.

Chhonkar PK. 2003. Organic farming: science and belief. Journal of the Indian Society of Soil Science, 51(4): 365-377.

Collins HP, Rasmussen PE and Douglas, CL. 1992. Crop Rotation and Residue Management Effects on Soil Carbon and Microbial Dynamics, 56 (3): 783-788.

Dhaliwal and Pathak, 1993. *Pesticides* in the developing world: a boon in bane, In: *Pesticides*. Their Ecological *Impact* in developing countries 1-29.

Dwivedi PD, Mukul Das and Khanna SK. 2004. Need for upgrading food quality monitoring capacities in the new globalized market scenario. Current Science, 87(6): 729-730.

FAO. 2001. A report of the round table on organic agriculture and climate change.

Gaur AC, Nilkantan S and Dargan KS. 2002. Organic Manures, ICAR, New Delhi, India.

Ghayur Alam 2000. A study of biopesticides and biofertilisers in Haryana, India.

Gianessi L and Reigner N. 2007. Barriers to widespread conversion from chemical pest control to non-chemical methods in U.S. Agriculture.

Gopalan C. 1997. Diet related non-communicable diseases in South and South East Asia. Lessons from Contrasting Worlds. London: John Wiley and Sons, pp. 10–23.

Gopinath KA, Saha S, Mina BL, Kundu S, Pande H and Gupta HS. 2008. Influence of organic amendments on growth, yield and quality of wheat and soil properties during transition to organic production. Nutrient Cycling in Agro-ecosystems, 82: 51-60.

Gupta RK and Abrol IP. 2000. Salinity build-up and changes in the rice-wheat system of the Indo-Gangetic plains. Experimental-Agriculture, 36(2): 273-284.

Håkansson S. 2003. Weeds and Weed Management on Arable Land – An Ecological Approach. CABI Publishing. Wallington, Oxon, UK.

Hameeda B, Rupela OP, Wani SP and Reddy G. 2006. Indices to assess quality, productivity and sustainable health of soils receiving low-cost biological and/ or conventional inputs. International Journal of Soil Science, 1(3): 196-206.

Handa SK. 1995. The pesticide industry. Kothari's Desk Books, New Delhi. 383-388.

He X, Logan T, Traina S. 1995. Physical and chemical characteristics of selected U.S. municipal solid waste composts. Journal of Environment Quality, 24: 543–552.

Hemalatha M, Thirumugan V and Balasubramanian R. 2000. Effect of organic sources of nitrogen on productivity and quality of rice (*Oryza sativa*) and soil fertility. Indian Journal of Agronomy, 45(3): 564-567.

IFOAM 2002. Organic Agriculture and Food Security.

Jacks CV, Brind WD and Smith R. 1955. Mulching Technology. Common Wealth. Bulletin of Soil Science 49, 118.

Jat SK and Maitry SK. 2006. Effect of farmyard manure and biofertilizers at differential levels of nitrogen and phosphorus application on growth and productivity of wheat (*Triticum aestivum*) in lateritic tract of West Bengal. National Symposium on Conservation Agriculture and Environment. BHU, Varanasi, India.

Johnson HJ, Colquhoun JB, Bussan AJ, and Rittmeyer RA. 2010. Feasibility of organic weed management in sweet corn and snap bean for processing. Weed Technology, 24: 544-550.

Khalil AI, Beheary MS and Salem EM. 2001. Monitoring of microbial populations and their cellulolytic activities during the composting of municipal solid wastes. World Journal of Microbiology and Biotechnology, 17: 155- 161.

Knezevic SZ, Datta A, Bruening C and Gogos G. 2012. Propane-Fueled Flame Weeding in Corn, Soybean, and Sunflower. University of Nebraska, Lincoln Extension Service.

Krishna Moorthy PN. 2005. Principles of Insect pest and diseases management for organic crop production. Invited Paper, National Seminar on Organic Farming – Current Scenario and Future Thrust, pp. 65-70.

Kumar SR and Subbarayudu B. 2004. Weather–insect pest cyclic pattern and biotic stress management in sorghum. Poster paper presented at the Natioanl Seminar on 'Organic farming – Prospects and Challenges in the Millennium.

Kumar SR, Shinde GG, Wanjari SS, Thakur NS, Singh YK and Pushpendra S. 2003. Methods to minimize chemical fertilizer usage in rainfed sorghum – a multilocation study. Paper presented at the National Seminar on "Organic products and their future prospects.

Lewis WJ, van Lenteren JC, Sharad P and Tumlinson JH. 1997. A total system approach to sustainable pest management, 94: 12243-12248.

Lundkvist A, Salomonsson L, Karlsson L and Dock-Gustavsson A. 2008. Effects of organic farming on weed flora composition in a long term perspective. European Journal of Agronomy, 28(4): 570-578.

Mamo M, Rosen CJ and Halbach R. 1999. Nitrogen availability and leaching front soil amended with municipal solid waste compost. Journal of Environmental Quality, 28(4): 1075- 1082.

Moldes A, Cendon Y and Barral MT. 2007. Evaluation of municipal solid waste compost as a plant growing media component, by applying mixture design. Bioresource Technology, 98: 3069-3075.

Neelima TL, Bhanu Murthy VB, Ramanjaneyulu AV. 2007. Response of rice to N fertilizer and sunhemp green manuring. Research on Crops, 8(3): 533-536.

Onduru DD, Diop JM, Van der Werf E and De Jager A. 2002. "Participatory on-farm comparative assessment of organic and conventional farmers' practices in Kenya." Biological Agriculture and Horticulture, 19(4): 295–314.

Panda RK and Behera S. 2003. Non-point source pollution of water resources: problems and perspectives. Journal of Food, Agriculture and Environment, 1(3/4): 308-311.

Ramanjaneyulu AV, Giri G and Shivay YS. 2007. Productivity and sustainability of fodder sorghum-Mustard cropping system as influenced by N, P, *Azotobacter* and PSB. International Journal of Tropical Agriculture, 25 (1-2): 307-315.

Ramanjaneyulu AV, Spandana Bhat P. and Neelima TL. 2012. Sustainable Agriculture. RARS (ANGRAU, Palem 509 215, Mahabubnagar district, Andhra Pradesh, p. 13.

Reddy RM, Reddy AN, Reddy YTN, Reddy NS. and Anjanappa M. 1998. Effect of organic and inorganic sources of NPK on growth and yield of pea. Legume Research, 21(1): 57– 60.

Richard T and Woodbury P. 1992. The impact of separation on heavy metal contaminants in municipal solid waste composts. Biomass Bioenergy, 3(3–4): 195–211.

Roy Rabindra N, Misra Ram V and Montanez A. 2002. Decreasing reliance on mineral nitrogen – yet more food. AMBIO: A Journal of the Human Environment, 3(2): 177-183.

Rupela OP, Gowda CLL, Wani SP and Ranga Rao GV. 2005. Lessons from non-chemical input treatments based on scientific and traditional knowledge in a long-term experiment.184-196 in the Agricultural Heritage of Asia: Proceedings of the International Conference (YL Nene ed.). 6-8 December 2004, Asian Agri-History Foundation, Secunderabad-500009, AP, India.

Salvador V, Garibay and Katke J. 2003. Market Opportunities and Challenges for Indian Organic Products. Research Institute of Organic Agriculture, pp. 184-196.

Samawat SA, Lakzian and Zamirpour A. 2001. The effect of vermicompost on growth characteristics of tomato. Agricultural Science and Technology, 15(2): 83–89.

Scialabba N and Müller-Lindenlauf M. 2010. Organic agriculture and climate change. Renewable Agriculture and Food Systems, 25(2): 158–169.

Sheppard AW, Shaw RH and Sforza R. 2006. Top 20 environmental weeds for classical biological control in Europe: a review of opportunities, regulations and other barriers to adoption. Weed Research, 46(2): 93-117.

Signor D and Carlos Eduardo Pellegrino Cerri. 2013. Nitrous oxide emissions in agricultural soils: a review. Pesquisa agropecuária tropical, 43(3): 322-338.

Singh S, Singh B and Mishra B. 2012. Microorganisms in Sustainable Agriculture and Biotechnology Springer Netherlands, pp. 127-151.

Singh, Ravindra and Agarwal, SK. 2004. Effect of organic manuring and nitrogen fertilization on productivity, nutrient-use efficiency and economics of wheat (*Triticum aestivum*). Indian Journal of Agronomy, 49(1): 49-52.

Sofia PK, Prasad R. and Vijay VK. 2006. Organic farmingtradition reinvented. Indian Journal of Traditional Knowledge, 5(1): 139–142.

Srijita D. 2015. Biopesticides: an ecofriendly approach for pest control. World Journal of Pharmacy and Pharmaceutical Sciences, 4(6): 250- 255.

Stockdale EA, Lampkin NH and Hovi M. 2001. Agronomic and environmental implications of organic farming systems. Advances in Agronomy, 70(3): 261–327.

Subbalakshmi Lokanadhan P, Muthukrishnan and Jeyaraman S. 2012. Neem products and their agricultural applications. Journal of Biopesticides, 5: 72-76.

Suman Gupta and Dikshit AK. 2010. Biopesticides: An ecofriendly approach for pest control. Journal of Biopesticides, 3(1): 186-188.

Thakore Y. 2006. The biopesticide market for global agricul tural use. Industrial Biotechnology, Fall, pp. 194-208.

Tomar VK, Bhatnagar RK and Palta RK 1998. "Effect of vermicompost on production of brinjal and carrot," Bhartiya Krishi Anusandhan Patrika, 13 (3-4): 153–156.

Vikas G, Manisha R, Omkar G and Babita K. 2013. Bio-fertilizers- increasing soil fertility and crop productivity. Journal of Industrial Pollution Control, 56(2): 89-95.

www.myenjoyzone.com

www.organicfarming.com.au

Yadav. 2010. Organic Agriculture. National Project on Organic farming. Dept. of Agriculture and Cooperation, Govt. of India.

Yadav S, Yogeshwar MK, Yadav B, Subhash and Kalyan S. 2013. Effect of organic nitrogen sources on yield, nutrient uptake and soil health under rice (*Oryza sativa*) based cropping sequence. Indian Journal of Agricultural Sciences, 83(2): 170–175.

Chapter 4

Vermicompost: A Viable Resource in Organic Farming

T.L. Neelima[1], A.V. Ramanjaneyulu[2], S.R. Kumar[3], K.A. Gopinath[4] and M. Venkata Ramana[5]

[1]Agricultural Polytechnic, [2,5]Regional Agricultural Research Station, Palem – 509 215, Telangana
[3]ICAR-Indian Institute of Millets Research, Hyderabad – 500 030, Telangana
[4]ICAR-Central Research Institute for Dryland Agriculture, Hyderabad – 500 059, Telangana
E-mail: avr_agron@rediffmail.com

1. INTRODUCTION

"Nobody and nothing can be compared with earthworms in their positive influence on the whole living Nature. They create soil and everything that lives in it. They are the most numerous animals on earth and the main creatures converting all organic matter into soil humus providing soil's fertility and biosphere's functions: disinfecting, neutralizing, protective and productive."

– ***Anatoly M. Igonin***

Aristotle called worms the "intestines of the earth" and Charles Darwin wrote a book on worms and their activities, in which he stated that there may not be any other creature that has played so important a role in the history of life on earth (Bogdanov, 1996) and called them as 'unheralded soldiers of mankind' and 'friends of farmers'.

Rapid growth of urbanization and industrialization has led to generation of large quantities of wastes. Millions of tons of municipal solid wastes (MSW) generated from the modern society are being added to landfills every day, creating extraordinary economic and environmental problems for the government to manage and monitor them for environmental safety. Construction of secured engineered landfills incurs 20-25 million U.S. dollars before the first load of waste is dumped (Sinha, 2010). Continuous escalation of socio-economic and environmental cost of dealing with current and future generation of mounting municipal solid wastes (MSW) was wintnessed. Besides, emission of greenhouse gases (GHG) like methane (CH_4) and nitrous oxides (N_2O) resulting from the disposal of MSW either in the waste landfills or from their management by composting is another major environmental problem.

A third of increase in world cereal production in the 1970s and 1980s has been attributed to increased fertilizer use (FAO, 2003). Similarly, the global usage of pesticides has increased considerably during the second part of the 20th century. In India, chemical agriculture triggered by widespread use of agro-chemicals in the wake of 'green revolution' of the late 1960s came as a 'mixed-blessing' rather a 'curse in disguise' for mankind. It dramatically increased the food grain production but severely reduced its 'nutritional quality' and also the 'soil fertility' over the years. The soil has become addict and increasingly greater amount of chemical fertilizers are needed every year to maintain the soil fertility and food productivity at the same levels. Increased use of agro-chemicals has virtually resulted into 'biological droughts' (decline in beneficial soil microbes and earthworms) in soils. Soil and water pollution due to seepage and drainage especially after heavy rainfall were other ill-effects on farmlands following the use of agro-chemicals. The farmers today are caught in a 'vicious circle' of higher use of agrochemicals to boost crop productivity at the cost of declining soil fertility. This is also adversely affecting the economy of the countries as the cost of agrochemicals has been rising all over the world.

Besides, agricultural development has contributed to global warming, reduction in biodiversity and soil degradation. Due to adverse effects of agro-chemicals, interest has been stimulated for the use of alternate sources of plant nutrition (Follet *et al.*, 1981). In view of the above, in recent years, much attention has been paid to manage different organic waste resources at low-input as well as eco-friendly basis. Integrated waste management envisages 'reduce' (optimization of usage), 'reuse' (One person's trash is another person's treasure) and 'recycle'. The fourth one is 'rot' which refers to recycling food waste and other organic materials through composting or vermicomposting.

2. Vermicomposing and Vermicompost

2.1 Waste to Wealth

Vermiculture is the production or culture of earthworms. On the other hand, it is the process by which worms are used to convert organic materials (usually wastes) into humus like material known as vermicompost. In case of vermiculture, population density of worms will be less, while, maximum worm population density is maintained in Vermicomposting. Vermicomposting differs from composting in several ways (Gandhi *et al.*, 1997). It is a mesophilic process, utilizing microorganisms and earthworms that are active at 10-32°C. The process is faster than composting because the material passes through the earthworm gut, whereby the resulting earthworm castings (worm manure) are rich in microbial activity and plant growth regulators and fortified with pest repellence attributes as well. In short, earthworms, through a type of biological alchemy, are capable of transforming 'garbage' into 'gold' (Vermi Co, 2001, Tara Crescent, 2003). Further, it is odorless process, reducing composting time by more than half and the end product is both 'disinfected', 'detoxified' and 'highly nutritive'. It is also the best possible alternative is to reduce the potential pollutants (Latifah Abd Manaf *et al.*, 2009).

The importance of earthworms in the breakdown of organic matter and the release of the nutrients has been known for a long time (Darwin, 1881). Earthworms are popularly known as 'environmental engineers'. They form the basis for 'circular economy in agriculture' *i.e.*, using food wastes of the society to produce food for the society again. As environmental engineers, earthworms are both 'protective' and 'productive' for environment and society. In nature, all organic matter eventually decomposes. But in vermicomposting, the worms speed up the process of decomposition and give a richer end product called "worm castings". It is an eco-biotechnological process that transforms energy rich and complex organic substances into a stabilized humus-like product (Benitez *et al.*, 2000). Vermicompost is like a 'brown gold' to produce 'green gold' (food crops). In some countries mainly Canada, United States, Australia, France and some Southeast Asia countries, earthworms have been used for waste stabilization for many years. Thus, vermicomposting is an 'economically viable', 'environmentally sustainable' and 'socially acceptable' enterprise. The use of vermicompost has long been recognized as an effective means of increasing crop yields through improved soil physical, chemical and biological properties (Gopinath *et al.*, 2010). As mineral fertilizers are not allowed in organic farming, there has been renewed interest in recycling wastes into valuable organic manure through vermicomposting and its subsequent utilization to nourish the crops grown organically.

2.2 Composition of Vermicompost

2.2.1 Chemical Properties

Werner and Cuevas (1996) reported that most vermicomposts contained adequate amounts of macronutrients and trace elements but were dependent on the sources of feedstock used. Businelli *et al.* (1984) reported that the highest elemental values were recorded in vermicomposts from cattle and horse manure mixture with 38.8 per cent organic carbon, 2.7 per cent total N and 1080 mg/kg NO_3-N. At the same time, the lowest elemental concentrations were recorded in the municipal waste compost with only 9.5 per cent organic carbon, 1.0 per cent total N and 503 mg/kg NO_3-N. Further, Atiyeh *et al.* (2000) found that conventional compost was higher in 'ammonium' and while vermicompost was rich in nitrate nitrogen which is easily available form of nitrogen. Vermicompost is a nutritive 'organic fertilizer' with 2-3 per cent nitrogen, 1.55-2.25 per cent phosphorus and 1.85-2.25 per cent potassium besides micronutrients. According to Suhane (2007), VC is 4 times more nutritive than cattle dung compost. Few authors such as Kale and Bano (1986), reported as high as 7.37 per cent N and 19.58 per cent P_2O_5 in worms' vermicast. Edwards and Burrows (1988) reported that vermicomposts, especially those from animal waste sources, usually contained more mineral elements than commercial plant growth media. Phosphorus was 64 per cent higher in vermicomposts than in the original organic material which was due to increase. When analyzed soil before and after application of vermicompost (after harvesting of groundnut crop), they found increase in organic carbon from 0.25 per cent (initial) to 0.48 per cent with papaya leaf litter, 0.60 per cent with *glyricidia* loppings, 0.56 per cent with *Celosia*, 0.51 per cent with *Parthenium* and 0.49 per cent with redgram stalks. It was due to release of CO_2 from decaying organic matter by the micro-organisms (Ravichandran *et al.*, 2001). Likewise, they also found increase in N availability by 11.8 to 27.8 per cent over initial level. Similarly, available potassium increased tremendously over initial value (105 kg/ha) with maximum increase found due to addition of *glyricidia* loppings (240 kg/ha) (Vijaya Sankara Babu *et al.*, 2008). Nutrient contents in vermicompost are often much higher than traditional garden compost (Table 4.1).

Table 4.1: Chemical Characteristics of Garden Compost and Vermicompost

Parameter	*Garden Compost*	*Vermi-compost*	*Reference*	*Vermi-compost*	*Reference*
Organic Carbon (per cent)	12.2	9.8-13.4	Nagavallemma *et al.* (2004)	NA	
pH	7.8	6.8	George (1994)	NA	
EC (*dS/m*)	3.6	7.1	George (1994)	NA	
Nitrogen (per cent)	0.8	0.51-1.61	Nagavallemma *et al.* (2004)	1.5-2.5	Thiruneela-kandan and Subbulakshmi (2014)
Phosphorus (per cent)	0.35	0.19-1.02		1.2-1.8	
Potassium (per cent)	0.48	0.15-0.73		1.5-2.4	
Calcium (per cent)	2.27	1.18-7.61		0.5-1.0	
Magnesium (per cent)	0.57	0.093-0.568		0.2-0.3	
Sodium (per cent)	<0.01	0.058-0.158		NA	
Zn (ppm)	0.0012	0.0042-0.110		0.05-0.1	Thiruneela-kandan and Subbulakshmi (2014)
Copper (ppm)	0.0017	0.0026-0.0048		0.0022-0.0036	
Fe (ppm)	1.1690	0.2050-1.3313		0.8-1.5	
Mn (ppm)	0.0414	0.105-0.2038		NA	
Boron (per cent)	0.0025	0.0034	George (1994)	NA	
Al (per cent)	0.7380	0.7012	George (1994)	NA	
Sulphate (per cent)				0.4-0.5	

2.2.2 Biological Properties

Vermicomposts have many outstanding biological properties. They are rich in bacteria, actinomycetes, fungi (Edwards, 1983; Werner and Cuevas, 1996) and cellulose-degrading bacteria (Werner and Cuevas, 1996). Earthworm castings obtained after sludge digestion are found to be rich in bacteria. Nair *et al.* (1997) compared the microorganisms associated with vermicomposts with those in traditional composts. The vermicomposts had much larger populations of bacteria (5.7×10^7), fungi (22.7×10^4) and actinomycetes (17.7×10^6) compared with those in conventional composts. The outstanding physico-chemical and biological properties of vermicomposts makes them organic amendment (Gopinath *et al.*, 2010) and excellent material as additives to greenhouse container media, organic fertilizers for various field and horticultural crops.

2.3 Rate of Food Intake and Multiplication of Earthworms

Adhikary (2012) while reviewing the literature on Vermicompost reported that earthworms consume various organic wastes and reduce the volume by 40-60 per cent and each worm weighs about 0.5 to 0.6 g, eats waste equivalent to its body weight and produces cast equivalent to about 50 per cent of the waste it consumes in a day. On the other hand, Visvanathan *et al.* (2005) found that most earthworms consume, at the best, half their body weight of organics in the waste in a day. *Eisenia foetida* can consume organic matter at the rate equal to their body weight every day. Earthworm participation enhances natural biodegradation and decomposition of organic waste from 60 to 80 per cent. As the worms double their population every 60-70 days, the process becomes faster with time. Under the optimum temperature (20-30°C) and moisture (60-70 per cent), about 5 kg of worms (approximately 10,000 numbers) can vermiprocess 1.0 ton of waste into vermicompost in just 30 days. An acre of land can have as many as five lakhs of earthworms, which can recycle as much as five tons of soil or more per year.

A population density of 2.5 to 5.0 kg/m^2 is required for optimum reproduction. Biomass of worms can be doubled for every 60 days at this rate under optimum growing conditions. Vijaya Sankara Babu *et al.* (2008) produced vermicompost by using various starter materials like papaya leaf litter, *glyricidia* loppings, *Celosia*, *Parthenium* and redgram stalks. At lower densities, slow or no reproduction rate is recorded. While at higher densities, increased competition for food and space slows down the reproduction rate (Bogdanov, 1996).

2.4 Mechanism of Worm Action in Vermicomposting

Earthworms promote the growth of 'beneficial decomposer aerobic bacteria' in waste biomass and also act as an aerator, grinder, crusher, chemical degrader and a biological stimulator (Binet *et al.*, 1998). Earthworms host millions of decomposer microbes in their gut (Singleton *et al.*, 2003). The number of bacteria and actinomycetes contained in the ingested material increased up to 1000 fold while passing through the gut (Edward and Fletcher, 1988). A population of worms numbering about 15,000 will in turn foster a microbial population of billions in a short time. Under favorable conditions, earthworms and microorganisms act symbiotically and synergistically to accelerate and enhance the decomposition of organic matter in the waste. It is the microorganisms which break down the cellulose in the food waste, grass clippings and the leaves from garden wastes (Morgan and Burrows, 1982). The waste feed materials ingested is finely ground into small particles to a size of 2-4 microns and passed on to the intestine for enzymatic actions. The gizzard and the intestine work as a 'bioreactor'. The worms secrete enzymes such as proteases, lipases, amylases, cellulases and chitinases in their gizzard and intestine which bring about rapid biochemical conversion of the cellulosic and the proteinaceous materials in the waste organics. The final process in vermi-processing and degradation of organic matter is the 'humification' in which the large organic particles are converted into a complex amorphous colloid containing 'phenolic' materials. Only about one-fourth of the organic matter is converted into humus. Passing through the gut of the earthworm, recycled organic wastes are excreted as castings, or worm poop or worm manure, an organic material which looks like fine textured soil. This is fertile and rich in nutrients and readily available to plants (Hansen, 2007).

2.5 Types of Earthworms Used in Vermicomposting

Nearly 1800 species of earthworms are estimated to be present worldwide (Edwards and Lofty, 1972). Six earthworm species have been identified as potentially useful species to break down organic wastes. These are Tiger worm or brandling worm (*Eisenia foetida*), Red tiger worm (*E. andrei*), *Dendroba enaveneta* and redworms or red wigglers (*Lumbricus rubellus*) from temperate regions and African night crawler (*Eudrilus eugeniae*), Indian blue worm (*Perionyx excavates*) and *Perionyx hawayana* from the tropics (Sinha *et al.*, 2002). Of these, *Eisenia foetida* (compost worm/manure worm/redworm/red wiggler) and *Eudrilus eugeniae* are ideal. Most Indian species are not suitable for the purpose.

2.6 Requirements for Vermicomposting

Any material that provides the worms with a relatively stable habitat is called as bedding. It should have high absorbency, porous for better aeration, low protein and/or nitrogen content (*high carbon: nitrogen ratio*). The ideal moisture content range for vermicomposting is 70-90 per cent (Dominguez *et al.*, 1997; Georg, 2004). Bedding material should feel like a wrung out sponge with 3:1 ratio of water to bedding by weight. However, it should not be soupy or too dry. They prefer a temperature of 20°C but above 35°C, the worms will leave the area. The vermi-beds/bins must be protected from snakes and squirrels by erecting a mesh around.

Worms can survive in a pH range of 5 to 9 (Edwards, 1998), but some researchers say near to neutral condition (Latifah AbdManaf *et al.*, 2009) and few say 7.5 to 8.0 is optimum (Georg, 2004). In general, the pH of worm beds will come down over time. If the food source is alkaline, the effect is a moderating one, tending to neutral or slightly alkaline pH. If the food source or bedding is acidic (coffee grounds, peat moss), then the pH will go down to below 7.0. This condition may encourage pests such as mites. The pH can be adjusted upwards by adding calcium carbonate and downwards by adding peat moss. Earthworms are very sensitive to salts, hence, prefers salt content of less than 0.5 per cent (Gunadi *et al.*, 2002).

Earthworms prefer mostly organic food of plant or animal origin. Manures are the most commonly used. Dairy and beef manures are generally considered as the best natural food with exception of rabbit manure (Gaddie and Douglas, 1975). Horse manure is suitable for the growth of earthworms and does not need any pretreatment and can be applied directly as a feed. They especially enjoy vegetable and fruit peelings, grains, coffee grounds and filters and newspaper. They should not be overloaded with high citrus diet as orange peels can introduce d-limonene which gives moldy fruit stand. The 'sewage sludge' from the municipal water and wastewater treatment plants and the punch waste materials (gut contents of slaughtered ruminants) from abattoir also make good feedstock for earthworms (Edwards, 1988; Edwards, 1998; Fraser-Quick, 2002; Sinha *et al.*, 2009a). Livestock rearing waste *e.g.* cattle dung, pig and chicken excreta makes excellent feedstock for earthworms.

Solid waste from paper pulp and cardboard industry, food processing industries including brewery and distillery, vegetable oil factory, sugarcane industry, aromatic oil extraction industry, sericulture industry, logging and carpentry industry also make excellent feedstock for vermicomposting (Edwards, 1988; Edwards, 1998; Kale, 1998). Worms can also compost 'fly-ash' from coal power plants. Some trees such as cedar and fir having high levels of tannins, when used a feed, it can harm worms and even drive them from the beds. Pre-composting of wastes can reduce or even eliminate most of these threats (Gunadi *et al.*, 2002). The meat, dairy and oil based products shouldn't be added.

Ants, bacteria, beetle, centipedes, collembola, fruit flies, fungus gnat, mite, millipede, mold, pill bug or roly poly, slug, snail, soldier fly, soldier fly larva, sow bug, spider, springtail, moles and birds and white worm are some of the enemies of earthworm (Mark Leary, 2004). Moles and birds can be prevented by barriers such as wire mesh while, centipedes can be destroyed manually (Sherman, 1997). Avoiding sweet foods in the vermin beds and maintaining pH above 7.0 will control the ants. White and brown mites compete with worms for food only but red mites are parasitic thus suck blood or body fluid from worms as well as cocoons (Sherman, 1997). Maintaining the moisture level below 85 per cent, addition of calcium carbonate to keep pH stays at neutral or above can control mite. Worms may be affected by 'Sour crop or protein poisoning' disease (swollen clitellum'or worm crawl aimlessly around on top of the bedding) which can be solved by avoiding protein rich foods/ overfeeding and keeping pH at neutral or above.

2.7 Vermicomposting Systems

Though there are three basic types of vermicomposting systems *viz.*, windrows, beds or bins and flow-through reactors are followed in other countries, but, in India, the following two very simple methods are popular for preparation of vermicompost as narrated by Adhikary (2012).

2.7.1 Vermicomposting from Household Wastes

A wooden box of 45 × 30 × 45 cm or an earthen/plastic container with broad base and drainage holes should be used for this purpose. A plastic sheet with small holes should be placed at the bottom of the wooden box. 3 cm layer of soil and a 5 cm layer of coconut fiber for draining of excess moisture below it is kept inside the box. A thin layer of compost along with worms as inoculums are to be placed above it. About 250 worms are sufficient for the box. Vegetable wastes should be added in layers daily on top of the inoculums on daily basis. The top of the box should be covered with a piece of gunny bag to provide dim light inside the box. When the box is full, the box should be left undisturbed for a week. When the compost seems to be ready, the box should be kept in light for 2 - 3 hours so that the worms go down to the lowermost coconut fiber layer. The composted materials should be removed from the top of the box and gradually down and sieved for use in the urban or intensive horticultural and agricultural systems.

2.7.2 Vermicomposting of Farm Wastes

Pits of sizes 2.5 m × 1 m × 0.3 m (length, breadth and depth) are to be taken in thatched sheds with sides left open. The roof of the shed can be constructed with cement sheets for longevity. The length of pits may vary depending on the availability of space. The bottom and sides of the pit are made hard by compacting with a wooden mallet. At the bottom of the pit a layer of coconut husk is spread with the concave side upward to ensure drainage of excess water and for proper aeration. The husk is moistened and above this, biowaste (crop residue or farm waste) mixed with cow dung in the ratio of 8:1 has to be spread up to a height of 30 cm above the ground level and water is sprinkled daily. After the partial decomposition of wastes for 7 to 10 days, the worms are introduced @ 500 to 1000 in numbers per pit. The pit is covered with jute bags. Water has to be sprinkled over the bed to maintain moisture and a temperature of 20°C-30°C. At higher temperature, the worms will aestivate and at lower temperature, they will hibernate. When the compost is ready, water has to be withheld so that the worms will move to bottom of the heap. After one or two days, the compost from the top of the heap can be harvested. The undecomposed residues can be reused with worms for further composting.

Figure 4.1: A Model Vermicompost Shed and Beds at RARS, Palem.

Another method is mobile vermicomposting. In this method, preparation of vermicompost is similar to vermicomposting of farm wastes, but, vermi beds are replaced by plastic beds as shown in the Figure 4.2. As this plastic bed can moved to the places of interest based on the need, this is called as mobile type of vermicomposting.

Figure 4.2: Mobile Type of Vermicomposting at RARS, Palem.

2.8 Methods of Harvesting Worms

Harvesting of worms means separation of worms from bedding and vermicompost which is done by three methods as detailed below (Bogdanov, 1996). This is mainly done to sell the worms commercially.

2.8.1. Manual Methods

These are performed by small scale growers who sell the worms on a small scale. It involves handsorting or picking the worms directly from the compost by hand. Generally, worms avoid light. When the vermicompost containing worms is dumped on a flat surface in a light shade, the worms will quickly go below the surface. Then, harvesting of vermicompost can be started, but, when worms become visible again, it will be stopped. This process is repeated several times until the worms are totally harvested. These worms can be quickly scooped into a container, weighed and prepared for delivery.

2.8.2 Self-Harvesting (Migration) Methods

This method is based on the concept that worms migrate to new regions either to find new food or to avoid undesirable conditions such as dryness or light. A box is constructed with a screen bottom. The mesh is usually of size ¼″ to $^{1}/_{8}$″. There are two different approaches. In downward-migration system, worms are forced downward by strong light. The worms go down through the screen into a prepared, pre-weighed container of moist peat moss. Then, the compost in the box is removed and a new batch of worm-rich compost is kept. The process is repeated until the box with the peat moss has reached the desired weight. While, in case of upward-migration system, the box with the mesh bottom is placed directly on the worm bed. It is filled with damp peat moss and chicken mash, coffee grounds or fresh cattle manure. The advantage of this system is that the worm beds are not disturbed. The main disadvantage is that the harvested worms are in material that contains a fair amount of unprocessed food, which can be avoided by removing food and allowing the worms to consume what is left before packaging. This system is used extensively in Cuba with the difference that large onion bags are used instead of boxes (Cracas, 2000).

2.8.3 Mechanical Methods

Mechanical harvesting is the quickest and easiest method for separating worms from vermicompost. Mechanical harvester with a rotating cylinder are composed of screen material of different mesh sizes. The cylinder is rotated by a small electric motor mounted on one end of the cylinder. As the cylinder rotates, the castings fall through the screen. The worms 'ride'the entire distance of the tromme land pass through the lower end into a wheel barrow.

2.9 Methods of Harvesting Vermicompost

2.9.1 Cone Method

The contents of the vermi beds/bins can be placed in the shape of cones on a taurpalin or plastic sheet and later the worms are allowed to burrow down. The harvested compost will be transferred to a separate storage container and the worms are transferred to new vermi beds.

2.9.2 Migrating Method

This is a handy method of harvesting the vermicompost quickly. The compost which is ready should be pushed to one side. Simultaneously, prepare new bedding on another side of the vermi bin/bed and place the waste. Now, addition of food material has to be stopped for one side and added on another side. Likewise, harvesting sides can be alternated on a continual basis.

2.9.3 Scoop Method

This is a perfect method for people who only need a small amount of compost at a time. Worms are allowed to burrow away by gentle exposure to the light by opening the vermi bins/beds. This can also be done by stirring the surface. Then, after about 10 minutes, top layer of castings can be scooped off. Worms in the compost if any, have to be removed. This process can be repeated.

2.10 Vermicompost Stability

It depends on the factors detailed below:

2.10.1 Carbon : Nitrogen (C : N ratio)

The C:N ratio is one of the most widely utilized parameters to follow the development of material undergoing a composting or vermicomposting process and varies based on the type feedstocks used. The C:N ratio of a mature compost or vermicompost should ideally be around 10, but this is hardly ever achieved due to the presence of recalcitrant organic compounds, or materials that decompose poorly.

2.10.2 Humic Substances

During the maturation process, humic substances evolve qualitatively with an increasing predominance of humic acids over fulvic acids. The ratio between these two is considered an important index of compost maturity. It should be >1 in a mature compost (Saviozzi *et al.*, 1988).

2.10.3 Absence of Plant Inhibitors

Absence of bio-inhibitory aliphatic acids and phenolics is the best indicator of a vermicompost's maturity.

2.10.4 Absence of Human Pathogens

Vermicompost should be considered hygienic if 100 g of sample do not contain *Salmonella*, human viruses, infective parasitic helminthic eggs and not more than 5 $\times 10^4$ fecal coliforms and 5 $\times 10^5$ fecal streptococci.

2.10.5 Other Criteria

Organic nitrogen mineralization is a useful parameter to determine readily biodegradable nitrogen compounds and therefore it can be correlated inversely with compost stability.

3. Economics of Vermicomposting

Cost of production by vermiculture is significantly low by more than 60-70 per cent and is simply insignificant as compared to chemical fertilizers. The food produced will be a 'safe, chemical-free' for the society. It is a 'win-win' situation for both producers (farmers) and the consumers (feeders). As vermicompost also helps the crops to attain maturity and reproduce faster, it shortens the 'harvesting time' (Sinha *et al.*, 2007), this further cuts on the cost of production and also adds to the economy of farmers as they can grow more crops every year in the same farm plot. People are earning from **Rs** 5 to 6 lakhs every year from sale of both worms and their vermicompost to the farmers.

4. Advantges of Vermicompost

4.1 Increased Level of Beneficial Microorganisms

Vermicompost may be 1000 times microbially active than conventional compost (Edwards, 1999; Edwards, 2000). Many beneficial microbes are killed and nutrient especially nitrogen is lost during conventional aerobic composting process which is thermophilic (temperature rising up to 55°C). Earthworms create aerobic conditions in the waste materials by their burrowing actions, inhibiting the action of anaerobic micro-organisms which release foul-smelling hydrogen sulfide and mercaptans. The earthworms release coelomic fluids that have anti-bacterial properties and destroy all pathogens in the resulting compost (Pierre *et al.*, 1982). Further, the end product is more homogenous, richer in plant-available nutrients and humus and significantly low contaminants (Appelhof, 1997; Lotzof, 2000). Vermicompost is rich in beneficial micro flora such as P- solubilizers, cellulose decomposing micro-flora *etc.*

4.2 Less Nutrients Run-off

Vermicompost, like conventional compost, binds nutrients both in the bodies of microorganisms and through their actions, thus, less nutrient run-off and increases the use efficiency of chemical fertilizers. This is an extremely important environmental benefit of both composting and vermicomposting.

4.3 Soil Condition

Frequent addition of vermicompost make the soil 'highly porous' with greater 'water holding capacity'thus, helps to increase the ability of soil to retain moisture postponing the moisture stress during dry spells. Vermicompost is able to retain more soil moisture thus reducing the demand of water for irrigation by nearly 30-40 per cent. It helps to maintain better soil structure suited for crop growth (Lotzof, 2000). thus prevents soil erosion.

4.4 Level of Plant Available Nutrients

According to Atiyeh *et al.* (2000), compost was higher in ammonium, while, vermicompost was higher in nitrates. 'Vermicomposted manure' has higher N availability than conventionally composted manure on a weight basis. Supply rate of several nutrients, including P, K, S and Mg was increased by vermicomposting as compared with conventional composting (Short *et al.*, 1999). Vermicompost treated plots registered high availability of K compared to FYM treated plots. Bhaskar *et al.* (1992), inferred that earthworm increases the availability of K by shifting the equilibrium among the forms of K from relatively unavailable forms to more available forms.

4.5 Nutrient Uptake

Increased availability of nutrients in enriched vermicompost especially P would have enhanced root proliferation which helped in more uptake of K. Higher amounts of Ca and other bases present in worm casts have been reported by Vasanthi and Kumaraswamy (1996). Maximum Ca and Mg uptake was recorded by plants treated with enriched vermicompost. The higher content of these cations present in plants treated with enriched vermicompost may be due to increased uptake through enhanced availability from the soil. When compared to FYM treated plots, vermicompost treated plots showed an enhanced micronutrient uptake (Sailaja Kumara and Usha Kumari, 2002).

4.6 Ability to Repel Pests

Worm castings sometimes repel hard bodied pests (Arancon, 2004; Edwards and Arancon, 2004) due to production by the worms of the enzymes like chitinase, which breaks down the chitin in the insects' exoskeleton. Decreases in arthropod (aphid, mealy bug, spider mite) populations and subsequent reductions in plant damage in tomato, pepper and cabbage trials with 20 per cent and 40 per cent vermicompost additions (Edwards and Arancon, 2004). They also found significant suppression of plant parasitic nematodes in field trials with peppers, tomatoes, strawberries and grapes. Further research is required, before vermicompost can be considered as an alternative to pesticides or non-toxic methods of pest control.

4.7. Ability to Suppress Diseases

High levels of beneficial microorganisms in vermicompost protect the plants by out-competing pathogens for available resources. Edwards and Arancon (2004), reported significant suppression of incidence of the diseases *viz., Pythium* on cucumbers, *Rhizoctonia* on radishes in the greenhouse and *Verticillium* on strawberries and *Phomopsis* and *Sphaerotheca fulginae* on grapes by vermicompost applications. Vermicomposting could indeed be used as a method for destroying pathogens, with a success rate equal to conventional composting (Eastman, 1999; Eastman *et al.*, 2000). Something inside the worm destroys the pathogens, leaving the castings pathogen free (Appelhof, 2003). Vermicompost spread on farm land will not result in pathogen contamination of ground or surface waters.

4.8. Mitigate Climate Change

Agriculture is a significant contributor to climate change, through the release of carbon from soils and generation of methane gas from livestock and their manure. Composting or vermicomposting helps in locking carbon in organic matter and organisms within the soil. Nearly five to seven times as much fertilizer could be displaced per unit of vermicompost, thus decreases GHG emissions proportionately. Vermicompost has higher levels of nitrogen than compost which implies that the process is more efficient at retaining nitrogen, probably because of the greater microbial load. This in turn implies that less nitrous oxide is generated and/or released during the process. Since N_2O is 310 times as potent a GHG as CO_2, this could be a significant benefit. Vermicomposting emitted minimum of (N_2O) @ 1.17 mg/m^2/hour, as compared to 1.48 mg/m^2/hour from aerobic composting and 1.59 mg/m^2/hour from anaerobic composting. However, large-scale vermicomposting processes may in fact be a significant producer of NO_2 (Thiruneelakandan and Subbulakshmi, 2014).

4.9. Increases Biodiversity

Biodiversity is declining rapidly worldwide, heading for a mass extinction. Earthworms have an extremely important role to play in counteracting the loss of biodiversity. Worms increase the numbers and types of microbes in the soil by creating conditions under which these creatures can thrive and multiply. The earthworm gut has been described as a little "bacteria factory", spewing out many times more microbes than the worm ingests. Addition of vermicompost and cocoons to soil helps in tremendous increase in microbial activity. This below-ground biodiversity is the basis for increased biodiversity above ground.

4.10. Reduces Use of Chemical Pesticides and Input Costs

Widespread use of chemical pesticides became an important requirement for the growth of high-yielding varieties of crops which was more susceptible to pests and diseases. Continued application of chemical pesticides induced 'biological resistance' in crop pests and diseases and much higher doses are now required to eradicate them. There has been considerable evidence in recent years regarding the ability of vermicompost to protect plants against various pests and diseases either by suppressing or repelling them or by 'inducing biological resistance' in plants to fight them or by killing them through pesticidal action. Pesticide spray was significantly reduced where earthworms and vermicompost were used in agriculture (Suhane, 2007). Studies also indicate that use of vermicompost help in disease control by almost 75 per cent. This significantly cut down on the cost of food production.

5. Impact of Vermicompost on Growth and Yield of Major Crops/Trees

Vermicompost plays a major role in improving growth and yield of different field crops, vegetables, flower and fruit crops. Vermicomposts have consistently improved seed germination, enhanced seedling growth and development and increased plant productivity. In many field studies across India, replacement of 25 per cent of the nutrients provided to the plants by vermicompost produced significant greater increases in plant height and marketable crop yield than those produced by 100 per cent inorganic fertilization. The maximum benefit from vermicompost is obtained when it constitutes between 10 and 40 per cent of the growing medium. It appears that levels of vermicompost higher than 40 per cent do not increase benefit and may even result in decreased growth or yield (Arancon, 2004). Vermicompost has been found to have beneficial effects when used as a total or partial substitute for mineral fertilizer in peat-based artificial greenhouse potting media and as soil amendments in field studies. The specific effect of vermicompost in different crops is summarized hereunder.

5.1 Cereals and Millets

Vermicompost stimulates the growth of a wide range of plant species including cereals such as sweet corn (Lazcano *et al.*, 2011) and rice (Reddy and Ohkura, 2004; Sunil *et al.*, 2005). Gopal Reddy *et al.* (2012) reported that vermicompost (5 t/ha) performed equally with Neem cake (5 t/ha) but both were significantly superior over FYM and control in terms of rice grain yield. They also confirmed significantly low infestation by thrips and low incidence of blast and sheath rot in vermicompost amended plots as compared to control plots. Desai *et al.* (1999) stated that the application of vermicompost along with fertilizer N gave higher dry matter (16.2 g/plant) and grain yield (3.6 t/ha) of wheat (*Triticum aestivum*) and higher dry matter yield (0.66 g/plant) of the following coriander (*Coriandrum sativum*) crop in sequential cropping system. Suhane *et al.* (2008) reported exclusive application of 2.5 t/ha improved wheat grain yield by 17.6 per cent over mineral fertilizer besides reducing the water demand by 30-40 per cent and pests and disease incidence (Sinha *et al.*, 2009b). Sudhakar (2000) he reported grain yield enhancement in rice to the tune of 26, 35.7 and 54.4 per cent due to application of vermicompost (5 t/ha) prepared from rice straw, sugarcane trash and water hyacinth, respectively.

Gopinath *et al.* (2008) reported reduction in wheat grain yield by 28-41 per cent due to vermicompost application during transition period under organic farming. Similarly, Ramesh *et al.* (2005) reported less protein content and 12.6 per cent reduction in grain yield of wheat due to vermicompost application as compared to mineral fertilizers. Though vermicompost alone gave 39 per cent lower grain yield of rice, its' combination with *Azolla*, blue green algae and farm yard manure application gave on par yields with that of inorganic fertilizer (Singh *et al.*, 2007). Singh (2009) found that in the farm plots where vermicompost was applied in the 2nd, 3rd and 4th successive years, the growth and yield of wheat crops increased gradually over the years at the same rate of application of vermicompost *i.e.* at 20 q/ha. In the 4th successive year the yield was 38.8 q/ha which was very close to the yield (40.1 q/ha) where vermicompost was applied at 25 q/ha. Vermicompost also increases the nutritional quality of corn. Patil and Sheelavantar (2000), Reddy and Ohkura (2004) and Sunil *et al.* (2005) reported increased growth of sorghum due to vermicompost application.

5.2 Pulses and Oilseeds

Positive effects of vermicompost include stimulated seed germination (Karmegam *et al.*, 1999), increased dry matter production and nutrient content in green gram (Rajkhowa *et al.*, 2000; Nagavallemma *et al.*, 2004). Vermicompost application gave 35.1 and 48 per cent higher seed yield of greengram over recommended dose of N applied as FYM and urea, respectively. Bhaskaran *et al.* (2009) observed 56.14 per cent higher protein content, lowest fibre content (8.5 per cent), highest keeping quality (6.3 days) and palatability score (3.7) due to vermicompost application in cowpea. According to Ramanjaneyulu *et al.* (2013), vermicompost when applied for top dressing along with application of farm yard manure (FYM) as basal, resulted in stimulation in growth of organically grown redgram and enabled the plants to withstand mid season dry spell. They further stated combined application of FYM (5 t/ha) + vermicompost (2.5 t/ha) resulted in 18.4 per cent higher mean redgram seed yield over conventional inorganic farming. They also felt that organic products need to be given premium price to make organic farming as a profitable enterprise. They noticed superiority of organic redgram seed to inorganic and control plots in terms of protein and sugars. Gangaiah *et al.* (2013) reported that during conversion period from inorganic to organic, inorganic mode of production gave 189.7 and 68.4 per cent higher seed yield as compared to organic mode of production without amendments (0.49 t/ha) and with FYM/vermicompost (0.84 t/ha) application. There was significant build up of organic carbon content and available P in the soil amended with FYM and vermicompost application. The fresh and dry matter yields of cowpea (*Vigna unguiculata*) were higher when soil was amended with vermicompost than with biodigested slurry (Karmegam *et al.*, 1999; Karmegam and Daniel, 2000). Positive effects of vermicompost application were recorded on growth of sunflower (Devi and Agarwal, 1998; Devi *et al.*, 1998).

5.3 Vegetables

Positive effects of vermicompost include stimulated seed germination growth in several plant species such as tomato plants (Gutiérrez-Miceli *et al.*, 2007; Zaller, 2007). Application of vermicompost at 5 t/ha significantly increased yield of tomato (*Lycopersicon esculentum*) (5.8 t/ha) in farmers' fields in Adarsha watershed, Kothapally, Andhra Pradesh compared to control (3.5 t/ha). Similarly, greenhouse studies at Ohio State University in Columbus, Ohio, USA have indicated that vermicompost enhances transplant growth rate of vegetables. Amendment of vermicompost with a transplant grown without vermicompost had the highest amount of red marketable fruit at harvest. In addition, there were no symptoms of early blight lesions on the fruit at harvest. The yield of pea (*Pisum sativum*) was also higher with the application of vermicompost (10 t/ha) along with recommended N, P and K than with these fertilizers alone (Reddy *et al.*, 1998). Vermicompost may also increase the nutritional quality of tomatoes (Gutiérrez-Miceli *et al.*, 2007), Chinese cabbage (Wang *et al.*, 2010), spinach (Peyvast *et al.*, 2008) and lettuce (Coria-Cayupán *et al.*, 2009) and strawberries (Singh *et al.*, 2008), sweet corn (Lazcano *et al.*, 2011). Mean soluble and also insoluble solids were found to be significantly higher in vermicmpost applied plots as compared to control in tomato crop (Rakesh Joshi and Adarsh Pal, 2010). Application of enriched vermicompost (with *Azospirillum* and P-solubilising organisms) resulted in improvement in biometric observations like no. of leaves/plant, shoot/root ratio and plant height in chillies (Padmavathiamma *et al.*, 2008).

5.4 Fruits

Positive effects on fruit crops such as banana and papaya (Acevedo and Pire, 2004; Cabanas-Echevarria *et al.*, 2005) and strawberry (Arancon *et al.*, 2004) were reported. In strawberry, more number and yield of marketable fruits besides no. of runners and flowers on vines were observed. In a study in Australia, Webster (2005) found that vermicompost increased yield of 'cherries' for three years after 'single application'. Yield was much higher when the vermicompost was covered by 'mulch'. Vermicompost treated vines produced 18 per cent more bunch numbers and 23 per cent more grapes in the vineyards. In case of papaya, 25 per cent yield reduction was noticed due to vermicompost application even though there was not much difference in yield parameters as compared to mineral fertilizer (Ray *et al.*, 2008)

5.5 Spices, Condiments, Medicinal and Aromatic Crops

Vermicompost stimulates the growth of several crops like pepper (Arancon *et al.*, 2005) and garlic (Argüello *et al.*, 2006). It also showed positive effects on some aromatic and medicinal plants (Prabha *et al.*, 2007).

6. Impact of Vermicompost Application on Soil Properties

Role of vermicompost in improving the physical, chemical and biological properties is a well proven fact. Vermicompost application increases soil porosity in the range of 30-50 μm and 50-500 μm size thus aeration (Lunt and Jacobson, 1994), decreases particle and bulk density (Singh *et al.*, 2007; Gopinath *et al.*, 2008; Parthasarathy *et al.*, 2008), improvement in hydraulic conductivity by 8 per cent and six fold increase in water holding capacity (Lee, 1985), development of good soil structure, increase in infiltration and resistance to erosion (Rose and wood, 1980) as compared to inorganic fertilizer treatments.

Drinkwater *et al.* (1995), Gopinath *et al.* (2008) and Bhaskaran *et al.* (2009) reported significant increase in soil pH in vermicompost added soils as compared to FYM or mineral fertilizer added soils. Gopinath *et al.* (2008) reported that vermicompost amended soils do get benefitted due to buildup of organic matter (C-sequestration). Bhaskaran *et al.* (2009) noticed perceptible increase in cationic exchange capacity, C/N ratio and available N, P and K in vermicompost amended soils.

Vermicompost provides good nourishment for the growth of several microorganisms. Thus, it is used as a carrier medium for biofertilizers. Its' incorporation into agricultural fields enhanced the soil microbial biomass carbon (Singh *et al.*, 2007; Ramanajaneyulu *et al.*, 2013), enzymatic activity of phosphatase and dehydrogenase (Gopinath *et al.*, 2008; Bhaskaran *et al.*, 2009; Ramanajaneyulu *et al.*, 2013), population of actinomycetes and bacteria (Singh *et al.*, 2007) and higher activity of acid phosphatase, urease and β glucoside (Saha *et al.*, 2008; Tajeda *et al.*, 2009).

7. Effect of Humic Acid Extracts from Vermicompost on Crops

Humic acids (HA) extracted from vermicompost were found to promote plant growth (Atiyeh *et al.*, 2002) and lateral root emergence and root area (Canellas *et al.*, 2002). They enhance plant growth by binding the plant growth hormones present in the soil. Application of HA showed encouraging results in marigold, pepper, straw berries and tomatoes in terms of germination, growth, flowering and yields, similar to that of indole-3-acetic acid (IAA) (Arancon *et al.*, 2006).

8. Effect of Vermicompost Leachates

8.1 Vermicompost "Teas"

It is an aqueous extract ("teas") from vermicompost. It can be produced either by passing water through vermicompost or keeping vermicompost in water (1-7 days). First, vermicast is collected from vermibed and then kept in mesh bag and soak it in a water with a ratio of 1:10 (vermicast:water). The contents have to be stirred intermittently to make the Vermicompost "teas" ready after 24 hours. It is known to suppress plant pathogens such as *Plectusporium*, *Verticillium* and *Rhizoctonia*, significantly in laboratory and greenhouse experiments.

8.2 Vermiwash

Vermiwash is a liquid plant growth regulator (leachates) derived from vermicompost. It contains high concentration of plant nutrients (Tejada *et al.*, 2008), humic acids (Atiyeh *et al.*, 2002; Ordonez *et al.*, 2006), high amount of enzymes, vitamins and hormones like auxins and gibberellins. Thus it regulates nutrient absorption. Its' positive effects were reported by Guiterrez-Miceli *et al.* (2008) in sorghum, Tejada *et al.* (2008) in tomatoes, Singh *et al.* (2010) in strawberries and farmers of North India in brinjal. Vermiwash effectively reduced the pest and disease incidence thus use of chemical pesticides (Sinha *et al.*, 2009c). It can be applied in two ways. Spray the mixed solution of 1 litre of vermiwash with 7-10 litres of water on the leaf (upper and lower side) in the evening on the growing crop or 1 litre of vermiwash + 1 litre of cow urine + 10 litres of water (mix thoroughly and keep it over night) on one bigha of land to control various crop diseases.

9. Promotional Policies

Vermicomposting of diverse organic wastes and use of vermicompost in agriculture is being commercialized all over the world by developed countries like U.S., Canada, U.K., Australia, Russia and Japan to developing countries like India, China, Chile, Brazil, Mexico, Argentina and the Philippines (Bogdanov, 1996; Sherman, 2000).

U.S. has some largest vermicomposting plants in world and is encouraging people for 'backyard vermicomposting' to divert wastes from landfills. (Bogdanov, 1996; Bogdanov, 2004). The 'American Earthworm Company' started a 'vermicomposting farm' in 1978-79 with 500 t/month of vermicompost production (Edward, 2000). 'Vermicycle Organics' produced 7.5 million pounds of vermicompost every year in high-tech greenhouses. The sale of vermicompost grew by 500 per cent in 2005. 'Vermitechnology Unlimited' has doubled its business every year since 1991 (NCSU, 1997; Kangmin, 1998).

A large scale vermicomposting plant has been installed in Canada to vermicompost municipal and farm wastes (Georg, 2004). In UK, large 1000 metric ton vermicomposting plants have been erected in Wales to compost diverse organic wastes. In France, 20 tons of mixed household wastes are being vermicomposted everyday using 1000 to 2000 million red tiger worms (*Elsenia andrei*) in earthworm tanks (Visvanathan *et al.*, 2005).

The 'Envirofert Company' of New Zealand is vermicomposting about 5-6 thousand tons of green waste every year. They are also planning to vermicompost approximately 40,000 tonnes of food wastes from homes, restaurants and food processing industries every year to produce vermicompost which would eventually replace chemical fertilizers in farm production in New Zealand (Frederickson, 2000).

Vermicomposting is being done on large scale in Australia as a part of the 'Urban Agriculture Development Program' utilizing the urban solid wastes (Lotzof,2000). Vermicomposting of sludge from the sewage and water treatment plants is being increasingly practiced in Australia and as a result it is saving over 13,000 m^3 of landfill space every year in Australia (Komarowski, 2001; Dynes, 2003).

India has also launched vermicomposting program of MSW. In recent years it is growing as a part of 'village and farm waste management', 'sustainable non-chemical agriculture' combined with 'poverty eradication' program. It has enhanced the lives of poor in India and generated self-employment opportunities for the unemployed. In several Indian villages, NGOs are freely distributing cement tanks and 1000 worms and encouraging men and women to collect waste from villages and farmers, vermicompost them and sell both worms and vermicompost to the farmers (Suhane, 2007; NIIR, 2009). Department of Agriculture is giving 50 per cent subsidy for farmers towards construction of vermisheds and beds to encourage organic farming through vermicomposting.

10. Conclusions

Keeping in view the increasing interest towards organic farming across the globe, vermicompost has been recognized as one of the potential organic sources of nourishing the plants in Agriculture, Horticulture and Floriculture. Vermicompost is rich in macronutriuents, besides micronutrients, plant growth promoting hormones, beneficial soil microbes and mycorrhizal fungi. In view of ever increasing demand for organic foods across the world and its pivotal role in maintaining the soil health and productivity, vermicompost is definitely going to be a viable component of organic farming. Switching over to sustainable agriculture by 'Vermiculture Revolution' can truly bring in 'economic prosperity' for the farmers and the nations with 'environmental security' for the earth. Further, earthworms do have multifarious uses in Agriculture and pharmaceutical industries.

11. Future Thrust

Though research work on impact of humic acid extracts and other leachates like vermiwash has been initiated across the world but still the information on economical dose of application in various crops is meager. There is a need to explore the possibility of use of vermicompost for top dressing to meet the needs of long duration field crops. Further, efforts must be made to encourage organic farmers to manage vermicomposting units in their farms so as to efficiently utilize the crop residues and animal excreta.

REFERENCES

Acevedo IC and Pire R. 2004. Effect of vermicompost as substrate amendment on the growth of papaya (*Carica papaya* L.). Interciencia, 29: 274–279.

Adhikary S. 2012. Vermicompost, the story of organic gold: A review. Agricultural Sciences 3(7): 905-917 http: //dx.doi.org/10.4236/as.2012.37110.

Appelhof M. 1997. Worms eat my garbage, 2nd Edition, Flower Press, Kalamazoo, Michigan, U.S. (http: //www.wormwoman.com).

Appelhof M. 2003. Notable Bits. In: *WormEzine*, Vol. 2(5). (Available at http: // www.wormwoman.com.

Arancon N. 2004. An interview with Dr. Norman Arancon. In Casting Call, 9(2), August.

Arancon NQ, Edwards CA, Bierman P, Welch C and Metzger JD. 2004. Influences of vermicomposts on field strawberries - 1: Effects on growth and yields. Bioresource Technology, 93: 145–153.

Arancon NQ, Galvis PA and Edwards CA. 2005. Suppression of insect pest populations and damage to plants by vermicomposts. Bioresource Technology, 96: 1137-1142.

Arancon NQ, Edwards CA, Lee S, Byrne R. 2006. Effects of humic acids from vermicomposts on plant growth. European Journal of Soil Biology, 42: S65-S69.

Argüello JA, Ledesma A, Nuriez SB, Rodriguez CH, Diaz Goldfarb MC. 2006. Vermicompost effects on building dynamics, non-structural carbohydrate content, yield and quality of Rosado Paraguayo garlic bulbs. Horticulture Science, 41(3): 589-592.

Atiyeh RMS, Subler CA, Edwards G, Bachman JD, Metzger and Shuster W. 2000. Effects of vermicomposs and composts on plan growth in horticultural container media and soil. Pedo biologia, 44: 579-590.

Atiyeh RMS, Lee S, Edward CA, Arancon NQ and Metzger JD. 2002. The influence of humic acids derived from earthworm processed organic wastes on plant growth. Bioresource Technology, 84: 7-14.

Beetz A. 1999. Worms for Composting (Vermicomposting). ATTRA-National Sustainable Agriculture Information Service, Livestock Technical Note, June, 1999.

Bhaskar A, Macgregor AN and Kirkmam JH. 1992. Influence of soil ingestion by earthworms on the availability of potassium in soil. An incubation experiment. Biology and Fertility of Soils, 14: 300-303.

Bhaskaran, Usha P and Devi K. 2009. Effect of organic farming on soil fertility, yield and quality of crops in the tropics. In: The proceedings of the International Plant Nutrition Colloquim XVI, UC Davis. Avaialble online: http: //www. escholarship.org/uc/item/7k12w04m.

Binet F, Fayolle L and Pussard M. 1998. Significance of earthworms in stimulating soil microbial activity. Biology and Fertility of Soils, 27(1): 79-84.

Bogdanov P. 1996. Commercial Vermiculture: How to Build a Thriving Business in Redworms. VermiCo Press, Oregon. pp 83.

Bogdanov P. 2004. The single largest producer of vemicompost in the world. In: Casting Call 9(3), October. http: //www.vermico.com

Benitez E, Nogales R, Masciandro G and Ceccanti B. 2000. Isolation by isoelectric focusing of humic–urease complexes from earthworm (*Eisenia foetida*) processed sewage sludges. Biology and Fertility of Soils, 31: 489–493.

Businelli M, Perucci P, Patumi M and Giusquiani PL. 1984. Chemical composition and enzymatic activity of worm casts. Plant and Soil, 80: 417-422.

Cabanas-Echevarría M, Torres–García A, Díaz-Rodríguez B, Ardisana EFH and Creme-Ramos Y. 2005. Influence of three bioproducts of organic origin on the production of two banana clones (*Musa* spp. AAB.) obtained by tissue cultures. Alimentaria, 369: 111–116.

Canellas LP, Olivares FL, Okorokova AL and Facknha RA. 2002. Humic acids isolated from earthworms compost enhance root elongation, lateral root emergence and plasma membrane H^+ - ATPase activity in maize roots. Plant Physiology, 130: 1951-1957.

Cooper E. 2009. New enzymes isolated from earthworms is potent fibrinolytic; ACAMIntegrative Medicine Blog; Oxford University Press Journal (UK).(http: //acam.typepad.com/blog/2009/04/index.html)

Coria-Cayupán YS, De Pinto MIS and Nazareno MA. 2009. Variations in bioactivesubstance contents and crop yields of lettuce (*Lactuca sativa* L.) cultivated in soils with different fertilization treatments. Journal of Agricultural and Food Chemistry, 57(21): 10122-10129.

Cordero CH. 2005. Earthworms can help dissolve blood clots for stroke patients(http: //www.thenewstoday.info/20051125/earthworms.can.help.dissolve.blood. clots.for. stroke.patients.html)

Cracas P. 2000. "Vermicomposting Cuban Syle", In: Worm Digest, Issue #25t.

Darwin C. 1881. The formation of vegetable mould through the action of worms withobservations on their habitats. John Murray, London, UK.

Desai VR, Sabale RN and Raundal PV. 1999. Integrated nitrogen management in wheat-coriander cropping system. Journal of Maharashtra Agricultural Universities, 24(3): 273–275.

Devi D, Agarwal SK and Dayal D. 1998. Response of sunflower (*Helianthus annuus* L.) to organic manures and fertilizers. Indian Journal of Agronomy, 43(3): 469–473.

Dominguez J, Edwards CA and Subler S. 1997. A comparison of vermicomposting and composting. Biocycle, 4: 57-59.

Drinkwater LE, Letourneau DK, Workneh F, Van Bruggen HC and Shennana C. 1995. Fundamental differences between conventional and organic tomato agroecosystems in California. Ecological Applications, 5: 1098-1112.

Dynes RA. 2003. Earthworms -Technology info to enable the development of earthworm production- Rural Industries Research and Development Corporation (RIRDC), Govt. of Australia, Canberra, ACT.

Eastman BR. 1999. Achieving pathogen stabilization using vermicomposting. Biocycle, 11: 62.

Eastman BR, Philip NK, Clive AE, Linda T, Bintoro G, Andrea LS and Jacquelyn RM. 2000. The effectiveness of vermiculture in human pathogen reduction for USEPA biosolids stabilization. Orange County, Florida: Environmental Protection Division.

Edwards CA. 1983. Earthworms, organic waste and food. Span shell chemical Co., 26(3): 106-108.

Edwards CA and Lofty JR. 1972. Biology of Earthworms. Chapman and Hall Ltd. London. pp. 1-288.

Edwards CA and Arancon N. 2004. Vermicomposts suppress plant pest and disease attacks. In: *REDNOVA NEWS,* http: //www.rednova.com/display/?id=55938.

Edwards CA. 1988. Breakdown of animal, vegetable and industrial organic wastes by earthworms. In: *Earthworms in Waste and Environmental Management* (CA Edwards and E F Neuhauser, Eds.), SPB Academic Publ. Co., the Hague, the Netherlands. pp. 21–31.

Edwards CA, Dominguez J and Neuhauser EF. 1998. Growth and reproduction of *Perionyx excavates* (Per.) (Megascolecidae) as factors in organic waste management. Biology and Fertility of Soils, 27: 155–161.

Edwards CA. 1999. Interview with Dr. Clive Edwards. In: *Casting Call,* (Peter Bogdanov, Ed.), VermiCo, Merlin, Oregon, 4(1), June Issue.

Edward CA. 2000. Potential of vermicomposting for processing and upgrading organic waste, Ohio State University, Ohio, U.S.

FAO, 2003. World Agriculture – Towards 2015/2030 – An FAO perspective. Available online http//www.fao.org documents/show_cdr.asp?url_file=/docrep/005/y4252e/y4252e00.htm

Follet R, Donahue R and Murphy L. 1981. Soil and soil amendments. Prentice hall, Inc., New Jersey.

Fraser-Quick G. 2002. Vermiculture–a sustainable total waste management solution, What's new in Waste Management, 4(6): 13-16.

Frederickson J and Ross-Smith S. 2004. Vermicomposting of pre-composted mixed fish/shellish and green waste. The Worm Research Centre, SR566, July. Available at http: //www.wormresearchcentre.co.uk.

Frederickson J. 2000. The worm's turn; Waste Management Magazine, August, UK.

Gaddie RE (Sr.) and Duglas DE. 1975. Earthworms for Ecology and Profit. Volume 1: Scientific Earthworm Farming. Bookworm Publishing Company, Cal. p 180.

Gandhi M, Sangwan V, Kapoor KK and Dilbaghi N. 1997. Composting of household wastes with and withoutearthworms. Environment and Ecology, 15(2): 432–434.

Gangaiah B, Ahlawat IPS, Shivakumar BG and Prasad Babu MBB. 2013. Impact of organic mode of production on performance of pigeonpea (*Cajanus cajan* l.) during conversion from conventional to organic production. Indian Journal of Dryland Agricultural Research and Development, 28(2): 27-31.

Georg 2004. Feasibility of developing the organic and transitional farm market for processing municipal and farm organic wastes using large-scale vermicomposting. Publication of of Good Earth Organic Resources Group, Halifax, Nova Scotia, Canada (Available on http: //www.alternativeorganic.com).

Gopal Reddy B, Ram Gopal Verma N, Jagadeeshwar R, Madhavi A, Narsimha Reddy P, Sreedhar N, Sureder Raju CH and Vani Sreee S. 2012. Organic farming – Research Progress, ANGRAU Technical Bulletin-2/Rice Section/Telugu/2012/1000.

Gopinath KA, Saha S, Mina BL, Kundu S, Pande H, Gupta HS. 2008. Influence of organic amendments on growth, yield and quality of wheat and soil properties during transition to organic production. Nutrient Cycling in Agroecosystems, 82: 51-60.

Gopinath KA, Bandi Venkateswarlu, Banshi L. Mina, Nataraju KC and Konda Gayatri Devi 2010. Utilization of vermicompost as a soil amendment in organic crop production. Dynamic soil Dynamic plant, 4(Special issue I): 48-57.

George WD. 1994. Vermicomposting, Guide H-164, College of Agriculture and Glenn Munroe. Manual of On-Farm Vermicomposting and Vermiculture, Organic Agriculture.

Gunadi B, Charles B and Clive AE. 2002. The growth and fecundity of *Eisenia fetida* in cattle solids pre-composted for different periods. Pedobiologia, 46: 15-23.

Gutierrez-Miceli FA, Santiago-Borraz J, Montes Molina JA, Nafate CC, Abdud-Archila M, Oliva Llaven MA, Rincon-Rosales R and Deenodoven L. 2007. Vermicompost as a soil supplement to improve growth, yield and fruit quality of tomato. Bioresource Technology, 98: 2781-286.

Hansen D. 2007. Vermicomposting: innovative kitchen help. [Online] Available: http: //www.dnr.mo.gov/env/swmp/docs/vermicomposting.pdf (April, 12 2008).

Hwang CM, Kim Di and Huh SH. 2002. *In vivo* evaluation of lumbrokinase extracted from earthworms *Lumbricus rubellus* in a prosthetic vascular graft. Cardiovascular Surgery, 43: 891-894.

Jin L, Jin H, Zhang G and Xu G. 2000. Changes in coagulation and tissue plasminogen activator after the treatment of cerebral infarction with lumbrokinsae (from earthworms). Clinical Hemorheol Microcirc, 23: 213-218.

Kale RD. 1998. Earthworms: nature's gift for utilization of organic wastes. In: *Earthworm Ecology* (CA Edward, Ed.), St. Lucie Press, NY, ISBN 1-884015-74-376.

Kale RD and Bano K. 1986. Field trials with vermicompost: An organic fertilizer. In: Proceedings of national seminar on "Organic waste utilization by vermicomposting". GKVK, Bangalore, India, pp. 151-160.

Kangmin Li. 1998. Earthworm case, 4th ZERI World Congress, Windhoel, Namibia (Also in Vermiculture industry in Circular Economy; *Worm Digest* (2005) (http: //www.wormdigest.org/content/view/135/2).

Karmegam N, Alagermalai K and Daniel T. 1999. Effect of vermicompost on the growth and yield of greengram (*Phaseolus aureus* Rob.). Tropical Agriculture, 76(2): 143–146.

Karmegam N and Daniel T. 2000. Effect of biodigested slurry and vermicompost on the growth and yield of cowpea (*Vigna unguiculata* (L.). Environment and Ecology, 18(2): 367–370.

Komarowski S. 2001. Vermiculture for sewage and water treatment sludge. WATER – A publication of Australian Water and Wastewater Association, July 2001, pp. 39-43.

Lazcano C, Revilla P, Malvar RA, Dominguez J. 2011. Yield and fruit quality of four sweet corn hybrids (*Zea mays*) under conventional and integrated fertilization with vermicompost. Journal of the Science of Food and Agriculture, 91: 1244–53.

Latifah Abd Manaf, Latifah Abd Manaf, Mohd Lokman Che Jusoh, Mohd Kamil Yusoff Tengku Hanidza Tengku Ismail, Rosta Harun and Hafizan Juahir. 2009. Influence of Bedding Material in Vermicomposting Process. International Journal of Biology, 1(1): 81-91.

Lee KE. 1985. Earthwoerms, their Ecology and relationships with land use, Academic Press, Sydney, p. 411.

Li SL. 1995. Research on *di long's* (earthworms) effect in lowering blood pressure. Journal of Information, 12(3): 22-24.

Lotzof M. 2000. Vermiculture: an Australian technology success story; *Waste Management Magazine*; February 2000, Australia.

Lunt HA and Jacobson HG. 1994. The chemical composition of earthworm casts. Soil Science, 58: 367-375.

Mark Leary. 2004. California Integrated Waste Management Board (CIWMB), Office of Education and the Environment, at (916) 341-6769. www.ciwmb.ca.gov/ Schools/Curriculum/Worms/.

Moss R. 2004. Of enzymes, worms and cancer: the war on cancer (lumbrokinsae enzyme from earthworms); Worm Digest (http: //www.wormdigest.org/ content view/161/2/)

Mihara H, Sumi M, Mizumoto H, Yoneta T, Ikeda R and Maruyama M. 1990. Oral administration of earthworm powder as possible thrombolytic therapy, Recent Advances in Thrombosis and Fibrinolysis, Academic Press, NY, pp 287-298.

Mihara H, Maruyama M and Sumi M. 1992. Novel thrombolytic therapy discovered in oriental medicine using the earthworms. SE Asian Journal of Tropical Medicine and Health, pp. 131-140.

Morgan M and Burrows I. 1982. Earthworms/microorganisms interactions, Rothamsted Experimental Station Republic.

Munroe G. 2007. Manual of On-farm vermicomposting and vermiculture - A Publication of Organic Agriculture Centre of Canada, p. 39.

Nagavallemma KP, Wani SP, Lacroix S, Padmaja VV, Vineela C, Babu Rao M and Sahrawat KL. 2004. Vermicomposting - Recycling wastes into valuable organic fertilizer. Global theme on Agricosystems, A report No. 8, ICRISAT, Patancheru, Andhra Pradesh, India, p. 20.

Nair SK, Naseema A, Meena Kumari SK, Prabhakumari P and Peethambaram CC. 1997. Microflora associated with earthworm and vermicomposting. Journal of Tropical Agriculture, 35: 93-98.

NCSU. 1997. Large scale vermi-composting operations – data from Vermi-cycle Organics, Inc., North Carolina State University, U.S.

NIIR. 2009. The complete technology book on vermiculture and vermicompost, Publication of National Institute of Industrial Research, New Delhi, India, ISBN 8186623817.

Ordonez C, Tajeda M, Benetiz C, Gonzalez JL. 2006. Characterisation of a phosphorus-potassium solution obtained during a protein concentrate process from sunflower flour. Application on rye-grass. Bioresource Technology, 97: 522-528.

Padmavathiamma KP, Li YL and Kumari UR. 2008. An experimental study of vermin-biowaste composting for agricultural soil improvement. Bioresource Technology, 99: 1672-1681.

Parthasarathi K, Balamurugan M and Ranganathan LS. 2008.Influence of vermicompost on physic-chemical and biological properties in different types of soil along with yield and quality of the pulse crop-blackgram. Iranian Journal of Environmental Health Science and Engineering, 5: 51-58.

Patil SL and Sheelavantar MN. 2000. Effect of moisture conservation practices, organic sources and nitrogen levels on yield, water use and root development of *rabi* sorghum (*Sorghum bicolor* L.) in the vertisols of semi arid tropics. Annals of Agricultural Research, 21(21): 32–36.

Peyvast G, Olfati JA, Madeni S and Forghani A. 2008. Effect of vermicompost on thegrowth and yield of spinach (*Spinacia oleracea* L.). Journal of Food Agriculture andEnvironment, 6: 110-113.

Pierre V, Phillip R, Margnerite L and Pierrette C. 1982. Anti-bacterial activity of thehaemolytic system from the earthworms *Eisinia foetida*. Invertebrate Pathology, 40: 21-27.

Prabha ML, Jayraay IA, Jayraay R and Rao DS. 2007. Effect of vermicompost ongrowth parameters of selected vegetable and medicinal plants. Asian Journal of microbiology, Biotechnology and Environmental Sciences, 9(2): 321-326.

Rajiv K. Sinha, Krunal Chauhan, Dalsukh Valani, Vinod Chandran, Brijal Kiran Soni, Vishal P. 2010. Earthworms: Charles Darwin's 'Unheralded Soldiers of Mankind': Protective and Productive for Man and Environment. Journal of Environmental Protection, 1: 251-260. doi: 10.4236/jep.2010.13030.

Ramanjaneyulu AV, Madhavi A, Srinivas D, Neelima TL, Madhusudhan Reedy D and Dharma Reddy K. 2013. Organic farming in redgram on Alfisols. In: Compendium of abstracts of the First International Conference of Bio-resource and Stress Management, Feb 6-9, 2013, Kolkata, India, pp. 53-55.

Ray PK, Singh AK and Kumar A. 2008. Performance of Pusa Delicious papaya under organic farming. Indian Journal of Horticulture, 65: 100-102.

Rajkhowa DJ, Gogoi AK, Kandali R, Rajkhowa KM. 2000. Effect of vermicompost on greengram nutrition. Journal of Indian Society of Soil Science, 48: 207-208.

Reddy R, Reddy MAN, Reddy YTN, Reddy NS, Anjanappa N and Reddy R. 1998. Effect of organic and inorganic sources of NPK on growth and yield of pea (*Pisum sativum* L). Legume Research, 21(1): 57–60.

Rakesh Joshi and Adarsh Pal V. 2010. Effect of vermicompost on growth, yield andquallty of Tomato (*Lycopersicum esculentum* L.). African Journal of Basic and Applied Sciences, 2 (3-4): 117-123.

Ramesh P, Mohan Singh and Singh AB. 2005. Performance of macaroni and bread wheat varieties with organic and inorganic sources of nutrients under limited irrigated conditions of vertisols. Indian Journal of Agricultural Sciences, 75: 823-825.

Ravichandran C, Chandrasekaran GE and Christy Priyadharsini F. 2001. Vermicompost from different solid works using treated Dairy effluent". Indian Journal of Environmental Protection, 21: 538-542.

Reddy MV and Ohkura K. 2004. Vermicomposting of rice-straw and its effects onsorghum growth. Tropical Ecology, 45(2): 327-331.

Rose CJ and Wood AW. 1980. Some environmental factors affecting earthworms population and sweet potato production in the Tari Basin, Papua New Guinea, Papua New Guinea Journal 33: 1-13.

Saha S, Mina BL, Gopinath KA, Kundu and Gupta HS. 2008. Organic amendments affect biochemical properties of a subtemperature soil of the Indian Himalayas. Nutrient Cycling in Agroecosystems, 80: 233-242.

Sailaja Kumari MS and Ushakumari K. 2002. Effect of vermicompost enriched with rock phosphate on the yield and uptake of nutrients in cowpea (*Viggna unguiculata*). Journal of Tropical Agriculture, 40: 27-30.

Saviozzi A, Riffaldi R and Levi-Minzi R. 1988. Maturity evaluation of organic wastes. BioCycle, 29: 54–56.

Sherman R. 1997. Controlling mite pess in earthworm beds. North Carolina Cooperative Extension Service, Raleigh, NC. http: //www.bae.ncsu.edu/ people/faculty/sherma

Sherman R. 2000. Commercial systems latest developments in mid to large scale Vermicomposting. BioCycle, November 2000, p. 51.

Short JCP, Frederickson J and Morris RM. 1999. Evaluation of traditional windrow-compostingand vermicomposting for the stabilization of wase paper sludge (WPS). In: (DJ Diaz Cosin, JB Jesus and D Trogo, (Eds.), 6th International Symposium on Earthworm Ecology, Vigo, Spain, 1998. Pedobiologia. 1999. 43(6), pp. 735-743.

Singh R, Sharma RR, Kumar S, Gupta RK and Patil RT. 2008. Vermicompost substitution influences growth, physiological disorders, fruit yield and quality of strawberry (Fragaria xananassa Duch). Bioresource Technology, 99(17): 8507-8511.

Singh R, Gupta RK, Patil RT, Sharma RR, Asrey R, Kumar A and Jangra KK. 2010. Sequential foliar application of vermicompost leachates improve marketable fruit yield and quality of straw berry. Scientia Hortiulturae, 124: 34-39.

Singh PK, Rajiv Sinha K, Sunil Herat K, Ravindra S and Sunita K. 2009. Studies on Earthworms vermicompost as a sustainable alternative to chemical fertilizers for production of wheat crops. Collaborative Research on Vermiculture Studies, College of Horticulture, Noorsarai, Bihar, India and Griffith University, Brisbane, Australia.

Singh KP, Suman A, Singh PN and Srivastava TK. 2007. Improving quality of sugarcane growing soils by organic amendments under sub-tropical climatic conditions of India. Biology and Fertility of Soils, 44: 367-376.

Singleton DR, Hendrix BF, Coleman DC and Whitemann WB. 2003. Identification of uncultured bacteria tightly associated with the intestine of the earthworms (Lumricus rubellus). Soil Biology and Biochemistry, 35: 1547-1555.

Sinha RK, Heart S, Agarwal R, Asadi and Carretero K. 2002. Vermiculture technology for environmental management: study of action of earthworms *Eisenia foetida, Eudrilus euginae* and *Perionyx excavatus* on biodegradation of some community wastes in India and Australia. The Environmentalist, 22(2): 261 – 268.

Sinha RK, Bharambe G and Bapat PD. 2007. Removal of high bod and cod loadings of primary liquid waste products from dairy industry by vermi-filtration technology using earthworms. Indian Journal of Environmental Protection (IJEP), 27(6): 486-501.

Sinha RK and Chan A. 2009. Study of emission of greenhouse gases by Brisbane households practicing different methods of composting of food and garden wastes: aerobic, anaerobic and vermicomposting"; NRMA – Griffith University Project Report

Sinha RK, Herat S, Bharambe G and Brahambhatt A. 2009a. Vermistabilization of sewage sludge (biosolids) by earthworms: converting a potential biohazard destined for landfill disposal into a pathogen free, nutritive and safe bio-fertilizer for farms. Journal of Waste Management and Research, UK, (http: / /sagepub.com).

Sinha RK, Herat S, Valani D, Chauhan K. 2009b. Vermiculture and Sustainable Agriculture. American and Eurasian Journal of Agricultural and Environmental Sciences, 5(S): 01-55.

Sinha RK, Heart S, Bharambe G, Patil S, Bapat D, Chauhan K, Valani D. 2009c. Vermiculture biotechnology. The emerging cost effective and sustainable technology of the 21st century for multiple uses from waste and land management to safe and sustained crop production. Environmental Research Journal, 3: 2-3.

Suhane RK. 2007. Vermicompost (In Hindi); Publication of Rajendra Agriculture University, Pusa, Bihar, p 88 (www.kvksmp.org).

Suhane RK, Sinha RK and Singh PK. 2008. Vermicompost, cattle dung compost and chemical fertilizers: Impacts on yield of wheat crops. Communication of Rajendra Agricultural University, Pusa, Bihar, India.

Sunil K, Rawat CR, Shiva D. and Suchit KR. 2005. Dry matter accumulation, nutrient uptake and changes in soil fertility status as influenced by different organic and inorganic sources of nutrients to forage sorghum (Sorghum bicolor). Indian Journal of Agricultural Science, 75 (6): 340-342.

Sudhakar G. 2000. Investigations to identify crop waste/low land weeds as alternative source to organics to sustain the productivity of rice base dsystem. PhD thesis. Tamil Nadu Agricultural University, India

Tejada M, Gonzalez JL, Hernandez MT and Garcia C. 2008. Agricultural use of leachates obtained from two different vermicomposting processes. BioresourceTechnology, 99: 6228-6232.

Tejada M, Garcia Martinez AM and Parrado J. 2009. Effects of a vermicompost composted with beet vinasase on soil properties, soil losses and soil restoration. Catena, 77: 238-247.

Tara Crescent 2003. Vermicomposting. Development Alternatives (DA) Sustainable Livelihoods. (http: //www.dainet.org/livelihoods/default.htm).

Vasanthi D and Kumaraswamy K. 1995. Efficacy of vermicompost on the yield of rice and on soil fertility. National Seminar on Organic Farming and Sustainable Agriculture, October 9-11, UAS, Bangalore, Karnataka, India.

Vermi Co 2001. Vermicomposting technology for waste management and agriculture: an executive summary. (http: //www.vermico.com/summary.htm) PO Box 2334, Grants Pass, OR 97528, USA: Vermi Co. 9. Appelhof, Mary, 1997. Worms Eat My Garbage; 2nd (Ed.). Flower Press, Kalamazoo, Michigan, U.S. (http: //www.wormwoman.com).

Vijaya Sankara Babu M, Adinarayana G, Rama Subbaiah K, Balaguruvaiah D, Yellamanda Reddy T. 2008. Vermicompost with different farm wastes and problematic weeds. Indian Journal of Agricultural Research, 42(1): 52-56.

Visvanathan C, Trankler J, Jospeh K and Nagendran R. 2005. Vermicomposting as an Eco-tool in sustainable solid waste management. Asian Institute of Technology, Anna University, India.

Webster KA. 2005. Vermicompost increases yield of cherries for three years after a single application, EcoResearch, South Australia, (www.ecoresearch.com.au).

Wang ZW. 2000. Research advances in earthworms bioengineering technology. Medica, 31(5): 386-389.

Wang D, Shi Q, Wang X, Wei M, Hu J, Liu J. and Yang F. 2010. Influence of cow manure vermicompost on the growth, metabolite contents and antioxidants activities of chinese cabbage. Biology and Fertility of Soils, 46: 689-696.

Wang FC, Wang M, Li, Gui L, Zhang J and Chang W. 2003. Purification, characterization and crystallization of a group of earthworms fibrinolytic enzymes from Eisenia fetida, Biotechnology, 25(13): 1105-1109.

Wengling C and Jhenjun S. 2000. Pharmaceutical value and uses of earthworms; Vermillenium Abstracts, Flower field enterprises, Kalamazoo, MI (USA).

Werner M and Cuevas JR. 1996. Vermiculture in Cuba. BioCycle, June, 57-62.

Zaller JG. 2007. Vermicompost as a substitute for peat in potting media: Effects on germination, biomass allocation, yields and fruit quality of three tomato varieties. Science of Horticulture, 112: 191-199.

Chapter 5

Utilization of Locally Available Resources for Nutrient Management in Organic Farming

K.C. Nataraj[1], *Malleswari Sadhineni*[1], *K.A. Gopinath*[2] *and Ch. Srinivasa Rao*[3]

[1]*Agricultural Research Station, ANGRAU, Anantapur – 515 001, Andhra Pradesh*

[2]*ICAR-Central Research Institute for Dryland Agriculture, Hyderabad – 500 059, Telangana*

[3]*ICAR-National Academy of Agricultural Research Management, Hyderabad – 500 030, Telangana*

E-mail: natarajkc2014@gmail.com

1. INTRODUCTION

The International Federation of Organic Agriculture Movements (IFOAM) defines organic agriculture as "A whole system approach based upon a set of processes resulting in a sustainable ecosystem, safe food, good nutrition, animal welfare and social justice". Organic production therefore is more than a system of production that includes or excludes certain inputs. In this direction, the emphasis is placed upon minimizing the use of external inputs through the utilization of locally available organic resources for nutrient management in organic farming. The term organic does not explicitly mean the type of inputs used rather it refers to the concept of farm as an organism. Often organic agriculture has been criticized on the grounds that with organic inputs alone farm productivity and profitability might not be improved because the availability of organic sources is highly restricted (Chhonkar, 2003). True, organic resources availability is limited but under conditions of soil constraints and chemicals beggaries, organic inputs use have proved more profitable compared to agrochemicals (Huang *et al.*, 1993).

At field level, nutrient management is one of the main challenges facing the organic farmer. In the short term, the problem is supplying sufficient nutrients to the crop at the correct point in its development to achieve economically viable yields. In the long term, the challenge is to balance inputs and off takes of nutrients to avoid nutrient rundown. Both of these goals must be achieved through the management of organic matter. There are several doubts in the minds of not only farmers but also scientists whether it is possible to supply the minimum required nutrients to crops through organic sources alone and even if it is possible how are we going to mobilize that much of organic matter. At this juncture, it is neither advisable nor feasible to recommend the switch over from fertilizer use to organic manure under all agro-ecosystems. But for the present, the nutrient management under ongoing organic fields the best utilization of locally available organic resources needs the inventory.

According to a conservative estimate around 600 to 700 Mt agricultural waste is available in the country every year, but most of it is not used properly. We must convert our agricultural waste into wealth by mobilizing all the biomass in rural urban areas into bio energy to supply required nutrients to our starved soil and fuel to farmers (Veeresh, 1997). The use of organic manures is the oldest and most widely practised means of nutrient replenishment in India, even today. The proportion of cattle manure available for fertilizing purposes decreased from 70 per cent in the early 1970s to 30 per cent in 2015. The use of FYM has almost came down to 2 t/ha, which is much below the desired rate of 10-12.5 t/ha. As two thirds of all crop residues are used as animal feed, only one-third is available for compost making, which can add 2.5 Mt/year. The area under green manuring is about 7 M ha. Unlike fertilizers, the use of organic material has not increased much in the last two to three decades in India. The estimated annual available nutrient (NPK) contribution through organic sources is about 5 Mt, which could increase to 7.75 Mt by 2025.

India is bestowed with diverse agro ecosystems, producing different kinds of organic resources for the organic agriculture. The organic resources are plenty, they are region specific, cropping system and farming system specific. Thus, locally available organic resources have a significant role to play in nutrient management under organic farming in Indian conditions.

2. Major Sources of Locally Available Organic Resources

2.1 Crop Residues

Crop residues generated in the farm such as straw of cereals, oilseeds and stalks, stubbles and leaves of all most all the field crops are the potential organic sources. In regions where mechanical harvesting is done, sizeable quantities of residues are left in the field. Similarly major portion of the residues is used as animal feed and about 33 per cent are available for direct use. About 200 Mt of crop residues is produced from different crops annually. The potential of these has been estimated to be around 100 Mt annually for recycling in agriculture. The leguminous plant residues are degraded at a faster rate than cereal crop residues having wide C: N ratio. The uses of residues are generally most effective for water conservation when managed as surface mulch (ICAR-IISS, 2014). The recycling of crop residues has the advantage of converting the surplus farm waste into useful product for meeting nutrient requirement of succeeding crops. These crop residues are the food for soil microorganisms besides supply of organic carbon and plant nutrients in organic farming. Crop residue retention on the soil surface, substantially reduces run off, soil erosion, decrease soil evaporation and protect the soil fertility in organic farming (Lal, 1989).

2.1.1 Potential of Crop Residues as Source of Plant Nutrients in Organic Farming

In India, about 500-550 Mt of crop residues are being generated annually (MoA, 2012). As per the estimation of Ministry of New and Renewable Energy, Govt. of India, 2009, The generation of crop residues is highest in Uttar Pradesh (60 Mt) followed by Punjab (51 Mt) and Maharashtra (46 Mt). Among different crops, cereals generate maximum residues (352 Mt), followed by fiber crops (66 Mt), oilseeds (29 Mt), pulses (13 Mt) and sugarcane (12 Mt). The cereal crops (rice, wheat, maize, millets) contribute 70 per cent while rice crop alone contributes 34 per cent and wheat ranks second with 22 per cent of the crop residues. The Table 5.1 shows the major crop residues production annually and its nutrient potential.

Table 5.1: Major Crop Residues Quantity and its Nutrient Potential in India

Crop Residues	*Residues Quantity*	*N*	*P*	*K*	*Nutrient Total NPK*	*Potential utilizable NPK*
	(Mt)	*(Per cent)*			*(Mt)*	
Rice	110.5	0.61	0.18	1.38	2.17	0.77
Wheat	82.6	0.48	0.16	1.18	1.82	0.50
Sorghum	21.0	0.52	0.23	1.34	2.09	0.15
Maize	12.5	0.52	0.18	1.35	2.09	0.09
Pearl millet	15.6	0.45	0.16	1.14	1.75	0.09
Barley	2.5	0.52	0.18	1.30	2.00	0.02
Finger millet	5.3	1.00	0.20	1.00	2.20	0.04
Sugarcane	40.9	0.40	0.18	1.28	1.86	0.25
Potato	7.9	0.52	0.21	1.06	1.79	0.05
Pulses	13.7	1.60	0.51	1.75	3.86	0.18
Total	312.6	-	-	-	21.63	**2.15**

Source: Manna and Ganguly (1998).

If these crop residues are utilized properly in the field or made compost out of these residues and apply to the field will supplement the nutrients effectively in the organic farming. The Table 5.2 shows the various crop residues addition to the field will turn into an appreciable quantity of available plant nutrients for nutrient management.

Table 5.2: Quantity of Crop Residues Addition to Soil and its Potential Nutrient Supply

Crop Residues	*Quantity t/ha*	*N*	*P*	*K*
			(per cent)	
Rice straw	2.20	0.36	0.08	0.71
Rice husk	0.25	0.45	0.25	0.45
Wheat straw	1.8-2.5	0.53	0.10	1.10
Sugarcane trash	9.00	0.35	0.12	0.60
Groundnut shells	0.80	1.25	0.46	1.50
Pigeon pea stalks	1.5-2.0	1.10	0.58	1.28
Pearl millet straw	0.40	0.65	0.75	2.50
Cotton stalk	4.0-5.0	0.44	0.10	0.66
Maize straw	0.8-1.2	0.42	1.57	1.65
Mustard straw	1.8-2.2	0.45	0.53	1.48
Castor shell	0.50	1.01	0.35	1.86
Banana pseudo stem	9.00	0.61	0.12	1.00

Source: ICAR-IISS (2014).

However, the majority of this valuable resource is not utilised properly as a source plant nutrients, rather wasted or burned in the field (Table 5.3).

Table 5.3: Share of Unutilized Crop Residues Generated Annually in India

Crop Residue	*Unutilized Residues (per cent)*
Cereal crops	58.0
Fibre crops	23.0
Oil seed crops	7.0
Sugarcane	2.0
Pulses	2.0
Other crops	8.0

Source: MNRE, GOI (2009).

The residue generated is utilized mainly as industrial/domestic fuel, fodder for animals, packaging, bedding, in-situ incorporation, manuring, thatching and left in field for open burning. In case of combine harvesting almost all the residue generated is left in the field, which finally ends up in burning. Another estimate given by (Gupta *et al.*, 2002) shows that, the biomass of 10 t/ha crop (Particularly exhaustive crops like Maize, rice wheat *etc.*,) removes 730 kg NPK/ha from the soil that is often not returned to the soil. This may cause mining of soil for major nutrients leading to net negative balance and multi nutrient deficiencies in crops. Hence, this huge potential of nutrients from crop residues must be utilised effectively for better nutrient management.

2.1.2 Effective Utilization of Crop Residues as Plant Nutrients

The crop residues production and utilization varies from crop to crop and region to region, the leguminous crop residues are normally put into the compost pit or incorporated into the soil.

Incorporation of left over crop residues into the soil

Instead of burning or removing the crop residues from the field incorporate the residues into soil. This practice enhances the soil organic matter, N, P and K contents in the soil. Ploughing the land and mixing residue into the soil is the most efficient residue incorporation method. Crop residues may be incorporated partially or completely into the soil depending upon methods of cultivation or crops sown.

Making use of crop residues for the preparation of compost

The Left over crop residues are used for preparing the compost. In the cattle shed of the farmer, each kilogram of straw absorbs about 2-3 litres of urine, which is enriched with nitrogen. The residues of rice crop from one hectare land, on composting give about 3 tons of manure, which is rich in nutrients as equivalent to well decomposed farmyard manure or compost.

Fortification of crop residues for enriched nutrient content in the compost

The enriched compost can be prepared from the crop residues by way of special composting methods involving a combination of several bacterial species.

Use crop residues as surface mulch

Crop residues in excess are also used as surface soil mulch. This practice is most useful in arid and semi arid regions. It is suggested to leave the stubble part with few inches of above ground growth, either intentionally or as it happens automatically with mechanical harvests. The residue thus left out can be spread on the surface or ploughed back into the soil. During an aberrant rainfall year, applying organic matter as surface mulch is highly beneficial in terms of soil moisture conservation. Nevertheless, whenever crop residues are intended to act as manure there is a need to add them much before seeding to allow mineralization of immobilized nitrogen during the process of decomposition.

Crop residues fed to animals get recycled through FYM, whereas, residues placed in compost pits are recycled as rural composts. Nitrogen poor crop residues (having wide C: N ratio) such as residues of cereals and other non-leguminous plants are known to temporarily immobilise the soluble soil nitrogen and thus create a shortage of plant available nitrogen. If such residues are incorporated 3-4 weeks before the crop is sown, the nitrogen immobilisation effect is reduced.

The plant available nitrogen is immobilised maximum up to 25-30 days. Thus, skipping over of the maximum immobilisation period for sowing of crop or planting of crop would usually results in definite benefits even to the succeeding crop. However, under intensively cropped areas, there is hardly any time for residue incorporation and decomposition between two crops in multiple cropping systems. Hence, under such situations making use of crop residues for the compost preparation in the local farm itself saves the valuable organic resources.

In tropical and subtropical conditions under double cropping system, fallow periods of 1-2 months are generally available between two main crops in most of the situations. This period is important for incorporating the available crop residues to improve soil fertility. Residue incorporation along with sowing/planting of crop may not yield short term beneficial results. However, moisture conservation and temperature moderation is improved by using residues as surface mulches.

2.1.3 Effect of Crop Residues on Soil Physical, Chemical and Biological Health

Crop Residue management practices affect soil physical properties such as soil moisture content, aggregate formation, bulk density and soil porosity. Incorporation and/or retention of crop residues into the soils results in reduced bulk density and compaction of soils over a period of time (Bellakki *et al.,* 1998).The soil pH is the determining factor for nutrient availability and it is greatly influenced by the crop residues incorporated in the soil. Long-term straw application will build soil organic matter level and nitrogen reserves, and also increases the availability of macro and micronutrients. Residue characteristics as well as soil and crop management factors also affect residue decomposition. Under optimum temperature and soil moisture conditions, nitrogen immobilization can last 4-6 weeks. Soil microbial biomass is affected by the residue management practices. Many workers reported a decline in microbial biomass when residues are burnt. Residue incorporation results in more microbial activity than residue removal or burning. The plant nutrient availability depends upon soil microbial biomass and microbial activity. This in turn depends on the supply of organic substrates in soil. The population of soil flora and fauna is positively correlated with the plant biomass present in soil. Crop residues are also known to enhance nitrogen fixation in soil by asymbiotic bacteria (*Azotobacter chrococcum* and *Aspergillus agilis*). Due to increase in soil microbial population the activity of soil enzymes responsible for conversion of unavailable to available form of nutrient also increases.

2.1.4 Merits and Demerits of Crop Residues as Nutrient Source in Organic Farming

Merits

Crop residues are not purchased as like fertilizers, bio fertilizers or Farm yard manure.

- ☆ Crop residues do not cost the season for farmer as like green manure crop.
- ☆ Farmer need not transport the residues, they are generated in the farm and in-situ incorporation is the option.
- ☆ No contamination of any heavy metals or pollutants as in case of manure prepared from industrial/municipal wastes.
- ☆ Use of bio digester to decompose all the crop residues and even weeds in the farm. Supplies good amount of nutrients for organic farming.

Demerits

- ☆ Crop residue incorporation is the laborious process: During land preparation poses little problem, however with mechanised cultivation crop residues can be effectively incorporated
- ☆ Poor decomposition of residues: Due to wider C: N ratio and high lignin content some residues take quite long time to decompose fully.
- ☆ Soil moisture requirement for decomposition: In irrigated area moisture is sufficient to decompose but in rainfed, arid and semi arid regions takes long time to decompose
- ☆ Physical chopping of residues is difficult to fasten the decomposition process in the field
- ☆ Chances of pest and disease of the previous crop may carry to next crop. Hence, decomposition should be made quickly in the process.

2.2 Green Manures

Green manures are the plants which are incorporated into the soil to improve the soil fertility. This is an age old practice not new to Indian farmer. The function of a green manure crop is to add organic matter to the soil. Green manure crops are either grown in the same field where they are intended to be turned down or collected from outside and used in the stipulated field. These green manure crops are cover crops, legume crops and any crop which is succulent enough to decompose in the soil after its incorporation. To supplement the nutrient requirement under organic farming the green manures plays a very important role. These are the alternative to composted organic manures in the field.

2.2.1 Benefits of Green Manures

The green manures improve physical, chemical and biological health of the soil (Palaniappan and Annadurai, 2010). Besides that it helps in the reclamation of saline and alkali soils. The consistent use of green manures after decomposition releases a large amount of organic acid which neutralizes the soil reaction. It helps to maintain the organic matter status of arable soils builds up the soil structure and improves tilth. It improves soil structure by promoting the formation of crumb structure in heavy soils leading to better aeration and drainage. Vigorous root system of green manure keeps the soil particles bound together. It enhances the water holding capacity of light soils. Green manure crops prevent leaching of nutrients to lower layers.

The growth of green manure crops (especially, dhaincha and sunnhemp) is very fast. Within 40-50 days they accumulate 6-10 t/ha of biomass. However, decomposition is also very fast due to richness of nitrogen. Also serves as a source of food and energy for the soil microbial population which multiplies rapidly in the presence of easily decomposable organic matter. The enhanced activities of soil organisms not only cause rapid decomposition of the green manure but also result in the release of plant nutrients in available forms for use by the crops. In the process, it contributes nearly 40 to 80 kg nitrogen per hectare to the field. Besides supplying nutrients, green manure also prevents loss of nitrogen by leaching and erosion. For example the green manure crop, Dhaincha is found to mobilize soil phosphorus and potassium and other trace elements in the soil. Certain green leaf manure crops serve the dual purpose of nutrient supply and fodder supply. The leaves of perennial legume plants can be used for green leaf manuring. In the off season, green leaves can be fed to the cattle.

Green manure crops absorb nutrients from the lower layer of soils and leave them in the soil surface layer when ploughed in, for use by the succeeding crops. Leguminous green manure plants harbour nitrogen fixing bacteria, rhizobia, in the root nodules and fix atmospheric nitrogen. It also increases the solubility of lime phosphates; trace elements *etc.*, through the activity of organic acids during decomposition. Green manure crop on average fixes 60 -100 kg nitrogen per hectare in single season under favourable conditions. *Sesbania acculeata* (dhaincha), when applied to sodic soils continuously for four or five seasons, improves the permeability and helps to leach out the harmful soluble salts. Certain green leaf manures like Pongamia and Neem are reported to have anti insecticidal property and controls the soil borne insect pests and to some extent control diseases. Thus, green manures play an important role as a natural source of nutrients.

2.2.2 Problems with Green Manures

As green manure possess a narrow C:N (carbon and nitrogen) ratio, great care must be taken in planting of the main crop. The main crop has to be transplanted (*e.g.* rice) or sown (*e.g.* cereals) soon after the application of green manure crop. Under such condition an interval of one week under assured water supply will be the optimum time for the planting of the crop. It is often complained that a farmer loses a valuable season by growing the green manure crop. This problem can be easily solved by suggesting to grow green manure on bund or field boundaries is recommended. To supplement this, the green leaves from suitable plants can be collected and incorporated into the field directly.

2.2.3 Types of Green Manuring

In-situ green manuring

Growing of green manure crops in the field where its intended use is desired. The Crops for In situ green Manuring are: *Crotalaria juncea, Crotalaria mucronata, Crotalaria anagyroids, Sesbania aculeata, Sesbania speciosa, Phaseolus mungo, Phaseolus trilobus, Melilotus spp, Trifolium alexandrium, Cyamopsis tetragonoloba.*

Among these *Crotalaria juncea* is the most common green manure crop in India. It is grown in all states of the country. It is sown with onset of the monsoon. Grows very fast even in the poor soils and attains a height of 1-2 meter. The plant contains a large proportion of herbage and rapidly decomposable. About 20 to 25 ton of green matter can be obtained from one hectare in eight weeks. *Sesbania aculaeta* (dhaincha) is widely known green manure crop in India. It can tolerate water logging and alkalinity and also drought conditions. It is an ideal green manure crop for rice growing soils. In waterlogged fields, it grows to 1.5 to 1.8 meters. This can be used for the reclamation of saline and alkaline soil.

Besides these crops, certain other crops such as *Vigna mungo* grow very fast, establishes quickly in both rabi and kharif and make a very good green manure. *Lathyrus sps* is grown in North India as a cover crop. It is annual and mostly grown in winter and establishes itself even when sown in standing rice before harvest. Desmodium is sown as both cover and green manure crop. Cowpea which is mostly grown as a fodder crop is a promising green manure crop. Other leguminous plants such as *Dolichos biflorus* and Lupinus spp are used to certain extent as green manures.

Green leaf manuring

Refers to bringing in the green plants grown elsewhere for the purpose of incorporation into the soil. Green succulent leaves of leguminous or non-leguminous plants are used and incorporated by ploughing. This practice is called as green leaf manuring. The crops for green leaf manuring are: *Gliricidia maculate, Pongamia glabra, Calotropis gigantia, Tephrosia perpuria, Tephrosia candida, Indigofera teysmanuui, Sesbania speciosa, Ipomea carnea, Cassia tora* are the commonly grown green leaf manure crops. Among these crops, *Gliricidia maculate* is commonly used under farmers' situation. This plant grows well in variety of soils and grows to the height of about 4.5 meter. During a year two cuttings can be taken; first at the beginning of the monsoon and the second in December. The fresh weight of leaves contains 0.49 per cent nitrogen and 8.5 per cent carbon. The yield of leafy material per plant is about 22.5 kg after 10 years of planting. Leaves can be obtained for several years.

2.2.4 Green Manuring Techniques

Green manure crops can be grown on any type of soils; the sowing should be done with mango showers or with summer showers in rainfed areas. Green manure crops do not require any nitrogenous fertilizer as they are leguminous crops. However, application of 10-15 kg of nitrogen per hectare helps the crop to grow faster.

Time and method of incorporation

The success of green manuring depends on the right stage and time of incorporating green matter into the soil and giving sufficient interval before sowing or planting the crop (Thorup-Kristensen, 2003). For example the best results are achieved when a dhaincha crop of 45- 50 days old is incorporated into the soil just before transplantation of paddy. At this age dhaincha is in its pre-flowering stage and quite succulent. A 50-day old crop furnishes about 20-24 t/ha of biomass and supplies 95-115 kg of nitrogen in the same area. The crop should be ploughed in with the help of soil turning or disc plough. Ploughing buries the crop in 15-20 cm deep surface soil. Dhaincha for green manuring purpose should not be allowed to grow for more than 50 days or else the crop becomes fibrous. It takes longer time for a fibrous crop to decompose and liberate nutrients timely because of its high lignin content. A thumb rule for turning the crop of green manure is at the time when flowers have just started to emerge.

In case of sugarcane crop, the sunhemp is grown along with the main sugarcane crop and the green manure crop is incorporated after about 40 to 50 days growth at the time of earthling up. In case of plantation crops, green manure grown in the same field or brought from outside is incorporated for the decomposition in the field. Usually about six to eight weeks time is found to be sufficient for the decomposition. Whereas the green leaf manuring being a bulky one is usually applied as basal dressing before the main crop is raised in the field. In India, green manure is applied as basal dressing except for some perennial crops. After incorporation, sufficient time is allowed for decomposition to take place and afterwards only the main crop is sown or planted. However, the time will vary according to the crop and other agronomic practices followed.

The method of application varies from place to place depending upon other agronomic practices followed. In the case of green manuring when plants are grown in the same field where they are to be incorporated, the plants are cut at the proper stage to the ground level, placed in the furrows and covered by the next furrow. With the availability of labour saving implements like green manure trampler, the plants are trampled by working the implement and later on levelling the field. This practice is possible where rice is transplanted. In broadcast crop, a suitable modification is necessary and usually the green manure crop is incorporated during the first weeding. In the case of green leaf manuring, the plants are brought from other place where it is grown *e.g.* bunds. The leaves and tender twigs are incorporated in the soil.

Decomposition of green manures

Soil microorganisms play a major role in the decomposition of organic matter. The factors conducive to complete decomposition are the stage of maturity of the crop and the moisture level of the soil (Thorup-Kristensen, 2003). Decomposition, besides depending on moisture content of the soil, is also dependent on the composition of the green matter and the presence of available inorganic nutrients. In light soils, crop should be buried deeper than in heavy soils. Desired results will follow if moisture content is high in the beginning, producing semi-aerobic conditions and low afterwards for inducing aerobic condition under which nitrification can take place. In the normal well-drained soil, the end products are mainly carbon dioxide, nitrate, sulphate and other resistant residues. The green manuring stimulates the growth of the algae and the production of oxygen in the rice soils resulting in greater aeration or oxygenation of the roots. Green manure also produces some growth hormones and other bio-chemicals which stimulates the growth of paddy. Hence, green manuring represents a cheap and effective way of improving the soil fertility depending up on the moisture availability of soil. It improves the soil fertility in terms of nitrogen status in addition to other nutrients along with organic matter. For a green manure crop, a legume is preferable as it adds atmospheric nitrogen which is a distinct addition to the soil. Green manure crops serve as cover crop in the soil erosion areas and aids in conservation of moisture. It acts as a good amendment for the reclamation of problem soil like alkali soil. The success of green manuring is again depend on type of green manure; time, stage and depth of incorporation, rate of decomposition *etc.* Thus, it has the potential to improve the low fertility status of soil. A considerable reduction in the investment on fertiliser, the cost of which is increasing, could be achieved by green manuring. Green manuring can be an important component of low external input for organic agriculture.

2.3 Farmyard Manure (FYM)

Since early ages the value of FYM has been well recognised in India. Good quality FYM is considered as the prime organic resource for nutrient management under organic farming. It is the storehouse of all most all the plant nutrients. FYM is the mixture of dung and urine of farm animals mixed with litter and leftover material from fodder fed to the cattle available in the backyard of every farmer in Indian villages. Average quality FYM contains 0.5 per cent N, 0.2 per cent P_2O_5 and 0.5 per cent K_2O that is 12 kg nutrients per tonne of manure.

India has vast resource of livestock, which play a vital role in improving the nutrients for the organic farming. There are about 300 million bovines, 65.07 million sheep, 135.2 million goats and about 10.3 million pigs as per 19th Livestock Census in the country (Annual report, 2016-17). With this huge quantity of Farm yard manures available in the country, the potential and utilizable quantities are projected which are equivalent to fertilizer form of nutrients (Table 5.4).

Table 5.4: Potential and Utilizable Quantities of Farmyard Manure

Farmyard Manure	*Plant Nutrients (Million tonnes)*	
	Year-2011	*Year-2025*
Total	18.02	21.64
Utilizable	10.81	12.98
Fertilizer equivalent value	5.40	6.49

Source: Sekhon (1997).

However, in each tonne of dung about 8-12 kg of nitrogen,2-4 kg of phosphorus and 3-6 kg of potassium is available on an average. However the value of dung as manure is reduced because of leaching and volatilization loss of nitrogen. Cattle urine is generally not conserved and finds its ways into the drains. About 33.6 million or 17.5 per cent of all Indian homes use LPG as their primary cooking fuel, with 90 per cent of rural homes still dependent on some form of biomass (Antonette D'Sa and Narasimha Murthy, 2004) and thus is lost to agriculture. Hence, the need arises for the effective utilization of FYM for the effective nutrient management in organic farming.

2.3.1 Losses of Nutrients from Farmyard Manure during Collection and Storage

Losses during handling

FYM consists of two original components such as i) dung (solid portion) and ii) urine (liquid portion) which are subjected to different losses in different ways. It is most essential that both the parts of cattle manure are properly handled and stored:

- ✰ *Losses of dung*: Dung is valuable manure. Still larger portion is dried in dung cakes and burnt as fuel. Besides this large portion of cattle excrements is dropped outside the cattle shed, when the animals are grazing on the uncultivated lands. This can be used for preparation of FYM.

- *Losses of urine*: Urine contains nitrogen and potassium in large portions. But there is no good method of preserving the urine in our country. Most of the cattle shed have un-cemented or *kachha* floor and the urine gets soaked in the soils of *kachha* floor of the cattle shed and the large quantities of nitrogen are thus lost through the formation of gaseous ammonia.

Losses during preparation and storage

Cow dung and other farm wastes are collected daily and these are accumulated in manure pit in open space for months together. The manure remains exposed to sun and rain during this period. Due to this effect the nutrients are lost in following ways.

- *By leaching*: Nutrients of manures are water soluble and these are liable to get washed by rain water. The leaching loss of nutrients will vary with the surface exposed, the intensity of rain fall and the slope of the surface on which manure is heaped. The leaching loss may be prevented by erecting a roof over the pit.
- *By volatilization*: During storage, the urine and dung are decomposed and considerable amount of ammonia is produced. The ammonia combines with carbonic acid to form ammonium carbonate and bicarbonate, which are rather unstable and gaseous ammonia may be readily liberated and passes into atmosphere.

2.3.2 Ways to Minimize the Losses from FYM during Handling

- Adopt trench method as suggested by C.N. Acharya for handling of dung and urine.
- Use of Gobar gas plant/Biogas plant: 50 per cent of dung is made dung cakes and burnt as fuel for cooking. The use of cow dung in gas plant produces a combustible gas, methane used as fuel gas which is an improved method of handling FYM.
- Adopting covered method of storing FYM: Nutrients losses can be effectively controlled by this method.
- Adoption of BYRE (Cattle are stalled in a shed with a non-absorbent floor provided with necessary slope towards the urine drains.) system in collection of FYM.
- Proper field management of FYM: During spreading of FYM in the field in small heaps leads to loss of nutrients from it. It is advisable to spread the FYM before ploughing.

Vegetable crops like potato, tomato, sweet-potato, carrot, radish, onion *etc.*, respond well to the FYM. The other responsive crops are sugarcane, rice, napier grass and orchard crops like oranges, banana, mango and plantation crop like coconut. The entire amount of nutrients present in FYM is not available immediately. About 30 per cent of N, 60- 70 per cent P and 70 per cent K are available to the first crop and subsequently the next crop follows to absorb the nutrients. In addition to the plant nutrients supply in the organic farming, the FYM also improves the soil physical and biological properties over the continuous addition in the crop production enhance the soil fertility and productivity (Rana, 2011). However, it is practically impossible to check completely the losses of plant nutrients and organic matter during handling and storing of FYM. But to control the losses improved methods can be adopted to reduce such losses.

2.3.3 Improved Methods of Handling FYM

Trench method of preparing FYM

This method has been recommended by C.N. Acharya. Trenches of size 6 m to 7.5 m length, 1.5 m to 2.0 m width and 1.0 m deep are dug. All available litter and refuse is mixed with soil and spread in the shed so as to absorb urine. The next morning, urine soaked refuse along with dung is collected and placed in the trench. A section of the trench from one end should be taken up for filling with daily collection. When the section is filled up to a height of 45 cm to 60 cm above the ground level, the top of the heap is made into a dome and plastered with cow dung earth slurry. The process is continued and when the first trench is completely filled, second trench is prepared. The manure becomes ready for use in about four to five months after plastering. If urine is not collected in the bedding, it can be collected along with washings of the cattle shed in a cemented pit from which it is later added to the farmyard manure pit.

2.4. Farm Compost

The compost made from farm waste like sugarcane trash, rice straw, weeds and other plant material is called farm compost. Besides this, the animal refuse and vegetable wastes are also made to decompose and used as a valuable organic resource for the nutrient management in organic farming. The organic residues are piled up, moistened, turned occasionally to aerate and allowed adequate time to decompose partially and bring down the C: N ratio to about 30. The collected organic refuse may include straw, leaves, paddy husk, ground nut husk, sugarcane trash, bagasse, cattle dung, urine, vegetable wastes, hedge clippings, water hyacinth and all other residues counting organic matter. During composting under thermophilic and mesophilic conditions in windrows, heaps or pits method adequate moisture and aeration are essential.

The compost is like decomposed cattle manure in general appearance, more powdery and lighter in colour. Ordinary compost can be enriched with N through Azotobacter. Compost prepared by using N-fixing bacteria is called AZO- compost. Azocompost is the cheapest source of Nitrogen among all the organic manures (N-1.5 per cent). The farm compost prepared from waste like sugarcane trash, rice straw, weeds and other plant material contains 0.5 per cent N, 1.5 per cent P_2O_5 and 0.5 per cent K_2O. Farm compost prepared from Bangalore method is 0.80-1.24 per cent N, 0.40-0.59 per cent P_2O_5 and 2-3.3 per cent K_2O. This farm compost up on addition to soil in the organic farming replenishes plant nutrients, maintains soil organic matter content and helps in improving the physical, chemical and biological conditions of the soil. There are some improved methods of compost preparation (Krishan Chandra *et al.*, 2014) at farm level for the efficient utilization of farm wastes are given below:

2.4.1 Improved Methods of Compost Preparation

Chimney and Wall Method

For wall aeration two brick walls, 30 cm apart, 1 m high, 23 thick and having 40 holes of size 22 cm x 10 cm each are constructed in the centre of a 2 x 3 m pit. Chimney aeration is provided by constructing two one meter high chimneys, one metre apart on rectangular base of 23 cm. Each chimney has 40 holes of similar size as provided in wall aeration. Substrates are filled in layers with each layer consisting of three sub layers. Total quantity of substrate in 10 such layers in each pit was one tonne. After one month, the towers and chimney were sealed with dung and mud mixture. Chimney aeration yielded 60 per cent recovery with nitrogen ranging from 1.10 to 1.70 per cent. The biomass of obnoxious weed, *Parthenium histophorus* yields better quality compost than rice straw. However, before using the parthenium in compost preparation, it is to be ensured that the weeds are not viable. Enrichment of compost with Nitrogen Phosphorus and Potassium can be achieved by *Azotobacter, Azospirillum* and phosphate solubilizing and potassium mobilizing Biofertilizers. after the thermophilic phase is subsided at the rate of 2-4 kg each per tonne of substrate. The biofertilisers may be dissolved in 50 litres water and poured in holes previously made in compost piles. If bio fertilisers are not available while composting is in progress, enrichment can be done by mixing bio fertilisers with the harvested compost and heaping the treated compost in shed for at least two weeks. Inoculation of the substrates with celluloytic and lignolytic microorganisms like *Trichoderma harzianum, Aspergillus niger, Aterreus etc.* has been found to accelerate the decomposition during composting. Likewise, *Trichoderma harzianum* is the most effective organism for making compost from rice straw.

NADEP Method

NADEP method of compost making has been developed by a farmer, Narayan Rao Pandhari Pande, in Maharastra, India. This method is based on the principle of aerobic decomposition with natural flow of optimum air. The substrate is converted at the top by plastering with dung and soil to minimise the loss of moisture. To obtain 2 -2.5 tonnes of compost with 0.6 -1.0, 0.5-0.8 and 1.2-1.5 per cent nitrogen, phosphorus and potash respectively, the required raw materials are 1.4 to 1.5 tonnes of organic refuse, 900 -1000 kg of cattle dung, 1.7 to 1.8 tonnes of pulverised dry soil and 1500-2000 litres of water. These materials are filled layer by layer in a tank of 3 m x 2 m x 1 m size and made up bricks with holes (15-20 cm rectangular) on all 4 walls for easy entry and circulation of air. The tank is constructed above ground at high lying area to avoid entry of rainwater from surrounding place.

Internal surface of tank is painted with dung slurry. The tank is filled in layers. The first layer is made by spreading 100-110 kg of organic wastes on the floor of the tank. This is followed by a second layer containing slurry of 4-5 kg cattle dung in 125-150 litres water. Over the second layer, 50-60 kg of pulverised good quality soil is spread. The three layer combination is repeated till the tank is filled up to

45-50 cm above the brick level. The complete filling is done within two days, using materials of 10-11 layers. Topping of the tank is done by 5 cm plastering with a paste of dung and soil. Cracks are not allowed to develop on the heap to prevent the gas leakage. After 15-20 days, when the substrate shrinks down, a second filling is made in a way similar to that adopted in the beginning. The top is then arranged in a hut like shape and re-plastered. The moisture level of the mass is maintained at 15-20 per cent by sprinkling with water and dung slurry through holes. Normally the substrate takes 3-4 months to attain maturity without turning.

Padegaon Method

This method is recommended for composting resistant substrates like sugarcane trash and cotton stubbles. These materials are shredded into 30 cm size particles and trampled to make a 30 cm thick above ground layer. This layer is drenched with slurry consisting of wood ash, cow-dung and soil. Four or five such layers are added to the pile. The completed heap is about 1.5 m high, 2m wide as per necessary. Since the material is very resistant to decay, the heap is turned each month, re-trampled and sufficient water is added to keep it moist. The material is ready for use in about five months.

Indore Method

Sir Albert Howard (1924-26) at Indore, Madhya Pradesh, developed this method in which the conservation of cattle urine is effected by getting it absorbed in rice straw, straw dust and other organic wastes used as bedding in cattle shed. The urine soaked material along with fresh cow dung serves as major source of nitrogen for the microorganisms involved in composting. The material collected from cattle shed is spread evenly in a pit to form a layer of 10-15 cm thick. To this layer is add dung slurry made of 4.5 kg dung + 3.5 kg urine earth + 4.5 kgs of inoculums from 15 day old compost pit. Water is then sprinkled to achieve 100 per cent saturation. The layering is repeated to fill the pit within seven days. The material is turned three times, first two turnings in 15 days interval after filling the pit and third turning after one month of the first turning. To provide succulent biomass, seeds of sun hemp are grown on compost heaps and at the first turning, the green plants are turned in. During rainy season the piling of 20 cm carbonaceous material (leaves, hay, straw, saw dust, wood chips, corn stalks *etc.*) and 10 cm nitrogenous materials (fresh grass, weeds, digested sewage, sludge, poultry litter) in alternate layers is repeated until the pile is one metre high. The recommended size of the heap is 2.4 m square at the base and 2.1 m square at the top. However, the method is highly labour intensive and less suitable to those farmers who do not have enough cattle and irrigation facilities. The medium and big farmers with enough cattle and irrigation facilities can best of this method for compost preparation.

Bangalore Method

In this method, the disadvantages of Indore method are overcome by slowing down the rate of decomposition and avoiding the turnings. The substrates usually composted in this method are town refuse and night soil which are spread in alternate layers of 15 cm and 5 cm in trenches or pits. When the pit is filled to

15 cm above the ground level, it is sealed to prevent loss of moisture. After the initial aerobic decomposition for 8-10 days the material undergoes semi anaerobic decomposition. During this stage the rate of decomposition slows down taking about 6- 8 months for the compost to be ready. Often, the composting period is more than eight months due to high C:N ratio. Loss of organic matter and nitrogen is negligible and percentage recovery of compost is more. But, this method is not adaptable to heavy rainfall areas. In Indian conditions except coastal and high rainfall areas this method can be best used by the farmers to prepare good quality compost at the farm level.

2.5 Poultry Manure

The total poultry population in India is 729 million in number as per the 19th Livestock Census in the India (Annual report, 2016-17). Poultry industry plays a vital role in supplementing the nutrient requirement for the organic farming. In the states such as Andhra Pradesh, Punjab, Tamil Nadu, Karnataka and Maharashtra, poultry industry is well developed and viewed as potential source for the organic manures. Each poultry bird on an average produce 0.025 kg litter/day (excreta) dry weight basis, 2.8 Mt of poultry manure is produced annually in our country. The average manurial value of the poultry manure works out to be 27,553 tonnes of Nitrogen, 36,772 tonnes of phosphorus and 21,941 tonnes of potassium.

The poultry waste comprises of waste feed, solid and liquid dropping, litter, egg shell, diseased and dead birds, culled birds, feathers and the wastes from poultry sheds. Poultry manure is good source of nutrients for organic farming. The decomposition of the waste must be done with the use of fungus/bacteria to neutralise the effect of disease causing pathogens. The nutrient status of poultry manures is as detailed in Table 5.5.

Table 5.5: Nutrient Status in Poultry Manure

Source	*Nitrogen*	*Ammonia*	*Phosphorus*	*Potassium*
		(Per cent)		
Fresh chicken manure	3.7 – 8.8	0.4 – 1.1	1.2 – 2.0	1.2 – 2.7
Poultry manure	1.4 – 6.8	0.5 – 1.1	0.5 -3.5	1.0 - 2.7
Broiler litter	2.3 – 6.0	-	0.6 – 3.9	0.7 – 5.2

Source: Sims and Wolf (1994).

2.5.1 Loss of Nutrients on Storage

The solid and liquid excreta are excreted together resulting in no urine loss. Deep litter for poultry consists of groundnut shell, rice husk in a layer of 10-15 cm. When excreta are added, the litter becomes moist but remains aerobic. Aerobic fermentation occurs with the production of heat and loss of some CO_2 and ammonia. Deep litter containing more than 22 per cent moisture, when stored in open air rapidly loses its nitrogen due to high proteolytic activity. Hence, the immediate processing of poultry manure to prevent its rapid decomposition and save its nutrient properties is necessary.

If poultry manure is stored, nutrient losses occur and handling cost increases. Fresh poultry manure is difficult to handle because of its high water content and cannot be applied to crops due to caustic effects on foliage. Nitrogen in poultry litter is present in both organic and inorganic forms that are subject to volatilization, denitrification, immobilization, mineralization and leaching. Hence composting is necessary.

Composting is done to stabilize the poultry litter prior to land application. The process of composting produces a material with several advantages with respect to handling by reducing volume, mass of dry matter, odours, fly attraction and weed seed viability. Composting poultry manure under anaerobic conditions helps for greater recovery of final product and negligible loss of nutrients particularly nitrogen.

2.6. Sheep and Goat Manure

India has huge potential of sheep and goat manure particularly in rainfed regions. The dropping of sheep and goats contain higher nutrients than farmyard manure and compost. The goat with a body weight of 20-40 kg excretes 0.3 to 0.6 kg dung and 0.3 to 0.5 litres of urine every day. Whereas the sheep with a body weight of 25 – 40 kg excretes 0.3 -1.4 kg dung and 0.3 to 1.0 litres of urine per head per day. On an average, the sheep and goat manure contains 3, 1 and 2 per cent N, P_2O_5, K_2O respectively

The sheep and goat manures are applied to the field after sweeping the droppings of night halt shed in the backyard and then placed compost pits for decomposition and applied to field after its maturity but the nutrients present in the urine are wasted in this method. Even then sheep and goat manure are the valuable source of nutrients for organic farms.

2.6.1 Sheep Penning

It is the Indigenous technological knowledge (ITK), has its own importance as the practice stood the test of time, proved to be efficacious to local people and as the basis for their link with nature. Sheep penning is the traditional method of enhancing soil fertility. This practice involves camping of sheep in the crop fields over night after the crop harvest. The farmers prioritize the use of penning within the land patches with their production potential. For penning, the shepherds pool their sheep into a flock that numbers about 250-1,000 sheep. The location of the penning is changed every day. This practice helps to uniform distribution of manures. Normally, penning is performed for 7-15 days depending on the number of sheep and when excreta deposition on the fields is considered adequate, then the camp shift to other places (Nataraj *et al.*, 2016). High concentration of nitrogen in fresh sheep and goat excreta will be the major contributor soil fertility in rainfed regions.

2.7. Oil Cakes

The oil cakes are the concentrated organic manures used for the nutrient management under organic farming. The major oil cakes such as the castor, neem, madhuca, karanja, linseed, rape seed and cotton seed which are non-edible oil cakes may serve as useful organic manure as these contain high amounts of plant nutrients. Most of the non-edible oil cakes are valued much for their alkaloid contents which inhibit the nitrification process in soils. Neem cake contains the alkaloids - nimbin and nimbicidine which effectively inhibit the nitrification process. Similarly, Karanjin (*Pongamia pinnata*) and (*Madhuca butyracea*) is a potent nitrification inhibitor equal in efficiency to nitropyrin in retarding the nitrification process of ammoniacal nitrogen and increasing the yield, nitrogen uptake and grain protein content of rice. Not only in rice crop, oil cakes supplies the nutrients for all most all major crops grown under organic farming. Tapping and proper utilization of such locally available organic resources could provide substantial quantity of crop nutrients in organic farming.

2.8 Coir Pith Compost

The coir dust is a waste product of the coir industry and could be used as organic amendment. The coir waste accumulates in large quantities near the coir industrial units and it can be converted into valuable organic manure by proper composting with the aid of the mushroom fungus *Pleurotus sojorcaju*. The composting reduces the volume by 42 per cent besides narrowing the C: N ratio. Once the compost is matured, add potash mobilizing bacteria and phosphorus solubilising bacteria at the rate of 400 ml per tonne with coir pith compost.

2.9 Biogas Slurry

Biogas slurry is good source organic manure. The following are the different methods of applying bio digested slurry as manure:

- ✰ Air dried biogas slurry can be applied by spreading on the agricultural land at least one week before sowing the crop.
- ✰ The liquid slurry can be mixed directly with the running water in irrigation canal which will enable spreading of the slurry uniformly in the cropped area or in cultivation land.
- ✰ Biogas slurry can also be coated on the seeds prior to sowing. This acts as insecticide and prevents seeds or plants from insect attack. This helps in early germination and healthy growth of seedlings.
- ✰ The digested slurry is fed through the channel, flowing over a layer of green or dry leaves and filtered in the bed. The semi-solid slurry can be transported easily as it was in the consistency of fresh dung and used for top dressing of crops like sugarcane and potato.
- ✰ Bio digested slurry is also being used to fish culture, which acts as a supplementary feed.
- ✰ The digested slurry, if mixed with Azospirillum, Potassium mobilising bacteria (KMB) and Phosphate solubilising micro organism (PSM) at the rate of 200ml each per acre enhances the nutrient content.

2.10 Press Mud

Press mud is a by product from the sugar industry. It is also called as filter cake. Press mud is a soft, spongy, amorphous and brownish white to dark brown material containing sugar, fibre, coagulated colloids including wax, albuminoids, inorganic salts and soil particles. The composition and properties of press mud, however, vary depending upon the quality of cane and the process followed for the clarification of cane juice. Press mud is mainly used as (i) Source of plant nutrients (ii) Ameliorant for acidic and sodic soils (iii) Medium for raising sugarcane seedlings, and (iv) Carrier for legume bio-inoculants.

2.10.1 Utilization of Press Mud

The press mud is plough mixed in soils after it is broadcasted as done with FYM or compost. However, the banding of press mud in cane rows is more effective than broadcasting. Press Mud as a Source of Plant Nutrients contains all the macro and micro-nutrients, though in small amounts. All these nutrients become available to the growing plants after the degradation of press mud added to the soil. The use of press mud has benefited the growth and yield of many crops through direct, residual and cumulative effects. Hence, press mud can be one of the potential sources for the nutrient management in organic farming.

2.11 Vermicompost

Vermincompost is the important component of organic farming. Vermincompost is prepared from the vermicomposting technique. Vermicomposting is a composting process aided by earth worms. Vermicomposting is an appropriate technique for effective and efficient recycling of animal wastes like cattle dung, poultry manure, piggery excreta, crop residues, municipal soild wastes, effluents from agro industries by using earth worms. Among the various bio composting measures, the vermicomposting has been considered to be an advantageous treatment system for different wastes. It has been estimated that one tonne of moist organic matter can be converted into300 kg of compost by earth worms. The castings of earth worms are rich in nutrients (N. P. K, Ca and Mg), which are in readily available form. The main advantage of use of earth worms for composting of wide variety of organic residues is the reduced time required for their maturity. They can hasten the rate of decomposition, thereby reduce the time for composting and the product is ready for use within three months. There are as many as 3000 species of earth worms found in India. However, the most efficient epigeic (surface dwelling) species used for vermicomposting technology are *Eisenia foetida, Pheritima elonata, Eudrilus eugeniae* and *Peronux excavatus*. The earth worms can be utilized as "decomposer industry" to take care of the organic wastes of different origin and kinds. Hence, the vermicompost is one among the potential and major source of nutrients for organic farming.

2.12 Biofertilizers

The biofertilizers are the preparations of living microorganisms which are useful for promotion of plant growth through a variety of mechanisms like biological nitrogen fixation, solubilisation of insoluble phosphates and other nutrients, oxidation of sulphur, production of growth hormones and combating plant diseases. These include specific strains of bacteria, fungi and blue-green algae.

These microorganisms are capable of making unavailable form nutrients to plant available form

The biofertilizer resources are available as following groups:

- *Rhizobium*
- *Azotobacter* and *Azospirillum*
- Phosphate solubilizing microorganisms (PSBs) (*Bacillus polymyxa*, *Pseudomonas* and *Aspergillus*)
- Vesicular arbuscular mycorrhizae (VAM)
- Azolla
- Blue green algae
- Plant growth promoting rhizobacteria (PGPR)

2.13 Sugarcane Trash

Sugarcane trash is one of the real potential sources of organic matter in organic farming. The sugarcane trash removed during cropping to prevent lodging of the canes and the dead leaves collected at the time of cane harvest are generally burnt in the field itself. The trash may be converted into a very useful organic amendment by proper composting. But, the compost making from sugarcane trash has not become yet popular, because it decomposes at a very slow rate due to in wider C:N ratio and it is not economically viable as it requires land, labours and time for handling. However, ex-situ recycling and incorporation of cane trash and stubbles in the field is an age old practice but adopted on smaller scale. Burning of sugarcane trash results in potential loss of 40 to 50 kg N, 20 to 30 kg P_2O_5 and 3 to 4 tonnes of carbon/ha, while only 75 to 100 kg K_2O and mineral elements are left behind in the form of ash (Phalke *et al.*, 2017). Instead making use of the sugar cane trash will serve as an excellent source of nutrient management under organic farming.

2.14 Innovative Formulations As Nutrient Boosters Under Organic Farming

In recent days among farmers the interest is more on some innovative formulations developed by progressive farmers and different Non Government Organisations for soil enrichment and growth promotion of crops. These formulations serve as nutrient boosters for nutrient management under organic management. Some of the nutrient formulations are as follows.

2.14.1 *Sanjivak*

Used for enriching the soil with microorganisms and quick residue decomposition. The preparation procedure for this formulation involves the following steps: Mix 100-200 kg of cow dung with 100 L cow urine and 500 g of jaggary in 300 L of water in a 500 L closed drum. Allow it to ferment for 10 days. After 10 days dilute with 20 times water and sprinkle in one acre either as soil spray or along with irrigation water. Three applications are needed during the crop growth period. First application of sanjivak formulation is before sowing of the crop, second application 20 days after sowing and the third application 45 days after sowing.

2.14.2 *Jivamrut*

To prepare the jivamrut formulation take 100 litres of water in barrel and add 10 kg of cow dung plus 10 litre of cow urine. Mix well with the help of wooden stick add two kg of jaggary and two kg of gram or any pulse flour and mix this solution well with wooden stick. Then Keep this solution for fermentation for about 5 to 7 days. After that, shake the solution regularly for three times a day. The formulation can be used as soil application either by sprinkling or by applying through irrigation water. Three applications are needed during crop growth period. First application of Jivamrut formulation is before sowing of the crop. Second application is after twenty days of sowing. Third application is after 45 days of sowing.

2.14.3 *Amrit Pani*

The formulation can be prepared by mixing 10 kg of cow dung with 500 gm honey and mix thoroughly to form a creamy paste. Then add 250 gm of desi cow ghee and mix at high speed. After that, dilute the formulation with 200 litres of water. The prepared formulation can be sprinkled in one acre over soil or with irrigation water. After thirty days, apply second dose in between the rows of plant or through irrigation water.

3. Conclusions

Nutrient management under organic agriculture proved to be very effective under cropping systems approach rather than sole cropping. Crop rotation is the central tool that integrates the maintenance and development of soil fertility with different aspects of crop and livestock production in organic systems. Manures and crop residues are carefully managed to recycle nutrients in the system. The supply and management of nitrogen is more complex in organic than in conventional agriculture. The major challenge for nitrogen management in organic system is to synchronize the availability of nitrogen mineralized from manures and crop residues with crop demand. Most of the organic nutrient management practices are site specific and crop specific. Besides, locally available organic resources in India are not only for nutrient management under organic farming but also to restore the soil health, food and nutritional security, reducing environmental pollution and also for upholding the slogan of **"Swachh Bharat"**. Hence, the organic resources right from crop residues, green manures, farm compost, vermicompost and bio fertilizers to sugarcane trash needs skilful use of local organic resources for effective nutrient management under organic farming in India.

REFERENCES

Antonette D'Sa and Narasimha Murthy KV. 2004. LPG as a cooking fuel option for India. Energy for Sustainable Development, 8 (3): 91-106.

Bellakki MA, Badanur VP and Setty RA. 1998. Effect of long-term integrated nutrient management on some important properties of a Vertisol. Journal of Indian Society of Soil Science. 46: 176-180.

Chhonkar, PK. 2003. Organic Farming: Science and Belief' Dr.R.V. Tamhane memorial lecture at the 68 th Annual Convention of the Indian Soc. Soil Sci., 5 November, 2003, CSAU&T Kanpur.

Gupta RK, Shukla AK, Ashraf M, Ahmed ZU, Sinha RKP and Hobbs PR. 2002. Options for establishment of rice and issues constraining its productivity and sustainability in eastern Gangetic plains of Bihar, Nepal and Bangladesh. Rice-Wheat Consortium Travelling Seminar Report Series 4. New Delhi, India: Rice-Wheat Consortium for the Indo-Gangetic Plains, 36 p.

Huang, SS, Tai, SF, Chen, TC, and Huang SN. 1993. Comparison of crop production as influenced by organic and conventional farming systems. Taichung District Agricultural improvement station, Special Publication, 32: 109 -125.

ICAR-IISS. winter School mannual. 2014. Waste recycling and resource management through rapid composting techniques. ICAR-Indian Institute of Soil Science, Bhopal, Madhya Pradesh. December 3rd–23rd, 502 p.

Krishan Chandra, Greep S, Srivathsa RSH and Nath PR. 2014. Organic manures. In: Organic farming News letter, March 10(1): 1-12.

Lal R. 1989. Conservation tillage for sustainable agriculture: tropics versus temperate environments. Advances in Agronomy, 42: 85-197.

Manna MC and Ganguly TK. 1998. Recycling of organicwastes: Its potential turnover and maintainence in soil – A review. Agriculture Review, 19(2): 86-104.

MNRE. 2009. Ministry of New and Renewable Energy Resources. Government of India, New Delhi. www.mnre.gov.in/biomassrsources

MoA. 2009. Ministry of Agriculture. Government of India, New Delhi.2012. www.eands.dacnet.nic.in

MoA. 2017. Department of Animal Husbandry, Dairying and Fisheries. Annual report 2016-17, Ministry of Agriculture and Farmers Welfare, Government of India. Pp 1-162. www.dahd.nic.in/reports/annual-report-2016-17.

Nataraj KC, Vijaysankar Babu M, Narayanaswamy G, K. Bhargavi K, Sahadeva Reddy, B and Srinivasa Rao Ch. 2016. Nutrient Management Strategies in Groundnut-Based Crop Production Systems in Dryland Regions of Southern Andhra Pradesh. Indian Journal of Fertilisers, 12 (10): 58-75.

Palaniappan SP and Annadurai K. 2010. In: Organic Farming: Theory and Practices. Scientific Publishers, India, Jodhpur. 485 p.

Phalke DH. 2017. Effect of in-situ recycling of sugarcane crop residues and its industrial wastes on different soil carbon pools under soybean (Glycine max) - maize (Zea mays) system. Indian Journal of Agricultural Sciences 87 (4): 444–54.

Rana SS. 2011. Organic Farming. Department of Agronomy, College of Agriculture, CSK Himachal Pradesh Krishi Vishvavidyalaya, Palampur. 90 p.

Sekhon GS.1997 In: Plant nutrient needs, supply, efficiency and policy issues: 2000-2025. JS Kanwar and JC Katyal eds. NAAS, New Delhi. pp. 78-90.

Sims JT and Wolf DC. 1994. Poultry Waste Management. Agricultural and Environmental Issues. Advances in Agronomy, 52: 1-83.

Thorup-Kristensen K, Magid J and Jensen LS. 2003. Catch crops and green manures as biological tools in nitrogen management in temperate zones. Advances in Agronomy, 79: 227-284.

Veeresh GK. 1997. Organic Farming: Ecologically sound and economically sustainable. Paper presented at the International Conference on Ecological Agriculture: Towards Sustainable Agriculture, Punjab Agricultural University, November 1997. pp. 15-17.

Chapter 6

Microbial Resources: An Important Component of Organic Farming

Minakshi Grover[1]*, *Lata Nain*[1] *and Sreedevi Shankar*[2]

[1]*ICAR-Indian Agricultural Research Institute, New Delhi – 110 012*
[2]*ICAR-Central Research Institute for Dryland Agriculture, Hyderabad – 500 059, Telangana*
E-mail: minigt3@yahoo.co.in

1. INTRODUCTION

Indiscriminate use of synthetic fertilizers and pesticides in agriculture during the last few decades has caused pollution of soil and water, leading to soil nutrient imbalance, loss of soil biodiversity and reduced soil fertility. By 2020, 28.8 million tones of plant nutrients will be required to achieve the targeted production of 321 million tonnes of food grain, against the availability of 21.6 million tones only. The widening gap between nutrient removal and supplies, and the environmental hazards, are imposing threat to sustainable agriculture. Further increasing cost of chemical fertilizers due to dependence on fossil fuels, is making them unaffordable for the small and marginal farmers (Mishra *et al.*, 2013).

Soil microorganisms play key role in agriculture ecosystem functioning. Most of the biological activities in the soil are mediated by soil microflora. Sustainable agriculture depends upon the healthy communities of soil microbes that contribute toward crop productivity through nutrient cycling, promotion of plant growth and disease suppression. Thus, the productivity of soils can be restored by improving the biological health of soil through building high population of soil microflora. Thus a shift from chemical to natural resources seems essential for sustainable agriculture.

Organic farming has emerged as an important alternative in view of the growing demand for safe and healthy food and agricultural sustainability and environmental safety. Organic agriculture/farming is a production system which avoids or largely excludes the use of synthetic fertilizers, pesticides, growth regulators and live stock feed additives (Lampkin, 1990). Animal dung, crop residues, green manure, biofertilizers and bio-solids from agro-industries and food processing wastes are some of the potential sources of nutrients for organic farming. Microbial inoculants are essential component of organic farming and are the preparations containing live or latent cells of efficient strains of plant beneficial microorganisms which when used as seed coating or as soil application, exert positive effect on plant growth. The role and importance of microbial inoculants in sustainable crop production has been reviewed by several authors (Venkateswarlu *et al.*, 2008; Mishra *et al.*, 2013). Microbial cultures are used in organic farming for different functional traits. For example nitrogen fixing, nutrient solubilizing or mobilizing microorganisms are used as biofertilizers which provide nutritional benefits to the plants. Microbial cultures are also used for the biotic (pathogens and pests) and abiotic stress (drought, heat, salinity *etc.*) management. As compost hastners and enrichers they form important components of organic farming to generate nutrient rich compost by recycling crop residues. Besides above facts, the long term use of microbial inoculants is economical, eco-friendly, more efficient, productive and accessible to marginal and small farmers over chemical fertilizers.

2. Microorganisms in Organic Farming

All these soil microorganisms serve as biological inputs in organic farming resulting in sustainable practices for ecofriendly agriculture. They form important components of organic farming as biofertilizers, plant growth promoters, biocontrol agents, abiotic stress mitigators, compost hasteners and enrichers, green manures (Figure 6.1, Tables 6.1 and 6.2) and have potential application in sustainable agriculture.

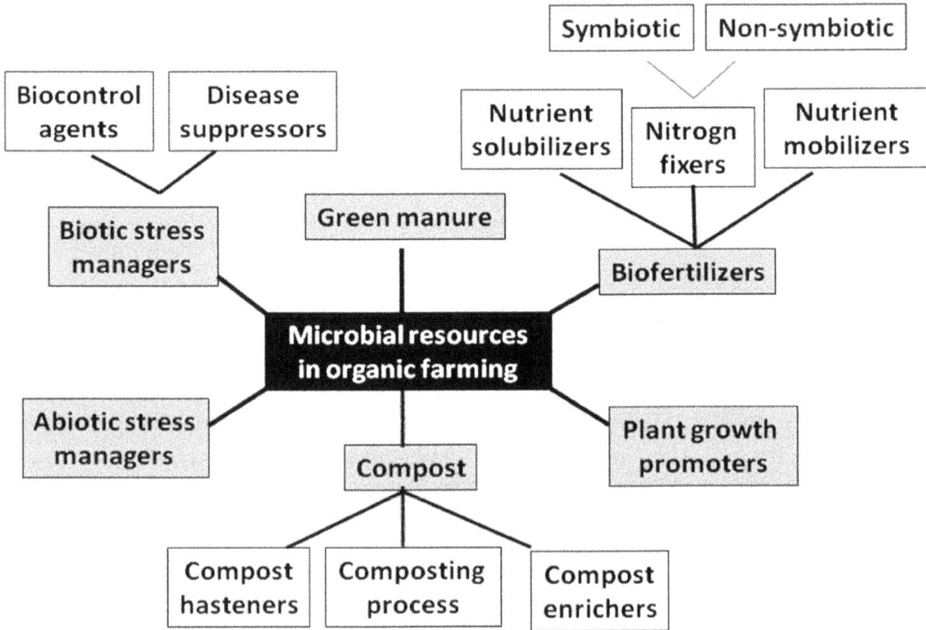

Figure 6.1: Microbial Resources as Inputs in Organic Farming.

Table 6.1: Microorganisms for Nutrient Management, Growth Promotion and Disease Management

Microbial Inoculant	*Crop*	*Observed Effects*	*Reference*
N-fixers			
Azotobacter	Pearlmillet	Increased plant biomass and grain yield	Venkataraman and Tilak (1990)
Azospirillum, Azotobacter, Klebsiella	Pearlmillet	Increased root and shoot N	Tiwari *et al.* (2003)
A. chroococcum Mac 68	Pearlmillet	Increased plant biomass and grain yield	Sangwan *et al.* (2011)
Rhizobium	Mungbean/ greengram	Enhanced plant biomass, nodulation and yield	Tripathi *et al.* (2012)
Rhizobium	Soybean	Increased plant biomass, yield and N uptake	Patra *et al.* (2012)
Azospirillum	Sorghum	Increased grain yield, water use efficiency.	Patil (2014)
Azotobacter strain Azo-8	Wheat	Increased plant biomass, seed weight, grain yield	Singh *et al.* (2013)
Nutrient solubilizers, mobilizers			
Bacillus megatherium var. *Phosphaticum* (PSB)	Sugarcane)	Improved available P in the soil, enhanced cane yield	Sundra *et al.* (2002)
PSB	Sorghum	Increased grain yield and P content in root and grain	Appanna (2007)
AM fungi	Maize-fingermillet cropping system	Increased plant biomass, yield, P uptake	Shrestha *et al.* (2009)
AM fungi + Pseudomonas	Sorghum	Increased nutrient uptake and plant biomass	Praveen Kumar *et al.* (2012)
AM fungi, *Azospirillium brasilense* and PSB	Finger millet	Increased P and N uptake and plant biomass	Ramakrishnan and Bhuvaneswari, (2014)
Pseudomonas + Rhizobium	Mung bean/ greengram	Enhanced growth and nutrient uptake	Praveen Kumar *et al.* (2015
Plant growth promoting rhizobacteria (PGPR)			
Bacillus sp. RM-2	Cowpea	Increased seed germination, plant biomass, and grain yield	Minaxi *et al.* (2012)
Pseudomonas fluorescens + Bacillus megaterium + Azospirillum brasilense	Maize	Root growth, nutrient uptake, kernel yield	Gulnaz *et al.* (2017)
Biocontrol agents			
Pseudomonas fluorescens	Pearlmillet	Biocontrol of *Sclerospora graminicola* (downy mildew). Enhanced germination, vigour, plant biomass, seed weight and yield.	Niranjan *et al.* (2004)

Microbial Inoculant	*Crop*	*Observed Effects*	*Reference*
Pseudomonas fluorescens	Rice	Biocontrol of *Rhizoctonia solani* (sheath blight)	Nagarajkumar *et al.* (2004)
Pseudomonas, Burkholde ria	Mungbean/ grrengram	Biocontrol of *Macrophomina phaseolina* (charcoal rot). Improved germination, plant biomass and yield	Minaxi and Saxena (2010)
T. viride	Mungbean/ grrengram	Biocontrol of *M. phaseolina* (root rot), yield improvement.	Leo Daniel *et al.* (2011)

Table 6.2: Microorganisms for Abiotic Stress Management in Plants

Microorganism	*Crop*	*Mechanism observed*	*Reference*
Drought stress			
Pseudomonas putida P45	Sunflower	EPS production and soil aggregation	Sandhya *et al.* (2009)
P. fluorescens Pf1	Greengram	Enhanced antioxidant enzymes status	Saravanakumar *et al.* (2011)
Pseudomonas spp.	Maize	Improved RWC and osmoprotectants level	Sandhya *et al.* (2010)
Bacillus spp.	Maize	Improved RWC, RAS/RT ratio and plant biomass	Sandhya *et al.* (2011)
Bacillus cereus	Mungbean/ greengram, chickpea, rice	Enhanced antioxidant enzymes status	Chakraborty *et al.* (2011)
Streptomyces spp.	Wheat	Improved plant growth	Yandigeri *et al.* (2012)
Bacillus spp.	Maize	Improved chlorophyll, RWC, soil moisture, plant biomass	Grover *et al.* (2014)
Citricoccus zhacaiensis	Onion	Improved germination and vigour	Selvakumar *et al.* (2015)
Pseudomonas putida	Chickpea	Improved growth, water status, membrane integrity, osmolyte level, antioxidant ability	Tiwari *et al.* (2015)
Heat stress			
Pseudomonas sp. AMK-P6	Sorghum	Improved biochemical parameters, induction of heat shock proteins	Ali *et al.* (2009)
Pseudomonas putida AKMP7	Wheat	Increased levels of osmoprotectants and membrane stability	Ali *et al.* (2011)
Cold stress			
Pseudomonas spp.	Wheat	Improved biochemical status, membrane stability and Na^{+}/K^{+} ratio	Mishra *et al.* (2011)
Salinity stress			
Pseudomonas fluorescens	Groundnut	Improved ACC deaminase activity	Saravanakumar and Samiyappan (2007)

Microorganism	Crop	Mechanism observed	Reference
Bacillus pumilus, Halomonas desiderata and *Exiguobacterium oxidotolerans*	Menthol	Enhanced antioxidant status and nutrient uptake	Bharti *et al.* (2014)
Bacillus licheniformis	Groundnut	Increase in fresh biomass, total length and root length	Goswami *et al.* (2014)
Pseudomonas simiae	Soybean	4-nitroguaiacol and quinoline promote soybean seed germination	Vaishnav *et al.*, 2016

2.1 Microbes as Biofertilizers

Biofertilizers are preparation that contains live microorganisms which, when applied on the seed, plant surface or soil, colonizes the rhizosphere and promote plant growth through increased supply of primary nutrients for the host plant (Kloepper *et al.*, 1980; Bhattacharjee and Jha, 2012). Their scope and importance can be realized from the fact that more than 43 million ha under paddy, 35 million ha under coarse cereals, 23 million ha under pulses, 25 million ha under groundnut and 4 million ha under soybean can be benefited by using one or other types of biofertilizers. Many soil microorganisms that provide nutrients to the plants through nitrogen fixation and nutrient soluilization and mobilization have been identified till date.

2.1.1 Microbes for Nitrogen Fixation

Biological nitrogen fixation is one of the important biological processes for environment sustainability. Since nitrogen is commonly the most limiting plant nutrient, biological nitrogen fixation (BNF) holds great promise in organic agriculture and can replace expensive chemical N fertilizer to great extent in many crops. Bacteria belonging to genera *Rhizobium, Bradyrhizobium, Allorhizobium, Sinorhizobium* and *Mesorhizobium* form symbiotic association with leguminous plants, and fix atmospheric nitrogen with the help of enzyme nitrogenase in the specialized structures called nodules. Frankia form symbiotic association with in non-leguminous trees. Besides, non-symbiotic nitrogen fixing bacteria (free-living, associative or endophytic) including cyanobacteria, *Azotobacter, Azospirillum and Acetobacter diazotrphicus, Herbaspirillum, Azoarcus* and *Bacillus* spp. can also be used for cereal crops. Biological nitrogen fixation contributes 180 × 10^6 metric tons N/ year globally, out of which symbiotic associations produce 80 per cent and the rest comes from free-living or associative systems (Graham, 1988). Nitragin was the first commericial biofertilizer which was granted US Patent in 1896. Since then, steady efforts on research related to plant beneficial microorganisms have continued, leading to selection of numerous microbial strains with diverse plant beneficial traits. At present, commercial formulations of many symbiotic and non symbiotic N fixers are available for different types of crops (Bhattacharjee and Dey, 2014).

2.1.2 Microbes for Nutrient Solubilization and Mobilization

Phosphorus (P) is major essential macronutrients for biological growth and development. Non-availability of P is also a limiting factor in biological nitrogen fixation. It is applied to soil in the form of phosphate fertilizers. However, a large portion of applied inorganic phosphate gets immobilized in the soil and becomes unavailable to plants. Many soil microorganisms are capable of solubilising the insoluble inorganic P of soil and make it available as plant usable forms. The rhizospheric phosphate solubilizing microorganisms (PSM) could be a promising source for P fertigation in organic agriculture. Bacterial genera like *Azospirillum, Azotobacter, Bacillus, Beijerinckia, Burkholderia, Enterobacter, Erwinia, Flavobacterium, Microbacterium, Pseudomonas, Rhizobium* and *Serratia* are reported as the most significant phosphate solubilizing bacteria (Sturz and Nowak, 2000; Sudhakar *et al.*, 2000; Mehnaz and Lazarovits, 2006). Many P solubilising microbial species have been reported as intimately associated with a large number of agricultural crops like potato, tomato, wheat, radish, pulses (Bhattacharyya and Jha, 2011). Many fungi including *Aspergillus* and *Penicillium* are also identified as potential P solubilizers. Formulations of some of these PSM are commercially available. The application of PSM inoculants can replace chemical phosphatic fertilizers in organic farming.

Potassium (K) is the third major essential macronutrient for plant growth. The concentrations of soluble potassium in the soil are usually very low and more than 90 per cent of potassium in the soil exists in the form of insoluble rocks and silicate minerals (Parmar and Sindhu, 2013). Potash solubilizing microorganisms can degrade silicate minerals releasing potassium and other elements for plant use. A wide range of bacteria namely *Pseudomonas, Burkholderia, Acidothiobacillus ferrooxidans, Bacillus mucilaginosus, Bacillus edaphicus, B. circulans* and *Paenibacillus* sp. has been reported to solubilize potassium from potassium-bearing minerals in soils such as micas, illite and orthoclases, by excreting organic acids. Beneficial effect of inoculating potassium solubilizing bacteria have been reported in many crops including cotton, rape, pepper, cucumber, sorghum, wheat and Sudan grass (Sheng, 2005; Liu *et al.*, 2012; Parmar and Sindhu, 2013). Inoculation of maize and wheat plants with *Bacillus mucilaginosus,* resulted in significantly higher mobilization of potassium from waste mica, which in turn acted as a source of potassium for plant growth (Singh *et al.*, 2010). Potassium solubilizing bacteria can be used to replace chemical source of potassium and thus have potential application in organic farming.

Zinc, an essential micronutrient is a constituent of various metabolic enzyme systems in plants. Several genera of rhizobacteria including *Pseudomonas, Bacillus, Acinetobacter, Burkholderia, Gluconacetobacter* (Iqbal *et al.*, 2010; Goteti *et al.*, 2013; Ramesh *et al.*, 2014) are known to solubilise zinc. Microbes solubilise the metal forms by protons, chelated ligands, and oxidoreductive systems present on the cell surface and membranes and by reducing the pH by releasing organic acids (Goldstein *et al.*, 1993; Hilda and Fraga, 1999; Goteti *et al.*, 2013). Positive effect of bacterial inoculation on plant growth and Zn uptake has been reported in many plants like maize, tomato, mungbean, soybean, wheat (Siddiqui and Shaukat, 2002; Iqbal *et al.*, 2010; Rana *et al.*, 2012; Goteti *et al.*, 2013; Ramesh *et al.*, 2014).

Mycorrhizae are the associations between plant roots and fungi. The extra radical mycelia present in mycorrhizal plants are adapted to explore the soil to a distance much longer than the non mycorrhizal plants and bring nutrients in soil solution. Thus mycorrhizae form channels between plant roots and soil to transport the solubilized nutrients from the soil to the plant roots. Mycorrhizae also improve soil structure by extending hyphae into the soil pores and secreting glomalin which acts as a binding agent for soil aggregation. The most studied beneficial effect of mycorrhizas on host plant is improved P nutrition. Additionally, the AM-fungi also increases the acquisition of micronutrients like Zn, Cu, Fe *etc.* by secreting the enzymes, organic acids which solubilize the fixed macro and micronutrients and make them available for the plant uptake (Mahdi *et al.*, 2010). The use of nutrient solubilising and mobilizing microorganisms as inoculants increases the nutrient uptake by plants (Chen *et al.*, 2006; Igual *et al.*, 2001; Sabannavar and Lakshman, 2009).

2.2 Microbes as Plant Growth Promoters

Besides improving nutrient acquisition many microorganisms improve plant growth directly by producing growth stimulating hormone as indole acetic acid, gibberlic acid, ethylene and abscisic acid or indirectly by suppressing plant pathogens. Plant growth-promoting rhizobacteria (PGPR) are free-living, soil-borne bacteria, isolated from the rhizosphere, that enhance the growth of the plant through direct and indirect mechanisms (Kloepper *et al.*, 1980; Glick, 2005) when applied to seeds or crops. Bacteria belonging to genera like *Pseudomonas, Azospirillum, Azotobacter, Klebsiella, Enterobacter, Alcaligenes, Arthrobacter, Burkholderia, Bacillus,* and *Serratia* have been reported to enhance the plant growth (Kloepper *et al.*, 1989; Okon *et al.*, 1994; Glick 1995; Vessey, 2003; Bhattacharyya and Jha, 2011; Nain *et al.*, 2010). Several PGPR inoculants have been currently commercialized which can be very useful in organic farming.

2.3 Microbes in Biotic Stress Management

Application of microorganisms for the management of different kinds of biotic stresses especially root-pathogens is not new. Plant growth promoting rhizobacteria colonizing the rhizosphere compete with plant pathogens for nutrients and space thus restricting their growth. Many rhizobacteria produce siderophores that have high affinity for iron and can chelate with iron making it unavailable to pathogenic microorganisms. PGPR may also synthesise anti fungal metabolites such as antibiotics and fungal cell wall lysing enzymes such as chitinase, glucanase, cellulose, protease, or produce hydrogen cyanide that suppress the growth of fungal pathogens causing root and plant disease. Systemic resistance induced by PGPR in host plant against pathogens has been termed 'induced systemic resistance' (ISR) (Kloepper *et al.*, 1992; Pieterse *et al.*, 1996). ISR is dependent on colonization of the root system by sufficient numbers of PGPR, and that can be achieved by coating seed with high numbers of bacteria or by adding bacterial suspensions to soil before sowing or at transplanting (Kloepper, 1996).

Pseudomonas fluorescens is one of the most extensively studied PGPR because of its antagonistic actions against several plant pathogens (Kavino *et al.*, 2007; Saravanakumar and Samiyappan, 2007; Harish *et al.*, 2009). *Pseudomonas* is a diverse genus containing a large number of species with a variety of catabolic and metabolic abilities. This highly diverse genus with a variety of catabolic and metabolic abilities can colonize an array of environmental niches. Numerous examples of plant growth stimulation by fluorescent *Pseudomonas* spp. have been reported (Kloepper *et al.*, 1980; Sivasakthi *et al.*, 2014). Biological control with fl uorescent pseudomonads offers an effective method of managing plant pathogens (Ramamoorthy *et al.*, 2001). These bacteria inhibit the fungal pathogens by producing antibiotics, lytic enzymes and by inducing resistance systemically in the plant by activating defense genes such as chitinase, β -1, 3-g1ucanase, peroxidase and phenylalanine ammonia lyase (Ramamoorthy *et al.*, 2001; Viswanathan and Samiyappan, 2001). Due to its ubiquitous distribution in the rhizosphere, fluorescent pseudomonads has broad spectrum of action in the suppression of fungi, bacteria and nematodes (Haas and Keel, 2003).

Bacteria belonging to genus *Bacillus* form another most commonly reported biocontrol agent (Compant *et al.*, 2005; Vessey, 2003). Certain species and strains of *Bacillus* are known to produce antibacterial or antifungal metabolites effective against many phytopathogenic microorganisms like *Macrophomina phaseolina, Sclerotium rolfsii, Fusarium oxysporum, Rhizoctonia solani, Phytophthora medicagenis, P. nicotianae, P. aphanidermatum or Sclerotinia minor. P. infestans etc.* (Asaka and Shoda, 1996; Grover *et al.*, 2009; Grover *et al.*, 2010). Formulations based on *B. subtilis* and other *Bacillus* species have been used as seed dresser in several crops for controlling plant disease (Schisler *et al.*, 2004). *Bacillus* strains have the ability to form endospores that confer them high stability as biofungicides or biofertilizers (Schisler *et al.*, 2004). The treatment of soybeans with *Bacillus cereus* has been shown to improve soybean yield in field (Osburn *et al.*, 1995). Systemic resistance in sugar beet was elicited by phyllosphere bateria *Bacillus mycoides* (Bargabus *et al.*, 2002).

Trichoderma is an antagonistic soil borne fungus that can effectively control soil-borne plant pathogens. The mode of action of its antagonism, are primarily by competing for substrates, chitinase production, production of antibiotics (trichodermin, viridin *etc.*) and mycoparasitism. Among several spp., *T. viride* is very popular as it has been accepted world over as a component in integrated disease management.

Besides suppression of pathogens, many microorganisms have been potentially used in the biocontrol of plant pests (insects, nematodes *etc.*). Microbial pesticides of bacterial (*Bacillus thuringiensis, Pseudomonas fluorescens*), fungal (*Entomophthora spp, Trichoderma polysporum, T. viridae, T. harzianum, Trichogramma chilonis Trichogramma braassiliensis, Beauvaria bassiana, Metarhizium anisopliae, Verticillium lecanii, Phascilomycetes*) and viral (nuclear polyhedrosis Virus, NPV), origins have been recommended for the biocontrol of various insects like *Helicoperva, Spodoptera,* borers, hairy caterpillars, mites, scales, white grubs, beetle grubs, *Heliothis*, lepidopteran, semiloopers, white flies, thrips, aphids, mealy bugs *etc.* in different crops like cotton, tomato, sugarcane, paddy, pulses, vegetables, fruits, sugarcane, groundnut, rice, potato *etc.* (http://agritech.tnau.ac.in/org_farm/IPM Booklet for OF).

These bio-control agents can protect the plant from different diseases and pests in an environment friendly manner. The phenomenon of biological control of soil borne- and foliar-pathogens through many microorganisms has been commercially exploited. With the increasing awareness about negative effects of chemical pesticides, biological control in India is now gaining significance as safe and economically viable option for controlling phytopathogens.

2.4 Microbes as Abiotic Stress Mitigators

Beneficial microorganisms are used in a variety of agroecosystems for augmentation of nutrient supply, biocontrol, bioremediation and rehabilitation of degraded lands (Vessey, 2003). Stressed ecosystems are however, the most challenging to realize optimum performance from Agriculturally important microorganisms (AIMs). Ecosystems with sub-optimal performance of soils and other resources in productivity are termed stressed ecosystems (Sehgal and Mandal, 1994). Major stress factors in India are drought or soil moisture deficit, which affect nearly two third area forming part of the arid and semi-arid ecosystems. The other important abiotic stresses are high temperature, soil salinity/alkalinity, low pH and metal toxicity.

Recently, role of microorganisms in imparting tolerance to plants against different abiotic stress has been reported in many crop plants. The term induced systemic tolerance (IST) has been used for microbial mediated abiotic stress tolerance in plants (Yang *et al.*, 2009). Reports have accumulated on enhanced tolerance of many crop plants including sunflower, maize, wheat, chickpea, groundnut, spices and grapes due to inoculation with bacteria like *Rhizobium, Bacillus, Pseudomonas, Pantoea, Paenibacillus, Burkholderia, Achromobacter, Azospirillum, Microbacterium, Methylobacterium, Variovorax, Enterobacter* under different abiotic stress conditions like drought, high and low temperature, salinity, flooding, nutrient deficiency (Ait Barka *et al.*, 2006; Arshad *et al.*, 2008; Kohler *et al.*, 2009; Grover *et al.*, 2011; Selvakumar *et al.*, 2012; Grover *et al.*, 2014). Role of mycorrhizae in imparting drought and salt tolerance in host plants is well documented (Busse and Ellis, 1985; Bothe *et al.*, 2012; Ruiz-Lozano *et al.*, 2012; Abdelmoneim *et al.*, 2014). Mycorrhizae are known to improve nutrient uptake, soil structure and plant water relations. Besides, the role of viruses and symbiotic fungi has been reported in abiotic stress alleviation in plants (Redmen *et al.*, 2002; Grover *et al.*, 2011).

Production of stimulatory phytohormones like indole acetic acid, gibberellins and some unknown determinants by PGPR, result in increased root length, root surface area and number of root tips, that enhance uptake of nutrients resulting in improved plant health under abiotic stress conditions (Egamberdieva and Kucharova, 2009). Production of cytokinin and antioxidants by microorganisms result in abscisic acid (ABA) accumulation and degradation of reactive oxygen species. Many aspects of plant life are regulated by ethylene levels and under stress conditions the plant hormone ethylene endogenously regulates plant homoeostasis resulting in reduced root and shoot growth. However, ACC deaminase producing bacteria can reduce the effect of stress ethylene by sequestering and degrading plant ACC (which is an immediate precursor for ethylene production) to get nitrogen and energy. Thus inoculation with ACC deaminase producing bacteria can help in, ameliorating plant stress and promoting plant growth (Glick *et al.*, 2007). The role of ACC deaminase producing PGPR in stress agriculture has been reviewed by Saleem *et al.* (2007). Inoculation with ACC deaminase containing bacteria induce longer roots which might increase water use efficiency of the plant due to enhanced uptake of water from deep soil (Zahir *et al.*, 2008). Certain microorganisms produce exo-polysaccharides which can bind soil particles to form microaggregates and macroaggregates. Plant roots and fungal hyphae fit in the pores between microaggregates and thus stabilize macroaggregates. Application of EPS producing microorganisms can help improve the soil structure thus improving water and nutrient retention (Sandhya *et al.*, 2009). EPS can also bind to cations including Na^+ thus making it unavailable to plants under saline conditions. Accumulation of osmoprotectants has been correlated with abiotic stress tolerance in plants. Proline as a compatible solute helps in maintaining osmotic turgor under stress, stabilizes macromolecules, acts as a sink of carbon and nitrogen for use after relief of stress, help in free radical detoxification *etc.* (Mohammadkhani and Heidari, 2008). Similarly *Rhizobium* mediated trehalose accumulation has been related with abiotic stress tolerance in legumes (Figueiredo *et al.*, 2008). Volatiles organic compounds (VOCs) like 2R, 3R-butanediol, salicylic acid (SA), and jasmonic acid emitted by microorganisms are reported to be involved in IST (Cho *et al.*, 2008). Table 2 summarizes reports, from Indian researchers, on use of microorganisms as bioinoculants in crop plants under stresses conditions. These researches open up new and exciting possibilities of utilizing microorganisms for enhancing tolerance of plants to abiotic stresses. Selection and application of efficient abiotic stress tolerant microorganisms can be an ecofriendly strategy to alleviate abiotic stresses in plants having potential application in organic farming.

2.5 Microbes in Green Manuring

Green manures are important components of organic farming. Cover crops, is a crop grown and then intentionally ploughed under in order to improve the soil nutrient status. Mainy leguminous crops alfalfa, clovers, soy beans, peas and sesbania are used as cover crops which harbor nitrogen fixing bacteria. Similarly, azolla harbors nitrogen fixing cyanobacteria, *Anabaena* and is used as green manure. Use of nitrogen gathering green manures add nitrogenous material to the soil along with other organic compounds thus helping in improving crop yields.

2.6 Microbes in Composting

Compost is a key component of organic farming as it improves physico-chemical and biological properties of soil besides improving organic matter content of the soil. Compost also introduces a variety of microorganisms into the soil that may assist in nutrient cycling and in the control of plant pathogens and pests. Thus compost can be an as an alternative for synthetic fertilizers and pesticides by the organic farmers. Conversion of undesired crop residues and biodegradable waste to nutrient rich mature compost is mediated through microorganisms. During composting, microorganisms use the organic matter as a food source, producing heat, carbon dioxide, water vapor, and humus as a result of their furious growth and activity. Composting process occurs in different stages having different temperatures and nutrient availabilities. The microflora during the composting keeps changing depending on the temperature profile of the compost pile. During initial stages of composting mesophillic bacteria predominate due to the availability of easily usable organic substances in the undigested feedstock. Heat released due to the activity of mesophilic bacteria raises the temperature (10°–40°C) within the compost heap. As the temperature increases thermophilic bacteria start dominating mesophilles and further raise the temperature of the pile (41°–65°C). This is called the active stage or most productive stage of composting as most of the organic matter is converted into carbon dioxide and humus, and the microorganism population grows. Spore forming bacteria *Bacillus, Clostridium,* and thermophilic bacteria *Thermus* have been isolated from the thermophilic stage. The high temperature during the active stage kills the pathogens and destroys the weed seeds present in the compost. Eventually, readily degradable organic substrate start depleting, activity of thermophiles decreases and the temperatures gradually return to the mesophilic range, and final curing stage begins. In the curing stage fungi and actinomycetes dominate and proliferate on recalcitrant organic matter such as chitin, cellulose and lignin by producing extracellular enzymes. *Cytophaga. Sporocytophaga* are dominant cellulolytic microorganisms in all types of composting processes. *Cellulomonas* and *Cytophaga* are the aerobic mesophilic bacteria able to degrade cellulose. Mesophilic aerobic and anaerobic forms of *Bacillus* spp are known to be cellulose and hemicellulose degraders. Thermophilic actinomycetes (*Thermoactinomyces, Streptomyces* and *Thermomonospora, Micromonospora* can solubilise cellulose and modify the lignin structure extensively. However, their ability to mineralize lignin is limited (Stutzenberger, 1972; Rao and Venkateswarlu, 1983; Eriksson *et al.,* 1990; Godden *et al.,* 1992; Saritha *et al.,* 2013). Several lignocellulolytic fungi like *Trichoderma harzianum, Pleurotus ostreatus, Polyporus ostriformis* and *Phanerochaete chrysosporium* play important role in composting of crop residues (Singh and Nain, 2014). Humic content gradually increases in the curing stage. The compost prepared under aerobic conditions with adequate curing can contribute to the soil fertility and plant growth in several ways.

2.6.1 Microbes in Composting

Bacteria constitute the majority of microorganisms in composting piles, with eubacteria and actinomycetes usually present in at least 100-fold greater numbers than fungi (Boulter *et al.,* 2000). Eubacteria or true bacteria, unlike fungi and actinomycetes thrive during all stages of composting. Composts support high population levels of bacteria with 78 per cent of them being gram-negative with *Pseudomonas* (28 per cent), *Serratia* (20 per cent), *Klebsiella* (11 per cent) and *Enterobacter* (5 per cent) being the dominant genera. Boulter *et al.* (2002) isolated all Gram positive bacteria identified as *Bacillus* spp. from compost. Gbolagade (2006) isolated bacteria involved in an outdoor single phase composting of sawdust wheat bran. The bacteria characterized were *Bacillus polymyxa, Enterobacter aerogenes, Micrococcus roseus, Citrobacter freundii, B. subtilis, Clostridium perfringens, B. licheniformis, P. aeruginosa, B. cereus and E. coli.*). These bacteria compete with pathogens for space and nutrients. Further they produce lytic enzymes and antibiotics (Csuzi, 1978; Dorherty and Preece, 1978) required for the suppression of pathogens. *Trichoderma viride,* a common fungal inhabitant of compost is known to to suppress the pathogens through production of antifungal antibiotics (Hoitink *et al.,* 1997). The effectiveness of compost in suppressing the growth of soil borne pathogens is now well known (Hoitink, 1980; Hoitink and Fahy, 1986; Schueter *et al.*, 1989). Studies have shown that composts that are sterilized by heating, irradiating with gamma radiation or taken from high temperature center of compost windrows are generally not suppressive (Chen *et al.*, 1988; Hadar and Mandelbaum 1986; Kuter *et al.*, 1988). The low level of suppression is correlated with low levels of microbial activity (Chen *et al.*, 1988). Addition of small amount of non-heated compost sand mixture to those heated can restore suppressive properties of compost. The elevated level of microbial activity result in increased competition between compost-inhabiting microbial population and plant pathogens for root exudates components essential for germination of fungal propagules. Bacilli are particularly attractive for practical use as biocontrol agents because they produce stable endospores, which can survive heat and desiccation during the process of composting (Turner and Backman 1991; Lumsden *et al.*, 1995; Osburn *et al.*, 1995.

2.6.2 Microbes as Compost Hasteners

The natural composting of agricultural residues rich in lignocellulose like paddy straw generally takes 180 days to obtain good and mature compost. High lignin content in the agrowastes restricts the enzymatic and microbial access to the cellulose. Use of some lignin degrading microorganisms in combinations with the cellulose-degrading microorganisms can hasten the composting process thereby overcoming the time constraint (Singh and Nain, 2014). Efficient lignocellulolytic fungi (*e.g., Trichoderma, Aspergillus awamori, Polyporous versicolor, Penicillium funiculosum, Phanerochaete chrysosporium etc.*), have been used as compost accelerators (Gaur *et al.*, 1982; Gaind *et al.*, 2005; Lata *et al.*, 2005). The consortium of four hypercellulolytic fungal cultures namely *Aspergillus nidulans, Trichoderma viride, Phanerochaete chrysosporium* and *Aspergillus awamori* was used for preparation of bioaugmented compost by using diverse agricultural waste *e.g.*, paddy straw, soybean trash, pearl millet, maize residues and mustard stover effectively (Gaind and Nain, 2010; Gaind *et al.*, 2009; Pandey *et al.*, 2009). Use of an efficient microorganism (EM) consortium (*Candida tropicalis, Phanerochaete chrysosporium, Streptomyces globisporous, Lactobacillus* sp. and enriched photosynthetic bacterial inoculums) along with compost inoculants further hastened composting of paddy straw and also improved the quality of the compost (Sharma *et al.*, 2014). Incorporation of poultry droppings and rock phosphate (1 per cent) resulted in generation of N-enriched phosphocompost within two month of composting.composting of paddy straw in perforated pits. A thermophilic fungal consortium of *A. nidulans, Scytalidium thermophilum* and *Humicola* sp. was effectively used in composting of soybean trash and paddy straw mixture (Kumar *et al.*, 2008). Similarly, a consortium of thermophilic microorganisms (*Scytalidium thermophilum, Humicola insolens* and *Sporotrichum thermophilum*) was used for production of compost at high temperature in tunnels to generate pathogen free compost within 10 days for mushroom cultivation. However efforts need to develop psychotrophic microbial accelerators for agrowaste management in temperate and hilly regions where biomass generation is tremendous due to rich vegetation and forests (Singh and Nain, 2014)

2.6.3 Microbes as Compost Enrichers

Exposure to heat during the thermophilic stage of composting is often responsible for killing plant and human pathogenic microorganisms. This heat also kills those beneficial microorganisms that cannot tolerate the high temperature. Although the re-inoculation of microorganisms occurs naturally in the compost after the thermophilic stage is over beneficial microorganisms can be inoculated into the compost to enhance their population over pathogenic ones. Bioinoculants such as P-solubilizers, nitrifiers and free-living N fixers and biocontrol agents can be employed to prepare enriched compost. Inoculation of PSM along with rock phosphate (12.5 per cent) and pyrite (10 per cent) can result into phospho compost. Addition of pyrite increases solubilization of rock phosphate by *A. awamori*. Compost enrichment with free living N fixers can help in maintaining the C: N ratio and can provide N nutrition to plants as well as non-nitrogen fixing microorganisms. Enrichment with biocontrol agents can improve the disease suppressiveness of compost. Application of *Trichoderma* treated compost could promote crop growth of lettuce, radish, tomato and okra (Cotxarrera *et al.*, 2002; Aldahmani *et al.*, 2005; Siddiqui *et al.*, 2008). Composts or compost tea (CT) amended with cultures of potential antagonistic cyanobacteria (*Anabaena oscillarioides* C12) or bacteria (*Bacillus subtillis* B5) exhibited enhancement of seed germination, seedling length and biomass of tomato crop in the presence of phytopathogenic fungi, a causal agent of damping off on tomatoes (Dukare *et al.*, 2011). Microbial-enriched compost tea (CT) is a water extract of compost that is amended with nutrient supplements during brewing to enhance their microbial diversity. Foliar application of microbial-enriched CT on improved the growth of muskmelon (*Cucumis melo* L.) and suppressed powdery mildew caused by *Golovinomyces cichoracearum* (Naidu *et al.*, 2013). Thus microbially enriched compost has great potential in organic farming.

3. Conclusions

Microbial inoculants as essential components of organic farming help in improving soil fertility and maintaining long-term sustainability. Different types of microbial inoculants can perform varied functions in organic farming. They can be used as biofertilizers for nitrogen fixation, solubilizing and mobilizing fixed macro and micro nutrients into forms available to plants, there by increases their efficiency and availability, as plant growth promoters by production of plant growth stimulatory substances, as biocontrol agent for suppressing the plant pathogens and pests, as compost hasteners and enrichers for producing quality compost, as soil conditioners for improving soil organic matter and soil structure. In context of environmental impact of chemical fertilizers and pesticides, organic farming is a viable option for sustainable agriculture. Microbial inoculants being important component of organic agriculture have enormous potential in sustainable agriculture in future.

4. Future Prospects

Microbial inoculants, which can fulfill diverse functions in organic farming, can provide promising solutions for a sustainable, environmentally friendly agriculture. Microbial inoculants as biofertilizers and biocontrol agents already exist, however their use in organic agriculture need to be promoted by generating awareness among the farmers. Microbes providing protection against abiotic stresses like salinity, drought, waterlogging, heavy metals, *etc.* need to be explored for their use in stress agriculture. Compost enrichment with growth promoting as well as stress alleviating microbial inoculants will be of emerging importance. The success of microbial inoculants in agriculture depends upon their ability to establish in soil and maintain an adequate population to exert beneficial effect to the host plants. The selection of efficient is the first and most important step to use microbial inoculants in agriculture. It takes lot of time and energy to develop one microbial formulation for a particular crop. Therefore, selection of the strain should be done with utmost care. The strain should be functionally efficient and highly competent post inoculation. Further it should be able to perform under different agro-climatic conditions. Next step is the mass multiplication of selected strains. There are some technical problems related to suitable carrier, economics and quality of the product produced *etc.* Future work should focus on development of quality product with extended shelf life by developing suitable multiplication protocols, and carrier systems. Role of extension workers is very important in the dissemination of the technology to the farmers. There is need to generate awareness among the farmers about the benefits as well as direction to use the microbial products. Another challenge in the use of microbial products in organic farming is that many plant-associated bacteria have been identified as opportunistic human pathogens (Berg *et al.,* 2005). For examples antagonistic species of the genera *Burkholderia, Enterobacter, Herbaspirillum, Ochrobactrum, Pseudomonas, Serratia, Staphylococcus, and Stenotrophomonas* that are root-associated bacteria that can interact with human hosts (Parke and Gurian-Sherman 2001; Ribbeck-Busch *et al.,* 2005; Egamberdieva *et al.,* 2008). Therefore it becomes necessary to evaluate the risk of each potential agriculturally beneficial microorganism.

REFERENCES

Abdelmoneim TS, Moussa TAA, Almaghrabi OA, Alzahrani HS and Abdelbagi I. 2014. Increasing Plant Tolerance to Drought Stress by Inoculation with Arbuscular Mycorrhizal Fungi. Life Science Journal 11(1) http: //ww w.lifesciencesite.com 10.

Ait bakra E, Nowak J and Clement C. 2006. Enhancement of chilling resistance of inoculated grapevine plantlets with a plant growth promoting rhizobacterium, *Burkholderia phytofirmans* strain PsJN. Applied and Environmental Microbiology, 72: 7246-7252.

Aldahmani JH, Abbasi PA, Sahin F, Hoitink HAJ and Miller SA. 2005. Reduction of bacterial leaf spot severity on radish, lettuce, and tomato plants grown in compost-amended potting mixes. Canadian Journal of Plant Pathology, 27: 186-193.

Ali SZ, Sandhya V, Grover M, Rao LV, Kishore VN and Venkateswarlu B. 2009. *Pseudomonas* sp. strain AKM-P6 enhances tolerance of sorghum seedlings to elevated temperatures. Biology and Fertility of Soils, 46: 45-55.

Ali SZ, Sandhya V, Grover M, Rao LV and Venkateswarlu B. 2011. Effect of inoculation with a thermotolerant plant growth promoting *Pseudomonas putida* strain AKMP7 on growth of wheat (*Triticum* spp.) under heat stress. Journal of Plant Interactions, 1-8.

Appanna V. 2007. Efficacy of phosphate solubilizing bacteria isolated from vertisols on growth and yield parameters of sorghum. Research Journal of Microbiology, 2: 550–559.

Arshad M, Sharoona B and Mahmood T. 2008. Inoculation with *Pseudomonas* spp. containing ACC deaminase partially eliminate the effects of drought stress on growth, yield and ripening of pea (*Pisum sativum* L.). Pedosphere, 18: 611-620.

Asaka O and Shoda M. 1996. Biocontrol of *Rhizoctonia solani* damping off of tomato with *Bacillus subtilis* RB14. Applied and Environmental Microbioloy, 62: 4081-4085.

Bargabus RL, Zidack NK, Sherwood JE and and Jacobsen BJ. 2002. Characterization of systemic resistance in sugar beet elicited by a non-pathogenic, phyllosphere-colonizing *Bacillus mycoides*, biological control agent. Physiological and Molecular Plant Pathology, 61: 289-298.

Berg G, Eberl L and Hartmann A. 2005. The rhizosphere as a reservoir for opportunistic human pathogenic bacteria. Environmental Microbiology, 7: 1673-1685.

Bharti N, Barnawal D, Awasthi A, Yadav A and Kalra A. 2014. Plant growth promoting rhizobacteria alleviate salinity induced negative effects on growth, oil content and physiological status in *Mentha arvensis*. Acta Physiologiae Plantarum, 36(1): 45-60.

Bhattacharyya PN and Jha DK 2012. Plant growth-promoting rhizobacteria (PGPR): emergence in agriculture. World Journal of Microbiology and Biotechnology, 28: 1327-1350.

Bhattacharjee R and Dey U. 2014. Biofertilizers, a way towards organic agriculture: A review. African Journal of Microbiology Research, 8: 2332-2342.

Bothe H. 2012. Arbuscular mycorrhiza and salt tolerance of plants. Symbiosis, 58(1-3): 7-16.

Boulter JI, Boland GJ and Trevors JT. 2000. Compost: a study of the development process and end-product potential for suppression of turfgrass disease. World Journal of Microbiology and Biotechnology, 16: 115-134.

Boulter JI, Trevors JT and Boland GJ. 2002. Microbial studies of compost: Bacterial identification and their potential for turfgrass pathogen suppression. World Journal of Microbiology and Biotechnology, 18: 661-671.

Busse MD and Ellis JR.1985. Vesicular-arbuscular mycorrhizal (*Glomus fasciculatum*) influence on soybean drought tolerance in high phosphorus soil. Canadian Journal of Botany, 63: 2290-2294.

Chakraborty U, Roy1 S, Chakraborty AP, Dey P and Chakraborty B. 2011. Plant growth promotion and amelioration of salinity stress in crop plants by a salt-tolerant bacterium. Recent Research in Science and Technology, 3(11): 61-70.

Chen W, Hoitink HAJ, Schmitthenner AF and Tuovinen O. 1988. The role of microbial activity in suppression of damping-off caused by *Pythium ultimum*. Phytopathology, 78: 314-322.

Chen YP, Rekha PD, Arun AB, Shen FT, Lai WA andYoung CC. 2006. Phosphate solubilizing bacteria from subtropical soil and their tricalcium phosphate solubilizing abilities. Applied Soil Ecology, 34: : 33-41.

Cho Sm, Kang BR, Han SH, Anderson AJ, Park JY, Lee YH, Cho BH, Yang KY, Ryu CM and Kim YC. 2008. 2R, 3r-butanediol, a bacterial volatile produced by *Pseudomonas chlororaphis* O6, is involved in induction of systemic tolerance to drought in *Arabdopsis thaliana*. Molecular Plant-Microbe Interactions, 21: 1067-1075.

Compant S, Duffy B, Nowak J, Clement C and Barka EA. 2005. Use of plant growth-promoting bacteria for biocontrol of plant disease: principles, mechanisms of action and future prospective. Appllied and Environmental Microbiology, 71: 4951-4959.

Cotxarrera L, Trillas-Gay MI, Steinberg C and Alabouvette C. 2002. Use of sewage sludge compost and *Trichoderma asperellum* isolates to suppress *Fusarium* wilt of tomato. Soil Biology and Biochemistry, 34: 467-476.

Csuzi S. 1978. The induction of a lytic enzyme in cultures of *Bacillus cereus*. Acta Biochimica Et Biophysica; Academiae Scientiarum Hungaricae, 13: 41-42.

Doherty MA and Preece TF. 1988. *Bacillus cereus* prevents germination of uredospores of *Puccinia allii* and the development of rust disease of leek. *Allium porrum* in controlled environments. Physiological and Molecular Plant Pathology, 22: 123-132.

Doherty MA and TF. Preece. 1978. *Bacillus cereus* prevents germination of uredospores of *Puccinia allii* and the development of rust disease of leek *Allium porrum* in controlled environments. Physiological Plant Pathology, 22: 123-132

Dukare AS, Prasanna R, Dubey SC, Nain L, Chaudhary V, Singh R and Saxena AK. 2011. Evaluating novel microbe amended composts as biocontrol agents in tomato. Crop Protection, 30: 436-442.

Egamberdieva D, Kamilova F, Validov S, Gafurova L, Kucharova Z and Lugtenberg B. 2008. High incidence of plant growth-stimulating bacteria associated with the rhizosphere of wheat grown on salinated soil in Uzbekistan. Environmental Microbiology, 10: 1–9

Egamberdieva D and Kucharova Z. 2009. Selection for root colonizing bacteria stimulating wheat growth in saline soils. Biology and Fertility of Soils, 45: 563-571.

Eriksson KEL, Blanchette RA and Ander P. 1990. Microbial and enzymatic degradation of wood and wood components. Springer Verlag, Germany: 60.

Figueiredo MVB, Burity HA, Martinez CR and Chanway CP. 2008. Alleviation of drought stress in common bean (*Phaseolus vulgaris* L.) by co-inoculation with *Paenibacillus polymyxa* and *Rhizobium tropici*. Applied Soil Ecology, 40: 182-188.

Gaind S, Pandey AK and Lata. 2005. Biodegradation study of crop residues as affected by exogenous nitrogen and fungal inoculants. Journal of Basic Microbiology, 45: 301-311.

Gaind S and Nain L. 2010. Exploration of composted cereal waste and poultry manure for soil restoration. Bioresource Technology, 101: 2996-3003.

Gaind S, Nain L and Patel VB. 2009. Quality evaluation of co-composted wheat straw, poultry droppings and oil seed cakes. Biodegradation, 20: 307-317.

Gaur AC, Sadasivam KV, Mathur RS and Magu, SP. 1982. Role of mesophilic fungi in composting. Agricultural Wastes, 4: 453-460.

Gbolagade JS. 2006. Bacteria associated with compost used for cultivation of Nigerian edible mushrooms *Pleurotus tuber-regium* (Fr.) Singer, and *Lentinus squarrosulus* (Berk.). African Journal of Biotechnology, 5: 338-342.

Glick BR, 1995. The enhancement of plant growth by free living bacteria. Canadian Journal of Microbiology, 41(Suppl 2): 109–114.

Glick BR. 2007. Promotion of plant growth by bacterial ACC deaminase. Critical Reviews in Plant Sciences, 26: 227-242.

Godden B, Ball AS, Helvenstein P, McCarthy AJ and Penninckx MJ. 1992. Towards elucidation of the lignin degradation pathway in actinomycetes. Journal of General Microbiology, 138: 2441-2448.

Goldstein AH, Rogers RD and Mead G. 1993. Mining by microbe. Biotechnology, 11: 1250-1254.

Goswami D, Dhandhukia P, Patel P, Thakker JN. 2014. Screening of PGPR from saline desert of Kutch: Growth promotion in *Arachis hypogea* by *Bacillus licheniformis* A2, In Microbiological Research, Volume 169, Issue 1, 2014, Pages 66-75, ISSN 0944-5013.

Gulnaz Y, Fathima PS, Denesh GR, Akhilesh Kumar Kulmitra and Shivrajkumar HS. 2017. Effect of Plant Growth Promoting Rhizobacteria (PGPR) and PSB on root parameters, nutrient uptake and nutrient use efficiency of irrigated maize under varying levels of phosphorus. Journal of Entomology and Zoology Studies 2017; 5(6): 166-169

Goteti PK, Emmanuel LDA, Desai S, and Shaik MHA. 2013. Prospective zinc solubilising bacteria for enhanced nutrient uptake and growth promotion in maize (*Zea mays* L.). Article ID 869697, 7 pages http: //dx.doi. org/10.1155/2013/869697.

Graham P. H. 1988. Principles and Application of Soil Microbiology. pp. 322–345

Grover M, Ali Sk Z, Sandhya V, Rasul A and Venkateswarlu B. 2011. Role of microorganisms in adaptation of agriculture crops to abiotic stress. World Journal of Microbiology and Biotechnology, 27: 1231-1240

Grover M, Madhubala R, Ali Sk Z, Yadav SK and Venkateswarlu B. 2014. Infuence of *Bacillus* spp. strains on seedling growth and physiological parameters of sorghum under moisture stress conditions. Journal of Basic Microbiology, 53: 1-11.

Grover M, Nain L and Saxena AK. 2009. Comparision between *Bacillus subtilis* RP24 and its antibiotic-defective mutants. World Journal of Microbiology and Biotechnology, 25: 1329-1335.

Grover M, Nain L, Singh SB and Saxena AK. 2010. Molecular and biochemical approaches for characterization of antifungal trait of a potent biocontrol agent *Bacillus* subtilis RP24. Current microbiology, 60: 99-106.

Haas D and Keel C. 2003. Regulation of antibiotic production in root-colonizing Pseudomonas spp. and relevance for biological control of plant disease. Annual Reviews of Phytopathology, 41: 117-153.

Hadar Y and Mandelbaum R. 1986. Suppression of *Pythium aphanidermatum* damping-off in container media containing composted liquorice roots. Crop Protection, 5: 88-92.

Harish S, Kavino M, Kumar N, Balasubramanian P and Samiyappan R. 2009. Induction of defense-related proteins by mixtures of plant growth promoting endophytic bacteria against banana bunchy top virus. Biological Control,51: 16-25.

Hilda R and Fraga R. 1999. Phosphate solubilizing bacteria and their role in plant growth promotion. Biotechnology Advances, 17: 319-359.

Hoitink HAJ. 1980. Composted bark, a lightweight growth medium with fungicidal properties. Plant Disease 64: 142-147.

Hoitink HAJ and Fahay PC. 1986. Basis for the control of soilborne plant pathogens with compost. Annual Review of Phytopathology, 24: 93-114.

Hoitink HAJ, Stone AG and Han DY. 1997. Suppression of plant diseases by composts. Horticultural Science, 32: 184-187.

Igual JM, Valverde A, Cervantes E and Velazquez E. 2001. Phosphate-solubilizing bacteria as inoculants for agriculture: use of updated molecular techniques in their study. Agronomie, 21: 561-568.

Iqbal U, Jamil N, Ali I and Hasnain S. 2010. Effect of zinc-phosphate-solubilizing bacterial isolates on growth of *Vigna radiata*. Annals of Microbiology, 60: 243-248.

Kavino M, Harish S, Kumar N, Saravanakumar D, Damodaran T, Soorianathasundaram K and Samiyappan R. 2007. Rhizosphere and endophytic bacteria for induction of systemic resistance of banana plantlets against bunchy top virus. Soil Biology and Biochemistry, 39: 1087-1098.

Kloepper JW, Leong J, Teintz M and Schroth MN. 1980. Enhanced plant growth by siderophores produced by plant growth promoting rhizobacteria. Nature, 286: 885-886.

Kloepper JW, Lifshitz R and Zablotowicz RM. 1989. Free-living bacterial inocula for enhancingg crop productivity. Trends in Biotechnology, 7: 39-43.

Kloepper JW, Rodriguez-Kabana R, Mclnroy JA and Young RW. 1992. Rhizospheric bacteria antagonistic to soybean cyst (*Heterodera glycines*) and root knot (*Meloidogyne incognita*) nematodes: Identification by fatty acid analysis and frequency of biological control activity. Plant and Soil, 139: 75-84.

Kloepper JW. 1996. Host specificity in microbe–microbe inter-actions. Bioscience, 46: 406-409.

Kohler J, Hernandez JA, Caravaca F and Roldan A. 2008. Plant growth promoting rhizobacteria and arbuscular mycorrhizal fungi modify alleviation biochemical mechanisms in water stressed plants. Functional Plant Biology, 35: 141-151.

Kumar A, Gaind S and Nain L. 2008. Evaluation of thermophilic fungal consortium for paddy straw composting. Biodegradation,19: 395-402.

Kuter GA, Hoitink HAJ and Chen W. 1988. Effects of municipal sludge compost curing time on suppression of *Pythium* and *Phytophthora* diseases. Plant Disease, 72: 751-756.

Lampkin, N H. 1990. Organic Farming. Farming Press: Ipswich

Lata, Gaind S and Pandey AK. 2005. Chemical characterization of composts prepared with diversified agro wastes. Indian Journal of Microbiology, 45(3): 245-247.

Leo Daniel AE, Praveen Kumar G, Desai S. and Mir Hassan ASK. 2011. *In-vitro* characterization of *Trichoderma viride* for abiotic stress tolerance and field evaluation against root rot disease in *Vigna mungo* L. Journal of Biofertilizers and Biopesticides, 2(3): 1-5.

Liu D, Lian B and Dong H. 2012. Isolation of *Paenibacillus* sp. and assessment of its potential for enhancing mineral weathering. Geomicrobiology Journal, 29: 413-421.

Lumsden RD, Lewis JA and Fravel DR. 1995. Formulation and delivery of biocontrol agents for use against soil borne plant pathogens. In: *Biorational Pest Control Agents: Formulation and Delivery*. HallFR and BarryJW (Eds) American Chemical Society, Washington DC. pp. 167-187.

Mahdi SS, Hassan GI, Samoon SA, Rather HA, Dar Showkat A and Zehra B. 2010. Bio-fertilizers in organic agriculture. Journal of Phytology, 2(10): 42-54.

Mehnaz S and Lazarovits G. 2006. Inoculation effects of *Pseudomonas putida, Gluconacetobacter azotocaptans*, and *Azospirillum lipoferum* on corn plant growth under greenhouse conditions. Microbial Ecology, 51(3): 326-335.

Minaxi, Nain L, Yadav RC and Saxena J. 2012. Characterization of multifaceted *Bacillus* sp. RM-2 for its use as plant growth promoting bioinoculant for crops grown in semiarid deserts. Applied Soil Ecology, 59: 124-135.

Minaxi and Saxena J. 2010. Disease suppression and crop improvement in moong beans (*Vigna radiata*) through *Pseudomonas* and *Burkholderia* strains isolated from semiarid region of Rajasthan. Biocontrol, 55(6): 799-810.

Mishra PK, Bisht SC, Pooja R, Selvakumar G, Joshi P, Bisht JK, Bhatt JC and Gupta HS. 2011. Alleviation of cold stress in inoculated wheat (*Triticum aestivum* L.) seedlings with psychrotolerant Pseudomonads from NW Himalayas. Archives of Microbiology, 193(7): 497-513.

Mishra DJ, Singh R, Mishra U and Shahi SK. 2013. Role of bio-fertilizer in organic agriculture: a review. Research Journal of Recent Sciences, 2: 39-41.

Mishra PK, Bisht SC, Bisht JK and Bhatt JC. 2012. Cold-tolerant PGPRs as bioinoculants for stress management. Bacteria in Agrobiology: Stress Management, 95-118.

Mohammadkhani N and Heidari R. 2008. Water stress induced by polyethylene glycol 6000 and sodium chloride in two maize cultivars. Pakistan Journal of Biological Sciences, 11: 92-97.

Nagarajkumar M, Bhaskaran R and Velazhahan R. 2004. Involvement of secondary metabolites and extra cellular lytic enzymes produced by *Pseudomonas fluorescens* in inhibition of *Rhizoctonia solani* the rice sheath blight pathogen. Microbiological Research, 159: 73-81.

Naidu Y, Meon S and Siddiqui Y. 2013. Foliar application of microbial-enriched compost tea enhances growth, yield and quality of muskmelon (Cucumis melo L.) cultivated under fertigation system. Scientia Horticulturae, 159: 33-40.

Nain L, Rana A, Joshi M, Jadhav SD, Kumar D, Paul S and Prasanna R. 2010: Evaluation of synergistic effects of bacterial and cyanobacterial strains as biofertilizers for wheat. Plant and Soil, 331: 217-230.

Niranjan S, Shetty NP and Shetty HS. 2004. Seed bio-priming with *Pseudomonas fluorescens* isolates enhances growth of pearl millet plants and induces resistance against downy mildew. Journal of Pest Management, 50: 41-48.

Okon Y and Labandera-Gonzalez CA. 1994. Agronomic applications of *Azospirillum*. In: Improving Plant Productivity with Rhizosphere Bacteria. Commonwealth Scientific and Industrial Research Organization, Adelaide, Australia, Edited by Ryder MH, Stephens PM, Bowen GD, 274–278.

Osburn RM, Milner JL, Oplinger ES, Smith RS and Handelsman J. 1995. Effect of *Bacillus cereus* UW85 on the yield of soybean at two field sites in Wisconsin. Plant Disease, 79: 551-556.

Pandey AK, Gaind S, Ali A and Nain L. 2009. Effect of bioaugmentation and nitrogen supplementation on composting of paddy straw. Biodegradation, 20: 293-306.

Parke, JL *and* Gurian-Sherman D. 2001. Diversity of the *Burkholderia cepacia* complex and implications for risk assessment of biological control strains. Annual Review of Phytopathology, 39: 22-258.

Patra RK, Pant LM and Pradhan K. 2012. Response of soybean to inoculation with rhizobial strains: effect on growth, yield, N uptake and soil N status. World Journal of Agricultural Sciences, 8(1): 51-54.

Patil S. 2014. Azosprillium based integrated nutrient management for conserving soil moisture and increasing sorghum productivity. African Journal of Agricultural Research, 9(23): 1761-1769.

Parmar P and Sindhu SS. 2013. Potassium solubilization by rhizosphere bacteria: Influence of nutritional and environmental conditions. Journal of Microbiology Research, 3(1): 25-31.

Pieterse CMJ, Van Wee SCM, Hoffland E, Van Pelt JA and Van Loon LC. 1996. Systemic resistance in *Arabidopsis* induced by biocontrol bacteria is independent of salicylic acid and pathogenesis-related gene expression. The Plant Cell, 8: 1225-1237.

Praveen Kumar G, Desai S, Leo Daniel AE and Pinisetty S. 2015. Impact of seed bacterization with PGPR on growth and nutrient uptake in different cultivable varieties of green gram. Asian Journal of Agricultural Research, 9 (3): 113-122.

Ramesh A, Sharma SK, Sharma MP, Yadav N and Joshi OP. 2014. Inoculation of zinc solubilizing *Bacillus aryabhattai* strains for improved growth, mobilization and biofortification of zinc in soybeanand wheat cultivated in vertisols of central India. Applied Soil Ecology, 73: 87-96.

Ramakrishnan K and Bhuvaneswari G. 2014. Effect of inoculation of AM fungi and beneficial microorganisms on growth and nutrient uptake of *Eleusine coracana* (L.) Gaertn. (Finger millet). International Letters of Natural Sciences, 8(2): 59–69.

Ramamoorthy V, Viswanathan R, Raguchander T, Prakasam V and Samiyappan R. 2001. Induction of systemic resistance by plant growth promoting rhizobacteria in crop plants against pests and diseases. Crop Protection, 20: 1-11.

Rana A, Joshi M, Prasanna R, Shivay YS and Nain L. 2012. Biofortification of wheat through inoculation of plant growth promoting rhizobacteria and cyanobacteria. European Journal of Soil Biology, 50: 118-126.

Rana A, Saharan B, Nain L, Prasanna R and Shivay YS. 2012. Enhancing micronutrient uptake and yield of wheat through bacterial PGPR consortia, Soil Science and Plant Nutrition, 58(5): 573-582.

Rao AV and Venkateswarlu B. 1983. Microbial ecology of the soils of Indian desert. Agriculture, Ecosystems and Environment, 10: 361-369.

Redman RS, Sheehan KB, Stout RG, Rodriguez RJ and Henson JM. 2002. Thermotolarance generated by plant/fungal symbiosis. Science, 298: 1581.

Ribbeck-Busch K., Roder A, Hasse D, de Boer W, Martínez JL, Hagemann M and berg G. 2005. A molecular biological protocol to distinguish potentially human-pathogenic strains of *Stenotrophomonas maltophilia* from non-pathogenic *S. rhizophila* strains. Environmental Microbiology, 7: 1853-1858.

Ruiz-Lozano JM, Porcel R, Bárzana G, Azcón R and Aroca R. 2012. Contribution of arbuscular mycorrhizal symbiosis to plant drought tolerance: State of the art. *In: Plant Responses to Drought Stress. From Morphological to Molecular Features.* Ricardo Aroca (Ed). Springer Berlin Heidelberg. pp. 335-362.

Sabannavar SJ and Lakshman HC. 2009. Effect of rock phosphate solubilization using mycorrhizal fungi and phosphobacteria on two high yielding varieties of *Sesamum indicum* L. World Journal of Agricultural Sciences, 5 (4): 470-479.

Saleem M, Arshad M, Hussain S and Bhatti AS. 2007. Perspective of plant growth promoting rhizobacteria (PGPR) containing ACdeaminase in stress agriculture. Journal of Industrial Microbiology and Biotechnology, 34: 635-648.

Sandhya V, Ali SkZ, Grover M, Reddy G and Venkateswarlu B. 2009. Alleviation of drought stress effects in sunflower seedlings by exopolysaccharides producing *Pseudomonas putida* strain P45. Biology and Fertility of Soil, 46: 17–22.

Sandhya V, Ali SZ, Grover M, Reddy G and Venkateswaralu B. 2010. Effect of plant growth promoting *Pseudomonas* spp. on compatible solutes antioxidant status and plant growth of maize under drought stress. Plant Growth Regulators, 62: 21-30. doi: 10.1007/s 10725-010-9479-4

Sandhya V, Ali SZ, Grover M, Reddy G and Venkateswaralu B. 2011. Drought–tolerant plant growth promoting *Bacillus* spp.: effect on growth, osmolytes, and antioxidant status of maize under drought stress. Journal of Plant Interactions, 6: 1–14.

Sangwan PS, Suneja S, Kukreja K and Raj M. 2011. Response of *Azotobacter chroococcum* in pearl millet under dryland farm conditions. Indian Journal of Dryland Agricultural Research and Development, 26: 97-100.

Saravanakumar D and Samiyappan R. 2007. ACC deaminase from *Pseudomonas fluorescens* mediated saline resistance in groundnut (*Arachis hypogea*) plants. Journal of Applied Microbiology, 102(5): 1283-1292.

Saritha M, Arora A, Singh S and Nain L. 2013. *Streptomyces griseorubens* mediated delignification of paddy straw for improved enzymatic saccharification yields. Bioresource Technology, 135: 12-17.

Scheuter C, Biala J and Vogtmann H. 1989. Antiphytopathogenic properties of biogenic waste compost. Agriculture, Ecosystems and Environment, 27: 477-482.

Schisler DA, Slininger PJ, Behle RW and Jackson MA. 2004. Formulation of Bacillus spp. for biological control of plant diseases. Phytopathology, 94: 1267-1271.

Sehgal J and Mandal DK. 1994. Concepts, assessment and monitoring in stressed ecosystems and sustainable agriculture. *In: Stressed ecosystems.* Virmani SM *et al.* (Eds), Oxford and IBH Publications, New Delhi, pp. 385-388.

Selvakumar G, Panneerselvam P and Ganeshamurthy AN. 2012. Bacterial mediated alleviation of abiotic stress in crops. In: Maheshwari DK (ed) *Bacteria in Agrobiology: Stress Management.*, Springer-Verlag, pp. 205-224.

Sharma A, Sharma R, Arora A, Shah R, Singh A, Kumar P and Nain L. 2014. Insights into rapid composting of paddy straw augmented with efficient microorganism consortium. The International Journal of Recycling of Organic Waste in Agriculture, 3: 54.

Sheng XF. 2005. Growth promotion and increased potassium uptake of cotton and rape by a potassium releasing strain of *Bacillus edaphicus*. Soil Biology and Biochemistry, 37: 1918-1922.

Shrestha G, Vaidya SG and Rajbhandari BP. 2009. Effects of arbuscular mycorrhiza in the productivity of maize and fingermillet relay cropping system. Nepal Journal of Science and Technology, 10: 51–55.

Siddiqui A and Shaukat SS. 2002. Zinc and glycerol enhance the production of nematicidal compounds *in vitro* and improve the biocontrol of *Meloidogyne javanica* in tomato by fluorescent pseudomonads, Letters in Applied Microbiology, 35(3): 212-217.

Siddiqui Y, Sariah M and Razi I. 2008. *Trichoderma*-fortified compost extracts for the control of *Choanephora* wet rot in okra production. Crop Protection, 27: 385-390.

Singh G, Biswas DR and Marwah TS. 2010. Mobilization of potassium from waste mica by plant growth promoting rhizobacteria and its assimilation by maize (*Zea mays*) and wheat (*Triticum aestivum* L.). Journal of Plant Nutrition, 33: 1236-1251.

Singh NK, Chaudhary FK and Patel DB. 2013. Effectiveness of Azotobacter bioinoculant for wheat grown under dryland conditions. Journal of Environmental Biology, 34: 927–932.

Singh S and Nain L. 2014. Microorganisms in the conversion of agricultural wastes to compost. Proceedings of the Indian National Science Academy, 80(2): 1-9.

Sivasakthi S, Usharani G and Saranraj P. 2014. Biocontrol potentiality of plant growth promoting bacteria (PGPR) - *Pseudomonas fluorescens* and *Bacillus subtilis*: A review. African Journal of Agricultural Research, 9: 1265-1277.

Sturz AV, Christie BR and Novak J. 2000: Bacterial endophytes: potential role in developing sustainable system of crop roduction. Critical Reviews in Plant Sciences, 19: 1-30.

Stutzenberger FJ. 1972. Cellulolytic activity of Thermomonospora curvata: nutritional requirments for cellulase production. Applied Microbiology, 24: 83-90.

Sudhakar P, Gangwar SK, Satpathy B, Sahu PK, Ghosh JK and Saratchandra B. 2000. Evaluation of some nitrogen fixing bacteria for control of foliar diseases of mulberry (*Morus alba*). Indian Journal of Sericulture, 39: 9-11.

Sundara B, Natarajan V and Hari K. 2002. Influence of phosphorus solubilizing bacteria on the changes in soil available phosphorus and sugarcane and sugar yields. Field Crops Research, 77: 43-49.

Tiwari S, Lata C, Chauhan PS and Nautiyal CS. 2015. *Pseudomonas putida* attunes morphophysiological, biochemical and molecular responses in *Cicer arietinum* L. during drought stress and recovery, Plant Physiology and Biochemistry, doi: 10.1016/j.plaphy.2015.11.001

Tripathi PK, Singh MK, Singh JP and Singh ON. 2012. Effect of rhizobial strains and sulphur nutrition on mungbean (*Vigna radiata* (L.) *wilczek*) cultivars under dryland agro–ecosystem of Indo–Gangetic plain. African Journal of Agricultural Research, 7(1): 34-42.

Turner JT and Backman PA. 1991. Factors relating to peanut yield increases after seed treatment with *Bacillus subtilis*. Plant Disease, 75: 347-353.

Vaishnav A, Kumari S, Jain S, Verma A, Tuteja N and Choudhary DK. 2016. PGPR-mediated expression of salt tolerance gene in soybean through volatiles under sodium nitroprusside. Journal of Basic Microbiology, 56: 1274-1288.

Venkataraman GS, Tilak KVBR (1990) Biofertilizers in sustainable agriculture. In: Kumar V, hrotriya GC, Kavre SV (Eds), Soil Fertility and Fertilizers. Vol. IV, Nutrient management and supply system for sustainable agriculture in 1990s, indian Farmers fertilizers Cooperative Ltd., New Delhi, India pp. 132–148

Venkatashwarlu B. 2008. Role of bio-fertilizers in organic farming: Organic farming in rain fed agriculture. Central Research Institute for Dryland Agriculture, Hyderabad, pp. 85-95.

Vessey JK. 2003. Plant growth promoting rhizobacteria as biofertilizers. Plant and Soil, 255: 571-586.

Viswanathan R and Samiyappan R. 2001. Antifungal activity of chitinases produced by some fluorescent pseudomonads against *Colletotrichum falcatum* Went causing red rot disease in sugarcane. Microbiological Research, 155: 309–314.

Yandigeri MS, Meena KK, Singh D, Malviya N, Singh DP, Solanki MK, Yadav AK and Arora DK. 2012. Drought-tolerant endophytic actinobacteria promote growth of wheat (*Triticum aestivum*) under water stress conditions. Plant Growth Regulators, 68: 411-420.

Yang J, Kloepper JW and Ryu CM. 2009. Rhizosphere bacteria help plants tolerate abiotic stress. Trends in Plant Science, 14: 1–4.

Zahir ZA, Munir A, Asghar HN, Arshad M and Shaharoona B. 2008. Effectiveness of rhizobacteria containing ACC-deaminase for growth promotion of peas (*Pisum sativum*) under drought conditions. Journal of Microbiology and Biotechnology, 18(5): 958-963.

Chapter 7

Weed Management in Organic Crop Production

J.S. Mishra[1]* *and V.P. Singh*[2]

[1]*ICAR-Research Complex for Eastern Region, Patna – 800014, Bihar*

[2]*ICAR-Indian Institute of Sugarcane Research, Lucknow – 226002, Uttar Pradesh*

E-mail: jsmishra31@gmail.com

1. INTRODUCTION

During the last century, agriculture has developed from more or less extensive subsistence farming to a highly intensive agricultural production. Production has become dependent on chemical pesticides and commercial fertilizers. However, because of the environmental, economic and food safety concerns, a change from a high-input and chemically intensive agriculture to a more sustainable form of agriculture is not only desirable, but has become a necessity. The cultivation of high yielding crop varieties responsive to fertilizers and irrigation, and the new intensive cropping systems have brought to the forefront the problem of weeds, which cause tremendous losses to crops in terms of productivity and quality. In general, impact of weeds on crop yields varies from high input to low input crop production systems. It has been estimated that weeds caused 5 per cent loss in agricultural production in most developed countries, while it caused 10 per cent loss in developed countries and 25 per cent loss in least developing countries (Bhowmik, 1998). Almost 34 per cent of attainable production is endangered by the weed competition worldwide as compared to 18, and 16 per cent by pests and pathogens in major field crops like wheat, rice, maize, potatoes, soybean and cotton (Oerke, 2006). Weeds offer serious threats to organic crop production, but very little attention has so far been paid to research on weed management in organic crop production (Barberi, 2002). Chemical intervention is not permitted for weed control purposes in organic farming systems. Concern about the potential increase in weed population without the use of herbicides has limited the uptake of organic farming (Bond and Grundi, 2001). However, as both public demands for organic produce and the profile of organic farming have increased in recent years, so too has the range of weed control options.

2. Weed Management Strategies

Weed control in organic crop production is a major problem because herbicides, the most effective tools, are not allowed for weed management, and hence the weed control is only based on cultural and mechanical methods. Manual weeding is the most laborious and uneconomical. Therefore, greater emphasis is being placed on weed prevention, cultural and mechanical methods including tillage and pre-and post- sowing inter-cultivation operations.

2.1 Weed Prevention

A majority of weeds were introduced from other countries and other areas of human emigration. On individual farms, weeds are introduced from neighboring farms along with grazing animals, field implements and machinery, from seeds, from road margin to field and from field to field through irrigation water, tractor and farm implements, *etc.* This process can be slowed, but not completely checked, by careful prevention and sanitation measures. The prevention of a weed problem is usually easier and less costly than control or eradication. The following measures can be suggested to prevent the introduction of weeds into non-inhabited field:

- ☆ Use 'clean' (weed seed-free) crop seed for planting
- ☆ Use organic manures only after thorough decomposition to kill weed seeds
- ☆ Clean harvesters and tillage implements before moving to non-weed infested area
- ☆ Avoid transportation or use of soil from weed infested area
- ☆ Inspect nursery stock or transplant for seed and vegetative propagules of weeds.
- ☆ Remove weeds that are near irrigation ditches, fence rows, right-of-way and other non-crop land
- ☆ Prevent reproduction of weeds
- ☆ Use weed seed screens to filter irrigation water.
- ☆ Restrict live stock movement in to non-weed infested area.

Other practices used to prevent and avoid potential weed problem at the state, regional or national level are weed laws, seed laws and quarantines. For example, sowing of wheat seed contaminated with *P. minor* seed is a major cause of spreading *P. minor* in wheat. It was reported that majority of farmer's seed drill boxes contained *P. minor* seeds in it (Chhokar *et al.*, 2012). Hence, use of clean and certified seed is desired. Combine harvesters moving from one field to another or even from one state to other states also contribute in *Phalaris* seeds dispersal. Therefore, it is also important to clean-up the farm machinery before introducing it in new areas/fields.

2.2 Cultural Methods

In organic farming practices, the approach to weed management involves the whole cropping system. The aim is not the eradication of weeds but to maintain a balance between crop plants and the weeds, with the grower adjusting the balance in favour of the crops whenever possible. Cultural practices that shift the balance of competition towards the crop usually will disfavour weed occurrence and improve crop yields. Crop rotation is at the heart of the organic system and although the method and timing of soil cultivations and choice of crop are usually linked, both contribute in different ways to manipulation of weed population. Any practice that provides vigorous uniform crop establishment usually will assist in reducing weed prevalence. Factors that improve crop competitiveness include:

2.2.1 Stale Seed-Bed Technique

The principle of flushing out germinabale weed seeds before cropping forms the basis of the stale or false seed bed technique, in which soil cultivation may take place days or weeks before planting or transplanting a crop. This depletes the seed bank in the surface layer of the soil and reduces subsequent weed emergence. Where light rains occur for an extended period before the onset of the monsoon, or irrigation is available, it may be possible to kill several flushes of weed growth before planting. To ensure success, cropping should be delayed until the main flush of emergence has passed. The emerged weed seedlings may then be killed by flaming or light cultivation. It is vital not to cultivate below the top 1-2 cm soil otherwise a further flush of weeds may emerge. The main advantage of the stale seedbed practice is that the crop emerges in a weed-free environment, with a competitive advantage over late-emerging weed seedlings. The practice of false seed bed technique may decrease weed infestation in crops by 80 per cent or more compared to standard seedbed preparation (Van der Weide *et al.*, 2002).

2.2.2 Tillage

Weed problem begins with weed seed in soil. It is the tillage which directly affects the weed seed bank by physical mixing of soil and in turn reducing the weed number. Tillage influences the vertical distribution of weed seeds in soil layer. Zero/minimum tillage favours weed seed build up in upper soil layers, while conventional tillage distributes the weed seeds in different soil layers. Weed seeds buried deep germinate but fail to emerge due to thick soil layer over it, resulting in death of weed seedling. Zero tillage reduces the emergence of *P. minor* because the seeds present in lower soil layer fail to germinate due to mechanical impedance. However, continuous use of ZT may shift the weed flora in favour of *Rumex dentatus* and *Malva parviflora* (Chhokar *et al.*, 2007a). Reduced tillage favours growth of *Agropyron repens*, *Cirsium arvense* and *Convolvulus arvensis* (Koch and Hess, 1980; Catizone *et al.*, 1990).

2.2.3 No Tillage and Night Tillage

A light requirement for seed to be released from dormancy is a wide spread mechanism found in many weed species. This requirement can be satisfied even after a brief exposure to light, such as that resulting from tillage. Thus, germination of weed species whose seed require light should be impeded if a no tillage system were implemented for crop production. Alternatively it is possible that night cultivation also could result in significant reduction in weed density (Scopel *et al.*, 1994). The problem of *P. minor* is comparatively less, as the weed seed lying at lower depths does not come to the surface as in case of conventional tillage. Infestation is also comparatively less due to non-availability of nutrients from well placed fertilizer near to the crop seed. For many buried seeds, it is the exposure to light during soil cultivation that stimulates germination. The exclusion of light during seedbed preparation has been shown to reduce weed emergence. Covering an implement to prevent light reaching the soil at the point of cultivation may be sufficient to reduce weed emergence by up to 70 per cent (Ascard, 1994). However, not all weed species have light-sensitive seeds. Welsh *et al.* (1999) observed the reduced density of *Chenopoduim album* by cultivating in darkness.

2.2.4 Furrow Irrigated Raised Bed Planting System (FIRBS)

It is a method where cultivation of crops in done on raised beds. In this system the problem of *Phalais minor* was reported to be very less as the weed seeds lying on top of the raised beds fail to germinate as the top of bed dry out quickly. This method also facilitates mechanical weeding as the area in the furrows can be easily cultivated and even manual weeding can be easily done (Chauhan *et al.*, 2003).

2.2.5 Crop Rotation and Cover Cropping

The composition and density of weed seed banks are frequently a reflection of long term crop rotations and management systems. Cropping sequences that provide varying patterns of resource competition, allelopathic interference, soil disturbance and mechanical damage provide an unstable environment that prevents the proliferation and dominance of a particular weed. The growth and reproduction of a troublesome weed species may be actively discouraged by introducing unfavorable practices in to a rotation. The inclusion of cover crops in the rotation at a time when land might otherwise lie uncropped will suppress weed development, yet maintain soil fertility and prevent erosion. In addition, the decomposing cover crop residues may release allele-chemical that inhibits the germination and development of weed seeds.

In monocropping system, numerous weed species persist and expand rapidly but crop rotation helps in interrupting life cycle of weeds and prevents any weed species to become dominant *e.g. Sorghum halepense* become predominant weed in continuous maize but can be controlled by rotating with cotton (Dale and Chandler, 1979). Certain weed species often associated with particular corps (eg. Barnyard grass in rice, wild oat and *Phalaris minor* in wheat, dodder in alfalfa *etc.*). Therefore, population of such weeds usually will increase whenever the crop is grown on the same field continuously for several years. Growing in sequence crops may discourage weed associations with crops that have sharply contrasting growth and cultural requirements. Incidence of *Phalaris minor* was higher in rice-wheat rotation than in maize-wheat rotation (Sen, 1981). Rotation of wheat with sorghum has been found quite effective in reducing wild oat seed reservoir in the soil (Martin and Felton, 1990). Rotating the wheat fields with other crops like sunflower, sugarcane or berseem helps in reducing the population of *P. minor*. Replacement of wheat by potato and vegetable pea in rice-wheat sequence can also help in *P. minor* management and improve system productivity (Chhokar *et al.*, 2012). Parasitic weeds can also be managed through crop rotation by rotating host crop with trap crops, as they induce germination of weed seeds but they themselves are not parasitized.

2.2.6 Intercropping

Intercropping suppresses weeds better than sole cropping and thus provides an opportunity to utilize crops themselves as tools of weed management. When two or more crops are grown together as intercrops, the total weed suppressing ability of a system will be higher cropping (Rao and Shetty, 1976). Weed suppressing ability of intercrops is dependent upon such factors as the component crops and cultivars selected, crop density, relative proportion of the component crops, their spatial arrangement and the fertility and moisture status of the soil. Many short duration pulses like, cowpea, greengram and soybean effectively smoother weeds without causing reduction in the yield of main crop. Patel *et al.* (1983) found that intercrops suppressed the weeds and increased the total productivity (Table 7.1).

Table 7.1: Effect of Intercropping in Pigeonpea on Weed and Yield of Crop

Crops	*Weed Dry Weight (g/m^2) at 60 DAS*	*Seed Yield (kg/ha)*	
		Pigeonpea	*Intercrop*
Pigeonpea sole	46	652	-
Pigeonpea + sorghum	18	593	514
Pigeonpea + maize	28	565	289
Pigeonpea + rice	32	597	125
Pigeonpea + blackgram	17	460	587
Pigeonpea + greengram	13	422	411
Pigeonpea + groundnut	29	592	592

2.2.7 Mulching

Mulch is a protective covering of materials maintained or placed on soil surface. Materials used for mulch includes crop residues, straw, leaves, paper plastic films, gravel and dry soil. Mulching the soil surface can prevent weed seeds germination or physically suppress seedling emergence, but is not effective against established perennial weeds.

Living mulch consists of a dense stand of low growing-species, established before or after the crops. In cereals, an under story of clovers (*Trifolium* spp.) has been shown to improve soil fertility, and reduce pest and disease problems in addition to suppressing weeds (George, *et al.*, 1997). A free-floating fern, *Azolla pinnata*, is used in low land and irrigated rice as a sort of companion crop. This fern has a symbiotic relationship with *Azolla anabenea*, nitrogen fixing blue green algae. This symbiotic relationship not only contributes up to 100 kg N/ha but also reduces weed growth by forming an *Azolla* blanket over the water surface in a rice crop. The

Azolla technique suppresses many annual weeds while leaving out certain annuals, which have strong culms (*e.g.* barnyard grass) and perennial weeds.

2.2.8 Crop Residue Management

Crop residues present on the soil surface can influence weed seed germination and seedling emergence by interfering with sunlight availability and creating physical impedance, as well as improving soil and moisture conservation and soil tilth (Locke and Bryson, 1997). In rice-wheat cropping system, management of crop residues left after combine harvest plays a major role on weed seedling emergence and efficacy of the herbicides. Burning of rice straw before wheat seeding is a common practice in northern India. Retaining crop residues on soil surface instead of burning will help in moisture conservation and weed suppression. Retaining a residue load of 5.0 and 7.5 t/ha reduced the weed infestation by 27.2 and 40.2 per cent in wheat (Chhokar *et al.*, 2009).

2.2.9 Competitive Crops and Cultivars

Crops differ in relative growth rate, spreading habit, height, canopy structure and inherent competitive character and accordingly differ in their weed suppressing ability. A quick growing and early canopy-producing crop would be expected to be better competitors against weeds than crops lacking these characters. Tall cultivars compete better for sun light than short cultivars. Using high yielding crop variety competitive against weeds in combination with other methods of weed control is one of the most economical approaches to attain optimal crop yield. Upland rice cultivars Vandana, Kalinga-III and RR-151-3 have shown better weed competitive ability and higher yield potential under sub-optimal weed management condition (ICAR, 2007). Tall wheat genotypes have greater weed suppression on the growth and development of weeds than do dwarf genotypes (Paul and Singh, 1979, Gill and Mehra, 1981).

2.2.10 Sowing Time

Planting time considerably influences the occurrence and manifestation of weed species. In timely sown chickpea, the weed population by 30 days is generally quite high to attract manual weeding whereas in late sown chickpea the buildup of adequate weed flora is only after 45 days. Delayed sowing of lentil and chickpea is also reported to reduce the infestation of *Orobanche* – a root parasite (Linke and Saxena, 1989). Malik *et al.* (1988) reported that in chickpea, the maximum emergence of most competitive weed, *Chenopodium album* L. occurred when crop was sown on November 5 and declined gradually with delay in sowing. However, in most winter pulses this cannot be a viable approach as delayed sowing invariably results in reduced yield. Sinha *et al.* (1988) reported that early sowing (10 August) and closer row spacing (30 cm) reduced the weed growth and increased the dry matter accumulation, LAI, NAR, CGR and grain yield of irrigated pigeonpea at Kalyani (West Bengal). Early planting of wheat encourages the problem of wild oat and broad-leaved weeds while late sowing increases the infestation of *P. minor*. Hence, the planting time should be so adjusted that it is unfavorable for the weed growth without reducing the crop yields.

2.2.11 Plant Geometry and Plant Density

Planting density and pattern modify crop canopy structure and in turn influence weed smothering ability. Narrow row spacing will bring variation in microclimate *viz*; light intensity, evaporation and temperature at soil surface. The increased shading of soil surface will smother weed growth. Increased population of maize (60,000 to 90,000 plants/ha) and decreased spacing were found to reduce the growth of weeds (Angiras and Singh, 1989). Bi-directional sowing and closer row spacing (15cm) were quite effective in suppressing the growth of *Phalaris minor* in wheat (Azad *et al.*, 1988). Mongia *et al.* (2005) reported lower weed density and higher wheat grain yield under closer row spacing (15 cm) and cross sowing (22.5 cm) as compared to normal sowing (22.5 cm).

2.2.12 Nutrient Management

Crops and weeds generally compete for the same nutrient pool. Increasing the levels of soil fertility can alter the competitive interactions between crops and weeds. Resource use by weeds often increases more rapidly with added nutrients, resulting in a greater ability of weeds to compete for other resources. Hence, nutrient management may be an important component of integrated weed management systems (Blackshaw *et al.*, 2005). Nutrient management in organic crop production involves crop rotation, cover cropping and addition of organic manures like compost, FYM, *etc.* Different organic sources of nutrients may change the weed dynamics and crop-weed interaction. Addition of organic manures as a source of nutrients can be a source of weed infestation. In most cases, addition of organic manures such as FYM and compost in fields leads to severe weed infestation in organic farming system (Dastgheib, 1989). Efthimiadou *et al.* (2012) reported the highest number and dry weight of weeds with addition of cow manure. The cow manure treatments promoted weed emergence and growth proportionally to the rate of application. Among the manures used, poultry manure has high N content and good weed control efficacy due to its phytotoxic character (Baig *et al.*, 2001). Haidar and Sidahmed (2006) have also reported that chicken manure was effective in reducing parasitic weed *Orobanche ramose.* Addition of crop residues has been found to reduce the weed biomass and density. Amanullah *et al.* (2006) have also reported the lowest weed dry matter for poultry manure followed by composted poultry manure (poultry manure and chopped sorghum straw). Mulching with barley residue reduced the weed biomass and density of redroot pigweed (*Amaranthus retroflexus* L.), common purslane (*Portulaca oleracea* L.) and prostrate knotweed (*Polygonum aviculare* L.) in sweet maize (Efthimiadou *et al.*, 2012). Use of soil amendments (cattle manure and potato compost and alternating years of legume green manure) substantially reduced the weed biomass, possibly by improving crop competitiveness (Gallandt *et al.*, 1998).

2.2.13 Water Management

Proper water management can reduce weeds in both wet and dry seasons. More weeds emerge earlier at high soil moisture than at low moisture level. Many weeds cannot germinate under flooded conditions. This is used as weed control method especially in rice crop. Maintaining 5 cm of water in rice field throughout the crop season creates anaerobic condition, which will be unfavourable for weed seed germination. Bhagat *et al.* (1999) concluded that irrespective of herbicide application, continuous shallow water ponding up to panicle initiation was effective in reducing weed diversity, density and biomass compared with saturated soil maintained throughout the life cycle of the crop. Balasubramanian and Krisharajan (2001) similarly recorded lower weed growth with continuous submergence in wet-seeded rice. Method of irrigation also influences the weed density. In drip irrigation the weed density is greatly reduced due to the wetting of soil near the plant base only. Rest of the area remains dry and free of weeds.

2.2.14 Soil Solarization

Soils act as a reservoir of weed seeds and therefore, the importance of soil solarization in destroying this seed reserve is increasingly being recognized. Solarization is a method of heating the surface by using plastic sheets placed on moist soil to trap the soil radiation. The process would raise the surface soil temperature by 8-10°C as compared to non-solarized soils. Duration of 4-6 weeks is sufficient to give satisfactory control of weeds. Many annuals, some perennials and parasitic weeds are sensitive to this treatment. Besides controlling soil borne pests, soil solarization is reported to enhance the availability of nutrients in soil and favour beneficial microflora, ultimately resulting in increased plant growth response in many crops. Chauhan *et al.* (1988) reported that solarization markedly decreased weed growth. Most of the annual weeds were effectively suppressed by solarization, however, the perennials such as *Cynodon dactylon, Cyperus rotundus* and *Convolvulus arvensis,* gradually recovered probably due to their deep rooted vegetative propogules in the soil. Variability in weed control due to soil solarization is influenced by soil type, temperature, moisture content, size and location of seeds or vegetative propagules in soil, besides the quality of polyethylene films. Sauerborn *et al.* (1989) reported significant decrease in *Orobanche* population in fababean and lentil with 20-50 days of solarization treatment.

2.3 Manual and Mechanical Weed Control

Manual weeding using traditional hand weeding tools is still a common method of weed control on small-scale crops in most of the developing countries. Hand weeding is feasible on small areas where sufficient work-force is available at relatively cheaper rates, and in high value crops. Use of mechanical weed control demands experience and knowledge of how to use the soil as a weed control agent. Climate and soil type play an important role in the possibilities for mechanical weed control. Monitoring the early development of weeds is necessary for timing of weed harrowing at the optimum stage. Mechanical weeders include cultivating tools such as hoes, harrows, times brush weeders, and cutting tools like mowers. The success of mechanical weeding depends upon the stage of weeds, crop geometry and climatic conditions. Hand weeding may also be used after mechanical inter row weeding to deal with weeds left in crop rows. In direct-seeded rice under upland conditions, weeds are controlled by tools like *Khurpi* or hand hoe/blade hoe/wheel hoe. In line sown/transplanted rice, rotary weeder or cono-weeder is also effective in controlling weeds. Mechanical weeding (20 DAS) followed by hand weeding (35 DAS) was found effective in weed management (Nair *et al.*, 2002). A dry land weeder (with a straight line peg arrangement) has shown excellent performance across a range of soil types with varying soil moisture levels and weed intensity, providing a labor saving of approximately 57 per cent compared with hand weeding (Subudhi, 2004).

2.3.1 Flame Cultivation

It is another successful alternative method that is catching up in the developed countries in the recent past. Flame cultivation involves the use of liquefied petroleum gas, mostly propane for pre-emergence and post-emergence weed control. Pre-emergence flaming is relatively problem-free, although the timing is critical. Selective flaming in the growing crop requires precision and carries a great risk of damage to the crop. Flaming equipments to burn off weeds has been developed in several counters including Germany, Holland, Sweden and Denmark (Holmoy and Netland, 1994).

2.4 Biological Control

The control of pests through biological control agents has been considered as the ideal method for managing pests under organic farming. No doubt, it is environmentally benign and ecologically and socially acceptable. However, their efficacy and economics are questionable. In weed management, their application is limited as the target weeds are many and the bio control agents are not expected to be effective on all of them. Moreover, control of one weed will result in substitution with other weeds, some of which could be more dangerous than the one that has been controlled. The use of insect pests as biocontrol agents has application in situations where only one weed is a problem such as in aquatic ecosystem, range lands, forestry or waste lands. The control of *Parthenium hysterophorous* with Mexican beetle *Zygogramma bicolorata* and the control of water hyacinth with weevil *Neochetina* spp. are a couple of successful examples of use insect pests for weed control. The Mexican beetle is very effective on *Parthenium* and excellent control of the weed has been reported from several places in India. However, the insect is active only during the rainy season and it hibernates in soil during the dry and winter months. Several species of fungi have been developed as mycoherbicides in developed countries for the management of few problematic weeds. However, they have a number of limitations and their use is limited to a very small area. *Colletotrichum gloesporioides* f. sp. *aeschynomene* is currently being tested as a mycoherbicide for the control of northern jointvetch (*Aeschynomene virginica*).

2.5 Green Manuring and Allelopathy

Green manuring is an alternative method of weed management. Green manures which are incorporated into the soil, manages the weeds through suppression and allelopathy. Green manure of *Brassica juncea* L. and *Sinapis alba* L. (Norsworthy *et al.*, 2005); sunflower *Helianthus annus* L. and *Sesbania aculeata* Poiret (Om *et al.*, 2002), or red clover (*Trifolium pratense* L.(Ohno *et al.*, 2000) were found effecting in weed management. The allelopathic properties of plants can be exploited successfully as a tool for weed reduction. Cereals have been reported to acquire allelopathic potential that affects the growth of other species (Dhima *et al.*, 2006; Zotarelli *et al.*,

2009). Bunna *et al.* (2011) reported that mulching of rice straw at 1.5 t/ha reduced weed biomass from 164 to 123 kg/ha and increased mungbean yield from 228 to 332 kg/ha. Moreover, Tabaglio *et al.* (2008) reported that rye mulches suppressed the emergence and growth of redroot pigweed by 40 per cent to 52 per cent and common purslane by 40 per cent to 74 per cent. Cover crops such as hairy vetch, ryegrass and oat could be used to reduce weed infestation (Isik *et al.*, 2009).

2.6 Integrated Weed Management

Although manual weeding is the most predominant practice followed in India, there is ample scope for reducing the labour requirement by integrating several preventive, mechanical and cultural methods of weed management. Integrated weed management (IWM) system is also more effective and ecologically more sustainable. However, this system is more knowledge intensive requiring better understanding of the weed biology and eco-physiology of crop-weed interaction. IWM practices should take in to account all aspects of the cropping system as a part of integrated crop management (ICM). The basic principle in IWM is to manipulate the crop-weed relationship in favour of the crop at the expense of the weed (Walker and Buchanan, 1982). Some of the following points may be considered while evolving the IWM practices for organic crop production:

- Greater emphasis on preventive methods such as weed-free crop seed, well-decomposed FYM, clean farm machinery *etc.*
- Summer tillage particularly in areas where perennial weeds are a problem
- Repeated shallow cultivations after rain or irrigation to create a weed-free seed bed
- Inclusion of green manure crops wherever possible
- Selection of competitive cultivars
- Maintaining optimum plant population
- Mechanical weeding wherever possible
- Growing green manure or fast growing cover crops particularly in orchards during rainy season
- Drip irrigation wherever possible
- Diversification and rotation of crops
- Mulching with organic or inorganic mulches
- Soil solarization in nurseries and in high value crops
- Transplanting in place of direct seeding wherever possible.

2.7 Utilization of Weeds

Weeds like Water hyacinth, Eupatorium, Lantana, Parthenium, Ipomoea, *etc.*, are widely spreading in farm environment, ponds, forest plantation, grassland and roadside at the cost of many other useful vegetation. Proper utilization of such biomass through appropriate technology may help in supplementing chemical fertilizers and adding organic matter to the soil. The tender and succulent portion of many of the dominant weeds with wider C:N ratio may also be utilized for vermicompost preparation which has enough potentiality for use in the production of organic tea as well as for vegetable farming.

3. Conclusion

Weeds offer a serious threat to organic crop production. However, very little effort has so far been made to research on weed management in organic agriculture. Weed management in organic crop production requires extensive planning and management. It must involve the many techniques and strategies, all with the goal of achieving economically acceptable weed control and crop yields. An integrated approach involving all cultural practices, aiming to optimize the whole cropping system rather than the weed control *per se*. It is stressed that manual and mechanical methods of weed control can only be successful where preventive and cultural weed management is applied to reduce weed emergence (*e.g.* through appropriate choice of crop sequence, tillage, smother/cover crops) and improve crop competitive ability (*e.g.* through appropriate choice of crop genotype, sowing/planting pattern and fertilization and irrigation strategies).

REFERENCES

Amanullah MM, Alagesan A, Vairapuri K, Sathyamoorthi K and Pazhanivelan S. 2006. Effect of intercropping and organic manure on weed control and performance of cassava (*Manihot esculenta* Crantz.). Journal of Agronomy 5, 589-594.

Angiras NN and Singh CM. 1989. Effect of weed control methods, cropping systems, plant population and soil fertility levels on nutrient removal and soil fertility status in maize. Indian Journal of Weed Science, 21: 69-74.

Ascard J. 1994. Soil cultivation in darkness reduced weed emergence. Acta Horticulturae 372, 167-178.

Azad BS, Singh H and Dubey AN 1988. Control of *Phalaris minor* in wheat by cultural and chemical methods. Pesticides, 22 (6): 41-42

Baig MK, Nanjappa HV and Ramachandrappa BK. 2001. Weed dynamics due to different organic sources of nutrients and their effect on growth and yield of maize. Research on Crops 2, 283-288.

Balasubramanian R and Krishnarajan J. 2001. Weed population and biomass in direct- seeded rice (*Oryza sativa*) as influenced by irrigation. Indian Journal of Agronomy, 46: 101-106.

Barberi P. 2002. Weed management in organic agriculture: are we addressing the right issues. Weed Research, 42 (3): 177-193.

Bhagat RM, Bhuiyan SI, Moody K and Estorninos LE. 1999. Effect of water, tillage and herbicide on ecology of weed communities in intensive wet-seeded rice ecosystem. Crop Protection, 18: 293-303.

Bhowmik PC. 1998. Herbicides in relation to food security and environment: a global perspective. Proceedings First International Agronomy Congress, Nov., 23-27, 1998, New Delhi, India. pp. 244-258.

Blackshaw RE, Molnar LJ and Larney FJ. 2005. Fertilizer, manure and compost effects on weed growth and competition in western Canada. Crop Protection, 24: 971-980.

Bond W and Grundy AC. 2001. Non-chemical weed management in organic farming systems. Weed Research, 41: 383-405.

Bunna S, Sinath P, Makara O, Mitchell J and Fukai S. 2011. Effects of straw mulch on mungbean yield in rice fields with strongly compacted soils. Field Crops Research, 124: 295-301.

Catizone P, Tedeschi M and Baldoni G. 1990. Influence of crop management on weed population and wheat yield. Proceeding of EWRS Symposium, Helsinki, Finland, 4-6 June.

Chauhan DS, Sharma RK and Chhokar RS 2003. Comparative performance of tillage options in wheat (*Triticum aestivum*) productivity and weed management. Indian Journal of Agricultural Sciences, 73(7): 402-406.

Chauhan DS, Sharma RK and Chhokar RS 2003. Comparative performance of tillage options in wheat (*Triticum aestivum*) productivity and weed management. Indian Journal of Agricultural Science, 73(7): 402-406.

Chauhan YS, Nene YL, Johansen C, Haware MP, Saxena NP, Sardar Singh, Sharma SB, Sahrawat KL, Burford JR, Rupela OP, Kumar Rao JVDK and Sithanantham S. 1988. Effect of solarization on pigeonpea and chickpea. Research Bulletin. No. 11, Patencheru, A.P. India, ICRISAT.

Chhokar RS, Sharma RK and Sharma I. 2012. Weed management strategies in wheat: A review. Journal of Wheat Research, 4(2): 1-21.

Chhokar RS, Sharma RK, Pundir AK and Singh RK 2007a. Evaluation of herbicides for control of *Rumex dentatus*, *Convolvulus arvensis* and *Malva parviflora*. Indian Journal of Weed Science, 39: 214-218.

Chhokar RS, Singh S, Sharma RK and Singh M 2009. Influence of straw management on *Phalaris minor* control. Indian Journal of Weed Science, 41: 150-156.

Dale JE and Chandler JM. 1979. Herbicide crop rotation for Johnson grass (*Sorghum halepense*) control. Weed Science, 27: 479-485

Dastgheib F. 1989. Relative importance of crop seed, manure and irrigation water as sources of weed infestation. Weed Research, 29: 113–116.

Dhima KV, Vasilakoglou IB, Eleftherohorinos IG and Lithourgidis AS. 2006. Allelopathic potential of winter cereal cover crop mulches on grass weed suppression and sugarbeet development. Crop Science, 46: 1682-1691.

Efthimiadou A, Froud-Williams RJ, Eleftherohorinos I, Karkanis A and Bilalis DJ. 2012. Effect of organic and inorganic amendments on weed management in sweet maize. International Journal of Plant Production, 6(3): 291-307.

Gallandt ER, Liebman M Corson S. Porter GA and Ullrich SD. 1998. Effects of pest and soil management systems on weed dynamics in potato. Weed Science, 46: 238-248.

George SK, Donaldson G and Clements RO. 1997. Cover cereal bi-cropping with reduced agrochemical inputs; the yield and costs. Optimising cereal inputs: its csientific basis, Aspects of Applied Biology, 50: 487-492.

Gill HS and Mehra SP. 1981. Control of wild oats with dichlofop methyl (Illoxon) in irrigated wheat. In: Proceedings of 8th Asian Pacific Weed Science Society Conference, Bangalore, pp. 17-22.

Haidar MA, Sidahmed MM. 2006. Elemental and chicken manure for the control of branched broomrape (*Orobanche ramosa*). Crop Protection, 25: 47-51.

Holmoy R and Netland J. 1994. Band spraying, selective flame weeding and hoeing in late white cabbage, part 1. Acta Horticulturae, 372: 223-234.

ICAR. 2007. Vision 2025.NRCWS Perspective plan. Indian Council of Agricultural Research, New Delhi, India.

Isik D, Kaya E, Ngouajio M and Mennan H. 2009. Weed suppression in organic pepper (*Capsicum annuum* L.) with winter cover crops. Crop Protection, 28: 356-363.

Koch W and Hess M. 1980. Weeds in wheat. Wheat Documenta, Ciba-Geigy. Ciba_Geigy Ltd., Basle, Switzerland, pp. 33-40.

Linke KH and Saxena MC. 1989. Proceedings of International Workshop on Orobanche Oberrenchal. Germany, Aug, 19-22,1989.

Locke MA and Bryson CT. 1997. Herbicide-soil interactions in reduced tillage and plant residue management systems. Weed Science, 45: 307-320

Malik RK, Balyan RS and Bhan VM. 1988. Effect of herbicides and planting dates of weeds in chickpea. Indian Journal of Weed Science, 20(2): 75-81.

Martin RJ and Felton WL. 1990. Effect of crop rotation, tillage practices and herbicide use on the population dynamics of wild oats. Proceedings of 9th Australian Weeds Conference, pp. 20-23.

Mongia AD, Sharma RK, Kharub AS, Tripathi SC, Chhokar RS, and Jag Shoran 2005. Coordinated research on wheat production technology in India. Karnal, India: Research Bulletin No. 20, Directorate of Wheat Research. p.

40.

Nair HG, Choubey NK and Srivastava GK. 2002. Influence of nitrogen and weed management practices on weed dynamics in direct-seeded rice. Indian Journal of Weed Science, 34: 134-136.

Norsworthy JK, Brandenberger L, Burgos NR and Riley M. 2005. Weed suppression in *Vigna unguiculata* with a spring-seeded Brassicaceae green manure. Crop Protection, 24: 441-447.

Oerke EC. 2006. Crop losses to pests: Centenary review. Journal of Agricultural

Sciences, 144: 31-43.

Ohno TK, Doolan K, Zibilske LM, Liebman M, Gallandt ER and Berube C. 2000. Phytotoxic effects of red clover amended soils on wild mustard seeding growth. Agriculture Ecosystems and Environment, 78: 187-192.

Om H, Dhiman S, Kumar S, Kumar H. 2002. Allelopathic response of *Phalaris minor*to crop and weed plants in rice-wheat system. Crop Protection, 21: 699-705.

Patel CL, Patel ZG and Patel HC. 1983. Non-cash technology for minimizing weed growth in pigeonpea. Indian Journal of Weed Science, 15(2): 143-46.

Paul S and Gill HS 1979. Ecology of *Phalaris minor* in wheat-crop ecosystem. Tropical Ecology, 20: 186-191.

Rao MR and, Shetty SVR. 1976. Some biological aspects of intercropping systems on crop-weed balance. Indian Journal of Weed Science, 8: 32-43.

Sauerborn J, Linke KH, Saxena MC and Kock W. 1989. Solarization: a physical control methods for weeds and parasitic plants in Mediterranean Agriculture. Weed Research, 29: 391-97.

Scopel AL, Ballare CL and Radosevich SR. 1994. Photo-stimulation of seed germination during soil tillage. New Phytolologist, 126: 145-152.

Sen DN. 1981. Ecological approaches to Indian weeds. Geobios International Publisher. Jodhpur. 301 p.

Sinha AC, Mandal BB and Jana PK. 1988. Physiological analysis of yield variation in irrigated pigeonpea in relation to time of sowing, row spacing and weed control measures. Indian Agriculturists, 32(2): 1777-85.

Subudhi ECR. 2004. Evaluation of weeding devices for upland rice in the Eastern Ghat of Orissa, India. International Rice Research Notes, 29: 79-81.

Tabaglio V, Gavazzi C, Schulz M and Marocco A. 2008. Alternative weed control using the allelopathic effect of natural benzoxazinoids from rye mulch. Agronomy for Sustainable Development, 28: 397-401.

Van Der Weide RY, Bleeker PO and Lotz LAP 2002. Simple innovations to improve the effect of the false seed bed technique. In: Proceedings of the 5th Workshop of the EWRS Working Group on Physical and Cultural Weed Control, Pisa, Italy, pp. 3-4.

Welsh JP, Philipps L, Bulson HAJ and Wolfe M. 1999. Weed control strategies for organic cereal crops. In: Proceedings 1999 Brighton Conference- Weeds, Brighton. U.K. pp. 945-950.

Zotarelli L, Avila L, Scholberg JMS and Alves BJR. 2009. Benefits of vetch and rye cover crops to sweet corn under no-tillage. Agronomy Journal, 101: 252-260.

Chapter 8

Management of Insect-Pests in Organic Crop Production

B. Shivanna

College of Agriculture, University of Agricultural Sciences, Bengaluru – 500 065, Karnataka
Email: shivannab@gmail.com

1. INTRODUCTION

Insects and pathogens are highly mobile and well adapted to farm production systems, pest control tactics and will continue to be major constraints to agricultural production throughout the world. In 1940's, Synthetic chemical pesticides were introduced and used widely on agricultural crops in the hope that they would control agricultural pests. It is now clear that, their use has some unfortunate consequences. Heavy agricultural reliance on synthetic chemical fertilizers and pesticides is having serious impacts on public health and the environment (Pimentel *et al.*, 2005). In the recent years, population of many pests has developed resistance to many commercially available pesticides (Ramakrishnan *et al.*, 1984; Rame Gowda, 1999). Many of these synthetic chemical pesticides are broad spectrum, killing not only arthropod pests but also beneficial organisms that serve as natural pest-control systems. Without benefit of the natural control that keep pest populations in check, growers become increasingly dependent on chemical pesticides to which pests may eventually develop resistance. In fact, pest resistance currently limits the efficacy of many insecticides, fungicides and herbicides and there are some pests for which no effective pesticides are available. Thus, there is an urgent need for an alternative approach to pest management that can complement and partially replace current chemically based pest management practices.

Though 'Green revolution' helped us to overcome domestic food deficits and ushered in an era of food security, sole reliance on an array of chemicals has led to several problems *viz.*, insecticide resistance, resurgence, residual hazards, lack of bio-diversity and replacement of natural enemies created imbalance in nature and resulted in outbreak of pests and diseases. To overcome all these problems, organic agriculture has emerged as a dynamic 'Alternate Farming System'.

In organic farming system, pest control strategies are largely preventive rather than reactive. It is a holistic system designed to optimize the productivity and fitness of diverse communities within the agro-ecosystem, including soil organisms, plants, livestock and people. The principal goal of organic production is to develop enterprises that are sustainable and harmonious with the environment.

Historically, organic wastes of animal origin were most commonly used for nutrition and plant protection. In addition, herbs processed in liquid excreta of animals were used for plant protection. Cow dung manure and liquid manure called Kunapajala were used universally. In this direction, scientific evaluation of biodynamic pesticides, botanicals and biorationals including indigenous technologies are considered very much essential to combat noxious pests of crop plants. Use of botanical pesticides for protecting crops from insect pests has assumed greater importance in recent years all over the world (Hiremath, 1994). Several organic technologies, if adopted in current conventional production systems, would most likely be beneficial. These include:

1. Employing off-season cover crops
2. Using more extended crop rotations, which act both to conserve soil and water resources and also to reduce insect, disease, and weed problems
3. Increasing the level of soil organic matter,which helps to conserve water resources and mitigates drought effects on crops and
4. Employing natural biodiversity to reduce or eliminate the use of nitrogen fertilizers, herbicides, insecticides, and fungicides. Some or all of these technologies have the potential to increase the ecological, energetic, and economic sustainability of all agricultural cropping systems, not only organic systems.

Integrated pest and nutrient management systems and certified organic agriculture can reduce reliance on agrochemical inputs as well as make agriculture environmentally and economically sound. The main aim of organic agriculture is to augment ecological processes that foster plant nutrition yet conserve soil and water resources. Organic systems eliminate agrochemicals and reduce other external inputs to improve the environment and farm economics. It is the most successful agri-environmental schemes, as humans benefit from high quality food, farmers from higher prices for their products and it often successfully protects biodiversity. However, there is little knowledge if organic farming also increases ecosystem services like pest control.

According to the organic standards, insect pest problems may be controlled through cultural, mechanical or physical methods; augmentation or introduction of predators or parasites of the pest species; development of habitat for natural enemies of pests; and non-synthetic controls, such as lures, traps, and repellents. When these practices are insufficient to prevent or control crop pests, a biological and botanical method of pest control would be beneficial. However, the conditions for using the material must be documented in the organic system plan. Pest management plans are site-specific. Farmers should develop their own strategies based on their knowledge, available time, and capital—the resources they can devote to pest management.

2. Pest Management in Organic Farming

Integrated Pest Management (IPM) practices are more crucial in organic farming systems as opposed to other farming systems because producers cannot use conventional insecticides for a quick fix.

On organic farms, where the focus is on managing insects rather than eliminating them, success depends on learning about three kinds of information:

1. Biological information: What the insect needs to survive can be used to determine if pest insects can be deprived of some vital resource.
2. Ecological information: How the insect interacts with the environment and other species can be used to shape a pest resistant environment.
3. Behavioral information: How the insect goes about collecting the necessities of life, can be manipulated to protect crops.

Organic farming systems rely on ecologically-based practices such as cultural and biological pest management, and virtually exclude the use of synthetic chemicals in crop production. Genetically modified crops are not allowed. Under organic farming systems, the fundamental components and natural processes of ecosystems, such as soil organism activities, nutrient cycling, and species distribution and competition, are used directly and indirectly as farm management tools and to prevent pest populations from reaching economically-damaging levels. For example, crops are rotated, planting and harvesting dates are carefully planned, and habitats that supply resources for beneficial organisms are provided. Soil fertility and crop nutrients are managed through tillage and cultivation practices, crop rotations, cover crops, and supplemented with manure, composts, crop waste material, and other allowed substances.

Pest management plans are site-specific. Farmers should develop their own strategies based on their knowledge, available time, and capital—the resources they can devote to pest management. It is the basic information on pest insects and the management tactics that organic farmers can use to keep pest insect populations at levels that do not pose an economic threat to crops. The following practices used together comprise an integrated pest management (IPM) approach.

2.1 Cultural Practices

The first key step toward managing pest insects is using cultural practices that suppress pest species and encourage their natural enemies.

Insect pest problems are influenced by three components of a farming system. Farmers can manipulate all these components to suppress pest species.

- ☆ The crop species and cultivar present a set of resources, growth habits, and structure.

- ✰ Production practices, such as rotation, timeliness of planting and harvesting, spacing of plants, fertility and water management, tillage, mulching, sanitation, and companion planting.
- ✰ Agro-ecosystem structure includes field borders, natural vegetation, and other crop production areas that resupply fields with pest insects and beneficial species when crops are replanted.

Protecting the Crop

Sanitary measures: Healthy planting material, clean seeds, clean tools *etc.*

Use of Resistant Varieties: Exploiting host resistance.

Cultural Measures

2.1.1 Higher seed rate

Higher seed rate will help to retain required plant population even after uprooting and destroying the infested plants by pests like, stem borer and shoot fly.

2.1.2 Planting distances

Difference in plant densities have many type of advantages in different cropping systems. For example, High plant density reduces necrosis of groundnut, low plant density reduces damping off in nursery and sorghum charcoal rot. Wider row sparing in rice reduces BPH in rice.

2.1.3 Crop rotation and intercropping

Intercropping, crop rotation (rotation of crops with non host plants) will help to reduce the incidence of many key pests in different ecosystems.

2.1.4 Use of trap crops

Trap crops have significant role to play in cropping systems with which many important or key or major pests can be managed. For instance, Castor is being recommended as trap crop against *Spodoptera* in groundnut and tobacco; Marigold as trap crop in tomato (16:1) against *Helicoverpa* will reduce the incidence. Similarly, Bhendi/Okra can be used as trap crop in cotton (10:1), which will help to trap the boll worms and stem weevils of cotton.

2.1.5 Use of fertilizers

Excessive use of fertilizer will result in succulent and vulnerable conditions for the attack of insect pests and diseases. Hence excessive use of N fertilizer in most of the crops will aggravate the pest problem; while use of organic manures will induce tolerance to ORA to pest and diseases.

2.2 Mechanical Methods

These methods involve collection and destruction of insect pests attacking crop plants. There are few instances where mechanical methods are being successfully deployed for managing the insect pests.

- Root grubs in adult stage can be collected during summer on the day of first rain at 7.30 to 9.00 pm with the help of petromax or fire and destroyed.
- The grown up caterpillars of *Helicoverpa, Spodoptera etc.* can be collected by hand picking and subsequently destroyed.
- Destruction of stubbles and crop residues - Stubbles of sorghum and maize known to give shelter to stem borer larva which continue their next generation after summer. Hence, destruction of such stubbles either by burning or by burying in the soil will kill the stem borer larvae or pupae.
- The egg masses and the early larval stages of *Spodoptera*, Bihar hairy caterpillar soon after hatching can be collected easily on tobacco, soybean, groundnut and other crops and destroyed to avoid the further spread of the pest.Similarly, theegg masses of stem borer of paddy and early shoot borer of sugarcane *etc.* can be removed by clipping the tips of the leaf and destroyed.
- Passing a rope across the paddy field will help to eliminate the leaf cases along with case worm larvae which will fall in water and die.

2.3 Biological Control

Biological control has a long history of use in pest management and has gained renewed interest because of problems encountered with the use of pesticides. The term 'biological control'has been used, at times, in a broad context to encompass a full spectrum of biological organisms and biologically based products including pheromones, resistant plant varieties, and autocidal techniques such as sterile insects. The historical and more prevalent use of this term is restricted to use of insect pathogens and natural enemies to manage populations of pest organisms (David *et al.*, 2005).

2.3.1 Using Insect Pathogens

Organisms that cause disease in insects can be exploited to help control pest populations by managing the environment to favor insect disease or by applying allowable purchased products. The use of insect pathogens to manage pests is called *microbial control.* Soil serves as a natural home and reservoir for many kinds of insect pathogens, including viruses, bacteria, protozoa, fungi, and nematodes. Pest control products based on insect pathogens are available commercially, and some products are allowed in organic production.

Major characteristics of insect pathogens:

- They kill, reduce reproduction, slow growth, or shorten the insect pests life
- Specific to target species or to specific life stages
- Their effectiveness may depend on environmental conditions or host abundance
- The degree of control by naturally occurring pathogens may be unpredictable

- ✰ They are relatively slow acting; they may take several days or longer to provide adequate control
- ✰ They may cause epizootics
- ✰ Microbial insecticides are compatible with the use of predators and parasitoids, which may help to spread some pathogens through the pest population. Beneficial insects are not usually affected directly because of the specificity of a microbial product, but some parasitoids may be affected indirectly, if parasitized hosts are killed. Insecticide applicators should note that although microbials are non-toxic to humans in the conventional sense, safety precautions should be followed to minimize exposure.

Insect-pathogenic Bacteria

Many insect diseases are caused by bacteria (Table 8.1). The most commonly used bacterial product available to organic growers is *Bacillus thuringiensis* (Bt). This bacterium produces an insecticidal protein that provides effective control for many pest insects and has very little effect on non-target insects and natural enemies. The commercial Bt products are powders containing a mixture of dried spores and toxin crystals.

Table 8.1: Some Important Bacteria and their Hosts

Bacteria	*Target Hroups*
Bacillus thuringiensis var. *kurstacki* (and others)	Caterpillars
B.thuringiensis var. *israelensis*	Aquatic flies
B.thuringiensis var. *tenebrionis*	Beetles
B.sphaericus	Mosquitoes
B. papillae	Scarab beetles
B. lentimorbus	Scarab beetles
B. larvae	Bees

Mode of action: Its activity is due primarily to endo (crystal protein) and exotoxin production which affects the insect gut. A high gut pH is needed to activate the crystal protein. Spores germinate after insect affected by toxins, then propagate. The insect dies as spores and gut bacteria proliferate in the body.

Insect-Parasitic Fungi

Fungi are considered to play vital role as biological control agent of insect populations. A very diverse array of fungal species are found from different classes that infect insects (Table 8.2). More than 900 species of entomopathogenic fungi belong to 100 genera are recorded and only 10 species have been commercially exploited of which *Beauveria bassina, Metarhizium anisopliae, Nomuraea rileyi, Lecanicillium lacanii, Hirsutella thompsoni,* and *Paecilomycis* sp. are popular in India. Specific fungal strains in commercial products target wide range of insects *viz.,* thrips, whiteflies, aphids, caterpillars, weevils, grasshoppers, ants, Colorado potato beetles, and mealybugs. Most fungi can cause natural outbreaks (epizootics) when environmental conditions are favourable.

Table 8.2: Some Important Fungi and Common Hosts

Fungi	*Target Host*
Entomophthora muscae	Flies
Erynianeo aphidis	Aphids
Beauveria bassiana	Many insects
Metarhizium anisopliae	Mainly coleopterans
Coelomomyces spp.	Mosquitoes and bugs
Hirsutella thompsonii	Mites
Verticillium lecanii	Sucking insects/pests
Nomurea rileyii	Lepidopteran caterpillars

Mode of action: Fungi invade insects by penetrating their cuticle or "skin." Once inside the insect, the fungus rapidly multiplies throughout the body. Death is caused by tissue destruction and, occasionally, by toxins produced by the fungus. The fungus frequently emerges from the insect's body to produce spores that, when spread by wind, rain, or contact with other insects, can spread infection.

Beauveria bassina

B. bassiana is one of the most widespread pathogens of insects. It has been exploited to many pests like root grubs, caterpillars, BPH, coffee berry borers, *Helicoverpa etc.* Several species have been developed as commercial products because of their ability to mass produced. Specific fungal strains in commercial products target thrips, whiteflies, aphids, caterpillars, weevils, grasshoppers, ants, Colorado potato beetles, and mealybugs.

Metarhizium anisopliae

A strain of *M. anisopliae* (Metschnikoff) found in Africa, is very infective to grasshoppers and locusts. By mixing the conidia with oil, infectivity of the pathogen is maintained even at low humidity, allowing the pathogen to be applied effectively in the field even in arid environments. It is known in Africa as green muscle disease. It can infect more than 300 insect hosts. The potential is proved beyond doubt in the management of different species of insects mainly coleopterans, grasshoppers, termites root grub, pyrilla of sugarcane, BPH in paddy, rhinocerous beetle, *etc.*

Lecanicillium lacanii

L. laccanii has been reported to be highly potent against sucking pests like aphids, thrips, scales, mealy bugs, hoppers, whiteflies, mites *etc.* This fungus could reduce the green bug infestation considerably in coffee plantation in South India.

Nomuraea rileyi

N. rileyi, an important fungal pathogen which causes natural mortality of few lepidopteran hosts like *Spodoptera, Helicoverpa,* semiloopers, cutworms, hairy caterpillars *etc.*, in high density cropping systemsin transitional tracts and high humid areas. These entomopathogens being facultative can easily and cheaply be mass multiplied on any carbon rich substrate especially on broken rice.

Insect Parasitic Nematodes

Some nematodes that are commercially available are *Steinernemacarpocapsae, S. feltiae, S. riobrave, Heterorhabditis bacteriophora* and *H. megidis* (Table 8.3). Treatment with these nematodes can be expensive, and they are most commonly used for managing soil insect pests in high-value crops, such as turf, nurseries, citrus, cranberries, and mushrooms, as well as in home lawns and gardens. As production technologies improve, however, the cost of using nematodes is falling, and it may be economical to use them in lower value crops.

Table 8.3: Some Appropriate Nematode Host Targets

Nematode Families and Species	*Target Groups*
Heterorhabditidae	
Heterorhabditis bacteriophora	Lepidoptera, Coleoptera
H. megidis	Coleoptera
H. marelatus	Coleoptera, Lepidoptera
Steinernematidae	
Steinernema carpocapsae	Lepidoptera, Coleoptera and Siphonaptera
S. feltiae	Diptera
S. glaseri	Coleoptera (Scarabaeidae)
S. kushidai	Coleoptera (Scarabaeidae)
S. riobrave	Lepidoptera, Orthoptera, Coleoptera (Curculionidae)
S. scapterisci	Orthoptera

Nearly 40 known families of nematodes parasitize and consume insects and other arthropods. Some are hunter-cruisers while others are ambushers. The most beneficial of these "entomopathogenic" nematodes belong to the Heterorhabditidae and Steinernematidae families. Both families are "obligate" parasites: their survival depends on their hosts and on the symbiotic relationships the nematodes have evolved with disease-causing *Xenorhabdus* and *Photorhabdus* bacteria.

Steinernema

It is the most widely used species for insect control. It is the most readily available for yard and garden use because it is easier to rear and handle. In field applications, *Steinernema carpocapsae* tend to be most effective against caterpillar larvae. In laboratory and field trials, it has controlled sod webworms, cutworms and certain borers (raspberry crown borer, carpenter worm). Another speices, *Steinernema feltiae* is effective against certain fly larvae. It is sometimes used to control fungus gnats and other soil dwelling insects of greenhouse-grown plants. They are less effective against white grubs, root maggots, rootworms and black vine weevil.

Heterorhabditis

It is less commonly available because it is more difficult to rear and more susceptible to environmental extremes. However, field trials consistently show that this genus outperforms *Steinernema* for control of white grubs. Several species are sold including *Heterorhabditis bacteriophora, H. megadis, and H. indica*. It is also effectively controls many nursery pests that feed in the root zone, such as black vine weevil and citrus-infesting root weevils.

Mode of action: These nematodes carry bacteria in their bodies that are toxic to insects. That's why they are called 'entomopathogenic' nematodes. Entompathogenic is the scientific term for 'insect-killing'. The nematodes and bacteria are always found together because they depend on each other. The bacteria need the nematodes to deliver them into the insects, and the nematodes need the bacteria for food and to create conditions in the insect that allow it to reproduce. The bacteria are safe to animals and have only been found in association with these nematodes and infected insects, never living freely in soil.

Insect-Parasitic Viruses

Insect viruses are obligate disease-causing organisms that can only reproduce within a host insect. They can provide safe, effective, and sustainable control of a variety of insect pests, although they are most effective as part of a diverse IPM program. All are highly specific in their host range, usually limited to a single type of insect. These currently include,

- Gemstar LC (Certis USA), a nuclear polyhedrosis virus of *Heliothis*and*Helicoverpa*spp. (Corn earworm, Tobacco budworm)
- Spod-X LC (Certis USA), a nuclear polyhedrosis virus of *Spodoptera*spp. (beet armyworm);
- CYD-X (Certis USA) and Virosoft CP4 (BioTEPP, Inc.), granulosis virus of *Cydiapomonella*, the codling moth; and
- CLV LC (Certis USA), a nuclear polyhedrosis virus of *Anagraphafalcipera*, the celery looper.

2.3.2 Natural Enemies using Insect

Farmers can manage their fields to provide habitats for species that eat and live on pest insects. This can be accomplished through conserving and augmenting beneficial populations. An important component of the biological environment is natural enemies or beneficiaries-the predators and parasitoids that dampen pest insect populations. Organic farmers often assume that withholding conventional pesticides will have a beneficial effect on population levels of species that weaken and kill pest insects. The absence of conventional pesticides will likely encourage the natural enemies of pest insects. But that encouragement may not be enough to provide substantive control of chronic pests without additional changes in the agro-ecosystem, which provide habitat for the pests *and* their natural enemies.

When considering strategies to enhance biological control, it may be helpful to look at them in two categories:

Conservation

These techniques *conserve* naturally occurring beneficial species. It involves managing crops and surrounding vegetation in a manner that encourages natural enemies and their negative impacts on pest species. This involves identifying anything that suppresses the natural enemies and modifying agricultural fields and surrounding vegetation to provide habitats for them. Some of the practices are followed to conserve natural enemies in the cropping area are intercropping, cover crops, field borders and hedgerows *etc.*

Augmentation

These techniques *augment* or supplement the actions of naturally occurring beneficial species. *Augmentation* involves adding something to an agro-ecosystem to regulate pest population densities. It usually involves either mass production of beneficials or genetic enhancement to improve their ability to survive negative influences. The most common example of augmentation is purchasing commercially available predators or parasitoids for release into greenhouses or commercial crop fields. If the natural enemy is released in large numbers (tens of thousands to hundreds of thousands), it is called an *inundative* release. If a few are released (hundreds to thousands) with the expectation they will reproduce and increase in number, then it is called an *inoculative* release.

Careful planning is critical to the success of a natural enemies release program. There are three important considerations when using natural enemies:

- ☆ The natural enemy selected should be correct for the specific situation and specific pest
- ☆ The quality or viability of the product purchased should be adequate
- ☆ The timing and rate of application should be correct

2.4 Pheromones and Other Attractants

Organic farmers can use chemical attractants to trap pest insects, disrupt their reproduction cycle, and monitor their population levels. Insects evolved many different ways of finding each other to mate. Some insects can make a sound as loud as a chainsaw; others have striking colors. Many insects find each other over long distances by emitting chemical signals or *pheromones* to attract individuals of the same species into an area so they can find each other to mate. Once the individuals get close together, visual cues-such as color, shape, and behavior-become more important. Entomologists have determined the chemical structure of pheromones for many pest species and duplicated them synthetically.

Insects also use other chemical messages. Chemical cues to the location of food can draw insects into a particular area where, once they get close enough, visual and tactile cues lead them to food sources. Pheromones and other chemical attractants can be used in several different ways:

- ☆ To monitor pests
- ☆ To disrupt mating and
- ☆ Mass Trapping

2.4.1 Pheromones to Monitor Insect Populations

Monitoring insect pests is one of the most effective uses for pheromones. They can be used to detect the first arrival of a pest insect species or an increase or decrease in populations. Attractants are used in several different types of traps (including sticky, wire, mesh, pan, and water traps) that all work on the same basic principle: Attract insects to the trap where they can be captured and counted. For the right species, with experience, pheromone traps can be effective and inexpensive monitoring tools.

2.4.2. Use of Pheromones to Disrupt Mating

Mating disruption works by permeating the crop environment with a chemical message that prevents pest insect adults from locating each other to mate. The logic of using pheromones to disrupt mating is simple: If enough males can be confused so they cannot find mates, then almost all females are unmated and fewer immature insects will result. Because there are fewer immature insects, crop damage is minimized or eliminated. In some cases this approach has worked well.

2.4.3. Use Pheromones for Mass Trapping

Pheromones can be so powerful that many or most of the adults insects in an area can be trapped and killed. It is easy to conclude that so many are caught that the number remaining cannot cause substantial crop loss. Insects, however, can mate repeatedly. So the remaining males will usually be sufficient to keep pest populations at economically damaging levels.

2.5. Natural Insecticides

Allowable organic and inorganic chemicals, insecticidal oils and soaps, microbial insecticides, particle films, and botanicals, when used in combination with the above pest management strategies, can help to suppress pest insect populations. They are produced by refining natural substances.

Natural insecticides fall under three major categories:

- ✰ Botanicals (Plant extracts)
- ✰ Biopesticides (Pesticides derived from natural materials as animals, plants, microorganisms and certain minerals)
- ✰ Biorationals (insect growth regulators/third generation insecticides)

2.5.1 Botanicals

Many plants and minerals have insecticidal properties; that is, they are toxic to insects. Botanical insecticides are naturally occurring chemicals (insect toxins) extracted or derived from plants or minerals (Table 8.4). They are also called natural insecticides. (Turney and McIlveen, 1983).

Pyrethrum and Pyrethrins

Pyrethrum is the powdered, dried flower head of the pyrethrum is daisy, *Chrysanthemum cinerariaefolium*. Most of the world's pyrethrum crop is grown in

Kenya. The term "pyrethrum" is the name for the crude flower dust itself, and the term "pyrethrins" refers to the six related insecticidal compounds that occur naturally in the crude material. They are extracted from crude pyrethrum dust as a resin that is used in the manufacture of various insecticidal products.

Mode of action: Pyrethrins exert their toxic effects by disrupting the sodium and potassium ion exchange process in insect nerve fibers and interrupting the normal transmission of nerve impulses. Pyrethrins insecticides are extremely fast acting and cause an immediate "knockdown" paralysis in insects. Despite their rapid toxic action, however, many insects are able to metabolize (break down) pyrethrins quickly. After a brief period of paralysis, these insects may recover rather than die. To prevent insects from metabolizing pyrethrins and recovering from poisoning, most products containing pyrethrins also contain the synergist, piperonylbutoxide (PBO). Without PBO the effectiveness of pyrethrins is greatly reduced.

Table 8.4: Botanicals Used as Natural Insecticides

Generic Name	*Oral LD50*	*Dermal LD50*	*Signal Word*
Pyrethrins	1,200-1,500	>1,800	Caution
Rotenone	60-1,500*	940-3,000	Caution
Sabadilla	4,000	—	Caution
Ryania	750-1,200	4,000	Caution
d-Limonene	>5,000	—	Caution
Linalool	2,440-3,180	3,578-8,374	Caution
Neem	13,000	—	Caution

Rotenone

Rotenone is insecticidal compound that occurs in the roots of *Loncho carpus* species in South America, Derris species in Asia, and several other related tropical legumes. Rotenone is extracted from cube roots in acetone or ether. Extraction produces a 2-40 per cent rotenone resin which contains several related but less insecticidal compounds known as rotenoids. The resin is used to make liquid concentrates or to impregnate inert dusts or other carriers. Most rotenone products are made from the complex resin rather than from purified rotenone itself. Alternatively, cube roots may be dried, powdered and mixed directly with an inert carrier to form an insecticidal dust.

Mode of action: Rotenone is a powerful inhibitor of cellular respiration, the process that converts nutrient compounds into energy at the cellular level. In insects rotenone exerts its toxic effects primarily on nerve and muscle cells, causing rapid cessation of feeding. Death occurs several hours to a few days after exposure. Rotenone is extremely toxic to fish, and is often used as a piscicide.

Sabadilla (Veratrine Alkaloids)

Sabadilla is derived from the ripe seeds *Schoenocaulon officinale*, a tropical lily plant which grows in Central and South America. When sabadilla seeds are aged, heated, or treated with alkali, several insecticidal alkaloids are formed or activated. These alkaloids are physiologically active compounds that occur naturally in many plants.

Mode of action: In insects, Sabadilla's toxic alkaloids affect nerve cell membrane action, causing loss of nerve cell membrane action, causing loss of nerve function, paralysis and death. Sabadilla kills insects of some species immediately, while others may survive in a state of paralysis for several days before dying.

Ryania

Ryania comes from the woody stems of *Ryania speciosa*, a South American shrub. Powdered Ryania stem wood is combined with carriers to produce a dust or is extracted to produce a liquid concentrate. The most active compound in ryania is the alkaloid ryanodine, which constitutes approximately 0.2 per cent of the dry weight of stem wood.

Mode of action: Ryania is a slow-acting stomach poison. Although it does not produce rapid knockdown paralysis, it does cause insects to stop feeding soon after ingesting it. Little has been published concerning its exact mode of action in insect systems.

Neem (Azadiractin)

Neem products are derived from the neem tree, *Azadirachta indica*, that grows in arid tropical and subtropical regions on several continents. The principle active compound in neem is azadirachtin, a bitter, complex chemical that is both a feeding deterrent and a growth regulator. Meliantriol, salannin, and many other minor components of neem are also active in various ways. Neem products include teas and dusts made from leaves and bark, extracts from whole fruits, seeds, or seed kernels, and an oil expressed from the seed kernel. The product known as "neem oil" is more like a vegetable or horticultural oil and acts to suffocate insects.

Mode of action: Neem is a complex mixture of biologically active materials, and it is difficult to pinpoint the exact modes of action of various extracts or preparations. In insects, neem is most active as a feeding deterrent, repellent, growth regulator, oviposition (egg deposition) suppressant, sterilant, or toxin.

Citrus Oil Extracts: Limonene and Linalool

Crude citrus oils and the refined compounds d-limonene and linalool are extracted from orange and other citrus fruit peels. Limonene, a terpene, constitutes about 90 per cent of crude citrus oil, and is purified from the oil by steam distillation. Linalool, a terpene alcohol, is found in small quantities in citrus peel and in over 200 other herbs, flowers, fruits, and woods.

Mode of action: The modes of action of limonene and linalool in insects are not fully understood. Limonene is thought to cause an increase in the spontaneous activity of sensory nerves. This heightened activity sends spurious information to motor nerves and results in twitching, lack of coordination, and convulsions. The central nervous system may also be affected, resulting in additional stimulation of motor nerves. Massive over stimulation of motor nerves leads to rapid knockdown paralysis. Adult fleas and other insects may recover from knockdown, however, unless limonene is synergized by PBO. Linalool is also synergized by PBO.

Products containing ingredients derived from plants are considered pesticides. However, products containing these active ingredients must be registered for use by the Environmental Protection Agency (EPA) and used in accordance with the provisions of the Federal Insecticide, Fungicide and Rodenticide Act (FIFRA) (Larson *et al.*, 1985).

2.5.2 Biopesticides

The term "biopesticides" usually refers to all biological materials and organisms that can be formulated and used as pesticide to manage obnoxious pests and diseases threatening the productivity of crops and animals. Biopesticides are inherently less harmful and have fewer environmental effects than conventional pesticides.

Biopesticides, are derived from such natural materials as animals, plants, bacteria, and certain minerals. As of early 2015 there were approximately 430 registered biopesticide active ingredients and over 1320 actively registered biopesticide products (Table 8.5). EPA generally requires less data to register a biopesticide than to register a conventional pesticide, thus the registration process is faster.

Table 8.5: Some of the Plant Products Registered as Biopesticides

Plant Product Used as Biopesticide	*Target Pests*
Limonene and Linalool	Fleas, aphids and mites,also kill fire ants, several types of flies, paper wasps and house crickets
Neem	A variety of sucking and chewing insect
Pyrethrum/Pyrethrins	Ants, aphids, roaches, fleas, flies, and ticks
Rotenone	Leaf-feeding insects, such as aphids, certain beetles (asparagus beetle, bean leaf beetle, Colorado potato beetle, cucumber beetle, flea beetle, strawberry leaf beetle, and others) and caterpillars, as well as fleas and lice on animals
Ryania	Caterpillars (European corn borer, corn earworm, and others) and thrips
Sabadilla	Squash bugs, harlequin bugs, thrips, caterpillars, leaf hoppers, and stink bugs

Categories of biopesticides include:

A. Microbial pesticides, in which a microorganism (*e.g.*, a bacterium, fungus, virus or protozoan) is the active ingredient;

B. Plant-Incorporated-Protectants, in which pesticidal substances are produced by crop plants as a result of genetic material being added to the plant (*e.g.*, *Bt* insecticidal protein), and

C. Biochemical pesticides, which are naturally occurring substances that control pests by non-toxic mechanisms, such as sex pheromones that interfere with mating and scented plant extracts that attract insect pests to traps. With plant-incorporated protectants, the toxin and its genetic material.

The major advantages of biopesticides in organic farming *i.e.*, advantages over chemical pesticides are

- ☆ Economical once method is developed
- ☆ Selective in action and hence no side effects
- ☆ Self propagating and self perpetuating as they are biological entitles
- ☆ No development of resistance
- ☆ Safe to non-target organisms
- ☆ Virtually permanent unless ecosystem is disturbed
- ☆ Effective against pests that are not accessible through chemical approaches and
- ☆ Compatible with other pest management tools except with broad spectrumtoxicants.

Successful utilization of biopesticides and bio-control agents in Indian agriculture include:

- ☆ Control of diamondback moths, *Helicoverpa* on cotton, pigeon-pea, and tomato by *Bacillus thuringiensis*
- ☆ Control of mango hoppers, mealy bugs and coffee pod borer by *Beauveria*
- ☆ Control of white fly on cotton by neem products
- ☆ Control of *Helicoverpa*on gram by N.P.V.
- ☆ Control of sugarcane borers by *Trichogramma* spp.
- ☆ Control of rots and wilts in various crops by *Trichoderma*-based products.

2.5.3 Biorationals/Insect Growth Regulators (IGRs)

Insect Growth Regulators (IGRs) are compounds which interfere with the growth, development and metamorphosis of insects. IGRs include synthetic analogues of insect hormones such as ecdysoids and juvenoids and non-hormonal compounds such as precocenes (Anti JH) and chitin synthesis inhibitors.

Ecdysoids

These compounds are synthetic analogues of natural ecdysone. When applied in insects, kill them by formation of defective cuticle. The development processes are accelerated bypassing several normal events resulting in integument lacking scales or wax layer.

Juvenoids (JH Mimics)

They are synthetic analogues of Juvenile Hormone(JH). They are most promising as hormonal insecticides. Juvenoids have anti-metamorphic effecton immature stages of insect. They retain *status quo* in insects (larva remains larva) and extra (super numary) moultings take place producing super larva, larval-pupal and pupal-adult intermediates which cause death of insects. Juvenoids are larvicidal and ovicidalin action and they disrupt diapause and inhibit embryogenesis in insects.

Example: Methoprene is a JH mimic and is useful in the control of larva of hornfly, stored tobacco pests, green house homopterans, red ants, leaf mining flies of vegetables and flowers

Anti JH or Precocenes

they act by destroying corpora allata and preventing JHsynthesis. When treated on immature stages of insect, they skip one or two larvalinstars and turn into tiny precocious adults. They can neither mate, nor ovipositand die soon.

Example: PBOs (PiperonylButoxide)

Table 8.6: Classical Biological Control Agents Used in India

Biocontrol Agent	*Imported from*	*Against*
Cryptolaemus montrouzieri	Australia	Mealy bugs
Rodolia cardinolis	USA	*Iceryapurchasi* in citrus, Casurina
Teleno musremus	New guinea	*Spodopteralitura* on tobacco
Eriborus trochanteratus	Srilanka	*Opisina arenosella*
Lepatomastix dactylopii	West Indies	Mealy bugs in citrus, coffee, guava
Curinus coerulus	Thailand	Subabul psyllid
Cephalonomia stephanoderis		Coffee berry borer
Cyrtobagous salvinae	Argentina	Gaint water fern *Eichhornia crassipes*
Neochaetina bruchi *N. eichhorniae* *Orthogalumna terebrantis*	Brazil	Water hyacinth
Zygogramma biocolorata	Mexico	Parthenium
Dactylopus tomentosus	Srilanka	Pricklypear

Chitin Synthesis Inhibitors

Benzoyl phenyl ureas have been found to have theability of inhibiting chitin synthesis in vivo by blocking the activity of the enzymechitinsynthetase. Two important compounds in this category are Diflubenzuron(Dimilin) and Penfluron. The effects they produce on insects include

- Disruption of moulting
- Displacement of mandibles and labrum
- Adult fails to escape from pupal skin and dies
- Ovicidal effect.

Chitin sysnthesis inhibitors have been registered for use in many countries andused successfully against pests of soybean, cotton, apple, fruits, vegetables, foresttrees and mosquitoes and pests of stored grain:

- Methropene for suppression of beetles, flies, mosquitoes, ants, and others on food and non-food crops, ornamentals, and livestock.

Table 8.7: Release Dosage of Parasitoids in different Crops

Crop	*Biotic Agent*	*Dosage/ha*	*Frequency of Application*	*Method of Application*	*Remarks*
Sugarcane					
Borers	*Trichogramma chilonis*	50,00/ha release	10 days interval 8 times at 30 DAT	Stapling	Pheromone trap catch data
Pyrilla perpusilla	*Epiricania melanoleuca*	2-3 egg masses or 5-7 cocoons in 40 selected spots/ha	Before onset of rainy season	Uniformly spread in 40 spots	Increase dosage by 10 times in endemic areas
Root grubs	*Metarhizium anisopliae*	10-12.5 kg/ha	June-July	Mix with 500 kg FYM spread uniformly in the field	Increase dosage to double in endemic areas
Sugarcane woolly aphid	*Micromusigorotus, Dipha aphidvora*	1000-1500/ha pupae 1000-1500/ha Larvae/ Pupae	Once during August-September	Place cocoons in dried leaves	Detrashing should be avoided
Rice					
Yellow stem borer *Scirpophaga incertalus*	*T. japonicum*	50,000/ha release	30,37 and 44 DAT	Stapling egg card bits	Pheromone traps
N. lugens (BPH)	*Cyrtorhinus lividipennis*	50-75 nymphs/m^2	After noticing the pest	Infested spots	If ratio is 1:4, No need
	M. anisopliae	200g 2x10 cfu/g	After noticing the pest	Infested spots – by Knapsack sprayer	
Cotton					
All Bollworms *H. armigera* *E. insulana*	*T. chilonis*	50,000/ha per release	6 times after 40 DAS	Stapling egg cards	Use selective pesticides
	Ha NPV	500 LE/ha	After noticing the pest	Spraying	Spray during evening time add jaggery and UV protectant

Crop	Biotic Agent	Dosage/ha	Frequency of Application	Method of Application	Remarks
Tobacco					
S. litura	*Telenomus remus*	50,000/ha per release		Release adults	Distribute uniformly in the nursery beds
	SI NPV 250 LE/ha	After noticing the pest	5 times at weekly interval	Spraying	Spray with Knapsack sprayer Add jaggery and UV protectants
Aphids	*C.carnea*	50,00/ha or 6/plant	4th week at weekly interval	2nd instar grubs	-
H.armigera	Ha NPV	250 LE/ha	Four times	Spraying	Spray with knapsack sprayer Add jaggery and UV protectants
Coconut					
Opisina arenosella (coconut black headed caterpillar)	*Goniozus nephantidis*	8-10/palm 3000-4000/ acre	Need based for each generation	Releasing adults	Release when larvae are noticed in the field
	Bracon brevicornis	8-10/palm 3000-4000/ acre	Need based for each generation	Releasing adults	do
	Bracon brevicornis	2:5 ratio	Need based for each generation	Releasing adults	do
	Elasmus nephantidis	2:5 ratio	Need based for each generation	Releasing adults	Release when larvae are noticed in the field
Oryctes rhinoceros (Rhinoceros beetle)	Baculovirus	10 beetles	One inoculative release	Release in the middle of the field at night	Wide coverage better results
	M.anisopliae	$4x10^6$ conidia/m^3	Once in rainy season	Mix spores in manure pits	Wide coverage better results
Arecanut/*Achnaspis longirostris*	*Chilocorus nigrita*	20 or 50 beetles/plant	Release after noticing the pest	Release adults beetles	Only to infested trees

Crop	Biotic Agent	Dosage/ha	Frequency of Application	Method of Application	Remarks
Coffee					
Mealy bug pseeudococcus	*Cryptolaemus mantrouzeri*	8-10 beetles/infested plant	After the blossom	Release adult beetles	Ant suppression should be adopted
Green scale	*M.anisopliae*	1 g/lit 2x10^{8}cfu/ml	-	-	Use knapsack sprayer for proper coverage
Coffee berry borer	*Beauveria bassiana*	2x10^{8}cfu/ml	120-150 days after blossom setting	Spray the conidia during evening time	Use knapsack sprayer for proper coverage
Apple					
Eriosoma lanigerum	*Aphelinus mali*	1000 adults/mummies tree	Once, as soon as infestation noticed	Releasing adults or placing mummies	Effective in valleys on aerial population
Quadraspidiosus perniciossus (Sanjose scale)	*Encarsia perniciosi*	1000 adults/tree	Once, in spring	Releasing adults or placing mummies	In endemic areas repeat release
	Chilocorus infernalis	20 adults or 50 grubs/ tree	Once in April-May	Releasing adults or placing mummies	In endemic areas repeat release
Cydiapomenella	*T.embryophagum*	2000 adults/tree	At weekly interval	Releasing adults	Pheromone catch
Citrus					
Icerya purchasi	*Rodolia cardinailis*	10 beetles/tree	Once on noticing adults	Releasing adults	Ant suppression
Planococcus citri (mealy bug)	*Cryptolaemus-montrouzieri*	10 beetles/tree	After the blossom	Releasing adults	Ant suppression
	Leptomastiz dactylopii	3000 adults/tree	Need based	Releasing adults	Ant suppression
Papiliodemolius(citrus butterfly)	*B.t.k*	1ml/lit 0.5 per cent ai	Single application for each generation		Spray with knapsack sprayer
Coccus viridis	*Verticillium lecanii*	2x10^{5}cfu/ml	Single application onset of monsoon	Spray with knapsack sprayer	Proper coverage

Crop	*Biotic Agent*	*Dosage/ha*	*Frequency of Application*	*Method of Application*	*Remarks*
Grapes					
Maconellicoccus hirsutus (mealy bug)	*Cryptolaemus montrouzieri*	2500-3000 beetles/acre 10 beetles/infested vine	Once as soon infestation is noticed	Release adults	Ant suppression
	Verticillium lecanii	$2x10^5$cfu/ml	Single application onset of monsoon	Release adults	Proper coverage
Guava					
Chloropulvinari apsidii (mealy bug)	*C. mantrouzieri*	10-20 beetles infested/ plant	Once as soon infestation is noticed	Release adults	Ant suppression
Mango					
Mealy bug	*C. mantrouzieri*	10-20 beetles infested/ plant	Once as soon infestation is noticed	Release adults	Ant suppression
Vegetables					
Beans *Tetranychus* spp.	*Phytoseialus persimilis*	10 adults/plant	Once in 30 days after germination	Adults Release	Release in brinjal and straw berry also
Cabbage *Plutella xylostella*	*Bt*	500g	Need based or weekly interval	Knapsack sprayer	Evening spray
	M.anisopliae	$2x10^8$ conidia/ml	Need based or weekly interval	Knapsack sprayer	Evening spray
	Beauveria bassiana	$2x10^8$ conidia/ml	Need based or weekly interval	Knapsack sprayer	Evening spray
S.litura *T.ni*	*N.rileyi* *N.rileyi*	$2x10^8$ conidia/ml	Need based or weekly interval	Knapsack sprayer	Evening spray
Tomato *H.armigera*	*Trichogramma brasiliensis*	50,00/ha	Weekly interval 6 times from 25 DAT or egg laying period	Stpling parasitized egg card uniformly	Pheromone trap catches
	Ha NPV	250LE/ha	Thrice during growth period	Knapsack sprayer	Jaggery, teepol, boric acid
	N.rileyi	$2x10^8$ conidia/ml	Thrice during growth period	Knapsack sprayer	Jaggery, teepol, boric acid

Crop	*Biotic Agent*	*Dosage/ha*	*Frequency of Application*	*Method of Application*	*Remarks*
Potato *Agrotis* spp. (cutworm)	*Stenernema carpocapsae*	5 billion infective juveniles/ha	On detection of cutworm larvae	Apply in soil thro irrigation water	Highly susceptible to descocation
Brinjal *Leucinodes orbanalis*	*T. chilonis*	50,00/ha	Weekly interval 6 times from 25 DAT	Staple parasitized egg card uniformly	Pheromone trap catches
Weeds *Eichhornia crassipes* (water hyacianth)	*Neochetina eichhorniae, N. bruchi*	100-5000 beetles/water body	Release weevils in water bodies at the onset of monsoon	Releasing adults or infested weed mats	Take up area wide operation
	Orthogalumna terebrantis	10,000 to 50,000 mites/ water body	Release mites in water bodies at the onset of monsoon	Releasing adults or infested weed mats	Take up area wide operation
Salvinia molesta (Giant water fern)	*Cyrtobaugous salviniae*	100-5000 beetles/water body	Release weevils in water bodies at the onset of monsoon	Releasing adults or infested weed mats	Take up area wide operation
Subabul *Heteropsylla cubana*	*Curinus coeruleus*	20-25 beetles/tree	Twice during July and October	Release adults	Inoculate as many places as possible
Parthenium hysterophorus	*Zygogramma bicolorata*	1000-2000 beetles/ha	During June and July	Release adults	Inoculate as many places as possible

- S-Kinoprene for whiteflies, gnats, aphids, mealybugs and scale ornamentals.
- S-Hydropene for cockroaches in food-handling establishments.

Advantages of using IGRs

- Effective in minute quantities and so are economical
- Target specific and so safe to natural enemies
- Bio-degradable, non-persistent and non-polluting
- Non-toxic to humans, animals and plants

3. Conclusions

Various organic agricultural technologies have been used for about 6000 years to make agriculture sustainable while conserving soil, water, energy, and biological resources. Recent quests for effective, safe, and lasting pest management programs have been targeted primarily toward development of new and better products with which to replace conventional toxic pesticides. We assert that the key weakness with our pest management strategies is not so much the products we use but our central operating philosophy. The use of therapeutic tools, whether biological, chemical, or physical, as the primary means of controlling pests rather than as occasional supplements to natural regulators to bring them into acceptable bounds violates fundamental unifying principles and cannot be sustainable. We must turn more to developing farming practices that are compatible with ecological systems and designing cropping systems that naturally limit the elevation of an organism to pest status. We historically have sold nature short, both in its ability to neutralize the effectiveness of ecologically unsound methods as well as its array of inherent strengths that can be used to keep pest organisms within bounds. If we will but understand and work more in harmony with nature's checks and balances we will be able to enjoy sustainable and profitable pest management strategies, which are beneficial to all participants in the ecosystem, including humans.

REFERENCES

Adrien R, Muriel VM, Jean PS and Jean RE. 2010. Biological Control of Insect Pests in Agroecosystems: Effects of Crop Management, Farming Systems, and Seminatural Habitats at the Landscape Scale: A Review of Advances in Agronomy, 109: 219-259.

David P, Hepperly P, James H, David D and Rita S. 2005.Environmental, Energetic and Economic Comparisons of Organic and Conventional Farming Systems. BioScience, 55: 7.

Feber RE, Firbank LG, Johnson PJ and Macdonald DW. 1997. The effects of organic farming on pest and non-pest butterfly abundance. A synthesis of current research Agriculture, Ecosystems and Environment, 64: 133-139.

Garratt MPD, Wright DJ and Leather SR. 2011. The effects of farming system and fertilizers on pests and natural enemies: A synthesis of current research Agriculture, Ecosystems and Environment, 141: 261– 270.

Hokkanen HMT. 1991. Trap Cropping in Pest Management. Annual Review of Entomology, 36: 119-138.

Kajimura T, Maeoka Y, Widiarta IN, Sudo T, Hidaka K and Nakasuji F. 1993. Effects of organic farming of rice plants on population density of leafhoppers and plant hoppers. I. Population density and reproductive rate. Japanese Journal of Applied Entomology and Zoology, 37: 137-144

Landis DA, Wratten SD and Gurr GM. 2000, Habitat management to conserve natural enemies of arthropod pests in agriculture. Annual Review of Entomology, 45: 175-201.

Larson LL, Kenaga EE and Morgan RW. 1985. Commercial and experimental organic insecticides. Entomological Society of America. pp. 105.

Lewis WJ, Van Lenteren JC, Sharad CP and Tumlinson, JH. 1997. A total system approach to sustainable pest management. Proceedings of the National Academy of Sciences, 94: 12243-12248.

Mohankumar S and Ramasubramanian T. 2014. Role of Genetically Modified Insect-Resistant Crops in IPM: Agricultural, Ecological and Evolutionary Implications. Integrated Pest Management, pp. 371-399.

Mzough N. 2011. Farmers adoption of integrated crop protection and organic farming: Do moral and social concerns matter. Ecological Economics, 70: 1536–1545.

Pickett CH and Bugg RL. 1998. Enhancing biological control: Habitat management to promote natural enemies of agricultural pests. University of California Press.

Rao A. 2002 Influence of chemical fertilizers and organic manures on the groundnut pests. Indian Journal of Plant Protection, 34: 30-33.

Roger CR, Vincent C and Arnason, JT. 2012. Essential Oils in Insect Control: Low-Risk Products in a High-Stakes World. Annual Review of Entomology, 57: 405-424.

Rosaih R. 2001. Performance of different botanicals against the pests complex in bhendi. Pestology, 25: 17-19.

Samantha MC, Zeyaur RK and John AP. 2007. The Use of Push-Pull Strategies in Integrated Pest Management. Annual Review of Entomology, 52: 375-400.

Schmutterer H. 1990. Properties and Potential of Natural Pesticides from the Neem Tree, *Azadirachtaindica*. Annual Review of Entomology, 35: 271-297.

Shelton AM and Badenes-Perez FR. 2005. Concepts and Applications of Trap Cropping in Pest Management. Annual Review of Entomology, 51: 285-308.

Turney HA and Mcilveen G. 1983. Insect control guide for organic gardeners. B-1252. Texas A and M AgriLife Research, pp. 12.

Urs KCD. 1987. Prospects of Indigenous plant products as future alternatives to conventional pesticides. In: *Proceedings of Symposium on Alternative Insecticides*, pp. 202.

Weisz R, Smilowitz Z and Christ B. 1994. Distance, rotation, and border crops affect Colorado potato beetle (Coleoptera: Chrysomelidae) colonization and populationdensity and early blight (*Alternaria solani*) severity in rotated potato fields. Journal of Economic Entomology, 87(3): 723-729.

Wyss E, Luka H, Pfiffner L, Schlatter C, Uehlinger G and Daniel C. 2005. Approaches to pest management in organic agriculture: a case study in European apple orchards. Organic Research, pp. 33-36.

Zehnder G, Gurr GM, Kuhne S, Wade MR, Wratten SD and Wyss E. 2007. Arthropod pest management in organic crops. Annual Review of Entomology, 52: 57-80.

Chapter 9

Non-Pesticidal Management: An Experience

G.V. Ramanjaneyulu, Zakir Hussain, G. Chandra Sekhar and G. Rajashekar

Centre for Sustainable Agriculture (CSA), Hyderabad – 500 017, Telangana
E-mail: gvramanjaneyulu@gmail.com

1. INTRODUCTION

Indian Agriculture is as old as human civilization. In this era of modernizing agriculture, the 'traditional cyclic methods which involved multiple approaches to solve a problem and pluralistic ways of understanding an issue or a phenomenon, are getting replaced by the 'linear industrial approaches' The frame work of their understanding always saw the indigenous knowledge as 'traditional', 'subsistence' and 'backward'. The mainstream agricultural institutions not only failed to understand the Indian socio-economic-political-ecological context but also the science behind the production practices or the cropping systems. The experiments with NPM in Andhra Pradesh and Telangana clearly showed that the contemporary innovations (CI) based on Indigenous technical knowledge (ITK) could have spread effectively if proper support is provided.

Pest management is seen as killing insects with toxins externally sprayed (Pesticides) over crops or genetically modified (GM) to produce internally in the plants. However, in the process, the ecological problems with pesticides and GM crops are completely ignored. The learning's from Integrated Pest Management (IPM) projects and Farmer Field Schools (FFS) worldwide should have led to research on the complex interaction between crop ecology, agronomic practices, insect biology and climate change to develop effective methods to manage disease and insect control strategies. Similarly the farmers' knowledge on using the local resources could have been captured and the principles could have been standardized. But, FFS mostly remained as a paradigm shift in agricultural extension: the training program that utilizes participatory methods "to help farmers develop their analytical skills, critical thinking, and creativity and help them learn to make better decisions". The agricultural research and extension system worldwide still continue to believe in chemical pesticide based pest management in agriculture. While the inevitability of pesticides in agriculture is promoted by the industry as well as the public research and extension bodies, there are successful experiences emerging from farmers' innovations call for a complete paradigm shift in pest management.

2. Shifting Paradigms: Non-Pesticidal Management

The economical and ecological problems due to pests and pesticides, respectively in agriculture gave rise to several eco-friendly innovative approaches which do not rely on the use of chemical pesticides. These initiatives involved rediscovering traditional practices and contemporary grass root innovations supplemented by strong scientific analysis mainly supported by non-formal institutions like NGOs. Such innovations have begun to play an important role in development sector. This also shows how collaborative work of diverse players like public institutions, civil society organizations and farmers can generate new knowledge and practice and can evolve more sustainable models of development.

Pest is not a problem but a symptom. Disturbance in the ecological balance among different components of crop ecosystem makes certain insects reach pest status. In fact, experience shows that most of the pest problems are pesticide induced (Ramanjaneyulu *et al.*, 2009). From this perspective, a concept 'Non Pesticidal Management' (NPM) was evolved. It is an 'ecological approach to pest management using knowledge and skill based practices to prevent insects from reaching damaging stages and proportions by making best use of local resources, natural processes and community action.

Non-pesticidal management is mainly based on:

- ☆ Understanding crop ecosystem and adopting suitable cropping systems and crop production practices. The type of pests and their behavior and also the natural enemies' composition varies with the cropping system.
- ☆ Understanding insect biology and behavior and adopting suitable preventive measures to reduce the pest numbers. Curative measures if needed, are based on using various products like made out of locally available resources. The measures coupled with natural processes effectively manage the pests.
- ☆ Building farmers knowledge and skills in preventive and curative measures. This is done through a farmer field school approach.

2.1 Backyard Biotechnology

Most of the curative practices are based on making and using products from the locally available plant and animal materials. Various processes like making solutions, concoctions, decoctions and fermented products. These products are locally made used. Hence, only the processes are standardised but not the products.

On one hand, farmers found them easy to make, cheaper and very useful. Even if some of them have highly toxic material (*e.g. Holarrhena antidysenterica* (*Pala Kodise*), people never faced a problem as they know how to handle them safely. The recipes of these products are more or less based on regular food preparations, hence, women find it easy to make.

2.2 Components of NPM

2.2.1. Good Quality Seed and Growing Healthy Crop

Growing healthy plants is the first step in NPM. Selection and use of good quality seed which is locally adopted either from traditional farmers' varieties or improved varieties released by the public sector institutions is important. Farmers are suggested to make their decision based on a seed matrix regarding suitability of the different varieties into their cropping patterns, based on the soil types, reaction to pest and diseases and their consumption preference. They maintain the seed in their village seed banks. This ensures farmers to go for timely sowing with the seeds of their choice. In rainfed areas, timely sowing is one critical factor which affects the health and productivity of the crop. The seed is treated with concoctions depending on the problem for example cow urine, ash and asafetida concoction provides protection against several seed borne diseases like rice blast, or *beejaraksha, beejamrut* to induce microbial activity in the soil and kill any seed borne pathogens. Similarly in crops like brinjal where there is a practices of dipping of seedlings in milk and dipping fingers in milk before transplanting each seedling was observed to prevent viral infections. Several such practices are documented and tested by the farmers.

2.2.2. Build Healthy Soils

Healthy soils give healthy crop. Chemical fertilizers especially nitrogenous fertilizer makes the plants succulent and increases the sucking pests like Brown Plant Hopper in rice. The practices such as use of crop residues or other biomass as surface mulch, using compost and green manures, intercropping of legumes in cropping systems, and biocontrol of insect pests and diseases, will help enhance yields and sustain soil fertility and health (Rupela *et al.*, 2007).

2.2.3 Enhancing the Habitat/Crop Diversity

Crop diversity is another critical factor which reduces the pest problems. Traditionally farmers have evolved mixed cropping systems, intercropping and crop rotation systems. These systems will create a better environment for nutrient recycling and healthy ecosystems. On the contrary, the monoculture of crops and varieties lead to nutrient mining and pest and disease buildup. Under NPM, farmers adopt mixed and intercropping systems with proper crop rotations. For *e.g.* in the crops like chillies, groundnut, cotton, sunflower where thrips are a major problem, sowing thick border rows of tall growing plants like sorghum or maize will prevent insects from reaching the crop. Farmers adopt marigold as a trap crop for the gram pod borer reduces the pest load on pigeonpea (Table 9.1). The flowers that have been oviposited by the female moths of Helicoverpa can be picked out and destroyed (KVK DDS, 2003).

Table 9.1: Trap Crops Used for Pest Management

Crop	*Pest*	*Trap Crop*
Cotton, groundnut	Spodoptera	Castor, sunflower
Cotton, chickpea, pigeonpea	Helicoverpa	Marigold
Cotton	Spotted bollworm	Okra

Source: KVK DDS (2003).

Other agronomic practices include several crop specific agronomic practices like making alley ways at every 2 m interval in rice will allow enough light and air to reach the bottom of the plant thus minimizes the BPH incidence and is suggested by the scientists (ANGRAU, 2007).

2.2.4 Understanding Insect Biology and Behavior

Insects damage the crop only in larval stage of the four stages in their life cycle. Every insect has different behavior and different weaknesses in each stage. They can be easily managed if one can understand the lifecycle and their biology.

In adult stage, light-community bonfires or light traps (electric bulbs or solar light) can be used to attract and kill RHC (red hairy caterpillar). Similarly adult insects of *Spodoptera* and *Helicoverpa* can be attracted by using pheromone traps. Normally pheromone traps are used to monitor the insect population based on which pest management practices are taken up. Natural Resources Institute, UK in collaboration with Tamil Nadu Agriculture University, Gujarat Agricultural University, Centre for World Solidarity, Asian Vegetable Research and Development Centre has evolved mass trapping method to control brinjal fruit and shoot borer and demonstrated on a large scale (http://www.nri.org, GAU, 2003). The adults of sucking pests can be attracted using yellow and white sticky boards.

In egg stage, some insects like *Spodoptera* lay eggs in masses which can be identified and removed before hatching. Insects also have preference for oviposition. *Spodoptera* prefers to lay eggs on castor leaves if available. Hence, growing castor as trap crop is practiced. By observing the castor leaves, farmers can easily estimate the *Spodoptera* incidence. *Helicoverpa* lays eggs singly, but has a preference towards okra and marigold (mostly towards plants with yellow flowers). Hence, marigold is used as a trap crop whereever *Helicoverpa* is a major problem. Rice stem borer lays eggs on the tip of the leaves in nurseries; hence, these tips have to be removed before transplanting (ANGRAU, 2007).

As larval stage is the most damaging they have to be identified and removed mechanically. In pupal stage, the larvae of RHC burrow and pupate in the soil. Deep summer ploughing exposes these larvae to hot sun which kills them. The larvae of stem borers in crops like paddy and sorghum pupate in the stubbles. So farmers are advised to cut the crop close to ground level and clear the stubbles. The larva of red hairy caterpillar (*Amsacta albistriga*) has a dense hair over the body hence no pesticide reaches it when sprayed. Therefore, it needs to be controlled in other stages of its life cycle. For any safe and economic method of pest management, one must understand how the pest live and die, where does it come from and when, where and how does it damage the crop. Knowledge of these biological attributes of pest will help farmers to use NPM methods successfully on a sustainable basis (GAU, 2003).

2.2.5 Understanding Crop Ecosystem

The pest and natural enemy complex are based on the crop ecosystem. The pest complex of cotton is completely different from that of sorghum. The pest complex in wet rice ecosystem differs from that of dry rice. Decision about any pest management intervention should take into account the crop ecosystem which includes cropping pattern, pest-predator population, stage of the crop *etc.* Similarly, the management practices followed in one crop cannot be practiced as such in other crops. For *e.g.* to manage *Helicoverpa* in pigeonpea, the farmers in Andhra Pradesh and Gulbarga shake the plants and collect the falling insects over a sheet and kill. Similarly, in paddy, there is a practice of pulling rope over the standing crop to control leaf roller.

2.2.6 Reactive Sprays for Control

Insect population may reach pest status if the preventive steps are not taken in time and also due to changes in weather conditions and insects coming from neighboring farmers' fields. Under these situations, based on the field observations, farmers can take up spraying botanical extracts and natural preparations (green sprays) instead of chemical pesticides. There are wide ranges of these preparations which are evolved by the farmers as detailed below (CSA, 2007).

Aqueous or Solvent Extracts

For example, Neem Seed Kernal Extract (NSKE) is very effective against many pests. For extracting 'Allenin' from garlic, kerosene is used as a solvent. After extraction, this solution is mixed with chilli extract and used against sucking pests (Prakash and Rao, 1997; Vijayalakshmi *et al.*, 1999, Prasad and Rao, 2006).

Decoctions

For *e.g.* plants like tobacco, *Nux Vomica etc.* contain volatile compounds which can be extracted by boiling them in water to get the decoction (Prakash and Rao, 1997, Vijayalakshmi *et al.*, 1999, Prasad and Rao, 2006).

Concoctions

These are mixtures. For *e.g.* aqueous extract of any five latex producing leaves is used to control pests in Tamil Nadu and other parts of South India (Prakash and Rao, 1997, Vijayalakshmi *et al.*, 1999; Prasad and Rao, 2006).

Fermented Products

Products made by fermenting the different botanicals with animal dung and urine. These products have rich microbial cultures which help in providing plant nutrients in addition to acting as pest repellents and pest control sprays. For *e.g.* cow dung urine-asafoetida solution is used to manage rice blast (Prakash and Rao, 1997, Vijayalakshmi *et al.*, 1999, Prasad and Rao, 2006). Tables 9.2 and 9.3 show various types of preparations which are used in NPM.

Table 9.2: Types of Preparations Used in NPM

Aqueous Solutions	*Decoctions*	*Solvent Extracts*	*Concoctions*	*Fermented Products*
Neem seed kernal extract	Vitex leaf	Chillic garlic extract	Chilli powder-garlic extract	Jeevamrutham
Pongamia seed	Tobacco leaf	Turmeric tuber, milk extract	Agniastram	*Panchagavya*
Seethaphal seed	Ipomea leaf	Egg, lemon solution	Brahmastram	Fish amino acid
Palakodise leaf	Bel leaf	Garlic, soap solution	*Panchapatra kashayam*	Rotten fruits solution
Palakodise stem	Wild basil		*Dasapatra kashayam*	Herbal tea
Mahua fruit	Turmeric leaf		*Aloe vera*-bougainvillea	Compost tea
	Connessi bark		Ginger-milk	Koonapa jalam
	mahua leaf			Dung-urine solution
	bel leaf and cow urine			Dung-urine, asafoetida solution
	Citronella leaf			
	Onion leaf			
	Pongamia leaf			
	Lantana leaf			
	Wild indigo			
	Acasia leaf			

Table 9.3: Nature of Action of Preparations and Target Pests

Chewing Insects	*Borers*	*Sucking Pests*	*Repellents*	*Broad Spectrum*
Neem seed kernel extract	Chilli garlic	Neem oil	Neem seed kernel extract	Jeevamrutham
Brahmastram	Agni astram	Vitex leaf decoction	Pongamia seed or leaf extract	Panchagavya
Neemastram	Neemastram	Tobacco leaf decoction	Cow urine	Compost tea
Panchagavya	NPV	Neem soap	Coriander as intercrop	Amrutajalam
cow dung urine solution		Neem seed kernel extract	Garlic as intercrop	
Neem oil		Compost tea		
Compost tea		Cattle urine hing		

Method of Preparation of Various Extracts

Method of preparations of one product in each of the category is given here. Same process can be followed for others.

Neem Seed Kernal Extract (NSKE)

Required materials: Neem seeds - 5 kg; soap powder - 100 gm

Method of preparation (Figure 9.1)

- ☆ Good quality 5 kg neem seeds to be dried under shade
- ☆ This powder can be packed in cloth and kept in 10 litres of water for 10-12 hrs
- ☆ Extract the decoction by pressing the cloth pack for 10-15 min
- ☆ Filter this solution through a thin cloth
- ☆ Add 100 gm of soap powder to the filtered solution
- ☆ Add 100 litres of water to the solution and spray it in 1.0 acre during evening time

Note: 5-10 kg of neem powder is required (depending on the crop stage and pest intensity); This solution cannot be stored; Increase the dosage depending on the crop

Figure 9.1: Method of Preparation of 5 per cent Neem Seed Kernel Extract.

stage and intensity; This solution can be used on all crops and nurseries, orchards to get better yields; Instead of soap powder, 500 gm of soap nut can also be used.

Vitex Leaf Decoction (Figure 9.2)

Presence of many alkaloids makes Vitex as an effective pesticide and fungicide.

Required material and procedure:

- ✰ Boil 5 Kg of Vitex in 10 litres of water for 30 min
- ✰ Stir the boiling solution regularly
- ✰ Cool the solution and filter through a thin cloth
- ✰ Add 100 gm soap powder to the decoction
- ✰ Add 100 litres of water to the decoction to spray in 1 acre
- ✰ Spray the decoction in the evening time

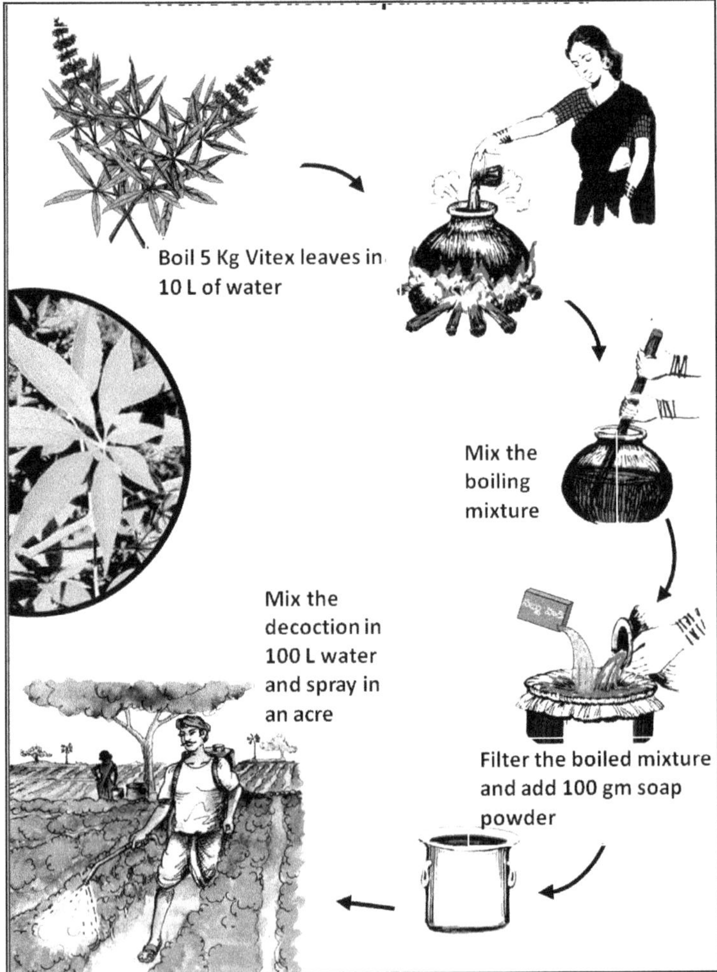

Figure 9.2: Preparation of Vitex Decoction.

Precautions: Tie a cloth to nose while making the decoction; Depending on the crop stage and pest intensity, this can be applied for two to three times; Never store the decoction

Chilli-garlic Solution

Required material: green chillies- 3 kg; garlic- 0.5 kg; kerosene- 250 ml; soap powder- 100 gm

Method of preparation (*Figure 9.3*)

- ☆ Grind the chillies after removing the petioles and add 10 liters of water to it. Keep this solution overnight
- ☆ Grind the 0.5 kg garlic and add 250 ml kerosene and keep overnight
- ☆ Next day morning filter the chilli solution through a thin cloth
- ☆ Do the same for garlic solution

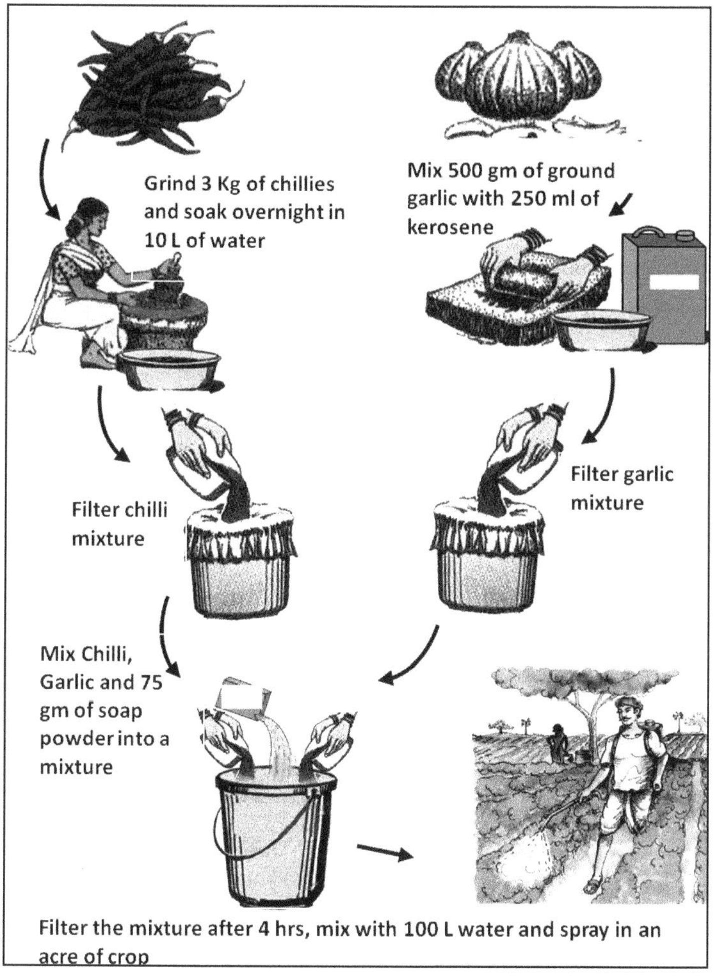

Figure 9.3: Preparation of Chilli-Garlic Solution.

- Mix chilli solution, garlic solution and soap powder thoroughly and make a mixture
- Add 100 litres of water to the above solution. This can be applied for one acre

Precaution: Apply oil to the body while preparing this decoction; Cover the body while spraying; Apply this solution only one or two times during the cropping season.

Concoctions: Concoctions are mixtures of compounds from plants. Concoctions are made when there is an incidence of pest complex. For *e.g. Panchapatra Kashayam* (5 leaves decoction), *dasapatra kashayam* (10 leaves decoction).

Panchapatra Kashayam

Required material: Neem leaves - 1 kg; Jatropha - 1 kg; Vitex - 1 kg; custard apple - 1 kg; mint - 1 kg; *Eucalyptus* leaves - 1 kg; Lantana – 1 kg

Method of preparation (Figure 9.4)

- Take any 5 types of aforementioned plant leaves
- Keep all these leaves in a pot and add 10 litres of water
- Keep them for whole night
- Next day boil the water
- Keep on stirring the solution and boil the water until the solution becomes 5 L decoction
- Cool the decoction
- Filter the solution and add 100 g of soap powder
- Mix this solution to 100 -150 litres of water to spray in one acre

Pests controlled: Sucking pests, small larvae, grass hoppers and fruit borers

How to use:

- This decoction can be applied to all crop types
- Based on the intensity of pest it can be used for one – two times in a crop period

Precaution:

- Cover your nose with a cloth while preparing the solution
- For effective results spray this solution during evening time
- Don't store the solution
- Don't spray this solution during early stages (30-45 days) of the crop
- Don't spray this solution on nurseries

2.3 Scaling Up of NPM

During 2004-05, when the farmers in Andhra Pradesh were experiencing a period of serious livelihoods crisis, the Society for Elimination of Rural Poverty

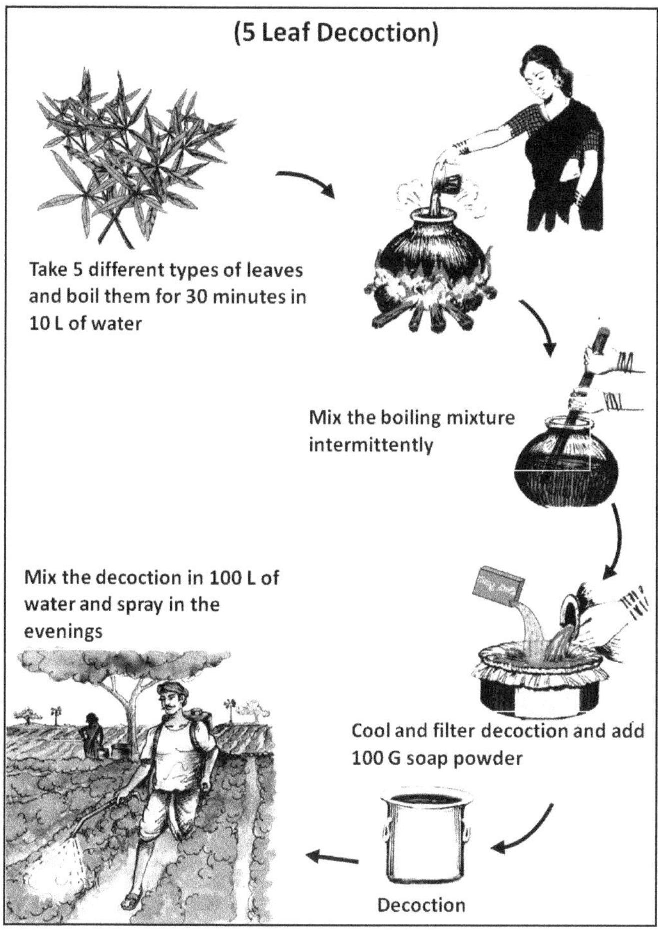

Figure 9.4: Preparation of *Panchapatra kashayam.*

(SERP), under the livelihood initiative of the Government of AP working with Federations of Women Self Help Groups, identified increasing cost of cultivation due to heavy dependency on external inputs as one of the main reasons for the growing indebtedness. Learning from the experiences of villages like Punukula and Enabavi in AP, SERP initiated pilot scaling up of Non-Pesticidal Management (NPM) in collaboration with a consortium of Civil Society Organizations led by Centre for Sustainable Agriculture in 2005-06 (Ramanjaneyulu *et al.*, 2009, Vijay Kumar *et al.*, 2009).

- ✰ Farmer Field School approach originally designed and promoted by FAO was suitably modified and established to train farmers (both women and men) regularly on NPM and other ecological farming practices.
- ✰ The program is implemented using experienced farmers as 'Community Resource Persons' (both women and men) and the Federations of Women Self Groups at the Mandal level (Block level) managed the entire program.

The program was supported by experienced local NGOs and Centre for Sustainable Agriculture as the nodal agency for technical support and project management till 2007-08.

- ☆ The initial success with NPM and SERP has cast its net wide across the country to identify best practices from the successful ecological farming models. Ecological/Natural Farming Master farmers like Sri. Bhaskar Save, Sri. Subash Palekar, Sri. Nammalwar have provided inspiration and necessary support to promote 'Polycrop' models, organic soil management practices, soil and water conservation, and *in situ* water harvesting practices.
- ☆ By 2007-08, the program spread to more than 2.8 lakh ha across the state, largely through the support of NGOs. It became evident during this period, that resource conserving, regenerative, sustainable agriculture practices which are largely based on local resources based solutions, farmers knowledge and skill in packaging them to suit their situations, bring in wide set of benefits that accrue to the practitioners and their farm ecology. In 2007-08, a state level Project Management Unit was setup to take over the roles of providing overall technical support and project management.
- ☆ As the scope of the intervention expanded, it was named as 'Community Managed Sustainable Agriculture' (CMSA). It represents a model of agriculture which is largely based on farmers' resources, knowledge and skills and the institutional systems for learning are managed by the Community.

2.3.1 Role of Community Institutions

One of the main objectives of this initiative was to establish a community managed learning and management system to build more accountability and ownership in the system.

a. *VO Sub Committee:* At the village level, a practicing farmer is identified as a Village Activist who is responsible for organizing and documenting FFS. The Village Organization (VO) is the federation of the all women SHG groups in the village and a subcommittee in the VO monitors the progress every month. Five such villages are grouped into a cluster and is supported by a Cluster Activist.

b. *MMS Sub Committee:* At the Mandal level (AP equivalent of blocks), the clusters are reviewed on a monthly basis by the Mandal Mahila Samakya (MMS) sub-committee. MMS identifies the villages where the agriculture program would be implemented based on the response of the members from villages which can be grouped as a cluster. MMS subcommittee can also enter into an agreement with any Resource NGO for providing technical support.

c. *Zilla Samakya Sub Committee:* At the district level, the program is reviewed on a monthly basis by the Zilla Samkya Sub Committee. The District Project Manager (DPM of SERP) provides the required administrative support in monitoring and documenting. *Zilla Samakya* Sub Committee identifies Mandals where the program would be implemented.

d. Community Resource Persons (*CRPs*): Successfully practicing farmers are selected as Community Resource Persons who help in supporting and promoting sustainable agriculture practices.

e. Non Governmental Organisations: Initially when the program was started, NGOs played an important role of providing handholding support to the women's SHGs for three years 2005-06 to 2007-08, both at grassroots and state level. Gradually, this role of project implementation has been taken over by SHGs and the capacity building by CRPs.

The funds for the program are released to the MMS and CRPs, Village and Cluster Activists and are paid by the MMS. Each participating farmer pays a registration fee of Rs. 30/year which would be deposited with the VO.

2.3.2 Implementation Strategy

The implementation process uses several important methods to ensure close community participation and learning and management involvement.

Involvement of Village

The villages are identified based on the expression of interest of the VO (Village Organization) members in the MMS (Mandal Mahila Samakya) sub-committee meeting. A village immersion program is organized where CRPs (Community Resource Persons), Cluster Activist and DPMs (District Project Manager) discuss about the agriculture situation and share learning from CMSA from other villages. The program identifies interested farmers and organizes them into FFS (Farmers Field School) groups. During the processes of immersion village, Resource Mapping is also done to identify locally available resources, cropping systems and local knowledge *etc.*

Farmers' Field School (FFS) forms the basic unit of learning. Each Farmer's Field School is a relatively homogenous group of farmers and the FFS meets every week in one of the member's field to learn, discuss and take decisions regarding actions to be taken managing their crops. Village activist and cluster activist will organize these field schools. Sometimes Community Resource Persons may also join.

ICTs for Information Sharing and Reviewing

Video Conferences to review progress are organized every fortnight. The cluster activists, MMS and ZS leaders, District Project Managers attend and share information. Similarly Mobiles phones are used to disseminate important alerts and suggestions, TV channels were used to share information regularly on production practices.

Further there are key sub-interventions which directly assist farmers gain greater role in the production process.

Community Seed Banks

In the identified villages, the seed requirements (both in terms of varieties and quantities) are mapped and breeder seed are procured for all the crops and farmers are trained to produce and use their seed. In crops like paddy, groundnut where seed requirement is high few farmers are identified at the village level to produce seed and make it available to other farmers at a price.

Custom Hiring Centers

Implements for ploughing, sowing, weeding *etc.*, are made available at the Village Organization on a custom hiring basis. This improved the access of small and marginal farmers to such implements.

NPM Shops

To promote micro enterprises which can supply ecological farming inputs NPM shops are promoted through PoP families. All NPM shop owners are trained on preparation of botanical extracts and loan is facilitated from MMS to establish NPM shops. So far 1944, NPM shops were established across the project implementation area. The income from NPM shops ranged from Rs.1500/per week- during the peak season to Rs.500/week- in lean season

The Community seed banks, Custom Hiring Centres for implements and NPM shops work as network to share surpluses with others when needed.

2.4 Reaching Out to the Poorest

Often agriculture related interventions tend to naturally work mainly with those farmers owning land and who have more resources and time and less afraid to take risks. Based on the learnings from the CMSA, two initiatives were developed to improve household food security of the poorest of the poor who form 34.7 per cent of the SHG groups promoted by SERP and to address rainfed areas which form 58 per cent of the cultivated land in the state.

2.4.1 Strategy to Support Poorest of the Poor (PoP)

With poorest of the poor, who have very few assets and form the lowest rung of the poverty ladder, being the main focus of the State, SERP initiated to facilitate PoP to adopt CMSA in at least 0.5 acre of land. Land leasing is facilitated for the landless. In this 0.5 acre land, SRI Paddy cultivation is taken up in 0.25 acre and a seven tier polycrop model (ranging from tuber crops to fruit crops, vegetables, pulses, serials *etc.*) in the remaining 0.25 acres (popularly called as 36 x 36 m model). During 2009-10, 251 models each of 0.5 acres (0.25 acre of multiple vegetables and 0.25 acre of SRI Paddy) were established under PoP strategy which gave an income ranging from Rs. 15,000 to 40,000 based on the cropping pattern and time of sowing. This model provides food and income round the year. Last two years data shows that a net income up to Rs.50, 000 in a year is possible along with improved household food and nutritional security. See Case study box below.

2.4.2 Rainfed Sustainable Agriculture (RFSA) in Convergence with Mahatma Gandhi National Rural Employment Guarantee Scheme (MGNREGS)

To make the most of possible synergies between generating employment, sustainable land improvement and drawing on existing government resources, comprehensive soil and moisture conservation works to improve the land of PoP were initiated converging with the MGNREGS from 2009-10. In the first year, it covered 1.28 lakh ha of 1.46 lakh SC/ST farmers 2009-10. The works includes farm ponds, conservative furrows, trenches, compost pits and vegetable mini kits and fruit plants. The total amount to be spent in one ha of land is Rs.1.00 lakh. Total out lay for the year 2010-11 is Rs.1630 crores covering 4 lakh ha in 22 districts. So far technical sanctions are accorded for Rs. 804 crores and administrative sanctions are accorded for Rs. 645 crores and RFSA works are grounded in 3,034 villages out of 3,332 villages. The final results are still awaited.

2.4.3 Case Study of Smt. M. Bojamma: Journey of an Ultra Poor Family from "Wage Seekers" to "Net Food Producers"

Smt. M.Bojjamma is a landless agriculture worker from Thadakanaplle village of Kallur mandal in Kurnool district. She hails from a Poor family. Her husband was working in a stone crusher unit. She got interested about the sustainable agriculture listening to the Cluster Activist in the VO meeting and wanted to try out.

Village Organization provided a loan amount of Rs.7, 000 for leasing in 0.5 acre of land with assured irrigation for one year (two seasons) in 2009. She attended all the training programs and enrolled for the farmer field school. She learned about sustainable agriculture methods, growing multistory cropping system and SRI – resulting in a diverse crop. During *Rabi* 2009, she took up only vegetables. The costs of cultivation were Rs. 3000 and she earned Rs. 26,500. The combined net incomes were Rs. 28,800 (*kharif*) + Rs. 23,500 (*rabi*) - Rs.7, 000 (land rental) =Rs. 45,300.

3. Impacts

The Community Managed Sustainable Agriculture is a comprehensive package and bringing together of several ecological farming practices. The enabling strength of the Women Self Help Group institutional platform has facilitated rapid spread of these practices. There have been a range of important impacts to farmers and poor rural households

- ✰ Reductions in costs of cultivation due to NPM are reported by all the farmers. The savings in costs range from Rs. 7500/ha in paddy and redgram upto Rs. 37500/ha in chillies
- ✰ A quick survey by SERP in three districts has shown that the number of cases of hospitalization due to pesticide poisoning has reduced from 242 cases/year to 146 cases/year a 40 percent drop in a year. In the villages which have adopted NPM the drop is 100 per cent
- ✰ Unlike the popular argument that pest outbreaks could happen, villages adopting NPM have not seen pest outbreaks caused due to ecological disturbance or pest resistance. Farmers could effectively manage rice

blast using a fermented solution of asafoetida, cow dung and urine), and sucking pests in cotton and chillies using similar methods

- Where organic soil management practices are adopted, the increased soil moisture conservation has helped to tide over drought spells for about 10 days or more. Encouraged by this, integrated effort to physically conserve water within the field by adopting conservation furrows, trenches and farm ponds was initiated
- Efforts to internalize the seed production at community level particularly in crops like paddy and groundnut have shown positive results
- AP produces and exports most of the chillies in the country. High pesticide residues often have led to the rejection of the export of chillies and products using chillies. The chillies produced in Guntur district adopting NPM practices were tested for pesticide residues and were found to be in accordance with EUROGAP standards
- During 2009-10, more than 7638 farmers (in addition to the 251 who come under PoP strategy) were supported to establish Intensive Farming System model in 36 x 36 m which produce food round the year with a combination of seasonal and perennial crops. The net incomes from these models ranged from Rs. 4000 to 12,000 per acre in addition to meeting the family food needs
- The CMSA approach enables bundling of various relevant services to farmer families, including credit access on the doorstep. Ultimately, the approach involves facilitating development of micro-credit plans for sustainable agriculture and linking farmers to commercial banks, especially where this related to marketing needs. Access to banks for farming reduced as the focus shifted to the poorest of the poor, who depended more on the group credit system. CMSA approach also facilitates the farmer's access to high quality inputs through a network of community seed banks and agricultural implements from Custom Hiring Centers run by the Federations of Women Self Help Groups.

3.1 Observations on Benefits and Strengths of the Program

- Getting out of pesticide poisoning is seen as a major benefit for the farmers. They clearly recognize and acknowledge that their health has significantly improved and the health costs have come down after adoption of NPM. While there is considerable scope to generally increase wider awareness on the benefits of appropriate use and correct application of pesticides and fertilizers, the NPM interventions have demonstrated the effectiveness of using local resources and preventive measures
- The ecological farming practices like NPM, organic soil management, multiple cropping models, SRI *etc.*, have been adopted by the farmers to a considerable extent. The Farmer Field School approaches to build the capacity and the confidence being part of the group have proven to be very useful in promoting such practices. The risk of failure of such practices is

also very low which make them easy to try. Even partial adoptions give benefit to the farmers

- ✰ Community management with FFS, CRPs, VOs and Mandal federations, has built in more ownership on the program - as explained above under the roles of the institutions of the poor
- ✰ Practices involving heavy earthen works like farm ponds, conservative furrows and trenches need more capacity building for the staff and labour involved. As the risk of failure of such models being high, more adaptation to local situations is needed and has a good capacity building plan for the people involved.
- ✰ Convergence of various interventions of IKP such as marketing, dairy along with CMSA will provide additional benefits.
- ✰ Convergence with line departments (Department of Agriculture, Department of Animal Husbandry) still seems to be a distant possibility due to rigid compartmentalization of their working. However discussions have with agencies have been undertaken to increase integration and linkages.
- ✰ The Community Resource Person based extension is working well for horizontal expansion of the program. Involving some more experienced resource organizations at state and district level will be important to strengthen the program.

Table 9.6: Pesticide Consumption (MT) during the Last Five Years (2005-06 to 2009-10)

States/UTs	*2005-06*	*2006-07*	*2007-08*	*2008-09*	*2009-10*
Andhra Pradesh	1997	1394	1541	1381	1015
Gujarat	2700	2670	2660	2650	2750
Haryana	4560	4600	4390	4288	4070
Jammu and Kashmir	1433	829	1248	2679.27	1640
Karnataka	1638	1362	1588	1675	1647
Kerala	571	545	780	272.69	631
Madhya Pradesh	787	957	696	663	645
Maharashtra	3198	3193	3050	2400	4639
Orissa	963	778	N/A	1155.75	1588
Punjab	5610	5975	6080	5760	5810
Rajasthan	1008	3567	3804	3333	3527
Tamil Nadu	2211	3940	2048	2317	2335
Uttar Pradesh	6671	7414	7332	8968	9563
West Bengal	4250	3830	3945	4100	NA
Total (India)	**39773**	**41515**	**43630**	**43860**	**41822**

Source: http://ppqs.gov.in/IpmPesticides.htm.

- ☆ Till now the program is implemented through the Women's Self Help Groups and their Federations. These institutions which are formed for thrift and credit, have shown that they can form an important platform for farming families. They can also then potentially form the springboard where by farming families can be organized into cooperatives for more focussed work along the value chain.
- ☆ AP s the only state which showed the reduction in pesticide use in the country in the last five years even when compared with the states growing Bt cotton or using low volume insecticides (Table 9.6). Cost of cultivation was brought due to adoption of NPM (Table 9.7).

Table 9.7: Reduction in Cost Due to Adoption of NPM

Crop	*Reduction in Cost Due to NPM (Rs/ha)*	*Reduction in Costs Due to Use of Organic Manures (Rs/ha)*	*Net Additional Income (Rs/ha)*
Rice	940	1450	5590
Maize	1319	2357	5676
Cotton	1733	1968	5676
Chillies	1733	1968	7701
Groundnut	1021	3462	10483
Vegetables	1400	390	3790

There has been a range of impacts of considerable importance to poor rural households in AP, as summarized below.

Economic Benefits	*Ecological Benefits*
Lower cost of production and substantial statewide savings	Better soil health, water conservation
Yield maintained or increased	Conservation of agro-biodiversity
Higher cropping intensity	Fewer pesticide related health problems
Higher household income	Smaller carbon footprint as a result of reduced use and production of inorganic fertilizers
Lower debt	
Lower risk perception and higher investment in agriculture	
Business innovation and new livelihood opportunities	

4. Conclusion

The experience by CSA shows that agro-ecological approaches like Non-Pesticidal Management (NPM) holds promise for bringing in ecological and economical sustainability in farming. The community managed extension system is the key for the success of such models and public investment has to be made to scale up the same.

5. Future Thrust

The research also should adopt the agroecological framework and assess and refine some of the practices which are evolved by farmers. The community managed extension system should be properly be studied and evolved into a model.

REFERENCES

ANGRAU. 2007. *Vyavasaya Panchangam* (Annual calendar of crop production practices in Telugu), Acharya NG Ranga Agricultural University, Hyderabad.

CSA. 2007. *Susthira Vyavasayam lo vividha padarthala Tayari* (Telugu), (Various preparations in sustainable agriculture), Society for Elimination of Rural Poverty, Government of Andhra Pradesh.

GAU. 2003. Non pesticidal management of brinjal fruit and shoot borer, technical bulletin, Department of Entomology, Gujarat Agriculture University.

KVK DDS. 2003. Annual Report. Unpublished report by Krishi Vigyan Kendra, Pasthapur.

Prakash A and Rao J. 1997. Botanical pesticides in agriculture. CRC Press. USA.

Prasad YG and Rao KV. 2006. Monitoring and evaluation: Sustainable cotton Initiative in Warangal District of Andhra Pradesh, Central Research Institute for Dryland Agriculture, Hyderabad, unpublished report.

Ramanjaneyulu GV, Chari MS, Raghunath TAVS, Zakir Hussain and Kavitha Kuruganti. 2009. Non Pesticidal Management: Learning from Experiences in Integrated Pest Management: Innovation-Development Process Volume-1. In: Rajinder Peshin and Ashok K Dhawan, Eds., Springer.

Rupela OP, Gowda CLL, Wani SP and Hameeda Bee. 2007. Evaluation of crop production systems based on locally available biological inputs in biological approaches to sustainable soil systems. In: Norman Uphoff, Ed. 2007.

Vijayalakshmi K, Subhashini B, Koul S. 1999. Plants in pest control: Pongamia, tulasi and aloe. Centre for Indian Knowledge Systems, Chennai, India.

Chapter 10

Management of Diseases in Organic Crop Production

Deeksha Joshi

ICAR-Indian Institute of Sugarcane Research,
Lucknow – 226002, Uttar Pradesh
Email: 14deeksha@gmail.com

1. INTRODUCTION

Organic farming as a cultivation system dates back to the prehistoric era. At the start of civilization, farming was predominantly based on the principles of organic farming and sustainable agriculture with humans co-existing in a close symbiotic association with nature. With growth, development and advent of chemical fertilizers and pesticides, the fine balance was disturbed resulting in various detrimental impacts both on the environment as well as human health. The World Bank report of 2007, ranks agriculture as one of the most hazardous occupations, alongside mining and construction, in developing and industrial countries mainly due to the large number of injuries caused by the improper/unsafe use of pesticides. Globally, pesticide residues in food, contaminated water as well as indirect and direct exposure to hazardous chemicals due to their unsafe/improper usage are responsible for >3 lakh deaths every year with majority of them occurring in developing countries like India. A study by FAO, WHO, and UNEP (2004) has broadly estimated that between 1 million and 5 million cases of pesticide poisoning occur each year. With more and more such reports corroborating the dangerous effects of chemical pesticides on environment and human health, there has been a continuously growing clamor amongst both the scientific community and public alike for a reversion to the older eco-friendly ways of farming like organic farming.

Organic farming emphasizes the use of management practices in preference to the use of off-farm inputs, taking into account that regional conditions require locally adapted systems. This is accomplished by using wherever possible, agronomic, biological and mechanical methods, as opposed to using synthetic materials, to fulfill any specific function within the system" (FAO, 1999). Fueled by a growing demand for organic foods, considerable area in the world has been converted to or is in the process of conversion to organic agriculture system. According to an International Federation of Organic Agriculture Movements (IFOAM) and Research Institute of Organic Agriculture (2014) report, there was 57.8 million hectares of organic agricultural land in the world by end of 2016 with Asia contributing around 4.9 million hectares. In case of India, National Centre for Organic Farming has reported that between the years 2003-04 to 2009-10, the country witnessed almost a 29 fold increase in area under organic agriculture. By 2016, the country had almost 1.5 m ha organic land with crops like cotton, fruits and vegetables accounting for the major share of organic agricultural land in the country.

As efforts are made to bring more and more area and crops under organic cultivation, effective management of diseases and insect pests becomes a major challenge facing the growers. Successful disease development in any crop occurs as a result of the interaction of three components *viz.* a virulent pathogen, a susceptible host and favourable environmental conditions also called as a 'disease triangle' (Agrios, 2005). Any effective disease management strategy should aim to break this triangle by targeting one or more of these components. For example, a susceptible host can be replaced with a resistant variety, pathogen can be eliminated by using pesticides *etc.* and unfavourable environmental conditions for disease development are created using various cultural methods. However, when shifting from conventional farming systems to organic cultivation, growers are faced with the problem that organic cultivation prohibits use of any synthetic chemical pesticide. Moreover, simply replacing the various pesticides with organically acceptable chemicals does not yield long term satisfactory results. In essence, effective and sustainable disease management under organic cultivation necessitates a holistic approach developed on the basis of thorough knowledge regarding target pathogens, their behavior, crop ecology, soil conditions, geographical region, environmental conditions *etc.* and requires the suitable amalgamation of multiple disease management tactics suited for the region. Also it should be kept in mind that organic farming does not aim at complete elimination or control of pathogens but rather focusses on suppressing these pathogens to levels not causing economic damage to crops.

2. Disease Managment Tools available and Approved for Use Under Organic Farming in India

2.1 Genetic

The first, most important and basic step for disease management under organic conditions is the use of resistant varieties. It is essentially an easy, safe and effective method for disease control which is compatible with all other disease management tools. Genetically modified crops like Bt cotton *etc.*, however, have been prohibited from use in organic farming. Resistant varieties constitute the first line of defence against attack by pathogens, but a major drawback with this option is the fact that it is not possible to develop resistant varieties against all pathogens infecting a crop. Breeding efforts are generally targeted towards developing resistant varieties against only a few economically important pathogens or against the prevalent races of a single pathogen. Moreover, with continuous cultivation of the same variety over several years, problems of resistance breakdown occur due to development of novel virulent pathotypes. In such conditions, growers have to continuously replace the existing variety with other newly developed resistant varieties. Also, it is a challenge to develop resistant varieties against soil borne pathogens like *Rhizoctonia solani, Sclerotium rolfsii, Sclerotinia sclerotiorum etc.*, which have a broad host range. Therefore, to manage such diseases against which resistant varieties do not exist, other available disease management tools have to be employed.

2.2 Cultural

Cultural control practices for diseases management primarily target to disrupt the environmental component of the disease triangle and also to some extent attempt to eliminate the pathogen by exclusion. It involves agricultural practices which either manipulate the environment to make it less suitable for pathogen survival and attack or preventive practices like sanitation which exclude the pathogen from the site. Quite often these simple modifications of a pathogens environment or habitat prove to be effective methods of disease control. Cultural control tactics are generally simple, low cost and compatible with a farmer's other management objectives (resistant varieties, chemicals *etc.*). One of the major aim of cultural control practices is to sustain or enhance the diversity of the ecosystems and develop a healthy soil. The more diverse the flora and fauna of a region, lesser are the chances of any one species becoming dominant and causing major disease outbreaks. For successful planning and implementation of cultural control for disease management, background information about the previous cropping systems followed in the field, prevalent pathogens and their life cycle, diseases outbreaks, time of appearance of the diseases in different crops of that region, soil type and nutrient status (organic carbon, macro and micronutrients) is essential. The major practices followed under cultural control for disease management are:

2.2.1 Pathogen Exclusion

Preventing contact between the host plant and pathogen is an important aspect of cultural control. This can be achieved by selecting healthy fields with no prior history of major disease outbreaks in the target crops, where possible, and/or by regularly changing the planting sites especially in case of diseases caused by soil borne pathogens. Certified disease free seeds should be used where available. Besides these methods, based on time of appearance and spread of various diseases in a given geographical area; modifications in planting time, maintenance of proper row to row distance *etc.* are some tactics which may help the host to evade the pathogen. However, it should be ensured that the certified seeds used are not treated with any fungicide.

2.2.2 Sanitation

Sanitation is a basic principle of cultural control and includes all activities which aim to eliminate or reduce the amount of inoculum present in the field, plant or storage area as well as to prevent the spread of inoculum to other healthy plants and plant parts. Agricultural practices like deep ploughing and burning of trash after harvest helps to remove or reduce pathogen inoculum in field. Incorporation of infected crop debris in soil exposes the pathogen to native soil microflora resulting in faster degradation and eradication of inoculum. Similarly, rouging of infected plants is an important practice which prevents secondary spread of diseases. Rouging involves regular field surveys for early spotting of infected plants and their tagging, collection, removal and burning as well as pruning of infected plants/ plant parts. The washing, cleaning, removal of soil and disinfestation of agricultural equipments at regular intervals helps to reduce spread of inoculum among different fields. Sanitation practices are especially useful for controlling secondary spread of various diseases like smuts (Agrios,4 2004).

2.2.3 Crop Rotation

Crop rotation is the practice of growing a series of dissimilar/different types of crops in the same area in sequential seasons. The major advantage of crop rotation in disease control is that it mitigates the build-up of pathogens and pests that often occurs under crop monoculture. Planting of non-host crops for a period of 3-4 years or planting of host crops in rotation with non host crops or leaving the field fallow for certain period of time especially during hot summer months helps in reducing the inoculum load in soil due to inability of the pathogen to survive in absence of a suitable host. Crop rotation is especially recommended for management of plant pathogens which survive in soil on crop debris for limited period of time. This practice can also improve soil structure and fertility by alternating deep-rooted and shallow-rooted plants as well as add various nutrients to the soil through the use of green manure in sequence with other crops. Efficacy of crop rotation can be further improved by combining it with different residue management, nutrient management and tillage practices (Brokus and Claassen 1992, Peters *et al.,* 2003; Honeycutt *et al.,* 1996). Another major advantage of crop rotation is that it improves the soil biodiversity resulting in build-up of native antagonistic microflora which further helps in suppression of pathogens (Lupwayi *et al.,* 1998, Garbeva *et al.,* 2004, Larkin 2008).

2.2.4. Intercropping

Intercropping refers to the cultivation of two or more crops, belonging to different families, simultaneously in the same field at the same time. This may involve either simultaneous planting of both crops or the planting of second crop after the first crop has completed its development. As a disease control tactic it relies on the rationale that crops belonging to different families are not likely to be attacked by the same group of pathogens thus hampering the active spread and buildup of pathogen inoculum. This practice may also result in the crops acting as a barrier for the spread of diseases. Indirectly, simultaneous cultivation of two different crops also affects changes in soil microflora and microbial dynamics as compared to cultivation of single crop reducing chances of inoculums buildup of any specific pathogen in the soil. Boudreau (2013) observed in his review that on comparing disease in monocrops and intercrops, especially foliar fungi, intercropping reduced disease in 73 per cent of more than 200 studies. Effective control of diseases like rust in leek (Theunissena and Schelling, 1996), Fusarium wilt in watermelon (Ren *et al.*, 2008), late blight of potato (Bouws and Finckh 2008), peanut bud necrosis (Gopal *et al.*, 2010) can be achieved through intercropping.

2.2.5. Water Management

Irrigation water serves as an important means for the secondary spread of pathogen inoculum within and between fields. The traditional method of irrigation *i.e.* flooding of fields, compounds this problem by carrying the inoculum from infected to non-infected plants as well as by creating water logged conditions in the field which are generally conducive for spread of most pathogens. Maintenance of proper drainage facilities in field and adoption of irrigation techniques like drip irrigation may help by restricting the spread of inoculum as well as contribute towards water saving (Feld *et al.*, 1990, Bell *et al.*, 1998, Blad *et al.*, 1978).

2.2.6 Soil Solarization

Soil solarization refers to a method of soil disinfestation in which moist soil is covered with transparent polythene mulch and exposed to sunlight during hot summer months. This causes heating of soil leading to the killing of various weeds as well as disease causing fungi and bacteria in the soil (Katan, 1981). The success of soil solarization is based on the fact that most plant pathogens are mesophylic, *i.e.* having a temperature threshold of approximately 37°C, beyond which the accumulation of heat effects over time are lethal. During summer months in sub-tropical/tropical conditions of India, the soil temperatures may reach as high as 55°C or more at 5 cm depth due to the solar heating of moist soil under transparent plastic films. At these high temperatures, various pathogens, nematodes and weeds are killed directly or indirectly by the temperatures achieved. Even thermo-tolerant and thermophilic microbes are weakened by the solarization process making them more susceptible to any microbial antagonist surviving the solarization process or applied to the soil after solarization (Katan, 1981). The success of soil solarization depends on the time of year when it is carried out, soil type, length of solarization and moisture to ensure maximum heat transfer. Soil moisture is a critical variable in soil solarization since the transfer of heat to weed seeds and plants and micro-organisms in soil is greatly increased by moisture. Similarly solarization is most effective when carried out during hot dry summer months. To further improve the efficacy of solarization, integrating it with other organically approved measures like seed or soil treatment with microbial antagonists has been shown to yield higher disease control (Gupta and Khosla, 2007, Gade *et al.*, 2007, Joshi *et al.*, 2009). Similarly incorporation of organic substrates like animal manures, crop residues *etc.* in moist soil prior to solarization also increases pathogen mortality as compared to solarization alone due to the combined effect of high temperatures along with the toxic effects of the volatile gases released by the organic substrates at these temperatures (Harendar, 2004, Kurt and Emir, 2004, Joshi *et al.*, 2009). Although soil solarization is the only effective method for disinfestation of soil under organic condition, the fact that it can be effectively used only in crops which are cultivated in warmer climates and during summer months imposes a severe limitation to its wide spread adoption. Moreover, covering large areas of soil with polythene mulch may not always be practical and cost effective. However, soil solarization can be an effective and feasible method to disinfest nursery beds and manage diseases like pre and post-emergence damping off, root rots and wilt complexes in different crops (Arora and Pandey 1989; Sofi *et al.*, 2007; Harendar, 2004, Gupta and Khosla, 2007; Gade *et al.*, 2007). Nevertheless, in organic farms, use of soil solarization to combat pests and diseases is restricted to circumstances where a proper rotation or renewal of soil cannot take place and it can be carried out only when permission has been given by the certification programme.

2.2.7 Compost Mediated Suppression and Balanced Plant Nutrition

Organic farming essentially relies on application of various organic substrates like composts, animal and green manures, crop residues *etc.* to meet the nutritional demands of the crop. An added advantage of the application of these organic substrates is that besides being the major source of nutrition for the plants; in some cases composts have been observed to suppress certain soil borne and foliar plant diseases. The first efforts in this direction were initiated during the 1950's when the nursery industry in the USA and Australia developed lower cost alternatives to peat from tree barks for use in potting mixes and ground beds for the production of woody ornamentals. Several growers observed that composted bark could suppress root rots caused by the pathogen *Phytophthora cinnamomi* against which effective resistant varieties or chemical control procedures were otherwise not available. Today studies have reported effective suppression of several pathogens like *R. solani, Fusarium, Pythium* sp. *etc.*, in field conditions (Hoitink 1997, 2001). Disease suppression by application of composts may be either due to the collective activity of the microbes present in the composts or in some cases has been attributed to the presence of specific antagonistic microbes like *Trichoderma* sp. in the applied compost (Hoitink *et al.*, 2006). The general microbial load present in the compost competes with plant pathogens for carbon, space, energy and nutrients like nitrogen (Hoitink, 1997, 2001). Moreover, specific antagonistic microbes may suppress the pathogen by direct antagonism or indirectly by inducing systemic resistance in plants. Soil application of composts from different sources (FYM, peat, bark compost, poultry manure, composted *Urtica*) has shown effective suppression of diseases caused by *Pythium* and *Phytophthora* sp., *R. solani, Fusarium* sp. and *S. rolfsii* (Kwok *et al.*, 1997; LaMondia *et al.*, 1999; Lewis *et al.*, 1992; Scheuerell *et al.*, 2005; Joshi *et al.*, 2008; Tuitrert *et al.*, 1998; Hoitink, 1997, 2001). Even though composts may serve the dual purpose of supplying nutrition to plants as well as acting as suppressors of disease, when using them in organic farming situations, it should be ensured that only composts produced on the organic farm itself or approved for use in organic farm should be used.

2.3 Microbial Pesticides for Biological Control of Diseases

Under such situations in organic farming where adoption of various cultural and genetic methods do not yield successful disease control, farmers have been permitted the use of various microbial pesticides. The use of such biopesticides for disease control can be undertaken only after the approval of and under guidance of an accredited certification body. Microbial biocontrol agents mainly work by altering the biotic and abiotic environment from one that favours disease/pathogen to one that discourages accumulation of infective or parasitic material and reduces the activity of the pathogen. As a strategy and philosophy for reducing crop losses from diseases, biological control is not new to agriculture. Practices of early farmers like rotation of crops, burrowing disease infested crop residues, fertilizing with organic manures *etc.* also indirectly resulted in enhancing population of beneficial antagonistic microbes which in turn suppressed disease causing pathogens. Essentially, biological control is based on the natural phenomenon that every biological entity has its adversary in nature. Darwin had said that in the struggle for life everything is born to eat and get eaten (Stone, 1980) and this maintains the biological balance under natural conditions. Amongst the numerous microbial antagonists explored for disease control till date, biopesticides based on *Trichoderma* sp., *Pseudomonas fluorescens* and *Bacillus* sp. are the most extensively used.

2.3.1 *Trichoderma* sp.

Trichoderma is a cosmopolitan fungi which is widely distributed all over the world and occurs in all types of soil and other natural habitats especially those containing high organic matter. Two species of *Trichoderma, viz. T viride* and *T. harzianum* are currently registered in India as biopesticides, for control of a wide range of pant pathogens causing diseases like wilt complexes, root rots, damping off, seedling rots, collar rots, white rots *etc.* (Sharma *et al.*, 2014). *Trichoderma* spp. can employ a diverse array of mechanisms to inhibit its target pathogen including direct mechanisms like mycoparasitism and/or antibiosis, competition for nutrient, oxygen or space or indirect mechanism like inducing disease resistance or tolerance in crop plants. Even in the absence of disease causing pathogen, this fungus exhibits growth promoting effects in a number of crops which manifests itself as improved germination, increased root growth, shoot growth and seedling vigour, changes in plant nutrient status due to better uptake of nutrients from soil which generally translates in higher yields. (Howell, 2003). *Trichoderma* based biopesticides have been classified as Class–IV which indicates low hazard (green label) indicating that they are extremely safe to mammals, man and other non-target organisms like pollinators, bees, fishes *etc.*, have no phyto-toxicity and are exempt from residual analysis (Chandra *et al.*, 2005).

2.3.2 *Pseudomonas fluorescens*

These are gram negative bacteria characterized as chemoheterotrophic motile rods with polar flagella. They have simple nutritional requirements, and this is reflected by their relative abundance in nature. Fluorescent *Pseudomonas* spp, can generally be distinguished from other pseudomonads by their ability to produce water-soluble yellow green pigment in specific media. These bacteria are credited with the potential of biological control of plant pathogens and growth promotion. They are known to produce a large number of secondary metabolites, which may affect the growth, and health of plants and/or may exert antagonistic activity against plant pathogens (Lugtenberg and Kamilova, 2009). As in case of *Trichoderma*, formulations of this bacteria are classified under Class-IV indicating low hazard, no phyto-toxicity and exempt from residual analysis (Chandra *et al.*, 2005).

2.3.3 *Bacillus* sp.

Bacillus subtilis, originally named *Vibrio subtilis*, is a gram positive, rod-shaped, bacteria which can form a tough, protective endospore, allowing it to tolerate extreme environmental conditions. The formulated product is generally developed from spores which are concentrated and dried into a powder. *Bacillus subtilis* may act by competing for space and nutrients with the pathogen or by exerting direct antagonistic activity by production of various antifungal compounds and enzymes (Lugtenberg and Kamilova, 2009; Cawoy *et al.*, 2011).

2.3.4 Application of Biocontrol Agents

Presently around 50 formulations based on the above three microbes have been registered under the Insecticides Act (1968) by the Central Insecticides Board and Registration Committee for management of various diseases of crop plants in India. These biopesticides are currently available in various dry as well as liquid formulations mainly wettable powders (WP), water soluble (WS) and aqueous suspension (AS). The formulations can be applied as seed and seedling treatment, soil application and/or foliar sprays depending on the target pathogen and crop. Table 10.1 lists the major biopesticides available in India, their formulations and diseases controlled by them in various crops.

Table 10.1: Commercial Formulations of Major Bioagents Available in India and Diseases Controlled

Microorganism	*Commercial Products Available in India*	*Major Diseases Controlled**
Bacillus subtilis	Biotilis ™ (Agri Life), Milastin K (Kan Biosys Pvt. Ltd.), Biotilis (Som Phytopharma Ltd.), Biosubtilin/Rog rakshak (Biotech International Ltd.)	*Verticillium* wilt in crops, damping off in vegetables, powdery mildew in mango, bacterial wilt of solanaceous crops, root rots in some vegetable crops, gray mold and bacterial spot of tomato, post harvest diseases of stone fruits
Trichoderma harzianum and *T. viride*	Bioderma-H/barrier and Bioderma/ Protector (Biotech. International Ltd.), Niprot (Pest Control India Ltd.), Bioprtotector *T. viride* (Manidharma Biotech Pvt. Ltd.), Shakti Trichoderma (Nivshakti Bioenergy PVt. Ltd.), Ecosom (Agri Life), Tricho Power (K N Biosceinces India Ltd.), Bioshield (Shree Biotech. and research Inputs), Fungiderma (Sai Agrotech)	Pre- and post-emergence damping off in various crops, root rots, white rot, collar rot in vegetables and pulses, banded leaf and sheath blight, anthracnose, smut of wheat, wilt complexes in pulse crops.
Pseudomonas fluorescens	Bio Protector (Manidharma Biotech Pvt. Ltd.), Shakti Pseudomonas (Nivshakti Bioenergy Pvt. Ltd.), Monas (K N Biosceinces India Ltd.), Ecomonas (P J Margo PVT. Ltd.), Su-mona (Pest Control India PVt. Ltd.), Biowin (Bio Agri Solutions), Biomonas/Shield Plus (Biotech International Ltd.)	Damping off in vegetable crops, bacterial wilt in solanaceous crops, blast and sheath blight of rice, leaf spot of wheat, dry root rot of pulses, antracnose and Fusarium wilt in vegetables.

**Source*: Cawoy *et al.* (2011); Cladwell *et al.* (2013); Sahrma *et al.* (2014), Ganeshan and Manoj Kumar (2005).

Seed Treatment

Seed coating with biocontrol agents is one of the easiest methods of delivering the biocontrol agents for the management of plant diseases. Besides protecting the germinating seed against various soil borne pathogens, seed treatment with bioagents also suppresses the existing seed borne inoculum, particularly the externally seed borne inoculum of diseases affecting both above ground and below ground plant parts. Even some internally seed borne disease like loose smut of wheat can be suppressed by seed treatment with bioagents (Singh and Maheshwari, 2001). When applied through seeds, these bioagents not only colonize spermoplane and spermosphere, but also rhizoplane and rhizosphere.

Seedling Treatment

In crops like vegetables and rice which involves raising of seedling in nursery followed by transplanting, dipping of seedling roots in aqueous suspension of biocontrol agents provides control of several soil borne pathogens and also may result in enhanced growth and vigour of crop in field.

Soil Treatment

Application of bio-agents directly to soil is the most effective method to suppress soil-borne pathogens like *Pythium* sp., *Phytophthora* sp., *R solani, S rolfsii, S sclerotiorum, Verticillium, Fusarium* spp. *etc.* which cause a wide array of diseases like root, stem and collar rots, damping off and wilts. In polyhouse cultivation or nurseries of crops, the bioagent formulations may be applied directly to soil. However large scale soil application of biopesticides in field is not always feasible since it requires bulk material and there may also be problem with uniform distribution of the biopesticide. In such cases, the biopesticides especially *Trichoderma* formulations, may be delivered to field through organic substrates like farmyard manure, vermicompost *etc.* In this method, the formulation is first mixed with moist FYM in a heap on farm itself @ 1 kg formulation/100 kg FYM. It is then covered with a polythene mulch and incubated for 15-20 days on the farm with turning as 5 days intervals to allow colonization by the bioagent. The colonized FYM is then distributed in field followed by irrigation. This methodology not only cuts down on cost but also provides a substrate for the bioagent to establish and proliferate.

Foliar Application

For control of foliar diseases like brown spot, blast and sheath blight of rice, spot blotch of wheat *etc.* foliar sprays of various bioagents have been recommended. However, success of an antagonist on leaf/sheath surface depends largely on the ability of the bioagent to colonize the leaf surface which may be deeply influenced by various environmental factors (humidity, temperature and sun light). As such foliar applications of bioagents is preferred during evening hours and tends to be more successful in rainy season crops.

2.4 Chemical Control

Under situations where all available and approved genetic, cultural and biological methods for disease management fail; farmers have been permitted the use of some non-synthetic, botanical or mineral pesticide products under organic farming. These include pesticides of mineral origin, particle film barriers, botanical pesticides made from plants, light mineral oils, plant extracts *etc.* However, some these chemicals/botanicals may exert a phytotoxic affect or have non-selective effects on the local ecosystem and as such their use has been "restricted". This implies that such chemicals should be applied only as a last resort and the conditions and the procedure for their use shall be set by the certification programme.

2.4.1 Materials of Mineral Origin

The two major fungicides of mineral origin permitted for used in organic farming for disease control are sulphur and copper salts and their various compounds.

2.4.1.1. Sodium Bicarbonate

Sodium bicarbonate, commonly known as baking soda has been used as a fungicide since the 1930's. Its use in organic farming for disease control has been permitted with restrictions. Sodium bicarbonate disturbs the potassium/sodium ion balance within the fungal pathogen cell causing collapse of cell wall (Cladwell *et al.,* 2013). Baking soda is generally applied as spray at a 1 per cent concentration and it's sprays have shown some efficacy in management of powdery mildews (Yildrim *et al.,* 2002), anthracnose, early tomato blight, leaf blight and spots and should be used in a preventive manner at the first indication of a disease (Cladwell *et al.,* 2013). For plants already affected with powdery mildew, it is better to wash/ hose down the plants so that maximum spores may be dislodged before spraying, thus giving better results. Besides water, some horticultural oil or soap may be added to the mixture to allow for better coverage of leaves and stems and may also enhance efficacy. However sodium bicarbonate may cause foliar burns on plants which appears as brown or yellow patches at the end of the leaves. In case of appearance of such symptoms, further dilution of the product is recommended to minimize the damage.

2.4.1.2. Quick Lime, Chloride of Lime and Calcium Hydroxide

These chemicals are permitted for use in organic farming but with restrictions. Although these compounds can be used directly as basic fungicides for control of some diseases like brown rot of peach cause by *Monilinia fructicola*; they are mainly used to produce other more effective fungicides. Calcium hypochlorite or chloride of lime is an inorganic compound with formula $Ca(ClO)_2$. It is generally marketed as chlorine powder or bleach powder for water treatment and can be used for sterilization of various agricultural implements so as to prevent spread of disease inoculum.

2.4.1.3. Clay (Bentonite, perlite, vermiculite, zeolite)

Use of clay is approved in organic farming system. It is a non-synthetic naturally occurring mineral which when prepared as a water suspension and sprayed, produces a dry white film layer on the plant surface. This particle film technology aims to control both arthropod pests and diseases of plants with a hydrophobic particle barrier (Cladwell *et al.,* 2013). It is primarily used as a dry, inert carrier material in various microbial pesticide and botanical preparations, soaps, oils *etc.* approved for use in organic farming. However, it also exerts direct impact as an insecticide by acting as a physical barrier preventing insects from feeding on the soft plant tissue and as a repellant by creating unsuitable surface conditions for egg laying by insects. As such it can be used for control of various sucking insect pests which transmit viral diseases.

2.4.1.4. Light Mineral Oils

A light mineral oil is any of the various colorless, odorless, light mixtures of higher alkanes derived from the distillation of petroleum. Light mineral oils have been used since 1860's for crop protection especially in management of insect pests. In plant disease management, petroleum oils have been predominantly explored for control of powdery mildews in grape vine and to some extent for control of diseases like banana leaf spot. The mode of action of these oils is probably the disruption of fungal membranes, interference with spore germination and attachment and induction of host plant resistance (Cladwell *et al.*, 2013). Besides their direct use as fungicides, light mineral oils are also frequently used as spray adjuvants and may improve efficacy of other products like soaps. However, they are incompatible with sulphur and copper based fungicides on some crops. The use of light mineral oils in organic farming is subject to the permission of the certification body.

2.4.1.5. Permanganate of Potash

Potassium permanganate ($KMnO_4$) is a violet coloured water soluble salt with fungicidal and bactericidal properties. Potassium permanganate acts by oxidizing all organic materials on contact. It has been used for management of powdery mildew caused by the fungi *Oidium*, downy and powdery mildew of grapes, Verticillium wilt of various solanaceous crops and Phomopsis leaf spot of grapevine. However, this salt is extremely phytotoxic if sprayed at higher doses (> 300 g/hl). Care should be taken not to mix this salt with organic substances (rotenone, Bt, *etc.*), due of its corrosive nature.

2.4.1.6. Diatomaceous Earth

Diatomaceous earth is comprised of the calcium carbonate skeletons of diatoms that developed in water bodies. Diatomaceous earth is primarily used for insect control and acts by causing the desiccation of insect body by interfering with the waxy layer on the body surface. Although it has no reported direct impact on plant diseases, it can be used for control of various insect vectors of viral plant diseases.

2.4.2 Material from Plant Origin

2.4.2.1 Plant Based Extracts

In Indian agriculture, the practice of using plant derivatives or botanical pesticides and fungicides for pest control dates back at least two millennia predating the discovery of all major classes of synthetic chemical insecticides and fungicides. Neem and its various derivatives (oil, seed extract, kernel extract, leaf extract) are the most common botanical pesticides currently being used in organic farming. Although these pesticides are mainly used for insect control their efficacy against plant pathogenic fungi, bacteria and nematodes has also been documented (Locke, 1995; Imtiaz, 2005). Soil application and sprays of neem oil have been shown to control phytopathogenic fungi like *Fusarium* sp. *R. solnai, S. rolfsii, Sclerotinia sclerotiorum, Botrytis cinerea, Puccinia, Magnaporthe grisea etc.* causing diseases like rusts, scabs, powdery mildew, blast and rots (Pasini *et al.*, 1997). Aqueous extracts of neem leaves and seed kernel extracts have shown effective control of diseases like leaf stripe of barley caused by *Drechslera graminea* (Paul and Sharma, 2002), groundnut leaf spots caused by *Phaeoisariopsis personata* and rust caused by *Puccinia arachidis* (Ganapathy and Narayansamy, 1990), fusarium wilt of brinjal and tomato (Hadian *et al.*, 2011) *etc.* The presence of compounds like nimbidin (a triterpenoid), isomeldenin, nimonol and azadirachtin have been implicated in the antimicrobial activity of neem.

In addition to neem and its products, garlic (*Allium sativum*, Linn.) extracts are also extensively recommended for disease control in organic farming in India (Yadav NCOF, Chandra *et al.*, 2005). Garlic extracts can suppress a number of plant diseases like powdery and downy mildew on cucumbers, late blight of potato, bacterial pathogens infecting tomato *etc.* (Russel and Mussa, 1977; Curtis *et al.*, 2004; Obagwu and Korsten, 2003; Slusarenko *et al.*, 2008). Garlic contains the antimicrobial substance allicin (diallylthiosulphinate), a volatile phytoanticipin which is produced in garlic when the tissue is damaged due to wounding and the substrate alliin mixes with the enzyme alliin-lyase. It works by crossing the cell membrane and undergoes thiol-disulphide exchange reactions with free thiol groups in cell proteins and these proteins then form the basis of its antimicrobial action (Slusarenko *et al.*, 2008). Eucalyptus (*Eucalyptus globulus*, Labill.), onion (*Allium cepa*) turmeric (*Curcuma longa*, Linn.), chili (*Capsicum annum*) and ginger (*Zingiber officinale*, Rosc) are some other plants whose extracts are also frequently recommended for disease control under organic farming conditions (Chandra *et al.*, 2005). However, it should be kept in mind that these extracts can be used only after approval by an accredited certification body. Also the extracts should preferably be prepared from plants grown under organic cultivation only.

2.5 Miscellaneous

In addition to microbial pesticides and some chemicals; the use of soft soaps, homeopathic, ayurvedic and some herbal and biodynamic preparations has also been permitted for disease control under organic farming situations. Most of these preparations can be made by the farmers themselves at their own farms using substances available at the farm.

2.5.1 Soft Soaps

Soft soaps, also called as fatty acid potassium salt, are obtained by mixing vegetal oils with alkaline substances such as soda and potassium hydroxide. The use of soft soaps for disease management in organic farming is permitted. Although soft soaps have been mainly exploited as insecticides against soft bodied insects or as additives with other crop protectant chemicals, they are reported to show some efficacy in management of powdery mildews (Mmbaga, 2002), sooty blotch and flyspeck of apples (Batzer *et al.*, 2002), black spot, canker, leaf spot, and rust diseases

(Ellis *et al.*, 1996). However, frequent sprays throughout the growing season with good plant area coverage are essential to attain effective disease control by soaps.

2.5.2 Hot Water Treatment/Moist Hot Air Treatment

Internal and externally seed borne inoculum is a major mode of transmission of several plant pathogens. Since seed treatment with fungicides is prohibited in organic farming, growers can resort to hot water or moist hot air treatment of seeds to eradicate the seed borne inoculum. Hot water treatment of seeds has been found to be especially successful in controlling seed-borne diseases like Alternaria leaf spot, black leg (*Phoma* spp.) and black rot in cole crops; Alternaria leaf spot of carrot; anthracnose (*Colletotrichum* sp.), Phomopsis blight and Verticillium wilt of brinjal; anthracnose and bacterial leaf spot of pepper; purple blotch and Stemphylium leaf blight of onion; early blight of tomato bacterial spot in tomato; tobacco mosaic and tomato mosaic in pepper; loose smut of wheat *etc.* (Doling 1965; Hermansen *et al.*, 1999; Nega *et al.*, 2003; Hossain *et al.*, 2009; Sridhar *et al.*, 2013). These diseases were considerably reduced following the treatment without any major impact on seed viability. For hot water treatment, water temperature is generally kept between 50-54°C, depending on the crop, and the treatment period may vary from 10 to 30 minutes. Moist hot air treatment of sugarcane setts at 54°C for 2.5 h has been observed to be one of the most effective methods for control of a number of sett borne diseases of sugarcane like grassy shoot disease, leaf scald and smut (Singh and Agnihotri, 1987).

2.5.3 Compost Teas

Compost tea or compost extracts is a term used to describe an aqueous extract made by steeping compost in water (Scheuerell, 2002; Ingham 2005). Based on the method of preparation, compost teas can be divided into two main categories *viz.* non-aerated fermented compost tea (NCT), and aerated fermented tea (ACT). NCT's originated as home made brews and are prepared by suspending compost in a container of water, in a ratio of 1:3 to 1:10 for up to 14 days, with intermittent stirring to extract the nutrients. The suspension is then filtered and applied to plants as sprays or soil drench near plant roots. ACT is generally brewed in mechanized systems for shorter periods of time (18 h to 3 days) and are supplemented with oxygen, nutrients and/or microbial starter cultures to enhance the biological activity of the tea (Scheuerell, 2004; Ingham, 2005; Naidu, 2010; Dukare *et al.*, 2011). Both NCT and ACT have been credited with suppressing a broad spectrum of soil borne and foliar pathogens. Soil drench and foliar sprays of compost teas has been shown to effectively control diseases like apple scab (Cronin *et al.*, 1996), powdery mildew disease on rose and tomato (Segarra *et al.*, 2009; Kone´ *et al.*, 2010); grey mold diseases on vegetables crops, strawberries, geranium, and tomatoes (Scheuerell and Mahaffee 2006; Kone *et al.*, 2010); damping off on cucumber seedling (Scheuerell and Mahaffee, 2004), bacterial spot of tomato (Al-Dahmani *et al.*, 2003); late blight of potato (Al-Mughrabi, 2007); common scab of potato tubers (Al-Mughrabi *et al.*, 2008); Choanephora wet rot on okra (Siddiqui *et al.*, 2009); and anthracnose diseases on pepper and cucumber (Sang and Kim, 2011). Compost teas mainly work by introducing a diverse range of microorganisms to agricultural systems and the degree of suppression accorded by the tea has frequently been correlated to the microbial dynamics of the tea. These microbes may directly antagonize the pathogen, inhibit spore germination or compete with pathogen for nutrients in the plant rhizosphere and phyllosphere. In addition, compost teas may also induce systemic resistance in plants against several pathogens (Scheuerell, 2002, 2004). Typically, compost teas can be prepared from composts derived from animal manures, landscape and agricultural plant material, bio-solids or food wastes; however, for use in organic farming systems, such teas should be prepared only from organic material produced on the organic farm unit itself or such composts which have been approved for use in the organic unit by accredited certification body.

2.5.4 *Panchgavya* and *Dashagavya*

Panchgavya refers to a blend of five products obtained from cows, namely fresh cow dung (5 kg), cow urine (3 l), cow milk (2 l), curd (2 l) and cow ghee (1 kg). *Dashagavya* is prepared by adding ingredients like sugarcane juice (3 l), tender coconut water (3 l), banana paste (12 fruits) and grape juice (2 l) to Panchgavya. To prepare the formulations, cow dung and ghee are initially mixed in a container and fermented for 3 days with intermittent stirring. The rest of the ingredients are added on the fourth day and allowed to ferment for 15 days with stirring twice daily. The final formulation is ready in 18 days. 500 g jaggery can be used in place of sugarcane juice while 100g yeast powder mixed with 100 g jaggery and 2 lit of warm water can be used in place of grape juice. These mixtures can be applied as seed treatment, soil drench (50 l/ha) or foliar sprays (3-4 l formulation diluted with 100 l water). Vedic literature has emphasized the role of panchgavya both in soil and plant health promotion as well as in according protection to plants against diseases (Sadhale, 1996). With the growing popularity of organic farming and organic foods, there has been a renewed interest in the use of such ancient tools for disease control and several studies are being carried out to scientifically validate the potential of Panchgavya against diseases and understand the underlying mechanisms. Sireesha (2013). observed that foliar sprays of panchgavya gave effective control of rice blast caused by *Magnaporthe grisea*. Similarly, seed treatment with panchgavya showed significant reduction in *Curvularia lunata* causing grain discoloration in rice while foliar sprays were found effective in reducing tip-over disease of banana caused by *Erwinia caratovora* (Nagaraja *et al.*, 2012), Phytophthora blight and anthracnose of bell pepper (Ashlesha and Paul, 2014) and early blight of tomato (Ramesh *et al.*, 2015). The presence of a diverse and high population of various fungal, bacteria, yeasts and other beneficial microbes is generally attributed to be the major factor contributing towards the fungicidal and bactericidal action of these mixtures. Sarkar *et al.* (2014) had observed enhanced production of defense related enzyme polyphenol oxidase in plants treated with panchgavya indicating that besides the direct antagonistic activity of the microbes, induced systemic resistance may also play a role in disease suppression.

2.5.5 Biodynamic Preparations

Biodynamic agriculture was originally developed by Rudolf Steiner (Steiner, 1924). and is based on the concept of "a holistic understanding of agricultural processes". This method shares considerable similarity with organic farming in as much as it primarily depends on use of various organic manures and composts and excludes the use of chemical fertilizers and pesticides. Under this method various biodynamic preparations were developed by Steiner with biodynamic preparation BD 508 reported to have prophylactic activity against diseases like mildews, blights and some other fungal diseases (Yadav, NCOF). BD 508 is made from fresh tissue of horse tail plant (*Equisetum arvense*). The tissue is boiled with water for 20 min to get a tea. The filtered tea can be stored in glass bottles, diluted and sprayed as per requirement.

2.5.6 Others

The various ingredients of panchgavya, alone and in combination with other substances, have been formulated to develop different formulations like jeevamrut and beejamrut, which can also be used as seed treatment and foliar sprays to protect plant against diseases in organic farming. Jeevamrut refers to a mixture of cow dung (10 kg), cow urine (10 l), jaggery (2 kg), any pulse grain flour (2 kg) and live forest soil (1 kg) mixed in 200 lit water. This mixture is fermented for 5 to 7 days with regular stirring thrice daily and applied in one acre with irrigation water. Seed treatment with jeevamrut has been shown to control anthracnose of chili, black scurf and brown rot of potato (Sridhar *et al.*, 2013; Yadav, NCOF). Beejamrut is another such mixture. It comprises of cow dung (5 kg), cow urine (5 l), cow milk (1 l) and calcium carbonate (250 g). It is used for seed treatment with seeds soaked in the solution and left overnight. Beejamrut is reported to protect seeds against various soil borne diseases especially fungal disease and the nitrogen rich urine also aids in seed germination (Yadav NCOF, Sridhar *et al.*, 2013). As in case of panchgavya and compost teas, the activity of jeevamrut and beejamrut may be due to the presence of beneficial microbes. Besides these formulations, some constituents of Panchgavya like cow urine are being used alone as seed treatment and foliar sprays for control of seed borne and foliar diseases in plants. Although there are few detailed scientific reports on the use and efficacy of these formulations, most Indian organic handbooks and resource material for organic farming today recommends use of cow urine, jeevamrut and beejamrut as seed treatment and foliar sprays for protection against a range of plant diseases (Sridhar *et al.*, 2013).

REFERENCES

Agrios GN. 2004. Plant Pathology, 5th Ed. Academic Press, San Diego.

Al-Dahmani JH, Abbasi PA, Miller SA and Hoitink HAJ. 2003. Suppression of bacterial spot of tomato with foliar sprays of compost tea under greenhouse and field conditions. Plant Disease, 87: 913-919.

Al-Mughrabi KI, Berthe´leme´C, Livingston T, Burgoyne A, Poirier R and Vikram A. 2008. Aerobic compost tea, compost and a combination of both reduce the severity of common scab (*Streptomyces scabiei*) on potato tubers. Journal of Plant Science, 3: 168-175.

Al-Mughrabi KI. 2007. Suppression of *Phytophthora infestans* in potatoes by foliar application of food nutrients and compost tea. Australian Journal of Basic and Applied Science, 1: 785-792.

Arora DK and Pandey AK. 1989. Effects of soil solarization on Fusarium wilt of chickpea. Journal of Phytopathology, 124: 13-22.

Ashlesha and Paul YS. 2014. Antifungal Bioefficacy of Organic Inputs against Fungal Pathogens of Bell Pepper. Indian Journal of Research, 3: 4-9.

Batzer JC, Gleason ML, Weldon B, Dixon PM and Nutter FW. 2002. Evaluation of Post-harvest Removal of Sooty Blotch and Flyspeck on Apples Using Sodium Hypochlorite, Hydrogen Peroxide with Peroxyacetic Acid, and Soap. Plant Disease, 86: 1325-1332.

Bell AA, Liu L, Reidy B, Davis RM and Subbarao KV. 1998. Mechanisms of subsurface drip irrigation-mediated suppression of lettuce drop caused by *Sclerotinia minor*. Phytopathology, 88: 252-259.

Blad BL, Steadman JR and Weiss A. 1978. Canopy structure and irrigation influence white mold disease and microclimate of dry edible beans. Phytopathology, 68: 1431-1437.

Bockus WW and Claassen MM. 1992. Effects of crop rotation and residue management practices on severity of tan spot of winter wheat. Plant Disease, 76: 633-636.

Boudreau MA. 2013. Diseases in Intercropping Systems. Annual Review of Phytopathology, 51: 499-519.

Bouws H and Finckh MR. 2008. Effects of strip intercropping of potatoes with non-hosts on late blight severity and tuber yield in organic production. Plant Pathology, 57: 916–922.

Cawoy H, Bettiol W, Fickers P and Ongena M. 2011. Bacillus-Based Biological Control of Plant Diseases In: Pesticides in the Modern World - Pesticides Use and Management (Eds.) Stoytcheva M. InTech China Shangha.

Chandra K, Greep S and Srivathsa RSH. 2005. Biocontrol agents and biopesticides. Retrieved from National Centre of Organic Farming http: //ncof.dacnet.nic.in/Training_Manual.html

Cladwell B, Sideman E, Seaman A, Shelton A and Smart C. 2013. Resorce guide for organic insect and disease management. New York State Agricultural Experiment Station (NYSAES) 630 West North Street Geneva, New York 14456.

Cronin MJ, Tohalem DS, Harris RF and Andrews JH. 1996. Putative mechanism and dynamics of inhibition of the apple scab pathogen *Venturia inaequalis* by compost extracts. Soil Biology Biochemistry, 28: 1241–1249.

Curtis H, Noll U, Störmann J and Slusarenko AJ. 2004. Broad-spectrum activity of the volatile phytoanticipin allicin in extracts of garlic (*Allium sativum* L.) against plant pathogenic bacteria, fungi and Oomycetes. Physiological and Molecular Plant pathology, 65: 79-89.

Doling DA. 1965. Single-bath hot-water treatment for the control of loose smut (*Ustilago nuda*) in cereals. Annals of Applied Biology, 55: 295-301.

Dukare AS, Prasanna R, Dubey SC, Nain L, Chaudhary V, Singh R and Saxena AK. 2011. Evaluating novel microbe amended composts as biocontrol agents in tomato. Crop Protection, 30: 436-442.

Ellis BW, Bradley FM, and Atthowe H. 1996. The Organic Gardener's Handbook of Natural Insect and Disease Control. Rodale Press, Emmaus, Pensylvania, pp.535.

Ellis MA, Ferree, DC. Funt RC and Madden LV. 1998. Effects of an apple scab-resistant cultivar on use patterns of inorganic and organic fungicides and economics of disease control. Plant Disease, 82: 428-433.

FAO. 1999. Organic Agriculture, Food and Agriculture Organization of the United Nations, Rome.

Feld SJ. Menge JA and Stolzy LH. 1990. Influence of drip and furrow irrigation on Phytophthora root rot of citrus under field and greenhouse conditions. Plant Disease, 74: 21-27.

Food and Agriculture Organization (FAO), United Nations Environment Programme (UNEP), and World Health Organization (WHO). 2004. "Childhood Pesticide Poisoning: Information for Advocacy and Action." UNEP, New York.

Gade R, Zote KK and Mayee CD. 2007 Integrated management of pigeon pea wilt using fungicide and bioagent. Indian Phytopathology, 60(1): 24-30

Ganapathy T and Narayanasamy P.1990. Effect of plant products on the incidence of major diseases of groundnut. International Arachis newsletter, 7: 20-21.

Ganeshan G and Manoj Kumar A. 2005. *Pseudomonas fluorescens*, a potential bacterial antagonist to control plant diseases, Journal of Plant Interactions, 1: 3, 123-134, DOI: 10.1080/17429140600907043

Garbeva P, van Veen JA and van Elsas JD. 2004. Microbial diversity in soil: Selection of Microbial Populations by Plant and Soil Type and Implications for Disease Suppressiveness Annual Review of Phytopathology, 42: 243-270.

Gopal K, Muniyappa V and Jagadeeswar R. 2010. Management of peanut bud necrosis disease in groundnut through intercropping with cereal and pulse crops. Archives of Phytopathology and Plant Protection, 43: 883–91.

Gupta AK and Khosla K. 2007. Integration of soil solarization and potential native antagonists for the management of crown gall on cherry rootstock colt. Scientia Horticulturae, 112: 51-57.

Hadian S, Rahnama K, Jamali S and Eskandari A. 2011. Comparing Neem extract with chemical control on *Fusarium oxysporum* and *Meloidogyne incognita* complex of tomato. Advances in Environmental Biology, 5: 2052-2057.

Haggag, WM and Saber MSM. 2007. Suppression of early blight on tomato and purple blight on onion by foliar sprays of aerated and non-aerated compost teas. Journal of Food Agriculture and Environment, 5: 302–309.

Harender R. 2004. Effect of solarization of farmyard manure amended soil for management of damping-off caused by *Pythium ultimum, Rhizoctonia solani, Scleroium rolfsii* in vegetable crop nurseries. Indian Journal of Agriculture Sciences, 74: 425-29.

Hermansen A. Brodal G and Balvoll G. 1999. Hot water treatments of carrot seeds: effects on seed-borne fungi, germination, emergence and yield. Journal of Seed Science and Technology, 27: 599-613.

Hoitink HAJ, Stone AG and Han DY. 1997. Suppression of plant diseases by composts. HortScience, 184-187.

Hoitink HAJ, Krause MS and Han DY. 2001. Spectrum and mechanisms of plant disease control with composts. In: Stoffella, P.J., Kahn, B.A. (Eds.), Compost Utilization in Horticultural Cropping Systems. Lewis Publishers, Boca Raton, Florida, pp. 263.

Hoitink HAJ, Madden LV and Dorrance AE. 2006. Systemic resistance induced by *Trichoderma* spp.: Interactions between the host, the pathogen, the biocontrol agent, and soil organic matter quality. Phytopathology, 96: 186-189.

Honeycutt CW, Clapham WM and Leach SS. 1996. Crop rotation and N fertilization effects on growth, yield, and disease incidence in potato. American Potato Journal, 73: 45-61.

Hossain MT, Hossain S, Islam A, Khan AR and Hossain MM. 2009. Awareness of the farmers on the use of hot water seed treatment device for controlling phomopsis blight of eggplant (*Solanum melongena* L.). Bangladesh Journal of Agricultural Research, 34: 723-727.

Howell CR. 2003. Mechanisms employed by *Trichoderma* species in the biological control of plant diseases: the history and evolution of current concepts. Plant disease. 87(1): 4-10.

Imtiaj A, Syed AR, Shahidul A, Rehana P, Khandaker MF, Sang-Beom K and Tae-Soo L. 2005. Effect of Fungicides and Plant Extracts on the Conidial Germination of *Colletotrichum gloeosporioides* Causing Mango Anthracnose, Mycobiology, 33(4): 200-205.

Ingham ER. 2005. The Compost Tea Brewing Manual, fifth ed. Soil Foodweb, Corvallis, OR.

Joshi, Deeksha, Hooda, KS and Bhatt JC. 2009. Integration of soil solarization with bio-fumigation and *Trichoderma spp.* for management of damping-off in tomato (*Lycopersicon esculentum*) in the mid altitude Region of north western Himalayas. Indian Journal of Agricultural Sciences, 79 (9): 754-757.

Joshi, Deeksha KS, Hooda JC, Bhatt BL, Mina and Gupta HS. 2009. Suppressive effects of composts on soil-borne and foliar diseases of French bean in the field in the western Indian Himalayas. Crop Protection, 28: 608-615.

Katan J. 1981. Solar heating (Solarization) control of soil borne pests. Annual Review of Phytopathology, 19: 211-236.

Koné SB, Dionne A, Tweddell RJ, Antoun H and Avis TJ. 2010. Suppressive effect of non-aerated compost teas on foliar fungal pathogens of tomato. Biological Control, 52: 167-173.

Kurt S and Emir B. 2004. Effect of solarization, chicken litter and viscera on populations of soilborne fungal pathogens and pepper growth. Plant Pathology Journal, 3: 118-24.

Kwok OCH, Fahy PC, Hoitink HAJ and Kuter JA. 1987. Interactions between bacteria and *Trichoderma hamatum* in suppression of *Rhizoctonia* damping-off in bark compost media. Phytopathology, 77: 1206-1212.

LaMondia JA, Gent MPN, Ferrandino FJ, Elmer WH and Stoner KA. 1999. Effect of compost amendment or straw mulch on potato early dying disease. Plant Disease, 83: 361-366.

Larkin RP. 2008. Relative effects of biological amendments and crop rotations on soil microbial communities and soilborne diseases of potato. Soil Biology and Biochemistry, 40: 1341-1351.

Lewis JA, Lumsden RD, Millner PD and Keinath AP. 1992. Suppression of damping-off of peas and cotton in the field with composted sewage sludge. Crop Protection, 11: 260-266.

Locke JE. 1995. Fungi. In: *The Neem Tree, source of Unique National Products for Integrated pest Management, Medicine, Industry and Other proposes*. (Ed.): H. Schmutterer VCH,Weinheim, Germany, p 118-125.

Lugtenberg B and Kamilova F. 2009. Plant-Growth-Promoting Rhizobacteria. Annual Review of Microbiology, 541-556.

Lupwayi NZ, Rice WA and Clayton GW. 1998. Soil microbial diversity and community structure under wheat as influenced by tillage and crop rotation. Soil Biology and Biochemistry, 30: 1733-1741.

Mmbaga MT and Sheng H. 2002. Evaluation of biorational products for powdery mildew management in *Cornus florida*. Journal of Environment and Horticulture, 20: 113–117.

Nagaraj MS, Umashankar N, Palanna KB and Khan ANA. 2012. Etiology and management of tip-over disease of banana by using biological agents. International Journal of Advanced Biological Research, 2: 483-486.

Naidu Y, Meon S, Kadir J and Siddiqui Y. 2010. Microbial starter for the enhancement of biological activity of compost tea. International Journal of Agricultural Biology, 12: 51–56.

Nega E, Ulrich R, Werner S and Jahn M. 2003. Hot water treatment of vegetable seed — an alternative seed treatment method to control seed-borne pathogens in organic farm. Journal of Plant Diseases and Protection, 110: 220-234.

Obagwu J and Korsten L. 2003. Control of citrus green and blue molds with garlic extracts. European Journal of Plant Pathology, 109: 221–225.

Pasini C D'Aquila F, Curir P and Gullino ML. 1997. Effectiveness of antifungal compounds against rose powdery mildew (*Sphaerotheca pannosa var. rosae*) in glasshouses. Crop Protection, 16: 251-256.

Paul PK and Sharma PD. 2002. *Azadirachta indica* leaf extract induces resistance in barley against leaf stripe disease. Physiological and Molecular Plant Pathology, 61: 3-13.

Peters RD, Sturz AV, Cartera MR and Sanderson JB. 2003. Developing disease-suppressive soils through crop rotation and tillage management practices. Soil and Tillage research, 72: 181-192.

Ramesh G, Ajithkumar K, Savitha AS and Patil SG. 2015. Integrated influence of organic manures in addition to inorganic fertilizers on growth, yield parameters and early blight disease of tomato (*Lycopersicon esculentum* l.). International Journal of Biological and Pharmaceutical Research, 6(6): 478-483.

Ren L, Su S, Yang X, Xu Y, Huang Q and Shen Q. 2008. Intercropping with aerobic rice suppressed Fusarium wilt in watermelon. Soil Biology and Biochemistry, 40: 834-844.

Russel PE and Mussa AEA. 1977. The use of garlic (*Allium sativum*) extracts to control foot rot of *Phaseolus vulgaris* caused by *Fusarium solani* f.sp. phaseoli. Annals of Applied Biology, 369-372.

Sahale N. 1996. Surpala's Vrikshayurveda (Science of life by Surpala). Agri History Bulletin No. 1. 104.

Sang MK and Kim KD. 2011. Biocontrol activity and primed systemic resistance by compost water extracts against anthracnoses of pepper and cucumber. Phytopathology, 101: 732–740.

Sarkar S, Kundu SS and Ghorai D. 2014. Validation of ancient liquid organics: *Panchgavya* and *Kunapajala* as plant growth promoters. Indian Journal of Traditional Knowledge, 13: 398-403.

Scheuerell S and Mahaffee W. 2002. Compost Tea: Principles and prospects for plant disease control. Compost Science and Utilization, 10: 313-338.

Scheuerell SJ and Mahaffee WF. 2004. Compost Tea as a container medium drench for suppressing seedling damping-off caused by *Pythium ultimum*. Phytopathology, 94: 1156-1163.

Scheuerell SJ and Mahaffee WF. 2006. Variability associated with suppression of gray mold (*Botrytis cinerea*) on geranium by foliar application of non-aerated and aerated compost teas. Plant Disease, 90: 1201–1208.

Scheuerell SJ, Sullivan DM and Mahaffee WF. 2005. Suppression of seedling damping-off caused by *Pythium ultimum, P. irregulare* and *Rhizoctonia solani* in container media amended with a diverse range of pacific northwest compost sources. Phytopathology, 95: 306-315.

Segarra G, Reis M, Casanova E and Trillas MI. 2009. Control of powdery mildew (*Erysiphe polygoni*) in tomato by foliar applications of compost tea. Journal of Plant PathololOgy, 91: 683–689.

Sharma P, Sharma M, Raja M and Shanmugam V. 2014 Status of *Trichoderma* research in India: A review. Indian Phytopathology, 67: 1-19.

Siddiqui Y, Sariah M and Razi I. 2008. Trichoderma-fortified compost extracts for the control of Choanephora wet rot in okra production. Crop Protection, 27: 385–390.

Singh D and Maheshwari DK. 2001. Biological control of loose smut of wheat. Indian Phytopathology, 54.

Singh RP and Agnihotri VP. 1988. Thermotherapy of sugarcane for disease control. In: *Review of tropical plant pathology* (Eds.) Raychaudhuri SP and Verma JP. pp. 305-329.

Sireesha O. 2013. Effect of plant products, panchagavya and bio-control agents on rice blast disease of paddy and yield parameters. International Journal of Research in Biological Sciences, 3: 48-50.

Slusarenko AJ, Patel A and Portz D. 2008. Control of plant diseases by natural products: Allicin from garlic as a case study. European Journal of Plant Pathology, 121: 313-322.

Sofi TA, Tewari AK and Razdan VK. 2007. Soil solarization reduces damping-off caused by soil-borne fungal pathogens and improve vigour of cauliflower seedling. Journal of Mycology and Plant Pathoogy, 37: 68-71.

Sridhar S, Ashok Kumar S, Thooyavathy RA and Vijayalakshmi K. 2013. Seed Treatment Techniques. Centre for Indian Knowledge Systems (CIKS) Seed Node of the Revitalising Rainfed Agriculture Network, pp. 21.

Steiner R. 1924. Geisteswissenschaftliche Grundlagen zum Gedeihen der Landwirtschaft. Rudolf Steiner Verlag, Dornach.

Sumangala K and Patil MB. 2009. Panchagavya - an organic weapon against plant pathogens. India Journal of Plant Disease Sciences, 4: 147-151.

Theunissena J and Schelling G. 1996. Pest and disease management by intercropping: Suppression of thrips and rust in leek. International Journal of Pest management, 42: 227-224.

Tuitert G, Szcech Mand Bollen GJ. 1998. Suppression of *Rhizoctonia solani* in potting mixtures amended with compost made from organic household waste. Phytopathology, 88: 764-773.

Willer H and Lernoud J. 2018. The World of Organic Agriculture. Statistics and Emerging Trends (Eds.) Willer and Lernoud. Research Institute of Organic Agriculture (FiBL), Frick and International Federation of Organic Agriculture Movements (IFOAM), Bonn.

Wilson CL, Solar JM, El Ghaouth A and Wisniewski ME. 1997. Rapid evaluation of plant extracts and essential oils for antifungal activity against *Botrytis cinerea*. Plant Disease, 81: 204-210.

World Bank. 2007. World Development Report 2008: Agriculture for Development. Oxford: Oxford University Press for The World Bank.

Yadav AK. Organic agriculture: Concept scenario, principles and practices. National Centre for Organic farming, Department of Agriculture and Cooperation, GOI.

Yildirim I, Onogur E and Irshad M. 2002. Investigations on the Efficacy of some Natural Chemicals against Powdery Mildew [*Uncinula necator* (Schw.) Burr.] of Grape. Journal of Phytopathology, 150: 697-702.

Zwieten MW, Stovold G and Zwieten LV. 2007. Alternatives to Copper for Disease Control in the Australian Organic Industry. RIRDC Publication No 07/110 RIRDC Project No DAN-208A.

Chapter 11

Eco-Friendly Management of Vertebrates in Agriculture

V. Vasudeva Rao

All India Network Project on Vertebrate Pest Management, PJTSAU, Rajendranagar, Hyderabad – 500 030, Telangana
E-mail: vasuvaidyula@yahoo.com

1. INTRODUCTION

Agricultural production in India is mainly affected by insect pests, plant diseases and weed plants to a greater extent. In the recent times, avian fauna and mammalian fauna with special reference to rodents, wild boars, blue bull and monkeys started gaining pest status and in certain cases a huge damage is being encountered due to some of these vertebrate pests. Among them, depredatory birds and wild boar have become regular menace for farmers in major crops resulting into enormous damage. Unlike other pests, which generally attack during a particular stage of the crop, wild boar cause damage right from seedling to till the maturity of the crop (Roberts, 1977; Shafi and Khokhar, 1986).The basic reason for such unexpected abrupt increase in their populations can be attributed to escalating rate of deforestation, which otherwise are the natural habitats of those species. Deforestation also resulted in the decline of Tigers, Panthers, Wild dogs, Wolf, and Jackal which are the natural predators for wild boars (Moreira *et al.*, 1997), there by indirectly contributing to the phenomenal raise in the wild boar populations. Over exploitation of forest resources by the mankind forced birds and higher vertebrates are out of their natural habitat and compelled them to depend on cultivated crops such as rice, maize, sorghum, pulses, oil seeds, fruits and vegetables. The damage caused by birds and higher vertebrates are more alarming than their actual feeding in the crop. Vertebrate damage is more pronounced in crop fields which are in close proximity with adjoining forests areas. Vertebrates in general moves in flocks/ groups and their activity is more during early morning and evening, noticeably active at dawn and dusk than in the actual day period, they have also a unique feature of identifying cropped areas through their structural appearance of seeds in case of birds and smell sensory mechanism in wild boar (Roberts, 1977; Syamsunder Rao, 2002).

However, today in India, views of the administrators and agricultural research scientists towards the depredatory birds have certainly changed. Attitude of the farmers towards birds and wild animals is changing at slower rate but they are still tolerant as they are under the influence of religious preaching. Under present economical crisis and modernization of agricultural ecosystem, there is a need to manage all the pests, including birds and higher vertebrates for sustainable productivity.

An investigation on both basic and applied aspects of birds as well as wild boar has generated valid information that has helped to evolve technologies for their management for increasing crop production. A total of 63 species of birds belonging to 19 families have been identified damaging several crops (Table 11.1). The number of bird species that affected various crops was: cereals-52, pulses-14, oilseeds-15 and fruits-23. Amongst the 46 species of beneficial birds, which devoured insects and rodent pests, all fed on insects while six of them also consumed rodents (Table 11.2). Twenty eight species of birds that inflicted damage to crops and fifteen of the beneficial species were omnivorous. Omnivorous birds have a dual role in our agro-ecosystem. The extent of damage in different stages of the crops by depredatory birds is given in Table 11.3.

Table 11.1: Important Depredatory Birds of Agricultural Crops

Family and Species	*Crop*	*Damage Status*
Family: Threskiornithidae		
1. Black Ibis (*Pseudibis papillosa*)	C, OS	VL
Family: Anatidae		
2. Greylag Goose (*Anser anser*)	C, V	VL
3. Bar headed Goose (*A. indicus*)	C, V	VL
4. Lesser Whistling Teal (*Dendrocygna javanica*)	C, V	L
5. Ruddy Shelduck (*Tadorna ferruginea*)	C	L
6. Pintail (Anas *acuta*)	C	VL
7. Common Teal (*A. crecca*)	C	VL
8. Gargeny (*A. querquedula*)	C	VL
Family: Phasianidae		
9. Grey Partridge (*Francolinus pondicerianus*)	C,P,OS,W,F,V	H
10. Common Peafowl (*Pavo cristatus*)	C,P,OS,F,V	M
Family: Gruidae		
11. Common Crane (*Grus grus*)	C,P,OS,W,F,V	L
12. Sarus Crane (*G. antigone*)	C,OS,W	VL
13. Demoiselle Crane (*Anthropoides virgo*)	C,P,OS,W	VL

Family and Species	*Crop*	*Damage Status*
Family: Rallidae		
14. Purple Moorhen (*Porphyrio porphyrio*)	C,W,V	VL
Family: Charadriidae		
15. Black tailed Godwit (*Limosa limosa*)	C	VL
16. Ruff and Reeve (*Philomachus pugnax*)	C	VL
Family: Columbidae		
17. Blue Rock Pigeon (*Columba livia*)	C,P,OS,W	H
18. Ring Dove (*Streptopelia decaocto*)	C,P,OS,W	M
19. Red Turtle Dove (*S. tranquebarica*)	C,P,OS,W	L
20. Spotted dove (*S. chinensis*)	C,P,OS,W	VL
21. Little Brown Dove (*S. senegalensis*)	C,P,OS,W	VL
Family: Psittacidae		
22. Large Indian Parakeet (*Psittacula eupatria*)	C,P,OS,W,F	VL
23. Rose-ringed Parakeet (*P. krameri*)	C,P,OS,W,F,V	H
24. Blossomheaded Parakeet (*P. cyanocephala*)	C,P,OS,F,V	VL
25. Slatyheaded Parakeet (*P. himalayana*)	F	VL
26. Bluewinged Parakeet (*P. columboides*)	F	VL
Family: Cuculidae		
27. Koel (*Eudynamys scolopacea*)	F,V	VL
Family: Capitonidae		
28. Great Hill Barbet (*Megalaima virens*)	F	VL
29. Small Green Barbet (*M. viridis*)	F,V	VL
Family: Picidae		
30. Scalybellied Green Woodpecker (*Picus squamatus*)	F	VL
31. Lesser Goldenbacked Woodpecker (*Dinopium bengalensis*)	F	VL
Family: Alaudidae		
32. Short-toed Lark (*Calandrella cinerea*)	C,W	VL
33. Skylark (*Alauda arvensis*)	C,W	VL
Family: Oriolidae		
34. Golden Oriole (*Oriolus oriolus*)	F	VL
35. Blackheaded Oriole (*O. xanthornus*)	F	VL
Family: Sturnidae		
36. Starling (*Sturnus vulgaris*)	C	VL
37. Rosy Pastor (*S. roseus*)	C,F	M
38. Common Myna (*Acridotheres tristis*)	C,F,V	M
39. Bank Myna (*A. ginginianus*)	C,F	VL
Family: Corvidae		
40. Redbilled Blue Magpie (*Cissa erythrorhyncha*)	F	VL
41. Indian Tree pie (*Dendrocitta vagabunda*)	F	VL
42. House Crow (*Corvus splendens*)	C,P,OS,F,V	M

Family and Species	*Crop*	*Damage Status*
43. Jungle Crow (*C. macrohynchos*)	C,P,OS,F,V	VL
Family: Pycnonotidae		
44. Whitecheeked Bulbul (*Pycnonotus leucogenys*)	C,F	VL
45. Redvented Bulbul (*P. cafer*)	C,F	VL
Family: Muscicapidae		
46. Common Babbler (*Turdoides caudatus*)	C,W	VL
47. Large Grey Babbler (*T. malcolmi*)	C,W	VL
48. Jungle Babbler (*T. striatus*)	C,W	VL
49. Streaked Laughing thrush (*Garrulax lineatus*)	F	VL
Family: Nectainiidae		
50. Purple Sunbird (*Nectarinia asiatica*)	F	VL
Family: Ploceidae		
51. House Sparrow (*Passer domesticus*)	C,W,V	H
52. Yellowthroated Sparrow (*Petronia xanthocollis*)	C,W,V	M
53. Baya (*Ploceus philippinus*)	C,W	VL
54. Blackthroated Weaverbird (*P. benghalensis*)	C,W	VL
55. Streaked Weaverbird (*P. manyar*)	C,W	VL
56. Whitethroated Munia (*L. malabarica*)	C,W	VL
57. Whitebacked Munia (*L. striata*)	C,W	VL
58. Spotted Munia (*L. punctulata*)	C,W	VL
59. Blackheaded Munia (*L. malacca*)	C,W	VL
60. Black headed Bunting (*Emberiza melanocephala*)	C	VL
61. Common Rose Finch (*Carpodacus erythrinus*)	C	VL
62. Redheaded Bunting (*E. bruniceps*)	C	VL
63. Crested Bunting (*Melophus lathami*)	C	VL

Food: C: Cereals, P: Pulses, OS: Oilseeds, V: Vegetables, F: Fruits, W: Weeds

Damage status: VL: Very limited, L: Limited, M: Moderate, H: Heavy,

Source: Survey reports of coordinated centres of AINP on Agricultural Ornithology.

Table 11.2: Major Beneficial Birds of Agricultural Importance

Family and Species	*Food*	*Beneficial Status*
Family: Ardeidae		
1. Pond Heron (*Ardeola grayii*)	I	L
2. Cattle Egret (*Bubulcus ibis*)	I	M
3. Smaller Egret (*Egretta intermedia*)	I	L
4. Little Egret (*E. garzetta*)	I	L
Family: Treskiornithidae		
5. White Ibis (*Threskiornis aethiopica*)	I	VL
6. Black Ibis (*Pseudibis papillosa*)	I	VL

Family and Species	*Food*	*Beneficial Status*
7. Glossy Ibis (*Plegadis falcinellus*)	I	VL
Family: Accipetridae		
8. Blackwinged Kite (*Elanus caeruleus*)	I,R	VL
9. Pariah Kite (*Milvus migrans*)	I,R	VL
Family: Charadriidae		
10. Redwattled Lapwing (*Vanellus indicus*)	I	VL
11. Yellowwattled Lapwing (*V. malabaricus*)	I	VL
Family: Laridae		
12. Whiskered Tern (*Chlidonias hybrida*)	I	L
13. Gullbilled tern (*Gelochelidon nilotica*)	I	VL
Family: Strigidae		
14. Barn Owl (*Tyto alba*)	I,R	VL
15. Spotted Owlet (*Athene brama*)	I,R	VL
Family: Alcedinidae		
16. Whitebreasted Kingfisher (*Halcyon smyrnensis*)	I	VL
Family: Meropidae		
17. Little Green Bee-eater (*Merops orientalis*)	I	L
Family: Coraciidae		
18. Indian Roller (*Coracias benghalensis*)	I	VL
Family: Upupidae		
19. Hoopoe (*Upupa epops*)	I	VL
Family: Hirundinidae		
20.Common Swallow (*Hirundo rustica*)	I	VL
21. Redrumped Swallow (*H. daurica*)	I	L
Family: Laridae		
22. Grey Shrike (*Lanius excubitor*)	I	VL
23. Baybacked Shrike (*L. vittatus*)	I	VL
24. Rufousbacked Shrike (*L. schach*)	I	VL
Family: Dicruridae		
25. Black Drongo (*Dicrurus adsimilis*)	I	L
Family: Artamidae		
26. Asy Swallow-shrike (*Artamus fuscus*)	I	VL
Family: Sturnidae		
27. Brahminy Myna (*Sturnus pagoderma*)	I	VL
28. Rosy Pastor (*S. roseus*)	I	M
29. Starling (*S. vulgaris*)	I	VL
30. Pied Myna (*S. contra*)	I	VL
31. Common Myna (*Acridotheres tristis*)	I	L
32. Bank Myna (*A. ginginianus*)	I	VL

Family and Species	*Food*	*Beneficial Status*
Family: Corvidae		
33. House Crow (*Corvus splendens*)	I,R	M
34. Jungle Crow (*C. macrorrhynchos*)	I,R	VL
Family: Pycnonotidae		
35. Whitecheeked bulbul (*Pycnonotus leucogenys*)	I	VL
36. Redvented Bulbul (*P. cafer*)	I	VL
Family: Muscicapidae		
37. Common Babbler (*Turdoides caudatus*)	I	VL
38. Large Grey Babbler (*T. malcolmi*)	I	VL
39. Jungle Babbler (*T. striatus*)	I	VL
40. PlainWren-Warbler (*Prinia subflava*)	I	VL
41. Ashy Wren-Warbler (*P. socialis*)	I	VL
42. Tailorbird (*Orthotomus sutorius*)	I	VL
Family: Motacillidae		
43. Yellow Wagtail (*Motacilla flava*)	I	VL
44. Yellowheaded Wagtail (*M. citreola*)	I	VL
45. Grey Wagtail (*M. cinerea*)	I	L
46. White Wagtail (*M. alba*)	I	VL
Family: Ploceidae		
47. Hose Sparrow (*Passer domesticus*)	I	L
48. Baya weaver (*Ploceus philippinus*)	I	L

I: Insects, R: Rodents, VL: Very limited, L, Limited, M: Moderate.

Source: Survey reports of coordinated centres of AINP on Agricultural Ornithology.

2. Damage Caused by Depredatory Birds to Various Crops

Pearlmillet

A total of 24 depredatory species were recorded in pearlmillet crop from five states. Rose-ringed Parakeet (*Psittacula krameri*), Rosy Pastor (*Pastor roseus*), House sparrow (*Passer domesticus*) and Baya (*Ploceus philippinus*) were the predominant species that damaged the crop in Northwestern India. White eared Bulbul (*Pycnonotus leucotis*) was reported for the first time to feed on the crop in large numbers in the arid and semi-arid zones. Estimation of bird damage varied highly in different states. It was highest in Gujarat (0.3 to 40 per cent) followed by Andhra Pradesh (1.5 to 9 per cent), Punjab (45 per cent) and Delhi (60 per cent) during *Kharif* season. In Gujarat, bird damage to the summer crop was significantly less (0.2 to 2.1 per cent), due to the absence of migratory birds especially the Rosy Pastor, synchronization of crop cultivation also resulted in better yield compared to *kharif* season (Table 11.3).

Table 11.3: Extent of Damage by Depredatory Birds in different Stages of the Crops

Crop	*Stage of Crop*	*Extent of Damage (per cent)*	*Bird Depredators*
Paddy (*Oryza sativa*)	Nursery	0.6-5	Baya, House sparrow, White throated Munia, Spotted Munia, Blackheaded Munia, Black throated Weaver, Streaked Weaver, Purple Moorhen, Blue Rock Pigeon, Ring Dove, Common Myna, House Crow
	Ripening	1.4 - 22.9	House sparrow, Baya, Black throated Weaver, Streaked Weaver, Whitethroated Munia, Spotted Munia,Blackheaded Munia, Rose-ringed Parakeet, Large Grey Babbler, Common Babbler, Jungle Babbler, Brahminy Duck, Lesser Whistling Teal, Gargany, Common Teal, Pintail, Purple Moorhen, Sarus Crane
Wheat (*Triticum aestivum*)	Sowing and seedling	0.5-6	House Crow, Jungle Crow, Blue Rock Pigeon, Ring Dove, Peafowl, Grey Partridge, Ruff, Blacktailed Godwit, Demoiselle Crane, Common Crane, Sarus Crane, Black Ibis
	Ripening	0.6-37	House Sparrow, Rose-ringed Parakeet, Whitethorated Munia, Spotted Munia, Large Grey Babbler, Jungle Babbler, Indian Peafowl, Demoiselle Crane, Common Crane, Sarus Crane, Short- toed Lark, Rufous tailed Finch Lark, Blackheaded Bunting, Redheaded Bunting, Ring Dove
Maize (*Zea mays*)	Sowing and seedling	10-20	Rock Pigeon, Ring Dove, Spotted Dove, House Crow, Jungle Crow, Common Myna, Rose ringed Parakeet, Indian Peafowl, Grey Partridge.
	Ripening	6-39	Rose-ringed Parakeet, House Crow, Large Indian Parakeet
Pearlmillet (*Pennisetum glaucum*)	Sowing and seedling	0.5 - 6	Rock Pigeon, Ring Dove, spotted Dove, Red Turtle Dove, Little Brown Dove, Baya, Blackthroated Weaver, Streaked Weaver, House Sparrow, Grey Partridge, Jungle Babbler, Large Grey Babbler
	Ripening	0.5 - 26	House Sparrow, Baya, Black throated Weaver, streaked Weaver, Whitethroated Munia, Blackheaded Munia, Spotted Munia, Whitebacked Munia, Red Munia, Redheaded Bunting, Black headed Bunting, Common Myna, Redvented Bulbul, Whitecheeked Bulbul, Large Grey Babbler, Common Babbler, Jungle Babbler, Whiteheaded Babbler, Rosy Pastor, House Crow, Rose-ringed Parakeet, Ring Dove
Sorghum (*Sorghum vulgare*)	Sowing and seedling	0.5 - 8	Blue Rock Pigeon, Ring Dove, Spotted Dove, Little Brown Dove, Red Turtle Dove, Indian Peafowl, Grey partridge, Large Grey Babbler, Jungle Babbler, Common Babbler
	Ripening	0.7 - 44	Rose-ringed Parakeet, House Sparrow, Baya, Black throated Weaver, Streaked Weaver, Whitethroated Munia, Spotted Munia, Whitebacked Munia, Black headed Munia, Rose Finch, Blackheaded Bunting, Redheaded Bunting, Ring Dove, Rosy Pastor, Common Myna, Bank Myna, House Crow, Jungle Crow, Blossom-headed Parakeet

Crop	Stage of Crop	Extent of Damage (per cent)	Bird Depredators
Blackgram (*Vigna mungo*)	Sowing and seedling	0.2-3	Blue Rock Pigeon, Ring Dove, House Crow, Grey Partridge
	Ripening	0.5-2.5	Rose-ringed Parakeet, House Crow, Blue Rock Pigeon, Ring Dove, Grey Partridge, Peafowl
Greengram (*Vigna radiata*)	Sowing and seedling	0.3-3	House Crow, Blue Rock Pigeon, Ring Dove, Grey Partridge
	Ripening	0.5-2	Rose-ringed Parakeet, House Crow, Blue Rock Pigeon, Ring Dove, Grey Partridge, Peafowl
Chickpea (*Cicer arietinum*)	Sowing	0.5-3	House Crow, Blue Rock Pigeon, Ring Dove, Grey Partridge, Peafowl, Sarus Crane, Jungle Crow
	Ripening	0.5-2	Blue Rock Pigeon, Ring Dove, Indian Peafowl, House Crow, Grey Partridge, Rose Ringed Parakeet
Pigeonpea (*Cajanus cajan*)	Sowing and seedling	0.2-2	Blue Rock Pigeon, Ring Dove, Indian Peafowl, House Crow, Jungle Crow, Grey Partridge
	Ripening	0.1-1	Rose-ringed Parakeet, House Crow, Jungle Crow
Cowpea (*Vigna unguiculata*)	Sowing and seedling	0.5-3	Blue Rock Pigeon, Ring Dove, House Crow, Grey Partridge, Indian Peafowl
	Ripening	10-40	Rose-ringed Parakeet, Ring Dove, Blue Rock Pigeon, Grey Partridge, Indian Peafowl, House Crow, Small Green Barbet
Sunflower (*Helianthus annuus*)	Sowing and seedling	0.5-3	House Crow
	Ripening	10-40	Rose-ringed Parakeet
Groundnut (*Arachis hypogaea*)	Sowing and seedling	0.5-36	Blue Rock Pigeon, Ring Dove, House Crow, Rose-ringed Parakeet, Indian Peafowl, Black Ibis
	Ripening	1-15	House Crow, Jungle Crow, Demoiselle Crane, Common Crane, Sarus Crane, Black Ibis, Indian Peafowl
Safflower (*Carthamus tinctorius*)	Sowing	2-15	House Crow
	Ripening		Rose-ringed Parakeet
Mustard (*Brassica nigra*)	Ripening		Rose-ringed Parakeet, Blue Rock Pigeon, Ring Dove, Grey Babbler
Soybean (*Glycine max*)	Sowing and seedling	2-15	Rose-ringed Parakeet, Blue Rock Pigeon, Grey Partridge
	Ripening		Rose-ringed Parakeet, Indian Peafowl, Grey Partridge, Blue Rock Pigeon

Source: Survey reports of coordinated centres of AINP on Agricultural Ornithology.

Wheat

Bird damage to wheat varied from 0.2 to 41 per cent in different parts of the country. The damage was significantly high in Rajasthan as compared to that in

Gujarat and Punjab. About 13 species of birds damaged standing wheat. Rose-ringed Parakeet, Ring dove and Baya weaver were the most common species damaging the crop in most of the fields. Large flocks of Demoiselle Crane (*Anthropoides virgo*), Common Crane (*Grus grus*) and Short-toed Lark (*Calandrella brachydactyla*), heavily damaged wheat fields in Bhal area of Gujarat. Ruff (*Philomachus pugnax*) and Black winged godwit (*Limosa limosa*) are reported for the first time to damage wheat at sowing and seedling stages in the coastal areas of Gujarat (Table 11.3).

Paddy

Although paddy is cultivated in several states of the country in vast areas, it was prone to heavy damage by birds under congenial ecological conditions. Damage to paddy was highest in Punjab (0.1 to 6.5 per cent), followed by Kerala (1.5 to 6 per cent), Andhra Pradesh (1.5 to 3 per cent) and Gujarat (0.1 to 1 per cent). Thirty-nine species of birds fed on the grains of standing crop in Gujarat. Baya weaver, House sparrow and Rose-ringed Parakeet were common and predominant depredatory birds in most parts of the state. Gargany Teal (*Anas querquedula*) and Lesser Whistling Teal (*Dendrocygna javanica*) damaged paddy crop in Kole area of Kerala (Table 11.3).

Sorghum

Bird damage to sorghum was highest in Rajasthan (2.3 to 48 per cent) as compared to that in Gujarat (0.4 to 18.6 per cent) and Telangana (0.5 to 16.6 per cent). This is one of the most preferred crops by the granivorous birds like Rose-ringed Parakeet, Rosy Pastor and Bank Myna. Total 26 species were recorded feeding on sorghum (Table 11.3).

Maize

Bird damage to maize was relatively less in Gujarat (0.3 to 2.5 per cent) as compared to Punjab (3.3 to 7.5 per cent), Telangana (3 to 9.1 per cent) and Rajasthan (0 to 20 per cent). Ten species of birds were recorded to feed on maize, of which the Rose-ringed Parakeet was most important in all the states. In Andhra Pradesh, the bird damage in maize ranged between 10 to 40 per cent and the damage was mainly caused by Rose-ringed Parakeet (Table 11.3).

Sunflower

Rose-ringed Parakeet and Common Crow (*Corvus splendens*) were the predominant depredators and caused 10 to 30 per cent damage in Telangana and 10 to 90 per cent damage in Rajasthan. In Punjab the mean percentage of damage in different years of study ranged from 5.7 to 29 per cent. Only because of the bird problem, the crop could not be introduced in Gujarat (Table 11.3).

Oil Palm

The damage was higher both in south Gujarat (10 to 27 per cent) and west Godavari district (3.3 to 30 per cent) of Andhra Pradesh. Common Crow, Jungle Crow (*Corvus macrorhynchos*), Rose-ringed Parakeet and Common myna (*Acridotheres tristis*) were the depredatory birds of this crop. For the first time, it was reported

Pariah kite damage in oil palm gardens at NCZ of Andhra Pradesh to the tune of 30-50 per cent (Table 11.3).

Groundnut

During sowing to sprouting stage, 3 to 33 per cent damage was done by ten species of birds in Saurashtra region of Gujarat. The migratory Demoiselle Cranes caused damage up to 10 per cent at the time of harvesting (Table 11.3).

Safflower

Rose-ringed Parakeet was the only species feeding mainly on the peripheral plants of the crop. The damage level was negligible in Telangana and Gujarat (Table 11.3).

3. Management Method for Depredatory Birds

3.1. Traditional Methods

3.1.1. Machan

A machan is erected amidst the maize crop. A semicircular mat made of bamboo splits is put on the machan to prepare a small hut for the shelter which is locally called *dhagla.* Sometimes, instead of semicircular mat, an umbrella type structure made of leaves of *Butea monosperma* (*Palash*) and bamboo sticks (locally called *dengcha*) is placed on machan. Loud calls are made from the *machan* to keep away the birds. Stones are thrown by locally made equipment called *gophana* (sling) to drive away birds.

3.1.2. Flagged Bamboos and Flagged Leader Shoots

Pieces of plastics and coloured clothes are tied on bamboo sticks which are erected amidst the crop in the field to keep away the birds. Sometimes these are placed at the periphery only. When the crop reaches the milky stage, flags of cloth are tied on leader shoots of some of the tall trees.

3.1.3. Use of 'Owl-rests'

Cushion owl-rests: A coiled mass of *Tectona grandis* leaves (teak) is wrapped at one end of a bamboo stick. Dozens of such sticks are erected in the field keeping the leaf mass upward. At night, owls are attracted to these perches and prey on night dwelling rats. This method is generally used in fields of wheat and chickpea.

Pole owl-rests: Poles of bamboo culms, 0.5 to 1.5 m long, are erected amidst chickpea, wheat and barley crops to provide perching stations to the owlets during night.

3.1.4. Pitcher-effigy (Scare Crows)

Pitcher-effigies (locally called *byawana* or *taoon*) are prepared by the farmers with locally available material. An old pitcher (terracotta vessel), having black outer surface due to use in kitchen for cooking purpose, is kept upside down on a vertically erected wooden pole of a man's height to symbolize the head of a man having black hair. Sometimes head is made by black cloth also. Then, a horizontal stick is tied to the vertical pole to resemble arms raised to shoulder level. An old shirt (*kurta*) is put on the wooden structure to make an effigy of a man working in field. These effigies are said to be effective in repelling raiding birds from crops (Sharma, 1994).

3.1.5. Hanging Crows

A hung dead crow is said to be very effective in repelling crows. This method is equally effective in houses as well as in fields.

3.1.6. Calls Made by '*Ghunku*'

'*Ghunku*' is a simple device made by locally available material. An earthen pitcher used in Persian wheel (to draw water from wells, locally called *ged*) is taken and a piece of goat skin having a hole in its centre is tied to the mouth of the pitcher. A tall feather of peacock is inserted in the hole and a knot is made at its lower end. This apparatus is held between the feet and then a massage like action is made on the feather with the thumb and the first finger. To make the action easy, few oil drops are also applied on the feather. A loud call is generated by this apparatus which is said to be effective in frightening nocturnal animals. This is also used during day-time to keep away birds.

3.1.7. Halan

This method is mainly practiced by *Saharias*. A string is tied loosely around or across the fields. Leaves of *Tectona grandis* are tied to the string in a series. This festoon of leaves is connected by another string at the mid-point. A man sitting on a machan pulls the string in jerky motions and dry, hanging leaves produce a typical buzz like sound which keeps away the birds and other animals. This indigenous device is called *halan* (*i.e.*, something which moves). Many variations of this system can be seen in the state. Sometimes bells instead of leaves are also used. Sometimes, single leaf *halan* (locally called *jalra*) made by bamboo pole and striated *Tectona grandis* leaves are also erected amidst the crop to keep away rodents and birds.

3.1.8. Drum Beating

Drums are beaten from some elevated places or machans to keep away the flocks of grain eating birds. This method is said to be effective against the raid of locusts also.

3.1.9. Use of Feathery Grass Inflorescence

Just after sowing of wheat and gram, feathery inflorescence of *Saccharum bengalense* are erected at random in the fields. This method is said to be effective to keep away the grain eating birds. This method is mainly practiced in eastern Rajasthan.

3.1.10. Methods to Protect Harvested Crop

Generally the harvested stems of sorghum and pearlmillet are bundled and then piled in a conical shaped heap, keeping all the ear ends upwards. These conical heaps are locally called *chhaurs*. Spiny bushes of *Zizyphus nummularia* are cut and placed on the top of the heap to keep away the bird flocks. This method is much practiced in eastern and central Rajasthan (Sharma, 1991).

3.2 Ecofriendly Management Methods

As a part of All India Net Work Project on agricultural Ornithology, multi-location trials for evaluation of different eco friendly management practices were carried out to determine their efficacy, feasibility and economical viability under different agro climatic zones. These methods were tested all over India, *i.e.*, Himachal Pradesh in the north to Kerala in the south with the varying cropping patterns stretched in different climatic regimes. As a result several eco-friendly economical management practices have been evolved for minimizing losses caused by depredatory birds which are elucidated as under:

3.2.1 Wrapping Method on Maize Crop

Covering maize cobs by wrapping adjacent green leaves around them reduced the damage to a negligible level by parakeets and crows, which were the major problem birds. Being camouflaged, the wrapped cobs escape detection by birds and thus the crop is protected. It is a very simple and effective method, which does not involve any material cost and is les laborious than scaring. This method does not have negative impact on the grain yield. All cobs need not be covered. Since parakeet damage is restricted to peripheral rows, covering of 50 per cent cobs at random on outer 3 rows of the field is sufficient to effectively reduce bird damage. Wrapping of Maize cobs by adjacent green leaves yielded 2202 kg/ha (*Kharif*) and 1901 kg/ha (*rabi*) against control 1195 kg/ha (*kharif*) and 1207 kg/ha (*rabi*) (Table 11.4).

3.2.3 Reflective Ribbon for Bird Scaring

Reflective ribbon is a polyester film with a shining metallic coating with red on one side and silver on the other. It is prepared by cutting along continuous polyester sheet in to strips of 1.5 cm width. Such strips, preferably 15 to 20 m long, are fixed parallel to the crop at 0.5 m height above the crop and at 5 m intervals using bamboo poles and strings. For better reflection, the ribbon should be fixed in north to south direction. During sunshine the reflection of sunlight and humming noise produced by the wind scares the birds from the field. The device is effective only for 15-20 days. After that the birds gets used to it. The ribbons do not scare birds under poor light condition and when the crop is grown in isolation. The technique of bird scaring by ribbons is very effective and easily acceptable to the farmers. Birds like rose ringed parakeet, House crow (*Corvus splendens*) and mynas on the crops like sunflower, maize, sorghum, pearl millet and orchards are scared by this devise. The reflective tapes have been proved to be effective against the Demoiselle cranes in groundnut and against depredatory birds in other cereals and fruit crops. The installation of reflective ribbon in maize crop resulted a yield of 2063 kg/ha (*kharif*) and 1807 kg/ha (*rabi*) when compared to control 1195 kg/ha (*kharif*) and 1207 kg/ha (*rabi*) (Table 11.4).

Table 11.4: Bird Management Methods and Yields of different Crops (2013-14)

Crop	*Treatment*	*Yield (kg/ha)*	
		Kharif	*Rabi*
Maize	Protected control (net)	1586	1505
	Fodder maize	1422	1341
	Sorghum fodder	1334	1361
	Ribbon	1389	1264
	Wrapping	1418	1352
	Control	847	756
	CD (5 per cent)	243	261
Sunflower	Ecodon (20 ml/l)	583	632
	Egg solution (20 ml/l)	708	1006
	Paper plates	938	1119
	Protected control (net)	1651	1339
	Ribbons	388	885
	Control	375	324
	CD (5 per cent)	127	157
Sorghum	Egg solution spray (20 ml/l)	2802	2648
	Neem oil spray	1466	1706
	Reflecting ribbon	2336	2135
	Protected control (net)	3180	2889
	Control	764	1190
	CD (5 per cent)	188	306

3.2.4 Screen Crop

Thick planting of sorghum (fodder) as well as of maize fodder significantly reduced parakeet damage in maize crop grown for grain production. Besides giving better yield, this practice also provided additional fodder. Using of fodder Maize as screen crop around Maize resulted a yield of 2150 kg/ha (*kharif*) and 2016 kg/ha (*rabi*) and fodder sorghum as screen crop resulted a yield of 1764 kg/ha (*kharif*) and 1605 kg/ha (*rabi*) against control 1195 kg/ha (*kharif*) and 1207 kg/ha (*rabi*) (Table 11.4).

3.2.5 Bio-acoustics

The acoustic equipment is very effective in driving away the birds from the crop fields. It consists of micro SD card with 20w amplifier, 1 speaker, one 12 v battery and one solar panel. Pre-recorded alarm, predatory and distress calls of birds are embedded and played. Depending on the intensity of bird activity, the frequency of play should be setup at regular time intervals. Broadcasting of such distress calls of depredatory birds keeps the birds away from sunflower, sorghum and also other crops. Installation of this equipment in sunflower resulted in higher yield 1540 kg/ha against control (594 kg/ha), while in sorghum gave 1263 kg/ha against control (743 kg/ha).

3.2.6 Automatic Mechanical Bird Scarer or Pyrotechnic Method

Automatic bird scarer can also be employed. This is a sound producing device which works continuously for a whole day with 1 kg of calcium carbide and water. One-hectare areas can be covered with this method and it is found effective in reducing crop losses by birds. Care must be taken about the frequency of firing and change of positions and directions to avoid bird getting habituated.

3.2.7 Botanical Repellents

Neem cake solution is prepared by soaking neem cake @ 200 or 300 g/l of water and kept for fermentation for 8-10 days. The fermented solution is than decanted and this solution is used as spray fluid. Neem cake solution @ 200 g for liter of water showed effective in controlling bird damage in Maize. Spraying of botanical formulation like BBR+ and Fortune Azar (Neem formulation) in the field, reduced number of visiting birds and resulted in higher yield (714 kg/ha) compared to control (532 kg/ha) in Andhra Pradesh. Spraying of BBR+ (5 per cent) and Fortune Azar (5 per cent) in the fields of pearlmillet reduced bird visits and significantly reduced bird damage (68 per cent) in Gujarat.

3.2.8 Habitat Manipulation

Creating continuous disturbances to the nesting sites of the depredatory breeding birds in and around the cropped areas that will force the birds to leave breeding grounds and shift to another area. For parakeets in addition to manual destruction of nests, closing the entrance of the nests proved effective reducing their population. Planting of some fruit bearing trees like Manila tamarind (*Pithecolobium dulce*), Flame of the forest (*Butea monosperma*), mulberry (*Morus alba*) and toothbrush tree (*Salvadora persica*) in and around cropped areas attract many granivorous birds during fruiting period and reduces the impact at vulnerable stage of the crop (maize).

3.2.9 Spraying of Egg Solution

Spraying of egg solution @ 25 ml/l of water was very effective in control of bird damage in safflower, maize, sunflower, sorghum, pearlmillet, and other food crops. This method is effective for 10 days; second spray is recommended if damage is seen after 10 days. Application of this innovative method the yields of different crops increased from 30-69 per cent. Egg solution spray @ 20 ml/l showed higher yields in sunflower during *kharif* (708 kg/ha) and *rabi* (1006 kg/ha) in Andhra Pradesh (Table 11.4).

3.2.11 Reflective Paper Plate

This method is very effective during the milky stage of the crop. In this, the paper plates are arranged on the stalk behind the flower such that the reflective surface faces outside, so that the sun rays will be reflected back. This prevents the birds' vision and approaching the crop. Birds may approach the crop, in case of the absence of sunlight may not find suitable perch on the flower due to slippery surface. Hence this is a cost effective method in minimizing the crop loss. Fixing of reflective paper plates on the stalks of sunflower showed good results against depredatory birds, which yielded 938 kg/ha (*kharif*) and 1119 kg/ha (*rabi*) which was higher than the control (375 kg/ha in *kharif* and 325 kg/ha in *rabi*) (Table 11.4).

4. Managment Methods for Wild Boar

Several ITKs are being employed by the farming community to ward off wild boar in different innovative ways. Some of such effective ways, practiced by local people were scientifically evaluated and validated for efficiency and economic feasibility. The following are the sum of such methods which are being recommended through AINP on Agricultural Ornithology.

4.1 Traditional Methods

4.1.1 Spraying of Dung Solution of Local Pigs

Territoriality is very high in wild boars which are being exploited under this method. The dung collected from local pigs will be made into solution and sprayed on soil to the width of 1 ft around the crop. This will confuse wild boars with a false assumption of entering into the territory of other pigs; there by their movement will be prevented to avoid territorial conflict. For sustained affectivity, it is desirable to go for 2-3 sprays with 7 days interval between each spray.

4.1.2 Human Hair as Respiratory Deterrent

Wild boar with poorly developed sight and hearing mechanism has to depend on its smell sensory mechanism only for movement as well as locating of food. In this process it moves from one place to other place only by a way of sniffing on the ground there by getting guided in to the desired routes. Spreading of human hair collected from local barber shops is an affective and low cost traditional method being followed by farmers. Technically this indigenous method do have scientific logic which clearly suggest that the human hair in the movement routs of the wild boar gets sucked through nostrils causing severe respiratory irritation. Due to this the wild boar gets totally disturbed and loses its track by making distress calls, which will ward off other wild boars entering into the cropped area. Several farmers are extensively practicing this method in different crops and controlling the damage caused by wild boar to the extent of 70-80 per cent.

4.1.3 Fixing of Used Colored Sarees

This method also is a farmers' innovation, which has a behavioural background as far as wild boar is concerned. By arranging used sarees of different colors around the crop will make wild boars to assume human presence in the area there by not preferring to enter into such areas. Even though, not feasible in all situations it has some marginal benefit in the areas of human movement. By using this, extent of damage by wild boar can be minimize to the level of 45-60 per cent

4.1.4 Burning of Dried Dung Cakes

The dried cakes made of dung of local pigs are burnt by placing them in earthen pots. This will ensure slow generation and spread smoke during dusk time. The smoke coupled with smell of local pig dung helps in sensitizing wild boar about the unexisting presence of local pigs. As a result, to avoid territorial conflict, the wild boars don't prefer to move in such areas.

4.1.5 Creation of Sounds and Light through Bornfire

To scare away the wild boars from damaging their crops, farmers employ methods such as using fire crackers, making sounds through local drums, empty tins, making bornfires and shouting. These methods have proved to be effective on community basis in protecting farmers' fields from the wild boars.

4.1.6 Use of Traditional Local Dogs for Scaring away Wild Boars

In endemic areas of wild boar attacks farmers do follow using of trained dogs on a community basis to scare away the approaching wild boars. In selected cases this method proved to be affective and sustainable.

4.2 Ecofriendly Management Methods for Wild Boar

4.2.1 Biological Barriers

To avoid additional expenditure in the name of physical barriers, extensive studies were undertaken by AINP on Agricultural Ornithology to utilize naturally occurring shrubs/plants/crops as physical barriers/bio-fencing. Bio-fences are useful not only as protective barriers but are also ecofriendly, giving additional income to the farmer through their produce.

4.2.2 Safflower Around the Crop

The practice of having 4-5 rows of safflower crop as border crop around groundnut was found to be most promising in preventing the damage by wild boar. Safflower crop being thorny will cause great amount of inconvenience and damage to wild boar especially when it is sown in closed spacing. In addition, safflower crop emits strong chemical odour effectively masking the odours emitted by groundnut crop. Due to this, wild boar at the first instance will fail in locating the groundnut crop, secondly even if it is located the thorns of the safflower plant causes mechanical injury or damage, thereby they will not try to enter into the groundnut field. Additional income realized through safflower crop comes as an added advantage to the farmer. This method was experimentally validated in different locations. During 2011-12, the experiment was conducted at ARS Thandur, Andhra Pradesh. The experimental plot yielded 1609 kg/ha against control (432 kg/ha). Similarly in the subsequent year (2012-13), the yield was higher (1560 kg/ha) against control (738 kg/ha) and during 2013-14, the recorded yield was 1313 kg/ha against control (862 kg/ha).

4.2.3 Castor around the Crop

This method is being widely popularized in maize, by planting 4-5 rows of castor with close spacing around the maize crop. Wild boars being capable of identifying maize only through smell can't do so owing to the strong odour emitted by the castor successfully masking the odour emitted by the maize crop. Damage in castor by wild boar is also not possible due to the non palatable nature of the plants with high amount of alkaloids and glycosides. Through this method, farmer is benefitted with additional income through castor. Usage of castor as border crop is practicable in both *kharif* and *rabi* seasons in crops like pulses and oilseeds. This method was experimentally validated in different locations. During 2011-12, the experiment was conducted at ARS Thandur, Andhra Pradesh. The experimental plot yielded 1846 kg/ha against control plot (355 kg/ha). Similarly in the subsequent year (2012-13), the yield was 5525 kg/ha against control (133 kg/ha) and during 2013-14, the experiment was conducted at college farm, Hyderabad and recorded the yield 4616 kg/ha against control (1641 kg/ha).

4.2.4 Planting of Thorny Bushes and Xerophytes around the Crop

Different xerophytic species like Cacti sp *Euphorbia caducifolia, E. meriifolia* and Opentia sp *Opuntia elatior, O. dillenii,* Zizipus sp *Ziziphus oenopolia, Z. mauritiana,* and agave sp *Agave americana, Caesalpinia cristata* can be planted on the bunds around the crop which will not allow the wild boars due to their thorny in nature. The wild boars after unsuccessful trail of entry get injuries and making alarming calls, which makes the other animals to flee the scene.

4.2.5 Planting of *Karanda* around the Crop

Planting of *karanda* (*Carrissa carandus*) around the crop as biofence does prevent effectively the entry of wild boars into the cropped area owing the thorny nature. Using *karanda* as a border crop gives enormous benefits to the farmer by giving value added products extract of medical important effective alternative to tamarind *etc.*, in addition to fulfilling of basic purpose of wild boar prevention.

4.2.6 Spraying of Egg Solution

By exploiting the habit of the wild boar using smell of the crop as criteria for identification, an extensive level of experiments were carried out to use spray of egg solution either on the border row of the crop or on the wet soil around the crop. The results has given a clear cut indication that spray of egg solution 20 ml/l of water was capable of successfully making the natural odour of the crop and thereby reducing the wild boar damage. Spraying of egg solution around the bunds of maize fields were capable of reducing wild boar damage, and the experiment during *Kharif* and *Rabi* yielded 807 and 917 kg/ha against control 573 and 558 kg/ha respectively.

4.2.7 Coconut Ropes Coated with Pig Oil

Preparation of solution with sufficient quantity of sulphur is mixed with local/domestic pig oil and the mixture is smeared on the coconut ropes and the ropes are erected in 1 feet width in three rows around the crop with the help of wooden poles. This mixture generates the typical smell there by repelling wild boars not to enter into the cropped area. For an effective use of this method two such applications should be done with ten days internal in between.

4.2.8 Barbed Wire Fence

Erecting of barbed wire around the field in three rows with first row is at the height of 1 foot from the ground. This is highly effective in preventing wild boars from entering into the cropped area. Use of barbed wire fence as border against wild boar around maize crop yielded 859 kg/ha during *kharif* and 717 kg/ha during *rabi*, which was predominantly higher than the control (Table 11.5).

Table 11.5: Efficacy of Wild Boar Management Methods on Maize Yield (2011-14)

Treatment	*Yield (Kg/ha)*					
	2011-12		*2012-13*		*2013-14*	
	Kharif	*Rabi*	*Kharif*	*Rabi*	*Kharif*	*Rabi*
Barbed wire	1939	1781	1049	-	859	717
Egg solution spray	1752	1354	-	1045	807	917
GI wire	2088	1941	-	-	781	708
Chain link fence	2057	2208	1066	1069	911	1008
Circular blade wire	2267	2383	1193	1195	964	1067
Control	823	909	786	780	573	558
CD (5 per cent)	213	354	151	165	370	314

4.2.9 Circular Razor Fence

The iron wire fixed with sharp razor blades at regular distance is kept 1 ft away from the cropped area as border by forming circular rings. The blades cause serious damage to the wild boar which try to enter into the field. This not only prevents the animal to enter into the field but also scares away other animals. The entangled animal makes alarm calls which deter away the other wild boars thereby saving the entire crop without any damage. Using of Circular razor fence as border against wild boar around maize crop yielded 964 kg/ha during *kharif* and 1067 kg/ha during *rabi*, which was predominantly higher than the control (553 kg/ha (*kharif*) and 558 kg/ha (*rabi*).

4.2.10 Chain Link Fence

It is most effective way of fixing a barrier which is more durable in nature. Chain link meshes of 3 feet height can be fixed around the crop by maintaining a distance of 1 ft away from the crop.

4.2.11 Solar Fence

This is a permanent type of physical barrier arranged around the cropped area which is gaining more popularity. This method is being widely practiced to prevent the damage by wild boar in high value crops. In this method a solar fence is fixed around the crop with 12 volts electricity being sent to the fence with the help of solar plates. The shock received by the animal during the contact will not be capable of killing the animal but certainly ward off not only the animal which comes in contact but also other following animals which will be scared due to the alarm calls of the shocked animal.

4.2.12 GI Wire Fence

A simple GI wire can also be use to create physical barrier for wild boars where three rows of GI wire fixed around the crop with the help of poles with a height of

1 feet from the ground level. This method is comparatively simple and economical as compared to other barriers. *4.2.13 Trenches method*

Digging of 2 ft wide and 1 ½ feet deep trench around the cropped area at a distance 1 ft from the crop keeps away the wild boars from the field. This method also helps as an excellent source for water conservation in the rain fed areas despite being effective method for wild boar management. This method gives additional advantage of preventing the damage of the crop by insect pests which are migratory in nature from one field to other field.

5. Conclusions

All the methods listed above are scientifically proved effective in controlling of depredatory birds and wild boar intrusion in different crop fields. Among all traditional and eco friendly methods, reflective ribbon, wrapping in maize, bio acoustics, screen crop, spraying of egg solution and habitat manipulation were proved effective and reduced crop losses by birds to the tune of 60-70 per cent. Where as in case of wild boar, bio acoustics, safflower as border crop around ground nut, castor as border crop around maize and sorghum, circular razor wire fencing are proved effective in reducing the damage to the extent of 70-80 per cent. Hence, large scale demonstrations and awareness to different stakeholders about these methods will certainly enhances the production and productivity of the crops and also minimizes man-animal conflict.

6. Future Thrust

Effective management of vertebrates mostly depends on the population density of the species, extent of crop area, crop types, season, climatic conditions, predatory pressure and physiological status of the animal. Therefore, the future thrust should be a holistic approach preferably multi disciplinary use of methods to reduce crop damage. The recommendations given by AINP on Vertebrate Pest Management are certainly aimed in this direction. Special attention should also be given to forecasting and forewarning models by using weather-crop interdependency and application of modern information technology techniques such as remote sensing, GIS (Geographical Information System) and GPS (Global Positioning System) to study different vertebrate distribution and habitat suitability. Efforts also should be intensified to formulate suitable agricultural policies by way of paying compensation and other benefits to reduce man-animal conflicts and enhance the environmental conservation in agricultural landscapes.

REFERNCES

Moreira LJ, Rosa L, Lourenço J, Barroso I and Pimenta V. 1997. *Projecto Lobo; Relatorio de Progresso, 1996.* (Cofinanciado pela U.E. – Programa Life). Ministério do Ambiente e dos Recursos Naturais; Instituto de Conservação da Natureza. Parque Natural de Montesinho, Bragança (Portugal), pp. 61.

Roberts TJ. 1977. The Mammals of Pakistan. I[st] ed. Ernest Benn Press, Limited., pp. 361.

Shafi MM and Khokhar AR. 1986. Some observations on wild boar (*Sus scrofa cristatus*) and its control in sugarcane areas of Punjab, Pakistan. Journal of Bombay Natural History. Society 83: 63.

Sharma SK. 1991. Plant life and weaver birds – with special reference to eastern Rajasthan. Ph.D. Thesis, University of Rajasthan, Jaipur.

Sharma SK. 1994. Of crows and sparrows. Newsletter on Birdwatch 34(2): 36-37.

Syamsunder Rao P. 2002. Research accomplishments of agricultural ornithology. All India Network Project On Agricultural Ornithology. (Parasharya BM, Vasudevarao V and Mathew KL, Ed.), Acharya NG Ranga Agricultural University, Rajendranagar, Hyderabad. Technical Bulletin–II.

Chapter 12

Non-Chemical Management of Insect-Pests of Stored Products

T.V. Prasad and M. Srinivasa Rao

ICAR-Central Research Institute for Dryland Agriculture, Hyderabad – 500 059, Telangana
E-mail: t.prasad@icar.gov.in

1. INTRODUCTION

It has been estimated that between one quarter and one third of the world grain crop is lost each year during storage (Muhammad Sarwar, 2015). Global loss in stored products, caused by insects, have been estimated to be between five and ten per cent. In the tropics it may reach up to 30 per cent. In general, warm grain temperatures at harvest and during storage, combined with grain moisture content of 12-13 per cent, are favorable to growth of insect populations (David Weaver and Reeves, 2004). There are a number of stored grain insect pests that infest food grains in farmer stores and public warehouses and massively surge due to un-controlled environmental conditions and poor warehousing technology used (Sarwar, 2009; Sarwar and Sattar, 2012). Many grain pests preferentially eat out grain embryos, thereby reducing the protein content of feed grain and lowering the percentage of seeds which germinate. Insects not only cause direct loss through consumption of kernels but also indirect losses which includes accumulation of frass, exuviae, webbing and insect cadavers. High levels of this insect detritus may result in grain that is unfit for human consumption. Most insect pests belong to the orders Coleoptera and Lepidoptera, which account for about 60 and 10 per cent, respectively, of the total number of species of stored-product insect pests (Atwal and Dhaliwal, 2008; FAO, 2009).

Storage losses occur at different stages in the movement of food from field to the consumer. There are 2 types of grain losses namely quantitative and qualitative losses. Quantitative losses include, loss of weight due to reduction in moisture content of the grain, loss due to grain depredation by insect, rodent and birds and loss due to grain mishandling. Qualitative losses include, discoloration of grains due to development of mould or faulty method of grain drying, loss due to deterioration in nutritional quality due to presence of insect fragments, excreta, urine *etc.*

2. Factors Affecting the Stroage of Grains

2.1 Abiotic Factors

2.1.1 Moisture

It is necessary to know the moisture content of the grains during storage. The grain can absorb moisture from the surroundings. If the temperature is more, moisture will be reduced. At 12 per cent moisture level, insect attack and multiplication are faster. When the moisture is reduced all the insects will die except *Trogoderma granarium*. It can survive at low moisture and also at high temperatures. If the moisture content is more than 14 per cent, the activity of fungus starts.

2.1.2 Temperature

Temperature is a determining factor in the development of all organisms and its effect is correlated with the amount of moisture present. When the temperature is increased the life cycle will complete in less time when grain temperature reaches 66°C the germination capacity is lost and nutritive value is decreased. Temperature below 15°C retard insect reproduction and development; if prolonged below 10°C will cause the death of most insects.

2.1.3 Oxygen

Like humidity and temperature, every insect species requires a certain percentage of oxygen for its development and survival. When the grains are stored in an air tight storage, the oxygen content gets reduced and development of insect ceases and it dies. Khapra beetle requires 16.8 per cent oxygen content for survival of eggs, therefore its multiplication can be checked by lowering the oxygen content below this level. Some insects can survive in absence of oxygen by reducing their metabolic rates and utilizing the oxygen in tissues.

2.1.4 CO_2

Excessive CO_2 causes varied reactions in different insects. Some insects can live in an atmosphere with high CO_2 for several days. However, an excess of this gas in the atmosphere causes growth retardation in many insects. Under high CO_2 concentration, the spiracles remain open leading to excessive water loss and insect dies due to desiccation.

2.2. Biotic Factors

2.2.1 Microorganisms

All types of stored grains are susceptible to attack by fungi. Majority of storage fungi grow quickly at some point in the temperature range of 20°C to 40°C and above 90 per cent relative humidity. Fungal infection affects germination and quality of the grains. During growth, some fungi produce chemicals which are toxic to man or domestic animals. They cause liver cirrhosis. Ex: *Aspergillus flavus* produces aflatoxin on groundnut.

2.2.2 Mites

A number of mites are associated with stored produces. Under suitable temperature conditions and high moisture content, they multiply rapidly to form dense populations. Ex: The most common mite occurring in food grains is *Acarus siro*.

3. Insect Pests of Stored Products and their Damage

The insect pests of stored gains are broadly divided into:

3.1 Primary Grain and Seed Feeders

These attack whole and undamaged grain. Larvae of these insects are confined within one seed during development.

3.1.1 Granary/Rice Weevil, *Sitophilus oryzae* (L.) (Coleoptera: Curculionidae)

It attacks rice, wheat and barley. The granary weevil is a small dark brown-black beetle about 4 mm long with a characteristic rostrum (snout) protruding from its head. An adult lays up to 450 eggs singly in holes chewed in cereal grains. Each egg hatches into a white, legless larva, which eats the grain from the inside. The larva pupates within the grain and the adult then chews its way out. Both grubs and adults cause the damage. *S. oryzae* and *S. zeamais* starts its attack in field itself. Adults cut circular holes. Heating takes place during heavy infestation, which is known as 'dry heating'.

3.1.2 Khapra Beetle, *Trogoderma granarium* Everts (Coleoptera: Dermestidae)

It attacks grains and oilseeds. Initial discoveries are often based on the presence of accumulation of cast larval skins. As larvae mature, they are able to attack whole grains and capable of destroying commodity completely. Larvae may enter diapause (dormant state) and remain inactive up to 8 years if conditions are unfavourable. It is initiated by cooler temperature, poor food quality or low population densities. Adults cannot fly and do not feed. The beetle prefers hot and dry climatic conditions.

3.1.3 Lesser Grain Borer, *Rhyzopertha dominica* (F.) (Coleoptera: Bostrichidae)

Adults are dark reddish brown and 3 mm long with distinctly shaped, loose 3-segmented club. Larvae are white and c-shaped. They are immobile at maturity stage with dark head capsule. Attacks all grains, especially wheat, barley, sorghum and rice. Adults and larvae also burrow through kernels and irregular shaped holes are formed in commodity. Adults and larvae feed on germ and endosperm reducing kernels to shells of bran.

3.1.4 Angoumois Grain Moth, *Sitotroga cerealella* (Olivier) (Lepidoptera: Gelechiidae)

The angoumois moth is yellow-brown with darker markings. Its host range is paddy, maize, jowar, barley and wheat. Larvae damage grains. It attacks both in fields and stores. In stored bulk grain, infestation remains confined to upper 30 cm depth only. Females lays white eggs on the surface of damp grains in stores or fields. Infestations impart an unpleasant smell and taste to the cereal.

3.1.5 Pulse Beetle, *Callosobruchus chinensis, C. maculatus* (L.) (Coleoptera: Bruchidae)

It attacks all whole pulses, beans and grams. Female lays eggs singly, glued to the surface of the pod (in fields) or on grains (in stores). Grubs eat up the grain kernel and make a cavity. Adults come out making exit holes. Adults are short lived, it is harmless and do not feed on storage produce at all.

3.1.6 Tamarind/Groundnut Bruchid: *Caryedon serratus* (Olivier) (Coleoptera: Bruchidae)

It is a major pest of tamarind and groundnut. Female lays eggs singly, glued to the surface of the pod (in fields) or on grains (in stores). Grub causes the damage. Circular hole on fruits and seeds of tamarind both in plantations and storage. Adult is brownish grey beetle with characteristic elevated ivory like spots near the middle of the dorsal side. Elytra do not cover the abdomen completely, which is called as pygidium.

3.1.7 Cigarette Beetle, *Lasioderma serricorne* (F.) (Coleoptera: Anobiidae)

Cigarette beetles are quite small, measuring about 2 to 3 mm and are reddish brown. Larval feeding causes direct damage to foodstuffs and non-food items. Beetles chewing through cardboard boxes, containers, and packaging cause indirect damage.

3.1.8 Drug Store Beetle, *Stegobium paniceum* (L.) (Coleoptera: Anobiidae)

They are similar to cigarette beetle larvae, but have shorter hairs and markings on the head which ends in a straight line across the frons just above the mouthparts. Antennae of the drugstore beetle end in a 3-segmented club and elytra (wing covers) have rows of pits giving them a striated (lined) appearance.

3.1.9 Sweet Potato Weevil, *Cylas formicarius* (Fabricius) (Coleoptera: Curculionidae)

Eggs are deposited in small cavities in the sweet potato root or stem. The female deposits a single egg at a time, and seals the egg within the oviposition cavity with a plug of faecal material. When the egg hatches the larva usually burrows directly into the tuber or stem of the plant.

3.1.10 Potato Tuber Moth, *Pthorimaea operculella* (Zeller) (Lepidoptera: Gelechiidae)

Female lays eggs singly on under surface of leaves and exposed tubers, larva is pale greenish. Larvae mine into leaves or bore into tender shoots and developing tubers. Larvae feed on potato leaves, stems, petioles, and more importantly potato tubers in the field and in storage. Rotting and foul smelling of damaged tubers.

Potato tuber eyes become pink due to deposition of silk and excrement by potato tuberworm infestation.

3.2 Secondary Grain and Seed Feeders

These cannot attack whole grains and feed mainly on broken or already damaged grains.

3.2.1 Red Flour Beetle, *Tribolium castaneum* (Herbst) (Coleoptera: Tenebrionidae)

It attacks stored grains, oilseeds, starchy materials, beans, peas, spices, dried plant roots, fruit, yeast and chocolate. Prefers damaged grain but also attack intact wheat kernels, feeding first on the germ and then on the endosperm. Releases a noxious secretion, when disturbed, resulting in a pungent odor in the infested commodity, rendering milled products unfit for consumption.

3.2.2 Saw Toothed Grain Beetle, *Oryzaephilus surinamensis* (L.) (Coleoptera: Silvanidae)

It affects oats (beetle is most often found here), wheat, barley, animal feed, flax and sunflower. It also affects milled and processed products, dried fruit, packaged foods. Severe infestations can cause grain to become overheated contributing to further damage. Both adults and larvae may feed on grain dust and larvae preferentially fed on the germ.

3.2.3 Rice moth, *Corcyra cephalonica* (Stainton) (Lepidoptera: Pyralidae)

Larvae are creamy white and web together particles of food and frass with silken threads into galleries in which they live and feed. The larval period is 20-30 days and during this period damage to sorghum grain occurs. Infested produce is densely webbed by the larvae with particles of food and frass with silken threads into galleries and thus the grain become unfit for consumption.

3.2.4 Fig or Almond or Warehouse Moth, *Ephestia cautella* (Walker) (Lepidoptera: Pyralidae)

The warehouse moth is a drab grey moth with a 10-12 mm wingspan. It usually infests the surface of stored grain. Moths live for only about two weeks, but during that time lay up to 200 eggs. These are distributed loosely on the grain surface. Larvae hatch out of the eggs and wander over the grain surface leaving a trail of silk which may form a thick mat covering the surface of the infested grain.

3.2.5 Flat Grain Beetle, *Cryptolestes pusillus* (Coleoptera: Cucujidae)

Eggs are laid throughout the stored grain and develop into tiny larvae with characteristic tail horns, biting mouth parts and three pairs of legs. They feed on damaged grain and wheat embryos. Pupation takes place in a cocoon. A complete life cycle takes from 4-5 weeks and adults may survive up to one year.

3.2.6 Indian Meal Moth, *Plodia interpunctella* (Hubner) (Lepidoptera: Pyralidae)

The female lays up to 200 eggs near the grain surface as it slowly passes from grain to grain spinning a silk thread. Severe infestations may form a surface web on the grain heap. Larvae attack the wheat germ, then pupate in a cocoon which may be found in cracks and crevices of buildings. It takes 4 weeks to complete one generation under warmer conditions.

3.3 Detritus and Mould Feeders

These do not attack clean and sound grains. Usually moulds are necessary in the diet to complete nutritional requirements.

4. Storage Structures

Grain storage is carried out for 3 purposes. They are: a) To retain a supply of food, b) To service a trading system, and c) To retain seed for planting in the following season.

4.1 Bamboo Structures

Bamboo indoor structures are called by different names in different states such as gade, duli tom or topa (Assam), thombai (Tamil Nadu), dholi (Orissa and Maharashtra) and galagi (Karnataka). Bukari and gola are the bamboo outdoor structures.

Advantages

- ☆ The building material and the technical know how is locally available
- ☆ The structure can be shifted because of light weight
- ☆ Suitable for storage of seeds and grains with excess moisture

Disadvantages

- ☆ The structures are not air tight and hence fumigation not possible
- ☆ Do not give protection against insects and rodents
- ☆ Short life and not fire proof
- ☆ Needs periodical maintenance

4.2 Timber Structures

Timber indoor and outdoor structures are called by different names in different states such as peti or bin (Uttar Pradesh), gutibharal (Assam) and pathayam (Kerala).

Advantages

- ☆ The building material and the technical know how is locally available
- ☆ The structure can be shifted because of light weight
- ☆ Suitable for storage of seeds and grains with excess moisture

Disadvantages

- Do not give protection against insects and rodents, termites and moisture
- The structures are not air tight and hence fumigation is not possible.

4.3. Paddy Straw Structures

These are made up of paddy straw alone or paddy straw mixed with mud. They are called puri (Andhra Pradesh), pura (Bihar, Bengal and Punjab), morai (Bihar and Bengal)

Advantages

- Low cost
- High moisture content grain can be stored as cross ventilation provides drying

Disadvantages

- The structures are not air tight and hence fumigation not possible
- Not moisture as well as fire proof

4.4. Mud and/or Brick Structures

These are called kothas or kothis (Bihar, Punjab and Rajasthan), mud bins, bharolas or bharolis when chimney shaped and petti when box shaped (Punjab), and mud, pukka and Pusa bin.

4.4.1 Pusa Bin

It is a modification of the ordinary mud storage structure commonly used in villages. To provide moisture-proof and airtight conditions, polyethylene film of 700 gauge thickness has been embedded at the top, bottom and on all the sides of the mud bin. The embedding process provides mechanical support and safety to polyethylene film. The construction of outer walls with burnt bricks up to 45 cm height makes the structure rat proof as well. The Pusa bin is constructed with unburnt bricks on burnt bricks or concrete floor to avoid rat burrowing. The bin can be constructed for storing different quantities of grain ranging from 0.4 to 3.2 tons. A round hole of 15 cm diameter is left at the bottom of the front wall to fix a pouch mate up of metal sheet or polyethylene pipe, for withdrawing the grain. All the sides of inner walls are plastered and 50 cm x 50 cm clear hole at the left hand corner of the front side is left far grain filling. On top wooden pole is spread at 45 cm distance for the support of the roof. The mud slabs (made with mud, paddy husk mixed with and bamboo split) are used as top cover. The top is also covered with 5 cm layer of mud. The structure is plastered and is allowed to dry thoroughly before filling of grain. Thoroughly dried grain should be filled to full capacity. The grain and seed both remain safe in the bin for more than 1 year.

4.4.2 Pusa Cubicle

This is a room like structure (3.5 m x 3.15.m x 2.60 m), a modification of Pusa bin to provide large storage capacity of 24 tons. On a platform of 3.73 m x 2.93 m x 0.07 m is made with unburnt bricks on a concrete floor (except 22 cm of outer sides with burnt bricks). A polyethylene sheet is placed on this platform and another platform of similar dimension is made with unburnt bricks. The 22 cm thick inner walls are constructed up to 2.60 m height. A wooden frame of 1.89 m x 1.06 m for door is fixed in the front side of 3.95 m wall. The roof can be made by wooden beam placed at 15 cm distance and covered with unburnt bricks. Use of coaltar painted bamboo splits or 1.5 cm thick wooden planks followed by plaster with 5 cm thick mud could be made. Structure is then allowed to dry. A polyethylene cover is placed from the top and two edges of bottom polyethylene and of the cover are then sealed by heating. Outer wall (22 cm thick) is now constructed, the first 45cm portion with burnt bricks and the rest with unburnt bricks. The roof is constructed with brick tiles or with 5 cm thick mud mixed with paddy straw. The door is made with either plywood or seasoned wooden planks by sandwiching of polyethylene sheet which extends 10 cm all around. This then folded on inside face of the door. A rubber gasket is fixed on this to provide complete air tightness.

Advantages

- ☆ These are cheap and handy
- ☆ These can withstand temperature fluctuations

Disadvantages

- ☆ These need regular maintenance
- ☆ Does not provide adequate locking facilities
- ☆ Not impervious to moisture, insect and fungus infestation

4.5 Underground Storage Structures

They are khatti (Uttar Pradesh, Punjab), peve (Maharashtra, Tamil Nadu, Karnataka, Madhya Pradesh), bahda, budu, khondia, koh or lathani (Madhya Pradesh), khani, banda (Rajasthan), khani (Orissa), bandha (Maharashtra)

Advantages

- ☆ Grains are free from seasonal fluctuations of temperature and humidity
- ☆ Low cost of construction compared to that of above-ground storage structures of similar capacity
- ☆ Ambient temperatures are relatively low and constant
- ☆ Hardly visible, and therefore relatively safe from thieves
- ☆ No need for continuous inspection
- ☆ Saves space and no danger of fire

Disadvantages

- ☆ Construction and digging are laborious

- Storage conditions adversely affect viability; the stored grains can only be used for consumption
- The grain can acquire a fermented smell after long storage
- Removal of the grain is laborious and can be dangerous because of the accumulation of carbon dioxide in the pit, if it is not completely full
- Inspection of the grain is difficult
- Risks of penetration by water
- No adequate protection against rodents and termites

4.6 Earthen Storage Pots

These are indoor structures of small capacities made from burnt clay. They are made from well kneaded mud on a potters wheel, sun dried and then baked in fire. They may have both inlets and outlets or only inlets. When filled with grain, the top is sealed with mud cover. The structures are placed either slightly above the ground or partly below it. They are used in Bihar (called bhauri or kundu), Uttar Pradesh and Punjab.

Advantages

- These are cheap and handy
- These can withstand temperature fluctuations

Disadvantages

- These are fragile and likely to break easily
- Not impervious to moisture, insect and fungus infestation

5. Grain Drying Methods and Aeration

5.1 Drying

It is essential to dry grain quickly and effectively after harvest and before storage to retain maximum quality, to attain a moisture content sufficiently low to minimize infestation by insects and microorganisms (bacteria, fungi, *etc.*), and to prevent germination. The major objective in drying grains is to reduce the moisture content so that spoilage will not occur before the grain is used.

Advantgaes

- It reduces loss of products due to weather, insects, rodents, birds, and natural shattering.
- It allows long-term storage with little deterioration
- It allows to have advantage of possible higher prices a few months after harvest
- It provides maintenance of seed viability
- It enables the production of a better-quality product

Drying Mechanism

In the process of drying heat is necessary to evaporate moisture from the grain and a flow of air is needed to carry away the evaporated moisture. There are two basic mechanisms involved in the drying process; the migration of moisture from the interior of an individual grain to the surface, and the evaporation of moisture from the surface to the surrounding air.

The rate of drying is determined

- ✰ Moisture content
- ✰ Temperature of the grain
- ✰ Temperature and relative humidity
- ✰ Velocity of the air in contact with the grain

Types of Drying Methods

5.1.1 Sun Drying

These make use of exposure of the wet grain to the sun and wind. Moist grains are spread in the open exposed to the sun for varying period depending up on the initial moisture and intensity of heat. Floor used for drying are either plastered with mud or bitumen or cemented.

Disadvantages

- ✰ Susceptible to pilferage by birds, cattle and human beings
- ✰ Requires large space
- ✰ It requires labour

5.1.2 Solar Dryers

An improved technology in utilizing solar energy for drying grain is the use of solar dryers where the air is heated in a solar collector and then passed through beds of grain. There are two basic types of solar dryer appropriate for use with grain: natural convection dryers where the air flow is induced by thermal gradients; and forced convection dryers wherein air is forced through a solar collector and the grain bed by a fan.

5.1.3 Drying with Heated Air

This can be applied either for batch or bulk drying. Incase of batch drying a definite quantity is move in heated atmosphere for a desired period. For bulk drying, grain is allowed to flow down in a current of hot air. Heating is done by either using furnace oil, or agricultural waste or stream

The rate of drying depends up on the rate of evaporation and diffusion when heat is applied for drying, heat is needed both to supply energy for evaporation and maintain a difference in vapour pressure to cause internal moisture to diffuse to surface.

5.1.4 Drying with Unheated Air

This involves forcing ordinary air through the wet grain. Efficiency and cost depend on the atmospheric conditions. This drying can also be carried out when grain is stored in bins, which is known as 'in bin drying'. Bins are provided with perforated floors to facilitate drying. Air is either blown through or sucked in. Blowing or sucking is arranged taking into consideration the environment conditions and initial moisture content. Efficiency and cost would depend up on the atmospheric conditions.

Advantages

- ☆ Equipment required is simple
- ☆ No hazards are involved

5.1.5 Mechanical Dryers

These employ the application of heat from combustion of fossil fuels and biomass resources, directly or indirectly, and in both natural and forced convection systems. Mechanical dryers commonly used are batch in bin dryers, bin layered drying, batch dryers and continuous flow dryers. Air serves two basic functions in grain drying. First, the air supplies the necessary heat for moisture evaporation; second, the air serves as a carrier of the evaporated moisture. Both functions are essential, regardless of the drying system used. The amount of moisture which can be removed from grain depends on the relative humidity and temperature of the drying air, the airflow rate, and the moisture content of the grain.

5.2 Aeration

Aeration involves forced movement of outside air through a bulk stored commodity to create a uniform temperature throughout the bulk. Cooling by aeration minimizes the growth of insect and mold populations and prevent spoilage and economic losses. In unaerated grain, the moisture from the grain is removed by air currents which condense leading to spoilage of grain.

Advantages

- ☆ Aeration is effective
- ☆ Aeration is safe
- ☆ Operation cost is low
- ☆ Grain quality is maintained
- ☆ No problem of resistance development

Disadvantages

- ☆ Aeration provides only short term protection
- ☆ Poor aeration can reduce quality
- ☆ Capital investment is high

- ☆ Process of aeration is complex
- ☆ Aeration can reduce moisture

6. Integrated Management of Storage Insect Pests

The pest control measures should be designed at the onset of grain storage. Firstly, locate the infestation source and then the source of infestation is eliminated properly. Traditional grain storage facilities may not offer protection, but promotion of the use of metal silos and resistant varieties for grain storage is an alternative approach to reduce losses (Adda *et al.*, 2002; Tadele *et al.*, 2011). The drying of the foods helps in reducing the moisture content to about 9-12 per cent in the drier areas, thus, minimizing the activities of storage insect pests and pathogens (Okunade *et al.*, 2001).

6.1 Monitoring

Pest monitoring is an important component in the IPM post-harvest practice for stored grain. Inspections should be done frequently, especially after first storage, to enable pest-management decisions to be made (Subramanyam and Hagstrum, 1995). Monitoring of stored grain insect pests with pit fall traps, delta pheromone traps and probe traps and UV light traps should be practiced.

6.2 Drying

Ensure proper drying to bring the moisture content lower than 10 per cent before storage.

6.3 Sanitation

- ☆ Cleaning of grains, removal and destruction of infested grains
- ☆ Cleaning of storage structures and receptacle thoroughly
- ☆ Removal of lose mud plaster and apply new one to fill cracks and crevices
- ☆ Use new gunny bags for storing the grains
- ☆ Both harvesting and threshing machines need to be thoroughly cleaned prior to use.
- ☆ Remove spilled grain outside the storage structure.
- ☆ Rat burrows to be closed with a mixture of broken glass pieces and plastered with cement.
- ☆ For additional protection, the inside and outside surfaces, foundations and floor of a storage facility should be treated with a residual insecticide, four to six weeks prior to harvest.

6.4 Physical Control

It is by exclusion, impact and removal. Exclusion by installation of proper doors of windows, preventing movement of transport vehicles in to the storage houses and use of pest exclusion meshes.

Impact: A machine used to disinfest cereals and other foods. The material is fed to the centre of a high-speed rotating disc carrying studs so that it is thrown against the studs; the impact kills any insects and destroys their eggs. The efficacy depends on speed, moisture and flow rate.

Treating the grains with diatomaceous earth or silicon dioxide which kills the insect by abrasion and desiccation. Diatomaceous earth (Insecto®) can be used effectively for on-farm storage. Diatomaceous earth (DE) is the remains of microscopic one-celled plants (diatoms). The insecticidal quality of DE is due to the razor sharp edges of the diatom remains. As the insects crawl through treated grain and dusted bins, the DE comes contact with the insects and the sharp edges punctures the insects exoskeleton. The powdery DE then absorbs the body fluids causing death from dehydration.

6.5 Botanicals

Botanicals such as neem possess repellent, anti-feedant and feeding deterrent properties against storage insect pests. Neem seed kernel powder at 4.0 per cent (w/w), neem seed oil at 1.0 per cent (v/w) and mahua oil at 1 per cent (v/w) proved repulsive and a potent oviposition inhibitor in checking damage by the pulse beetle, *C. chinensis* for up to 8 months in pigeon pea (Singal and Chouhan, 1997). Botanicals such as *Acorus calamus* (sweet flag) can be used for protection against storage pests. It has both killing properties and deterrent properties. Paranagama *et al.* (2003) reported that damage to grain was lower in *Oryza sativa* treated with the essential oils of *Cymbopogon citrates* (Stapf) and *Cymbopogon nardus* (Rendle) than in the control rice grains. *C. citrates* and *C. nardus* showed deleterious effects on oviposition and F1 adult emergence of the cowpea bruchid *Callosobruchus maculatus* (F.) compared with the control during no-choice tests.

6.6 Improved Storage Structures

Use of improved storage structures such as Pusa bin, Pucca kothi, will reduce the insect infestation and grain can be stored for long time.

6.7 Aeration

Aeration is used to either reduce the moisture or the temperature of grain. In both the situations the insect's development is arrested. Reducing moisture content through aeration is relatively easy even with out the use of heated air. It is a relatively a safe technique.

6.8 Extreme temperatures

6.8.1 High Temperatures

Exposure to high temperature for short time will eliminate all pests. Provide a super heating system by infrared heaters in the floor mills and food processing plants to obtain effective control of pests since mostly the stored produce insects die at 55–60°C in 10 to 20 minutes.

6.8.2 Low Temperature

Temperatures between 5 and 15°C delay insect development and then they die after a very long exposure. Temperature between -1 and 3°C can cause death in hours or days. Temperatures below -1°C can cause death more quickly. A temperature below 4°C results in death, particularly of the immature stages of almost all insect pests. Death occurs rapidly at freezing point. *T.castaneum* and *Oryzaephilus mercator* are highly susceptible to cold, whereas *Trogoderma* spp., *Ephestia* spp. and *Plodia interpunctella* are cold-tolerant species. In most field applications of a cold temperature to control insect pests, the insects are exposed to a temperature of 10–20°C for some days before being exposed to a lethal cold temperature.

6.9 Biological Control

Biological control has a limited scope in stored-grain pest management but is becoming an increasingly important part of an IPM approach. There are a number of insect predators and parasitic wasps that attack insect pests of stored grain. It is important to note that there are limited numbers of naturally occurring biological control agents. - *Anisopteromalus calandrae, Choetospila elegans* and *Lariophagus distinguendus* are parasitic wasp of grains. Warehouse pirate bug, *Xylocoris flavipes, Habrobracon hebetor* against Indianmeal moth (David Weaver and Reeves, 2004). Parasitoids *Theocolax elegans* (Westwood), a small pteromalid wasp attacks primary grain pests whose immature stages develop inside grain kernels, including the weevils (*Sitophilus* spp.), the lesser grain borer, the drugstore beetle, cowpea weevils and angoumois grain moth (Flinn *et al.*, 2006). *Trichogramma* spp. have also been evaluated against a variety of stored-product moths in bulk groundnut storage, bulk wheat storage and bakeries, as well as in warehouses and retail stores (Grieshop *et al.*, 2007). Stored-product moths commonly oviposit on packages and on shelves holding stored-product packages. *Trichogramma* spp. are especially promising as biological control agents on finished products because they attack the egg stage of the pests, thereby preventing invasion of products by first instars. *Dinarmus* spp. is a larval/pupal parasitoid of *Callosobruchus* spp., *Bruchus* spp., *Bruchidius atrolineatus* and *Acanthoscelides obtectus* in legume seed.

- ✰ Use of entomopathogens such as bacteria (*Bacillus thuringiensis*), fungi (*Beauveria bassiana* and *Metarrhizium anisopliae*), protozoa (*Nosema invadens* and *Farinosystis tribolii*) and NPV. *Bacillus thuringiensis* (Bt) formulations have proved effective for potato tuber moth control in various parts of world (Alvarez *et al.*, 2005).
- ✰ Use of egg parasitoids such as *Trichogramma* sp. and *Uscana* sp. and parasitoids such as *Anisopteromalus calandrae, Lariophagus distinguendus, Pteromalus cereallellae* and *Theocolax elegns*
- ✰ Release of predators such as *Xylocoris flavipes, Teretriosoma nigrescens* and *Thaneroclerus girodi*
- ✰ Use of parasitoids, *Copidosoma koehleri* and *Bracon gelechiae* Ashmead (Hymenoptera: Braconidae) against potato tuber moth (Symington, 2003, Alvarez *et al.*, 2005).

- ☆ PhopGV-based biopesticides (*Phthorimaea operculella* granulovirus (PhopGV) genus: Betabaculovirus of the arthropod-infecting Baculoviridae) are used to control *Phthorimaea operculella* (Zeddam *et al.*, 2013).

6.10 Radiation

Ionizing radiation sterilizes or kills insects by damaging cells and producing free radicals. Ionizing radiation that can be used for management of stored grain pest includes gamma rays and x- rays. Non ionizing radiation includes infrared and microwave radiation.

6.11 Controlled or Modified Atmosphere

Controlled or modified atmosphere (CA) refers to the process of altering the proportion of atmospheric gases oxygen, nitrogen and carbon dioxide (CO_2) to give an insecticidal gas. The advantage of the CA technique is that it provides a disinfestation method that is chemical-free and suitable for "organic" grain. The process of changing the atmosphere of a facility artificially by inducing CO_2 and N_2. Modify the storage atmosphere to generate low oxygen (2.4 per cent and to develop high carbon dioxide (9.0 – 9.5) by adding CO_2 to control the insects. Currently, the only practical method available to farmers is to introduce carbon dioxide from a gas cylinder into a gas-tight silo. A very high standard of gas-tightness is required, often with a supplementary bleed of gas, to hold at least 60 per cent carbon dioxide for at least 10 days or 30-40 per cent carbon dioxide for 14 days to kill all stages of the insects' life cycle.

Advantages

- ☆ Non polluting
- ☆ Safer to apply
- ☆ No harmful residues
- ☆ Requires continuous monitoring

Disadvantages

- ☆ Cost of application is high
- ☆ Duration of treatment is more
- ☆ CO_2 can not be used on some commodities because it forms carbonic acid which cause flavour deterioration

7. Conclusion

The effective management of storage insect pests may be ensured by drying the grains properly before storage, storing new grains in the clean godowns or receptacles and plugging all cracks, crevices and holes. The following points should be considered for safe storage of food grains: (i) proper selection of site (location); (ii) selection of storage structure; (iii) cleaning of storage structures; (iv) cleaning and drying of grains; (v) cleaning of bags; (vi) separate storage of new and old stock; (vii) cleaning of vehicles; (viii) use of dunnage; (ix) proper aeration; and

(x) regular inspection. Systems approach rather than a piecemeal one needs to be adopted. Manipulate the ecological factors like temperature, moisture content and oxygen through design and construction of storage structures/godown and storage to create ecological conditions unfavourable for attack by insects.Temperature above 42°C and below 15°C retards reproduction and development of insect while prolonged temperature above 45°C and below 10°C may kill the insects. Dry the produce to have moisture content below 10 per cent to prevent the buildup of pests. The existing post-harvest system has to be improved to cut post-harvest losses at the farm level where about 70 per cent of grains are stored and consumed as food and feed and for seed purposes.

8. Future Thrust

Traditional grain storage facilities may not offer protection, but promotion of the use of metal silos and resistant varieties for grain storage is an alternative approach to reduce losses. Need for development of low cost modified atmosphere storage facilities and heat treatment methods. More botanical pesticides may be screened which are effective even at low doses without causing any adverse effects to the seeds. Popularization and commercialization of biocontrol agents at farmers level. Efforts should be made for the use of infrared waves and low energy electrons for management of stored grain pests. New formulations and combinations of diatomaceous earth with other natural insecticides, such as *Beauveria, Spinosad, Metarhizium*, growth regulators, and vegetable essences, should be tested as insect control measures.

REFERENCES

Adda C, Borgemeister C, Biliwa A, Meikle WG, Markham RH and Poehling HM. 2002. Integrated pest management in post-harvest maize: A case study from Republic of Togo. Agriculture, Ecosystems and Environment, 93: 305-321.

Alvarez JM, Dotseth E and Nolte P. 2005. Potato tuberworm a threat for Idaho potatoes. University of Idaho Extension, Idaho Agricultural Experiment Station, Moscow.

Atwal AS and Dhaliwal GS. 2008. Agricultural Pests of South Asia and their Management. Kalyani Publishers, New Delhi, India.

David K. Weaver and Reeves A. 2004. Petroff pest management for grain storage and fumigation high. Plains IPM Guide, a cooperative effort of the University of Wyoming, University of Nebraska, Colorado State University and Montana State University.

FAO. 2009. Major Insect Pests of Stored Foods. Food and Agriculture Organization of the United Nations.

Flinn PW, Kramer KJ, Throne JE and Morgan TD. 2006. Protection of stored maize from insect pests using a two-component biological control method consisting of a ymenopteran parasitoid, *Theocolax elegans*, and transgenic avidin maize powder. Journal of Stored Products Research, 42: 218–225.

Grieshop MJ, Flinn, PW, Nechols JR and Schoeller M. 2007. Foraging success of three species of *Trichogramma* (Hymenoptera: Trichogrammatidae) in a simulated retail environment. Journal of Economic Entomology, 100: 591–598.

Muhammad Sarwar. 2015. Distinguishing and controlling insect pests of stored foods for improving quality and safety. American Journal of Marketing Research, 3: 201-207.

Okunade SO, Williams JO and Ibrahim MH. 2001. Survey of insect pests infestation of dried fruits and vegetables in Kano, Nigeria. Entomological society of Nigeria (ESN), 32nd Annual Conference Book of Abstracts, 8-11 October, 2001. 23p.

Paranagama P, Abeysekera T, Nugaliyadde L and Abeywickrama K. 2003. Effect of the essential oils of *Cymbopogon citratus*, *Cymbopogon nardus* and *Cinnomomum zeylanicum* on pest incidence and grain quality of rough rice (paddy) stored in an enclosed seed box. Journal of Food, Agriculture and Environment, 1: 139.

Sarwar M and Sattar M. 2012. Appraisal of different plant products against *Trogoderma granarium* Everts to protect stored wheat- A laboratory comparison. The Nucleus, 49(1): 65-69.

Sarwar M. 2009. Evaluating wheat varieties and genotypes for tolerance to feeding damage caused by *Tribolium castaneum* (Herbst) (Coleoptera: Tenebrionidae). Pakistan Journal of Seed Technology, 2: 94-100.

Singal SK and Chouhan R. 1997. Effect of some plant products and other materials on development of pulse beetle, *Callosobruchus chinensis* (L.) on stored pigeon pea, *Cajanus cajan* (L.) Millsp. Journal of Insect Science, 10: 196–197.

Subramanyam B and Hagstrum DW. 1995. Sampling. In: Subramanyam, B. and Hagstrum, D.W. (eds) *Integrated Pest Management of Insects in Stored Products*. Marcel Dekker, New York.

Symington CA. 2003. Lethal and sublethal effects of pesticides on the potato tuber moth, *Phthorimaea operculella* (Zeller) (Lepidoptera: Gelechiidae) and its parasitoid *Orgilus lepidus* Muesebeck (Hymenoptera: Braconidae). Crop Protection, 22: 513-519.

Tadele T, Stephen M and Paddy L. 2011. Effects of insect population density and storage time on grain damage and weight loss in maize due to the maize weevil *Sitophilus zeamais* and the larger grain borer *Prostephanus truncates*. African Journal of Agricultural Research, 6(10): 2249-2254.

Zeddam JL, Lèry X, Gómez-Bonilla Y, Espinel-Correal C, Páez D, Rebaudo F and López-Ferber M. 2013. Responses of different geographic populations of two potato tuber moth species to genetic variants of *Phthorimaea operculella* granulovirus. Entomologia Experimentalis et Applicata, 149: 138-147.

Chapter 13

Organic Farming in Rainfed Areas

K.A. Gopinath[1], V. Visha Kumari[1], M. Jayalakshmi[2] and P.S. Prabhamani[1]

[1]*ICAR-Central Research Institute for Dryland Agriculture, Hyderabad – 500 059, Telangana*
[2]*KVK-Banavasi, Acharya NG Ranga Agricultural University, Banavasi – 518 360, Andhra Pradesh*
E-mail: gopinath.icar@gmail.com

1. INTRODUCTION

Rainfed agroecosystems occupy a significant place in Indian agriculture covering about 80 M ha, in arid, semiarid, and subhumid climatic zones, constituting nearly 56 per cent of the net cultivated area. These areas contribute almost 100 per cent of forest products, 84–87 per cent of coarse grain cereals and pulses, 80 per cent of horticulture, 77 per cent of oilseeds, 60 per cent of cotton, and 50 per cent of fine cereals including rice, wheat, maize, sorghum *etc.* (Srinivasarao *et al.*, 2015). Further, rainfed regions support 60 per cent of livestock and 40 per cent of human population and contribute 40 per cent of food grains and several special-attribute commodities such as seed spices, dyes, herbs, gums *etc.* Due to yield plateuing in irrigated area in most of the crops, the second green revolution must therefore explicitly embrace rainfed areas with special focus on pulses and oilseeds to ensure nutritional security and agricultural sustainability. In rainfed regions, wide spectrum of agro-ecological conditions exists in different parts of country which are suitable for growing these crops throughout year.

A per the reports, still a vast majority of farmers in rainfed areas practice low or no external input farming which is well integrated with livestock, particularly small ruminants. The average use of chemical fertilizers in rainfed areas based on a survey of non-irrigated SAT districts was found to be 18.5 kg as against 58 kg in the irrigated districts (Katyal and Reddy, 1997). Based on several surveys and reports, it is estimated that up to 30 per cent of the rainfed farmers in many remote areas of the country do not use chemical fertilizers and pesticides. Thus, many resource poor farmers are practicing organic farming by default. The Government of India task force on organic farming and several other reviewers has identified rainfed areas and regions in north east as more suitable for organic farming in view of the low input use (GOI, 2001; Dwivedi, 2005; Ramesh *et al.*, 2005).

Even though the input uses are low, some crops receiving higher attention like cotton, hybrid sorghum and millets, groundnut, pigeonpea use relatively higher levels of chemical fertilizers and pesticides throughout the country. This is mainly because of the need due to varietal replacement and also greater accessibility to farmers for surface or ground water for protective irrigation. Thus it becomes essential to delineate different regions and crops within rainfed areas depending on the nature and level of input use, so that a proper research and policy initiative can be taken up for identifying prospective regions/commodities (Venkateswarlu, 2008).

2. Focus on Niche Areas and Commodities

Rainfed areas are reported to have relative advantage to go for organic farming primarily due to i) low level of input use, ii) shorter conversion period and iii) smaller yield reductions compared to irrigated areas. However, large scale conversions cannot be expected to happen because of several limitations particularly availability of organic amendments (Venkateswarlu, 2008).

Table 13.1: Selected List of Commodities with Potential for Organic Production in Rainfed Regions

Commodity	*Scope/Opportunity*	*Potential Area*
Cotton	Demand for organically produced lint to cut down on chemical use	Maharashtra, Andhra Pradesh, Karnataka, Gujarat
Sesame	Demand for organic sesame seed for medicinal and confectionery uses	Gujarat, Rajasthan
Niger	Demand for niger seeds produced organically for bird feed in Europe	Tribal areas of different states, in particular Orissa and Chhattisgarh
Lentil	Preference for Indian lentil in world markets; organic product to fetch price premium	Uttar Pradesh
Safflower	Growing market for safflower petals as natural food dye and herbal products	Maharashtra
Finger millet	Scope to export finger millet flour as health food ingredient	Karnataka, Orissa, Jharkhand
Medicinal herbs	Need for residue free crude drugs	All over India
Ginger/Turmeric	Demand for residue free spices/natural colours	Orissa
Groundnut	To produce residue/toxin free table varieties	Gujarat
Soybean	Demand for organically produced DOC for livestock feed	Madhya Pradesh

Source: Venkateswarlu (2008).

The inherent advantages of rainfed areas should be capitalized by encouraging organic farming in highly selected areas and commodities with edapho-climatic and price advantages. The primary focus should be on commodities which have export potential with price premiums. A list of such crops and the suggested areas are given in Table 13.1.

3. Organic Farm Management

Organic farm management is focused on the complete farm system including its integration with weather, environment, social, and economic conditions, rather than considering the farm as comprised of individual enterprises. In other words, it's the management of the entire surrounding system to ensure biodiversity and sustainability of the system. Crop production in organic systems is characterized by an increased diversity of cropping patterns in time and in space compared to intensive conventional crop production systems. The major objectives of such diversity are to operate a closed system for nutrients and organic matter and maintain crop health.

Adoption of soil and water conservation measures, a key component of rainfed farming is also one of the pillars of organic farming. Crop production with the use of alternative sources of nutrients such as crop rotation, residue management, organic manures and biological inputs, mulching or mulch cum manuring, green leaf manuring, cover cropping are the strategies followed that not only conserve moisture but also improve nutrient use efficiency. The use of FYM or other organic nutrient sources during aberrant rainfall years in particular have an additional advantage of protecting the crop from drought besides the nutritional benefits, so critical in drylands.

3.1 Selection of Crop Varieties

Crop improvement efforts during the second half of 20th century has been focused almost entirely on breeding for conventional farming systems. Conventional varieties have been developed with the aim of combining high productivity and standardized product quality under high-input conditions. The consequence is that organic farmers have to depend on varieties bred for cultivation with high external inputs. The amount of stress on the crop is expected to be more under organic management than under input-intensive conventional farming. The stress may be in the form of available nutrients, weed pressure, insect-pests and diseases. Hence, varieties that can respond to sub- optimum conditions (typical of organic farming conditions) are needed. Level of resistance to diseases and insect pests must be a criterion while selecting the variety for organic farming. Yield, quality, and market acceptability also have to be considered while selecting crop varieties.

Suitable varieties of different crops have been identified for rainfed organic farming in Northwestern Himalayan states (Table 13.2), based on field trials conducted at Viveanda Parvatiya Krishi Anusandhan Sansthan, Almora (Gopinath *et al.*, 2010).

Table 13.2: Varieties Suitable for Organic Cultivation in Rainfed Areas

Crop	*Varieties Evaluated*	*Suitable Varieties*	*Yield (q/ha)*
***Kharif* crops**			
Rice (upland)	VL 221, VL 163, VL 154 and VL 206	VL 154	22
Finger millet	VL 315, VL 146, VL 204, VL 149, VR 708 and PES 400	VL 149, VL 146 VL 315	23-25
Barnyard millet	VL 172, VL 29, VL 181, VL 201, PRJ 1 and K 1	All varieties except VL 181	19-22
Soybean	VLS 2, VLS 21, VLS 47, VLS 55 and local bhatt	VLS 47, VLS 2 VLS 21	17-20
***Rabi* crops**			
Wheat	VL 616, VL 829, VL 738, VL 719, VL 802, HS 240, HS 295, HS 420, VL 804, VL 832, HS 277 and Sonalika	VL 804, VL 829 VL 802	24-28
Field pea	VL Matar 1, VL Matar 3, VL Matar 40 and Rachana	VL Matar 40	12-13
Lentil	VL 1, VL 4, VL 103, VL 125, VL 126, VL 507, PL 4, PL 406 and PL 639	All varieties except VL 1	11-15

Source: Gopinath *et al.* (2010).

Ramesh *et al.* (2006) evaluated four varieties of pigeonpea under organic management at Indian Institute of Soil Science (IISS), Bhopal. Among the four varieties, ICPL-87119 recorded the highest number of pods/plant, 100-seed weight, seed and straw yield (Table 13.3). Jowahar-4 and Asha gave similar but significantly higher seed yields than BDN-2.

Table 13.3: Performance of Pigeonpea Varieties under Organic Management

Variety	*Pods/Plant*	*Seeds/Pod*	*100-Seed Weight (g)*	*Seed Yield (kg/ha)*	*Straw Yield (kg/ha)*
ICPL-87119	197	3.71	9.99	2524	4502
Jowahar-4	179	3.68	9.69	2138	4029
BDN-2	156	3.70	8.40	1807	3855
Asha	188	3.63	9.44	2078	4267
CD (P=0.05)	6.2	NS	0.49	172	232

Source: Ramesh *et al.* (2010).

3.2 Nutrient Management

The aim of nutrient management within organic farming systems is to work, as far as possible within a closed system. Organic farming practices aim to maximize the efficiency of nutrient cycling within the farm ecosystem (*e.g.* avoidance of losses from manure heaps, optimizing mineralization of soil organic N), and maximizing the fixation of atmospheric N by legumes. Organic standards minimize or eliminate use of synthetic or manufactured inputs and encourage maximum use of local natural resources. The inputs allowed as fertilizers in organic production are generally lower and more variable in nutrient content and plant-availability than commercial fertilizers. Therefore, they have to be applied at high rates to meet all the crop needs. Furthermore, there is greater likelihood of supplying some nutrients at excess rates, which may lead to increased risk of loss and negative environmental impact. There are a number of organic sources of nutrition and among them green manuring, composting, biofertilizers, vermicompost, panchakavya and biodynamics are important.

3.2.1 Rice-Toria-Blackgram System

Baishya *et al.* (2017) evaluated the effect of organic (alone), inorganic (alone) and integrated nutrient management on soil properties and productivity of rainfed winter rice–*toria*–blackgram system. The highest rice-equivalent yield (8.91 t/ha) was recorded with the application of 100 per cent NPK through organic sources plus intercropping with wheat during the *rabi* season and okra during the summer. Application of 100 per cent RDF through FYM, vermicompost and mustard seed-cake in the rice during the *kharif* season, followed by *toria* + wheat during *rabi* and blackgram + okra during the summer season resulted in the maximum average uptake of N, P and K (95.5, 20.1 and 84.9 kg/ha respectively). Bacteria, fungi and Actinomycetes counts as well as microbial biomass carbon in soil were the highest (253×10^4, 137×10^4, 195×10^4 and 163.95 µg/g respectively) in the organic management having 100 per cent supply of nutrients through 1/3 FYM + 1/3 vermicompost + 1/3 mustard seed cake along with 3.5 kg/ha each of *Azospirillum/ Azotobacter* and phosphate-solubilising bacteria. The highest dehydrogenase, phosphatase and urease activity in soil was also observed in the same treatment. Full dose of NPK through organic sources to rice–toria–blackgram sequence intercropping with wheat during the *rabi* and okra during the summer registered the highest earthworm count (25 numbers/m^3).

3.2.2 Barnyard Millet

Supply of nutrients to barnyard millet (*Echinochloa frumentacea*) exclusively through organic sources adequately supports the crop growth and yield, due to comparatively low nutrient requirement of the crop. In a field experiment at Ranichauri (Uttarakhand), various organic nutrient sources gave significantly lower productivity of barnyard millet than recommended dose of fertilizer (RDF) treatment in the initial two years (Table 13.4; Yadav and Malik, 2010). However, from third year onwards, organic treatments involving vermicompost (VC) + wild apricot-cake, WAC (50 per cent N from each source) with and without *Azotobacter* produced similar grain yields as that of RDF treatment. Among the treatments involving organic manures, VC + WAC (50 per cent N from each source) + *Azotobacter* on an average produced statistically similar yield (1.78 t/ha) as that of RDF treatment (1.92 t/ha). In the absence of any price premium for the organic produce, however, the net returns and benefit to cost ratio were highest with RDF followed by VC + WAC + *Azotobacter* treatment.

Table 13.4: Yield of Barnyard Millet and Economic Returns as Influenced by Different Sources of Organic Manures

Treatment	*Grain yield (t/ha)*				*B:C Ratio*	*Cost of Cultivation (Rs/ha)*
	2005	*2006*	*2007*	*2008*		
RDF (40-20-20 N-P_2O_5-K_2O)	0.62	2.17	2.49	2.39	1.59	9372
Vermicompost (VC) 4 t/ha	0.23	1.01	1.92	1.45	-0.15	24300
FYM 8 t/ha	0.21	0.60	1.31	1.55	0.40	8900
Wild apricot-cake (WAC) 4 t/ha	0.23	0.61	1.83	1.69	0.15	12300
VC + WAC (50 per cent N from each source) + *Azotobacter*	0.36	1.82	2.45	2.51	0.65	14345
Control	0.20	0.66	1.12	1.10	0.24	8300
LSD (P=0.05)	0.33	0.27	0.31	0.25	-	-

Source: Yadav and Malik (2010).

3.2.3 Maize-Wheat

Organic farming in the rainfed hill and mountain agriculture has great emphasis due to its high prospects for enhancing sustainable crop productivity and resource conservation. A four years field experiment carried out by Ghosh *et al.* (2012) at Central Soil and Water Conservation Research and Training Institute, Dehradun on a land having gentle slope (2 per cent) revealed that wheat equivalent yield was increased by 9 per cent with FYM 5 t/ha + vermicompost 1.0 t/ha + poultry manure 2.5 t/ha + minimum tillage + 3 weed mulch at 20, 40 and 60 DAS + Palmarosa vegetative barrier (T4) as compared to inorganic sources + conventional tillage + panicum as vegetative barriers (T_1). Four years average data showed reduction in runoff and soil loss by 44.8 and 36.8 per cent, respectively in T_4 treatment compared to inorganic treatment (T_1). On an average, T_4 treatment conserved soil moisture to the extent of 6.13 to 58.3 mm for wheat crop compared to T_1 with the reduction of micro-aggregates by 8-9 per cent which increased basic infiltration rate and water use efficiency. The nutrient use efficiency of all major nutrients (NPK) was also increased by 40.9 per cent in the same treatment compared to T_1 treatment. Even though resource conservation was higher in organic treatment but organic farming was not found profitable in the initial years as net profit was decreased by 13.0 per cent in T_4 treatment compared to T_1 after four years of cropping.

3.2.4 Sorghum-Pigeonpea Rotation

In a field experiment at CRIDA (Hyderabad), the performance of sorgum and pigeonpea were evaluated under six production packages *viz.* 1. High input chemical (HIC), 2. Low input chemical (LIC), 3. Integrated (INT), 4. Zero input (ZIP), 5. Low input organic (LIO) and 6. High input organic (HIO). The packages differed in respect of inputs and practices used for supplying nutrients to the crops and for controlling crop pests and diseases. The highest grain yield of sorghum (2853 kg/ha) was obtained with the high input chemical package (CRIDA, 2009). The next best treatment was low input chemical package (2663 kg/ha). The organic packages gave lower yield compared to chemical and integrated packages, even after four years under organic management. Considering only package specific labour and inputs, the high input chemical package was the most labour intensive (6 person days/ha) and input intensive (Rs 1732/ha) followed by the high input organic package (5 person days/ha and Rs 1600 inputs/ha).

In case of pigeonpea, the highest grain yield (1468 kg/ha) was obtained with the integrated package, followed by the high input organic package (1447 kg/ha). Marked differences in grain damage were observed between production packages. The lowest combined grain damage (13 per cent) by lepidopterans and pod fly was observed with the high input chemical package, followed by the integrated package (14.67 per cent). The high input organic package (Table 13.4) was most labour and cost intensive, requiring 12 person-days additional to routine crop production requirements, and Rs. 3312 for package related inputs per hectare. The integrated package, which gave the highest seed yield was far cheaper in terms of inputs (Rs 1206/ha), although the labour requirement was high (11 person-days/ha).

In a long-term experiment at ICRISAT, Hyderabad, crop yields (sorghum, pigeonpea, cotton, cowpea and maize) in seven out of nine years were similar or higher in the treatments involving low-cost and biological approaches than the treatment 'conventional agriculture' on a rain-fed Vertisol (Rupela *et al.*, 2005).

3.2.5 Lentil-Finger Millet System

Two sources of manures namely FYM and lantana were evaluated in lentil-finger millet cropping system (Gopinath *et al.*, 2010). After two years of study it was found that application of manures (FYM) based on P requirement of lentil (17.5 kg/ha) recorded similar grain yield as that of recommended NPK (Table 13.5). In finger millet, however, application of manures (FYM) based on N requirement of the crop (40 kg/ha) was found better compared to other treatments involving organic manures. Between the two sources of manures, FYM was found better compared to lantana. The results suggest that during initial years of organic farming, manures may be applied based on N requirement of cereals (finger millet) and P requirement of legumes (lentil).

Table 13.5: Performance of Lentil-Finger Millet Cropping sequence (q/ha)

Treatment	*Lentil*	*Finger Millet*
FYM for N	8.92	28.60
FYM for P	10.67	22.77
FYM for N/P	10.75	30.99
Lantana for N	7.92	26.95
Lantana for P	8.75	19.12
Lantana for N/P	8.67	25.22
Rec. NPK	11.34	34.20
Control	5.58	14.26

Source: Gopinath *et al.* (2010).

3.2.6 Sesame

Sesame has recently emerged as a valuable export crop, earning more than Rs. 1000 crores from the export of 2.5 lakh tonnes of sesame seed. Gopinath *et al.* (2009) evaluated the performance of sesame under organic vis-à-vis conventional production systems at Hyderabad. The organic production system involved application of FYM 3.7 t/ha + neem cake 900 kg/ha + ash 75 kg/ha + bone meal 75 kg/ha + elemental sulphur (ELS) 20 kg/ha + PSB 5 kg/ha (soil application) + *Azotobactor* 5 kg/ha, and pest management with *Trichoderma* (0.4 per cent) seed treatment + neem oil spray thrice at 15, 30 and 45 DAS, and *Azadirachtin* (0.03 per cent) spray at 30 DAS. The conventional system included application of recommended dose of chemical fertilizers (40:26.5:33.3 kg NPK/ha) and recommended pest management module. The grain and stalk yields were higher under conventional farming than that under organic farming and control in all the years (Table 13.6). The grain yield reduction under organic farming was 18 per cent in 2005, 22 per cent in 2006 and 20 per cent in 2007 compared with conventional farming.

Table 13.6: Sesame Yield Under Organic and Conventional Production Systems

Production System	*Grain Yield (kg/ha)*			*Stalk Yield (kg/ha)*		
	2005	*2006*	*2007*	*2005*	*2006*	*2007*
Conventional	656	247	416	1516	970	1848
Organic	535	193	333	1291	802	1584
Control	358	96	194	861	483	1136

Source: Gopinath *et al.* (2010).

3.2.7 Groundnut

Malligawad and Parameshwarappa (2006) evaluated the performance of groundnut under organic and inorganic production systems at Dharwad, Karnataka. The organic production system involved application of FYM @ 7.5 t/ha and Vermicompost @ 2 t/ha. Seed treatment with biofertilizers such as *rhizobium*, phosphate solubilzing bacteria (*Pseudomonas striata*) and plant growth promoting *rhizobacteria* (PGPR). Biopesticides such as neem seed cake @ 500 kg/ha (soil application), seed treatment with *Trichoderma harzianum* @ 5 g/kg seed and spraying crop with 5 per cent neem seed kernel extract were used for pest management. The inorganic production system involved seed treatment with Captan, application of recommended dose of inorganic fertilizers (25 kg N, 75 kg P_2O_5 and 25 kg K_2O/ha) and chemical pesticides.

Organic farming in groundnut produced 18.2 and 22.1 per cent higher dry pod yield and higher kernel yield over inorganic farming (2970 and 2345 kg dry pod and kernel yield/ha, respectively). Furthermore, use of organics in groundnut production also resulted in higher pod number/plant, dry pod weight/plant, double seeded pods, shelling per cent, 100-kernel weight and harvest index as compared to inorganic farming. Groundnut crop in organically amended plot did not show moisture stress during the period of dry spell of 38 days due to greater moisture conservation. On the contrary, groundnut in inorganic farming showed moderate to severe moisture stress during same initial dry spell period. The authors conclude that the groundnut productivity was not only higher in organic farming compared to inorganic farming during the first year, but would improve further and may become sustainable in the subsequent years of organic farming.

After three years of organic manure in groundnut revealed that application of FYM @ 10 t/ha recorded significantly higher groundnut pod yield of 2045 kg/ ha. overall improvement in soil fertility were observed when crop was fertilized with organic manures. Application of FYM @ 10 t/ha significantly increased the available nitrogen by 31.4, phosphorus by 27.3 and potassium by 62.8 per cent as compared to only application of 100 per cent RDF. The increase in soil nutrients after harvest of crop was due to addition of these nutrients through the application of FYM, farm waste and animal litter.

3.2.8 Soybean

In recent years, the farmers have switched over to cultivation of soybean under organic farming conditions. There is a global demand for organically grown soybean (Ramesh *et al.*, 2006). In a field experiment conducted at IISS, Bhopal, during the first year, conventional farming resulted in higher yield, gross income and benefit cost ratio compared to that of organic farming (Table 13.7). In the second year, both conventional and organic farming resulted in the comparable yields and almost similar economic returns. However, in the third year, organic farming recorded higher yields, gross income and benefit cost ratio compared to the conventional farming. Among different organic manures, application of poultry manure was found better, which gave 17.5 per cent higher yield than the chemical fertilizers.

Table 13.7: Economic Comparison of Organic vs. Conventional Soybean Production

Year	*Yield (kg/ha)*		*Gross Income** (Rs/ha)*		*Benefit-Cost Ratio*	
	*Organic**	*Conventional*	*Organic*	*Conventional*	*Organic*	*Conventional*
I	928	1027	11136	12324	1.30	1.36
II	863	851	10356	10212	1.21	1.12
III	1338	1243	16056	14916	1.87	1.64

*Mean of organic manure treatments.

** Income was calculated without the price premium for organic soybean

Source: Ramesh *et al.* (2010).

3.2.9 Cotton

In the rainfed tract of central India, cotton is grown on three million hectare (*i.e.*, 43 per cent of total area) of marginal lands where production is low due to poorly distributed rainfall, and eroded undulating nature of lands and low resources investment by farmers. Such soils require low-cost and low external input production systems to minimize cost on fertilizer and pesticides for imparting stability in production (Rajendran *et al.*, 2000). The authors recommend seed inoculation with *Azatobactor* or *Azospirillum*. FYM @ 15 t/ha must be added before preparatory tillage and mixed thoroughly. FYM should be well decomposed and should be preferably treated with composting organisms such as *Trichoderma viride*. The rate may gradually be brought down 5-10 t/ha, once the farm yield stabilises over a few years. Vermicompost @ 1-2 t/ha should be added supplementing FYM on the furrow lines on which sowing is done.

In situ green manuring with fodder cowpea and its burying at 40 days after sowing (DAS) will ensure a steady N supply during the grand-growth phase and flowering period, when the N demand peaks up in the crop. It hastens microbial activity in soil, reduces weed growth and enhances natural enemy build up. This provides around 400-500 kg dry matter per hectare with 2.5 per cent N and contributes 10-12 kg N/ha during squaring. Its additional benefits include smothering of weeds, controlling seasonal soil erosion and nurturing natural enemies of cotton pests.

In addition, dense stand of dhaincha (*Sesbania aculeaata*) can be raised around cotton field at a width of 2 m; its loppings cut and spread between cotton rows at 65-70 DAS. Its fast decomposing leaves provide N during early boll development period and stalks act as temporary mulch, preventing soil moisture evaporation.

From the studies conducted at Central institute for Cotton Research, Nagpur on the technology generation for sustainable organic cotton production and its evaluation with conventional farming with use of chemicals (fertilizers and insecticides), the following facts have emerged:

a. In a span of 3-4 years, there was gradual build up in soil productivity and seed cotton yield in organic farming was at par with non-organic farming.
b. In the initial year, the yield realization through organic system was about 40 per cent of that of conventional (non-organic) system, with comparable or slightly higher (7-10 per cent) production costs. This situation improved in subsequent years.
c. There was also a gradual improvement in soil fertility parameters such as organic carbon content and available P. The organic carbon content stabilized at 0.50-0.55 per cent after 4-5 years.
d. There was a good build up of natural enemies, such as egg, larval and pupal parasitoids, and also of various predators of cotton pests under organic farming.

India is one of the leading organic cotton producers in the world. To sustain the productivity of organic cotton, nutrition and plant protection play a very important role. In this context, the field experiment was carried out by Rudragouda *et al.*, 2015, to study the nutrient management practices on quality of cotton. The results of the two years revealed that combined application of compost (50 per cent) + vermicompost (50 per cent) equivalent to RDF recorded significantly higher values of ginning (34.26 per cent), uniformity ratio (45.62 per cent) and lower fiber fineness (3.71 g) over FYM @ 5 t/ha + RDF. Integrated application of compost (50 per cent) + vermicompost (50 per cent) equivalent to RDF + gliricidia green leaf manuring with surface application of jeevamrutha @ 500 l/ha recorded significantly more fiber length (35.62 mm), ginning per cent (34.60 per cent) and lower fiber fineness (3.63 g) over RDF alone and FYM + RDF. The combined use of organic manures mainly vermicompost, compost and liquid organic manures complimented each other and supplied all nutrients as required by crop to improve the quality of cotton and soil health.

3.2.10 Chilli + Cotton Intercropping System (2:1)

Chilli crop responds very well to organic production practices. The field trials conducted by Babalad *et al.* (2011) revealed that, two years conversion is required to stabilize the yield under organic system (Table 13.8). The third year yield showed significant superiority with organic system, compared to inorganic system and at par with integrated system. Combined application of enriched compost, poultry manure and green leaf manure (gliricidia) equivalent to recommended 100 kg N per ha produced significantly higher chilli equivalent yield and net returns as compared to inorganic system.

Table 13.8: Yield and Economics of Chilli + Cotton System as Influenced by Organic Nutrient Management Practices (Pooled 2004, 2005 and 2006)

Treatment	*Chilli Yield (kg/ha)*	*Cotton Yield (kg/ha)*	*Chilli Equivalent Yield (kg/ha)*	*Net Returns (Rs/ha)*
EC+GLM	385	540	608	15292
VC+GLM	436	553	628	14339
PM+GLM	466	554	691	17874
EC+VC+GLM	494	607	704	18023
EC+PM+GLM	515	601	688	19028
VC+PM+GLM	437	607	652	18652
Control	302	432	457	11896
LSD (0.05)	84.4	74.4	113.0	3652

EC: Enriched compost, GLM: Green leaf manuring, VC: Vermicompost, PM: Poultry manure.

3.2.11 Tuber Crops

Most of the tuber crops are grown by small and marginal farmers in rainfed and tribal areas and hence use of chemical fertilizers and insecticides are limited except in the case of cassava in the industrial production areas of Tamil Nadu (Salem, Dharmapuri, Namakkal, South Arcot districts) and Andhra Pradesh (Rajahmundry district). Tuber crops in general and aroids in particular, like elephant foot yam respond well to organic manures and there is considerable scope for organic production in these crops. Further tropical tuber crops are well adapted to low input agriculture. They are less prone to pest and disease infestations. Three years of experimentation has indicated that organic farming is a viable proposition in elephant foot yam (Suja, 2008). Of the three production systems tested in elephant foot yam (conventional, traditional and organic farming), organic farming proved promising with high yield (Table 13.9).

Table 13.9: Performance of Elephant Foot Yam under different Production Systems

Production Systems	Corm Yield (t/ha)	Dry Matter (Oer cent)	Starch (Per cent FW basis)	Oxalate (Per cent DW basis)	Crude Protein (Per cent FW basis)
Conventional	54.62	19.23	13.63	0.204	1.740
Traditional	51.71	20.49	16.62	0.189	1.876
Organic	64.48	21.37	16.57	0.172	2.160

Source: Suja (2008).

In another experiment, three species of yam/Dioscorea (*D. rotundata* (var. Sree Priya), *D. alata* (var. Sree Keerthi and *D. esculenta* (var. Sree Latha)) were tested under three production systems *viz.*, conventional, traditional and organic farming (Suja, 2008). *Dioscorea rotundata* produced higher yield in conventional practice (19.20 t/ha), which was on a par with organic farming (17.81 t/ha). In the case of *D. alata*, all the production systems were on par; traditional practice (20.64 t/ha) resulted in slightly higher yield than organic (19.46 t/ha) and conventional practices (19.04 t/ha). In *Dioscorea esculenta*, organic farming proved superior (8.58 t/ha).

3.2.12 Potato-French Bean

Application of the recommended dose of fertilizers to the potato and french bean crops (140 N: 120 P_2O_5: 60 K_2O kg/ha) through synthetic fertilizers resulted in 4.5 per cent higher system productivity than FYM application @ 30 t/ha along with inoculation biofertilizers (*Azotobacter* + *PSB*) (Yadav *et al.*, 2017). However, the maximum economic efficiency (Rs. 471/ha/day), net return (Rs.172.1 × 10^3/ha) and B:C ratio (1.3) were noticed by application of 30 t FYM/ha along with inoculation of biofertilizers. Increase in production efficiency due to application of biofertilizers was14.0 per cent over without biofertilizers with same level of FYM (30 t/ha). Either application 15 t FYM or 30 t FYM/ha with inoculation of biofertilizers generated more net returns than recommended dose of fertilizers due to premium on organic produce.

The organic sources of nutrients are the store house of nutrients and constant nutrient release pattern from organic sources to crop plants. Increase in rate of mineralization in organic manure resulting into nutrient transformations and their mobility into the plant system for longer period results in increases availability of nutrients to succeeding crops. Various levels of FYM and vermicompost applied to preceding popcorn showed significant increase in N, P and K content in the potato tubers as compared to inorganic nutrient @ 120 kg N/ha (Meena *et al.*, 2016).

3.2.13 Medicinal, Aromatic and Dye Crops

The organically grown medicinal plants fetch higher price in the international market and demand for organic product is increasing day by day with increase in awareness of the side effects of the synthetic drugs. In a field experiment conducted at CRIDA, organic fertilizers were applied by equating with recommended dose of N (50 kg/ha) in Senna, Andrographis and Ashwagandha. Farm yard manure, castor and neem cakes were applied 15 days before sowing. Vermicompost and NPK were applied at the time of sowing. Test crops were sown in the last fortnight of July. Highest Senna leaf, biomass of Andrographis and Ashwagandha root yields were recorded with the application of vermicompost, neem and castor cakes, respectively (Table 13.10). The application of castor cake also improved the quality of Andrographis with respect to andrographolide content in roots (CRIDA, 2008).

Table 13.10: Influence of Organic Nutrients on Yield and Quality of Medicinal Crops

Treatment	*Yield kg/ha*			*Andrographolide per cent*
	Senna (Fresh Leaves)	*Andrographis (Fresh Biomass)*	*Ashwagandha (Dry Root)*	
FYM	1517	1974	817	1.304
Vermicompost	1926	2666	632	1.032
Castor cake	1285	2258	1011	1.774
Neem cake	1750	2493	512	1.046
NPK	938	1192	451	1.664
Control	1314	747	332	1.008
LSD (P=0.05)	407.18	570.7	107.08	

Source: CRIDA (2008).

Table 13.11: Influence of Organics on Yield of Medicinal Plants

Treatment	*Yield kg/ha*		
	Senna (Fresh Leaves)	*Andrographis (Fresh Biomass)*	*Aswagandha (Dry Root)*
With tank silt	1777	1817	494
Without tank silt	1147	1089	369
LSD (P=0.05)	513	132	78
FYM	1205	1504	332
Vermicompost	1216	2128	462
Castor cake	1821	1114	344
Biomix	1325	1512	609
NPK	1943	1246	523
Control	1262	1214	331
LSD (P=0.05)	428	189	100

Source: CRIDA (2008).

In another experiment, application of tank silt increased the yields of senna and andrographis by 54.8 per cent, 67 per cent, and 33.8 per cent over no tank silt condition (CRIDA, 2008). Application of NPK, castor cake and bio mix (a mixture of *Trichoderma reseii, Phenerochyte chrysosporium, Bacillus subtilis* and *B. coagualans*) recorded higher yield in senna (Table 13.11). The increase in yields was observed with application of vermicompost, Biomix, FYM and NPK in andrographis. Biomix application recorded significantly higher root yields of ashwagandha over all the treatments followed by application of NPK. Lowest yields of all three crops were recorded in control and FYM treatments.

Aromatic plants, however, did not respond well to organic management (CRIDA, 2008). The biomass and oil yields of lemon grass and palmarosa were highest with application of inorganic fertilizers followed by vermicompost and castor cake application (Table 13.12). Highest 'citral A' content in lemongrass and 'geraniol' in palmarosa were recorded in vermicompost and castor cake application, and the lowest content in inorganic fertilizer.

Table 13.12: Influence of Inorganic Nutrients on Yield and Quality of Aromatic Plants

Treatment	*Lemongrass*			*Palmarosa*		
	Fresh Biomass (t/ha)	*Oil Yield (L/ha)*	*Citral A (per cent)*	*Fresh Biomass (t/ha)*	*Oil Yield (L/ha)*	*Citral A (per cent)*
Control	8.4	56	60	12.43	51	84
FYM	11.74	67	68	18.29	72	91
Vermicompost	13.79	80	78	21.2	113	84
Neem cake	12.53	77	57	17.99	57	79
Castor cake	13.74	78	63	19.05	84	69
NPK	16.96	92	49	23.08	87	88

Source: CRIDA (2008).

3.2.14 Spices

The All India Coordinated Research Project for Dryland Agriculture (AICRPDA) center at Phulbani, Odisha is closely working with farmers of Kandhamal District in cultivation of organic ginger and turmeric. About 61 farmer groups comprising 12,000 members have reconstituted themselves as KASAM (Kandhamal Apex Spice Association for Marketing). By adopting modern methods of organic farming, the Kandhamal's farmers produce over 10,000 tons of turmeric and 2000 tons of ginger a year, along with mustard, niger, tamarind, *etc.* These organic spices are being exported to Europe and other countries.

3.3 Pest Management

Pest control strategies under organic farming are largely preventive rather than reactive. The balance of cropped and uncropped areas, crop species, variety, temporal and spatial pattern of the crop rotation seek to maintain a diverse population of pests and their natural enemies and disrupt the life cycle of pest species. A combination of appropriate cultural techniques such as cultivation, crop rotation, smother crops, trap crops, irrigation, solarization *etc.* and the use of biological control agents are used to manage insect-pests, weeds and diseases.

Tropical countries are rich in tens or perhaps hundreds of plants (botanicals) with ability to help manage crop pests (both diseases and insect-pests). Also, microorganisms with the ability to kill/suppress crop pests occur in nature and some (*e.g. Bacillus thuringiensis*) are available commercially. In addition, each insect-pest has some natural enemies and a good number of them can be insects that are generally referred to as 'beneficial insects'. For example, as per a published report from ICRISAT, the most difficult insect-pest *Helicoverpa armigera* (also called legume pod-borer or cotton bollworm) has about 300 natural enemies, including beneficial insects (Sharma, 2001). Synthetic pesticides used in conventional agriculture would highly likely kill all beneficial insects that occur in a field while the options used in OF either do not affect them adversely or affect them in a relatively small way.

Creation of crop diversity by the introduction of one crop in to another crop is known as crop-crop diversity. Intercropping and mixed cropping systems are more popular forms of crop-crop diversity practiced in rainfed agriculture. These systems provide opportunities to create situations that are less prone to pests compared to the single crop situations or the monocultures (Srinivasa Rao *et al.*, 2006). In a study conducted at CRIDA, short and medium duration pigeonpea offered greater degree of protection against insect pest attack. Medium duration pigeonpea, in addition to being attacked by fewer pests, exhibited less flower shedding during dry spells, making it more suitable to dryland conditions. The diversity created by introducing sorghum, greengram and groundnut as intercrops in pigeonpea resulted in less congenial conditions for insect pests. Similarly, in another study the diversity created by introducing clusterbean or cowpea or greengram as intercrops in castor resulted in a buildup of natural enemies (*Micropliotis, Euplectrus*, coccinellids and spiders) of the major pests of castor and also resulted in less congenial conditions for insect pests (Srinivasa Rao *et al.*, 2007). Economic analysis also showed these intercropping systems to be more profitable than sole pigeonpea and castor.

The adoption of low external input integrated pest management module consisting of sequential application of neem seed kernel extract 5 per cent, neem oil 5 per cent, extract of *V.negundo* 1/10 w/w, pongamia oil 5 per cent, erection of bird perches and mechanical collection of larvae was found effective in managing/ controlling the pests in both intercropping systems.

3.3.1 Use of Bio-agents for Treatment of Seed and Planting Material

- ✰ Treat the seeds with talk-based formulations of *Pseudomonas flurescens* and *Trichoderma harzianum or T. viridae* each @ 10g/kg seed.
- ✰ Treat the nursery bed with any of the above formulations @ 50g/m^2. This treatment can be combined with the application of neem or pongamia cake @ 200g/m^2 before sowing.
- ✰ Treat the soil mixture (used for producing the nursery seedlings in polythene bags) with any of the above formulations @ 2kg/t soil mixture. In addition, add neem or pongamia cake @ 50 kg/t soil mixture.
- ✰ Spray the seedlings with *Bacillus thuringiensis* (BT) @ 1g/liter 10 days after germination.
- ✰ Before transplanting, dip the roots of seedlings in *Pseudomonas* solution for five minutes.

3.3.2 Main Field

- ☆ The bio-agent formulations can be used for enriching the FYM by applying 2 kg of *Pseudomonas flurescens* or *Trichoderma harzianum* and 50 kg of neem or pongamia cake to one tonne of FYM. This should be left for 15 days under the shade and then apply uniformly in the field.
- ☆ Apply neem cake @ 250 kg/ha to furrows at the time of planting and at 30 and 60 or 75 days after planting to reduce incidence of fruit borer, leaf miner and nematodes.
- ☆ Spray *Pseudomonas flurescens* (2 per cent) 2-3 times to induce disease resistance and control of foliar diseases.
- ☆ Spray Panchagavya @ 3ml/liter at 7-10 days interval to reduce pest incidence and boost the crop growth and production.
- ☆ Spray neem seed powder extract 4 per cent or neem soap (1 per cent) or pongamia soap (1 per cent) at 20-25 days after sowing and repeat 2-3 times at 15 days interval for managing sucking pests.
- ☆ spray Ha NPV @ 250 LE/ha at flower initiation stage and repeat 2-3 times at weekly intervals to reduce the fruit borer incidence.
- ☆ For, shoot borer, fruit borer, stem borer, hairy caterpillar and army worm *Andrographis paniculata* decoction 3 to 5 per cent or *Sida spinosa* decoction 5 per cent is found effective
- ☆ Erect pheromone traps @ 20/acre at flowering stage to trap shoot- and fruit-borers. Light trap and yellow sticky traps are also reported effective.

3.3.3 Other Methods of Pest Management

Sridhar *et al.* (2008) have reported that the following preparations are useful in management of certain insect-pests and diseases.

Ginger, garlic, chilli extract: For preparing the extract required for one hectare, 2.5 kg of garlic and 1.25 kg each of ginger and green chillies are taken and ground into a fine paste. This should be diluted with 18 liters of water and filtered. The concentration of the extract can be increased or decreased depending on the intensity of pest attack. This extract can be stored for a maximum of 3 days. This spray is reported to control shoot- and fruit-borer and Eplachna beetle in vegetables.

Five leaf extract: It is prepared using leaves of five different plants mentioned below. a) plants with milky latex, *e.g. Calotropis, Nerium, Cactus* and *Jatropa,* b) plants which are bitter, *e.g.* Neem, *Andrographis* and *Leucas,* c) plants that are generally avoided by cattle, *e.g. Ipomea fistulosa,* d) aromatic plants, *e.g. Vitex, Ocimum,* and e) plants that are not affected by pests and diseases, *e.g. Morinda, Ipomea fistulosa.* Any five of the above mentioned plant leaves should be collected in equal quantities (1 kg each) and pound well. Transfer this to a mud pot and add twice the quantity of water, one liter of cow's urine and 100g of asafetida. Tie the mouth of the pot tightly with a cloth. It should be mixed well daily in the evening and the filtrate can be used after a week. It is reported to give effective control of leaf eating caterpillar and leaf roller in vegetables.

Asafoetida solution: Take 100g of powdered asafoetida and mix in 5 liters of water. Dip the roots of seedlings (for transplanted crops) in this solution for 30 minutes and then transplant the seedlings in the mail field. This prevents soil borne fungal and bacterial diseases.

Cow dung solution: Mix 2.5 kg of cow dung in 25 liters of water and filter using a gunny cloth. Dilute the solution with 12.5 liters of water and filter again. This can be used for spraying one hectare area. This manages the bacterial diseases affecting the crop.

Fumigation combined with other organic methods: 200g/ac *Embelica ribes* is powdered well and put in a wide mouthed pot with burning charcoal and carried in the field in a direction opposite to the wind. After a week of this fumigation, 300 ml of *Acorus calamus* extract along with one liter of cow's urine is mixed with 8.7 litres of water is sprayed on the crop. This method has proved beneficial for wilting.

4. Conclusions

India is bestowed with a lot of potential to produce all varieties of organic products due to its various agro climatic regions and traditional knowledge. Due to low input demand rainfed area is known as the prime area and potential hotspots to produce organic products. As the export potential is increasing, overcoming the hurdles in rainfed area, developing cost effective production system, and adequate scientific research will help India emerge as a good exporter of organic products and increase farm net revenues.

5. Future Thrust

- ☆ Delineation of the potential areas/zones like hill and tribal areas, for rainfed organic farming by identifying contiguous blocks of areas with little or no chemical input use and where productivity can be enhanced by using permitted inputs to enable group certification to farmers.
- ☆ Carry out a country wide survey/inventorisation of areas in arid, semi-arid and dry sub humid regions about the level of chemical input use, productivity in selected commodities which have potential to fetch price premiums in international markets.
- ☆ Prepare an enlarged list of crops, herbs and livestock products which can be sourced from rainfed regions considering the international trade in organic food and allied products.
- ☆ Survey, documentation and critical evaluation of indigenous technological knowledge on organic farming.
- ☆ Inter-disciplinary and location-specific research has to be taken up for development of package of practices for organic farming. Organic production packages will be more location-specific than inorganic package of practices as the input use depends largely on locally available resources.
- ☆ Development of cost effective technology for on-farm organic manure production as well as large scale production of compost from domestic, agricultural, and industrial wastes.

- Generation of adequate scientific information on the yield, quality, economics and post-harvest aspects of various crops under different management levels and agro-climatic conditions.
- Extension machinery including State departments, NGOs, and other private organizations should develop intensive partnerships, at grassroots and policy levels in promoting organic farming.
- Development of quality standards, and setting up of effective marketing chains to sell organic produce in both domestic and international markets.
- Policy initiatives such as providing incentives to those farmers willing to shift to organic farming, providing market information, subsidized supply of inputs and group certification.

REFERENCES

Babalad HB, Patil RK, Math KK and Giraddi RS. 2011. Prospects of organic production system and health management in chilli. National symposium on spices and aromatic crops, 8-10 December, 2011. pp.124-134.

Baishya A, Bhabesh Gogoi, Hazarika J, Hazarika JP, Bora AS, DAS AK, Borah M and Sutradhar P. 2017. Comparative assessment of organic, inorganic and integrated management practices in rice (*Oryza sativa*)-based cropping system in acid soil of Assam. Indian Journal of Agronomy, 62(2): 118-126.

CRIDA 2008. Annual Report, 2007-08. Central Research Institute for Dryland Agriculture, Hyderabad, India. p. 119.

CRIDA 2009. Annual Report, 2008-09. Central Research Institute for Dryland Agriculture, Hyderabad, India. p. 130.

Dwivedi Vandana. 2005. Organic farming: Policy initiatives, Paper presented at the National Seminar on National Policy on Promoting Organic Farming, 10-11 March, 2005. pp.58-61.

GOI 2001. The report of the working group on organic and biodynamic farming, Planning Commission, Government of India. pp.1-25.

Gopinath KA, Balloli SS, Venkateswarlu S and Venkateswarlu B. 2009. Performance of sesame under organic and conventional production systems. Indian Journal of Oilseeds Research, 26: 376-378.

Gopinath KA, Saha S, Mina BL, Sushil SN, Hooda KS, Srivastva AK, Bhatt JC and Gupta HS. 2010. Organic farming research at VPKAS. Technical Bulletin, Vivekananda Parvatiya Krishi Anusandhan Sansthan, Almora, (In press).

Ghosh BN, Sharma NK, Pradeep Dogra and Dadhwal KS. 2012. Effect of integrated organic input management on resource use efficiency in maize - wheat cropping systems in sloping lands of the north-west Himalayas. Indian Journal of Soil Conservation, 40(1): 84-89.

Katyal JC and Reddy KCK. 1997. Plant nutrient supply needs: Rainfed food crops. In: *Plant Nutrient Needs, Supply, Efficiency and Policy Issues: 2000-2025* (Kanwar

JS and Katyal JC. Eds.), National Academy of Agricultural Sciences, New Delhi. pp. 91-113.

Malligawad LH and Parameshwarappa KG. 2006. Effect of Organics on the Productivity of Spanish Bunch Groundnut under Rainfed Farming Situations. 18th World Congress of Soil Science, July 9-15, 2006, Pennsylvania, USA.

Meena BP, Ashok Kumar, Shiva Dhar, Sangeeta Paul and Arvind Kumar. 2016. Productivity, nutrient uptake and quality of popcorn and potato in relation to organic nutrient management practices. Annals of Agriculture Research New Series, 37(1): 72-79.

Partap Tej and Vaidya CS. 2009. Organic farmers speak on economics and beyond. Westville Publishing House, New Delhi. 161p.

Ramesh P, Mohan Singh and Subba Rao A. 2005. Organic farming: Its relevance to the Indian context. Current Science, 88(4): 561-568.

Ramesh, P, Panwar NR, Ramana S and Singh AB. 2006. Organic soybean production. Extension Bulletin No. 1/2006, Indian Institute of Soil Science, Bhopal. 21p.

Rajendran TP, Venugopalan MV and Tarhalkar PP. 2000. Organic cotton farming in India. Technical Bulletin No. 1/2000, p. 39, Central Institute of Cotton Research, Nagpur.

Rudragouda F, Channagouda and Babalad HB. 2015. Impact of organic farming practices on quality parameters of cotton. Research on Crops, 16(4): 752-756.

Rupela OP, Gowda CLL, Wani SP and Ranga Rao GV. 2005. Lessons from nonchemical input treatments based on scientific and traditional knowledge in a long-term experiment. In: The Agricultural Heritage of Asia: Proceedings of the International Conference (Y L Nene ed.), December 6-8, 2004, Asian Agri-History Foundation, Secunderabad-500 009, AP. pp. 184-196.

Sharma HC. 2001. Crop protection compendium: Cotton bollworm/legume pod-borer, Helicoverpa armigera (Hübner) (Noctuidae: Lepidoptera): Biology and Management. Wallingford, Oxon OX10 8DE, UK: Commonwealth Agricultural Bureau International.

Sridhar S, Krithika P and Sridevi, R. 2008. Organic cultivation techniques- Tomato, Brinjal, Lady's finger and Chilli. Centre for Indian Knowledge Systems, Chennai. 29p.

Srinivasa Rao M, Rama Rao CA, Ramakrishna YS, Srinivas K, Sreevani G and Vittal KPR. 2006. Crop-crop diversity as a key component of IPM in pigeonpea. Research Bulletin 2006. Central Research Institute for Dryland Agriculture, Hyderabad. 24p.

Srinivasa Rao M, Rama Rao CA, Ramakrishna YS, Srinivas K, Sreevani G, Venkateswarlu B and Vittal KPR. 2007. Crop-crop diversity as a key component of IPM in pigeonpea. Research Bulletin 2007. Central Research Institute for Dryland Agriculture, Hyderabad. 20p.

Srinivasarao Ch, Lal R, Prasad JVNS, Gopinath KA, Singh R, Jakkula VS, Sahrawat KL, Venkateswarlu B, Sikka AK and Virmani SM. 2015. Potential and challenges of rainfed farming in India. Advances in Agronomy, 133: 113-181.

Suja G. 2008. Organic production of tropical tuber crops. In: *Organic farming in rainfed agriculture: Opportunuties and constraints.* (Venkateswarlu, B., Balloli, S.S. and Ramakrishna, Y.S. Eds.), Central Research Institute for Dryland Agriculture, Hyderabad. pp. 135-143.

Surekha K and Satishkumar YS. 2014. Productivity, nutrient balance, soil quality, and sustain-ability of rice (*Oryza sativa* L.) under organic and conventional production systems. Communications in Soil Science and Plant Analysis, 45: 415-428.

Venkateswarlu B. 2008. Organic Farming in Rainfed Agriculture: Prospects and Limitations. In: *Organic Farming in Rainfed Agriculture: Opportunities and Constraints.* (B. Ventateswarlu S.S. Balloli and Y.S. Ramakrishna, eds.). Central Research Institute for Dryland Agriculture, Hyderabad. pp. 7-11.

Yadav SK, Bag TK, and Srivastava AK. 2017. Effect of organic manure and biofertilizers on system productivity and profitability of potato (*Solanum tuberosum*)–French bean (*Phaseolus vulgaris*) cropping system. Indian Journal of Agronomy, 62(2): 155-159.

Yadav R and Malik N. 2010. Productivity of barnyard millet (*Echinochloa frumentacea*) in relation to organic nutrition under rainfed conditions of Western Himalaya region. Indian Journal of Agronomy, 55(2): 105-109.

Chapter 14

Organic Farming in Hill and Mountain Areas

Dibakar Mahanta, J.K. Bisht and J.C. Bhatt

ICAR-Vivekananda Parvatiya Krishi Anusandhan Sansthan, Almora – 263 601, Uttarakhand
E-mail: dibakar_mahanta@yahoo.com

1. INTRODUCTION

Sustainable agriculture is necessary to attain the goal of sustainable development. According to the FAO, sustainable agriculture is "the successful management of resources for agriculture to satisfy changing human needs while maintaining or enhancing the quality of environment and conserving natural resources" (Roychowdhury *et al.*, 2013). The problems posed by modern-day agriculture gave birth to the concept of organic farming, natural farming *etc.* (NAAS, 2005). Organic farming is found to meet the objectives of sustainable agriculture and is the best alternative to avoid the ill effects of chemical farming. The primary goal of organic agriculture is to optimize the health and productivity of interdependent communities of soil life, plants, animals and people. Many techniques used in organic farming are the traditional agriculture practiced in India. The origin of organic farming goes back, in its recent history, to 1940s, when, J.I. Rodale in the United States, Lady Balfour in England and Sir Albert Howard in India contributed to the cause of organic farming (Narayanan, 2005).

The hill and mountain regions of India covering the states of Jammu and Kashmir, Himachal Pradesh, and Uttarakhand in NW Himalayas and seven states of NE Himalayas have not yet been exposed to the use of chemical fertilizers and pesticides (Gopinath *et al.*, 2009; Das, 2012) and are endowed with vast natural resources. However, for the small and marginal farmers in Himalayas, the purchase of fertilizers and pesticides is and will continue to be constrained by their high costs and unavailability (Mahanta *et al.*, 2013). Again the per capita livestock population in Indian Himalayan states is very high than average of India (Anonymous, 2012;

Anonymous, 2011; Mittal *et al.*, 2008). The biological wastes produced from livestock, if not utilized as organic manure will add GHGs to the environment. This can fulfill the demand of organic manure to continue organic agriculture. Consumer demand for organic products is concentrated in North America and Europe; which comprise 90 per cent of global revenues, but only 34 per cent of global organic area. Hence, organic producer from other countries can export organic products to these countries with high price premium. The organic market expansion makes it possible for Himalayan farmers to emerge as major suppliers of organic products with high price premiums (Gopinath *et al.*, 2009). Hence the Indian hill and mountain ecosystem should be promoted for organic certification.

2. Current Status of Organic Farming in Indian Hill and Mountain Ecosystem

The soils of the hill region are much degraded owing to cultivation in steep slopes, negligible nutrient supplementation and biomass burning under traditional practices, especially in northeastern Himalayas. Cultivation of crops on sloping land without proper soil and water conservation measures exacerbate the degradation process, making the soil unfit for cultivation. Thus, the appropriate soil and nutrient management practices is the prime importance for sustainable hill agriculture. Organic farming is considered as one of the best options for protecting/sustaining soil health and produce healthy foods (Patel *et al.*, 2015). The hill ecosystem of Northeastern and Northwestern Himalayas is suitable for organic farming due to negligible use of fertilizer and agrochemicals, abundance of organic manure, especially plant biomass, and favourable climatic conditions for diverse crops (Patel *et al.*, 2015).

Table 14.1: State-wise Consumption of Plant Nutrients per Unit Area in the Indian Hill States during 2012-13

State	*Fertilizer Consumption (kg/ha)*			
	N	*P_2O_5*	*K_2O*	*Total*
Arunachal Pradesh	1.62	0.11	0.32	2.05
Assam	36.31	11.80	18.15	66.26
HP	35.53	7.19	7.51	50.23
JK	66.52	21.14	8.55	96.21
Manipur	26.09	3.68	1.38	31.15
Mizoram	12.33	0.68	0.23	13.23
Meghalaya	9.94	3.40	1.01	14.35
Nagaland	2.43	1.53	0.84	4.80
Sikkim	0.00	0.00	0.00	0.00
Tripura	39.09	20.66	12.94	72.69
Uttarakhand*	3.36	1.55	0.36	5.27
All India	84.54	33.44	10.36	128.34

*Only hill districts of Uttarakhand.

Source: Agricultural Statistics at a glance 2013; Chanda *et al.* (2013); Mittal *et al.* (2008).

In the past five decades, the traditional knowledge and practices of organic farming have been almost eroded from many parts of India due to influx of modern "green revolution" technologies (Gopinath *et al.*, 2009). However, many communities particularly in the hill and mountain regions have sustained this knowledge. Hence, most of the cultivated area in the Himalayas of India has largely remained organic by default. The total fertilizer use is quite low in all hill states than the national average of 128.3 kg/ha/year. There is no fertilizer use in Sikkim state (Table 14.1). Hence, the organic area in Indian hill states can be enhanced easily through certification process.

The organic area in the hilly states of India is about 0.09 m ha (Yadav, 2011). Among the hilly states of India, Sikkim was the first to be declared as an organic state. Nine States have drafted organic farming policies out of which Uttarakhand, Nagaland, Sikkim and Mizoram have declared their intention to go 100 per cent organic (Singh and Meena, 2013). The state of Mizoram has declared itself organic and has refused its allocation of chemical fertilizer in 2004 (Vidal, 2004). But the actual area under organic farming is more. Unfortunately India, especially hilly states of India are having very large organic areas than documented as little or no information about their land use is reported. The area under organic cultivation and number farmers practising it are highest in Uttarakhand among the hilly states of India (Table 14.2). The Uttarakhand state targeted for certification more than 0.1 million farmers over similar number of hectares by 2010 (Subrahmanyeswari and Chander, 2007), but the target was not achieved (Table 14.2).

Table 14.2: Area and Number of Farmers under Organic Certification Process in Indian Hill States of India

State	*Area (ha)*			*Number of Farmers*		
	Organic	*In Conversion*	*Total*	*Organic*	*In Conversion*	*Total*
Arunachal Pradesh	523.2	1374.3	1897.5	116	590	706
Assam	1598.2	3510.7	5108.9	479	2768	3247
HP	437.1	139.0	576.1	346	833	1179
JK	430.6	182.4	613.1	132	68	200
Manipur	1247.2	1924.2	3171.3	2066	2901	4967
Mizoram	18002.3	9857.6	27859.8	14177	13878	28055
Meghalaya	1366.0	1677.1	3043.1	823	2685	3508
Nagaland	3091.3	6554.4	9645.7	3459	15639	19098
Sikkim	2872.7	4521.5	7394.2	3130	4697	7827
Tripura	203.6	77.5	281.1	1	295	296
Uttarakhand	16158.9	14906.8	31065.6	20695	26484	47179
Total	**45931.0**	**44725.5**	**90656.4**	**45424**	**70838**	**116262**

Source: Yadav (2011).

3. Problems Limiting the Spread the Organic Farming in Indian Hills

3.1 Non-availability of Biopesticides

Non-availability of appropriate biopesticides, if available; also not in sufficient amount leads organic producers in hills of India at high risk (Pandey and Singh, 2012). There is no specific biopesticides for specific insect/pathogen of specific crop as chemical pesticide. Hence, it is very difficult to control once infested by pathogen/insect in organic farming.

3.2 Small Farmers Inaccessibility

In India, the Director General of Foreign Trade, New Delhi, permits the export of organic produce provided that these are produced, processed and packed under a valid organic certificate issued by a certification agency accredited by an accreditation agency designated by the Government of India. The Government of India has recognized USOCA, Tamil Nadu Organic Certification Department, Agricultural and Processed Food Products Export Development Authority (APEDA), Spice Board, Ministry of Commerce and Industry, Coffee Board and Tea Board for the purpose. However, lack of knowledge, capital and access to certification discourage small farm holders in Indian hills.

3.3 Social Acceptance

Majority of small farm holders in hills are still dependent on government incentives to meet the cost of input and are striving for a rational profit margin for their produce in indigenous market. Small farm holders in India therefore, are apprehensive in adopting organic farming for price premium and export. Major issues that constrains farmer's acceptance in Indian hills include: non-availability of organic supplements and lack of appropriate knowledge.

3.4 Lack of Local Market

Our country lack indigenous lucrative market for locally grown organic produce. Further, under conversion stage, economic viability depends on the status of the farm.

3.5 Forest Laws

Sikkim promoted organic cardamom crop, but discovered that forest laws do not allow cultivation on 'forest' lands, even though it is done without destroying forests.

3.6 Transition Period

Before producing marketable products, an organic farm has to have a transition period of 2 to 3 years depending upon the certifying agency's requirements (Escobar and Hue, 2007; Prasad, 2008; Wright *et al.*, 2013), and during this period, the farmers have to grow the crops as per standards set for organic farming. Yield declines during first year of conversion. Accordingly, there may be a deficit in net income, hitting small and marginal farmers (Prasad, 2008).

3.7 Organic Certification Complexity

Access to certification, cost involved and lack of knowledge often constrain farmers especially small land holders in Indian hills from adopting organic farming (Pandey and Singh, 2012).

3.8 High Labour Requirement

Organically produced foods must meet strict regulations (certification), intensive management and hence practised in small scale. Organic farming requires more labour to carry out manual and mechanical tasks essential to growing (Bello, 2008). The preparation for sale on the farm or on the market also involves more labour on organic holdings making organic product expensive.

3.9 Use of Crop Residues as Fodder

Crop residues not only enhance soil organic matter content but also strongly support soil biodiversity, especially earthworm activity to increase soil quality and C sequestration (Lal, 2004). In Indian hills, most of the crop residues are removed from the fields for use as fodder and fuel. A decrease in area under natural pasture creates tremendous pressure on crop residue to be used as fodder.

3.10 Animal Manure for Domestic Fuel

Some of rural populations in hills also use animal manure for domestic fuel. This further limits availability of animal manure for soil amendment. In a typical rural household, about 1500-2000 kg (dry weight basis) of cattle manure is used as domestic fuel annually. For maintaining soil fertility and meeting crop nutrient demands, large quantity of these organic supplements are needed (Pandey and Singh, 2012).

3.11 Loss of Urine during Collection from Cattle-shed

Under Indian hill conditions, the floor of the cattle-shed is usually uncemented or *kachha*. As such the urine passed by animals during night gets soaked into the *kachha* floor. Nitrogen from urine is lost through the formation of gaseous ammonia. Urine of cattle contains about 48 per cent N of total excreta. Hence, large quantities of N are lost through urine (Yawalkar *et al.*, 2002).

No special efforts are made in Indian hills to collect the liquid portion of manure. Saving the urine requires considerable investment in putting straw as bedding and construction of a concrete floor. The most serious loss of dung is through the open grazing of domestic animals and use of dung as fuel in some parts of Indian hills. It is often said that two-third of the manure is either lost during grazing or is utilized for making cakes (fuel).

3.12 Loss of Nutrients during Manure Preparation and Storage

Mostly, cattle dung and waste from fodder are collected daily in the morning and put in manure heaps in open space. Heaps go on accumulating dung and other materials for months together getting exposed to the sun and rain losing their nutrients as leaching and volatilization (Yawalkar *et al.*, 2002).

3.13 Lack of Grazing Land

Over 10.9 m ha of permanent pasture contains 42 animals per hectare for grazing against the threshold level of 5 animals per hectare. This leads to overgrazing and massive extraction of fodder. It also increases compaction and reduces infiltration and ultimately changes the soil quality also (Pandey and Singh, 2012).

3.14 Lack of Clarity About Organic Agriculture

Consumers were not always sure about what is really covered by organic farming, and the restrictions it implied. The situation was worsened by fraudulent use of labelling of organic products (Bello, 2008).The most important constraint felt in the progress of organic farming in Indian hill states is the inability of the government policy to take a firm decision to promote organic farming. There are number of firms in India which grow vegetables, fruits, plantation crops, spices and tea organically and export to other countries. For exports, an aggressive strategy demanding free access for certification needs to be adopted (Kaur and Kaur, 2014).

4. Options to Increase Area under Organic Farming in Hills

Despite the gain that can be availed by adopting organic farming, farmers in the hills are often reluctant to do so. The issues hindering the adoption are already discussed in the paper. Further, strategies to improve the adoption as discussed by Prasad and Gill (2009) and Kaur and Kaur (2014) are abridged below:

4.1 Clarity of Organic Farming vis-à-vis Conventional Agriculture

Farmers often lack clarity about the organic food production standard. Moreover, the do's and don'ts pertaining to the inputs to be used, pest management *etc.* are also not clear to them. Morever, complete knowledge about organic farming principles, practices and advantages accrued to grower is not filtered down to the small farmers, which should be the actual target and potential beneficiary of organic farming. Hence, awareness programmes may be necessary.

4.2 Scientific Composting and Vermicomposting

Preparation of compost, vermicomposting and biodynamic preparation *etc.* in scientific way can substitute and reduce the organic amendment requirement. There should be promotion for mass production and adoption of the bio-fertilizers, vermicompost, bio control agents and eco-friendly inputs.

4.3 Government Policy for Organic Farming

Government should take care of the complex and costly procedures of certification for small and marginal farmers. There should be provision for proper channel from domestic to global market and organic market infrastructure, development of post-harvest, processing and value addition facilities along with cold storage facilities. Domestic and international marketing of organic produce are to be facilitated through APMC, SMS with assistance from APEDA, SFAC, NABARD *etc.* Contract farming/buy back policy should be promoted for organics. Identification of potential areas/commodities within the region that could immediately be recognized for certification should be done. Regulating body in selected districts of the state to supervise and implement these policies is to be set up. Incentives to the growers in the initial years of shifting from traditional farming to organic farming should be provided.

4.4 Research

Research back up should be strengthened for developing location/commodity specific packages of practices.

5. Productivity and Soil Health in Organic Farming in Relation to Conventional System in Hills

5.1 Influence of Organic Farming on Crop Productivity and Soil Properties during Transition Period

Several studies have reported lower yields in organic conditions with comparison to chemical fertilizers (Gopinath *et al.*, 2009; MacRae *et al.*, 1990). The lower yields have been reported from the Indian Himalayas also (Tables 14.3 and 14.4) (Gopinath *et al.*, 2009; Saha *et al.*, 2007a). Initially lower yields on organic farms have been attributed to the negative effects of conventional practices on the soil microorganisms that mineralize soil organic matter, or that control soil-borne pests (Martini *et al.*, 2004).

Table 14.3: Cob Yield of Baby Corn (Mean of two years, experiment conducted for two years)

Treatment	*Cob Yield (kg/ha)*
Control	221
FYM 15 t/ha	925
Poultry manure 6 t/ha (97.8 kg N/ha)	823
Vermicompost 5 t/ha (35.3 kg N/ha)	501
Recommended NPK (120-25.8-33.3 kg N-P-K/ha)	1265

Source: Saha *et al.* (2007a).

But, these organic systems may lead to more biological activity and improve soil quality than industrial conventional systems (Castillo and Joergensen, 2001; Fließbach *et al.*, 2007; Garcia-Ruizet *et al.*, 2008; Gopinath *et al.*, 2009). Acid phosphatase, alkaline phosphatase, phosphodiesterase and inorganic phosphatase activities were studied with different organic sources (FYM, vermicompost, and lantana compost) and chemical fertilizer in wheat-baby corn-rajmash cropping system after three years. These enzymes mineralize P from organic and inorganic source of nutrients.

Chemical fertilizer treated soil showed less activity of acid phosphatase, alkaline phosphatase, phosphodiesterase and inorganic pyrophosphatase than all other organic source treatments, especially at higher levels. Application of earthworm casts lead to more acid phosphatase activity than others. The lower phosphatase activities in Lantana composts amended soil might be due to the fact that excess concentration of allelochemicals in those plots, which have hampered the growth of microbes.

Table 14.4: Effect of Treatments on Yields (t/ha) of Bell Pepper, French Bean and Garden Pea (Mean of two years)

*Treatment**	*Bell Pepper*	*French Bean*	*Gardenpea*
FYM 20 t/ha + biofertilizers (BF)	27.96	3.48	7.04
FYM 10 t/ha + PM 1.5 t/ha + VC 1.5 t/ha + BF	26.18	3.87	7.27
INM (FYM 10 t/ha + NPK)	37.40	5.49	7.08
Control	7.70	1.48	2.20

*PM = Poultry manure; VC = Vermicompost; INM = Integrated nutrient management.

Source: Gopinath *et al.* (2011).

All the treatments involving organic amendments registered significantly higher dehydrogenase, β-glucosidase, urease and acid phosphatase activity than did chemical fertilizer treatment, especially at higher levels (150 and 120 kg N/ha) (Gopinath *et al.*, 2008). The only exception, was the lower acid phosphatase activity in lantana compost applied plots at 150 kg N/ha. The incorporation of organic amendments into the soil influences the soil's enzymatic activities because the added material may contain intra and extracellular enzymes and may also stimulate microbial activity in the soil. The development of microbial populations, which is favoured by the root exudates of plants, may also be responsible for the stimulation of enzyme activity. FYM amended plots had the greatest urease activity, followed by vermicompost and lantana compost treatments. This may have been due to more simplified organic compounds in FYM, which leads to faster mineralization.

The soil bulk density was reduced significantly in all organic manure treatments in comparison with chemical fertilized plots (Table 14.5). The soil pH increased significantly to neutral range in all the plots treated with organic amendments, especially with FYM application compared with chemical fertilizer treatment after 2 years of wheat (Table 14.5). This illustrates the important role that organic manures in buffering the pH of the soil (Gopinath *et al.*, 2008). Similarly to pH, SOC was also significantly higher in manure amended plots than those under chemical fertilizer treatment. Among the organic amendments, application of FYM (150 kg N/ha) recorded the highest SOC compared with other treatments. Application of chemical fertilizer had higher levels of available N and P than most organic treatments, except FYM @ 150 kg N/ha (Table 14.5). Lower availability of plant nutrients in plots applied with organic amendments was expected, due to the slower release rates of organic materials, particularly during the initial years of transition to organic production.

Table 14.5: Effect of Source and Application Rate of Organic Amendments on Soil Properties

Treatment	BD (Mg/M³)	pH	SOC (Per cent)	Av N (mg/kg)	Av P (mg/kg)	Av K (mg/kg)
FYM @ 150 kg N/ha	1.16^{de}	6.5^{a}	1.88^{a}	184.8^{ab}	6.7^{b}	170.0^{a}
FYM @ 120 kg N/ha	1.16^{de}	6.3^{bc}	16.3^{bc}	164.0^{bcd}	4.7^{cd}	156.3^{ab}
VC @ 150 kg N/ha	1.14^{e}	6.4^{ab}	16.8^{b}	179.3^{bc}	5.2^{bc}	113.0^{cd}
VC @ 120 kg N/ha	1.22b^{cd}	6.3^{bc}	16.1^{bc}	166.5b^{cd}	5.0b^{cd}	108.5^{cde}
Lantana compost @ 150 kg N/ha	1.16^{de}	6.2^{cd}	14.5^{de}	158.0^{cd}	1.2^{g}	113.5^{cd}
Lantana compost @ 120 kg N/ha	1.21b^{cd}	6.2^{cd}	13.9^{ef}	153.5^{d}	2.2^{fg}	108.8^{cde}
Chemical fertilizer (120-60-40)	1.27^{ab}	5.7^{f}	11.9^{h}	201.9^{a}	9.6^{a}	107.5^{cde}
Control	1.33^{a}	6.0^{e}	10.6^{i}	129.0^{e}	2.9^{efg}	90.5^{f}

FYM: Farmyard manure; VC: vermicompost; BD: bulk density; SOC: soil organic carbon.

Means in the same column with different letters are significantly ($P < 0.05$) different.

5.2 Performance of Organic Farming Beyond Transition Period in Hills

When the experiments with organic manure were conducted for more than three years, the system was converted into organic. The productivity of crops and system completely change than what has been presented during transition period. Different nutrient sources were evaluated for six years on productivity and sustainability of gardenpea-french bean cropping system in the mid-Himalayas. All organic nutrient sources *viz.* poultry manure, farmyard manure and vermicompost provided higher mean pod yield than recommended NPK. Application of poultry manure @ 6 t/ ha produced 34 per cent more pod yield than the recommended chemical fertilizer with higher sustainability.

Similarly, organic source (FYM) was compared with chemical fertilizer with equal level of N, P and K application to both organic and chemical for 6 years in aromatic rice (Saha *et al.*, 2010). Organic source of nutrient provided higher mean grain yield than the chemical fertilizer with equal level of N, P and K. Interestingly, there was increase in grain yield during the third year in FYM-treated plots only. Further, yield reduction was observed in the chemical fertilized plots in third year. Within the sixth year, development of soil health was observed in organically fertilized plots, and yield decline was observed in chemical fertilized plots. Once nutrients in chemical fertilized soils get exhausted by the crop, it is difficult to sustain the productivity. Greater grain yield in FYM treated plots might be attributed to several factors, especially greater nutrient availability and less pest infestation, coupled with improvement in soil quality in these plots.

In another experiment, the highest productivity of garden pea equivalent yield of garden pea-french bean system was observed with application of 20 t FYM/ ha (FYM_{20}), which provided 54 and 29 per cent higher pod yield than NPK and INM treated plots, respectively (Mahanta *et al.*, 2013). Interestingly, the highest productivity was not achieved with chemical fertilizer (NPK) and the yields were significantly improved with FYM application (Table 14.6). The sustainability of garden pea-french bean cropping system with FYM application was enhanced considerably than the chemical fertilizer and integrated nutrient management after six years' of cropping in the Indian Himalayas (Table 14.9).

Table 14.6: Productivity and Sustainability of Garden Pea-French Bean Cropping System (Mean of six years) under different Nutrient Management Practices in the Indian Himalayas

Treatment	*Mean Pod Yield (t/ha)*			*Sustainable Yield Index*		
	Garden Pea	*French Bean*	*Garden Pea Equivalent Yield of Total System*	*Garden Pea*	*French Bean*	*Garden Pea - French Bean Cropping System*
Control	3.42	4.12	10.04	0.637	0.318	0.472
FYM*	8.90	14.44	31.47	0.705	0.517	0.606
NPK	5.96	9.14	20.49	0.615	0.381	0.525
INM	7.07	10.97	24.36	0.684	0.466	0.549
LSD (*P*=0.05)	1.06	2.16	3.08	-	-	-

*FYM = Application of 20 t FYM/ha; NPK = Recommended NPK application; INM = 50 per cent NPK + FYM 5 t/ha.

Source: Mahanta *et al.* (2013).

It was clearly proved from the probability analysis that the differences among treatments for yield and yield attributes widened as the experiment progressed in comparison with initial years (Mahanta *et al.*, 2015). The difference in yield between organic and chemical fertilizer increased significantly and continuously as the experiment progressed with higher sustainability in organic practices. The maximum net profit with higher sustainability index could be obtained through FYM application in comparison with chemical fertilizer in gardenpea-french bean system. The increase of net profit under FYM applied plots in comparison with chemical fertilizer was 1.52 lakh rupees per annum. It clearly dispelled the fear of profit in organic farming, and the profit is much more than the chemical fertilizer. The B:C ratio was 4.44 with FYM, whereas, it was 3.67 for chemical fertilizer during six years of gardenpea-french bean cropping system (Table 14.7).

Table 14.7: Mean Effect of Nutrient Management Practices on Economics of Garden Pea-French Bean Cropping System in the Indian Himalayas

Treatment	*Net Returns (× 1000 Rs/ha)*	*Net Returns per Rs Invested (B:C Ratio)*
Control	86	1.45
FYM 20 t/ha	392	4.44
NPK	240	3.67
INM	299	4.22
LSD at 5 per cent	54	-

Source: Mahanta *et al.* (2015).

It was observed that continuous application of FYM had a great beneficial effect on the population levels of total bacteria, fungi, actinomycetes and *Trichoderma* spp. over chemical fertilizer and INM in garden pea-french bean system (Table 14.8). There was at least 150 per cent increase in microbial population with addition of FYM over chemical fertilizer. Soil microorganisms govern the numerous nutrient cycling reactions in soils (Mäder *et al.*, 2002). Basically, the enzyme involved in intracellular microbial metabolism, *i.e.*, dehydrogenase and the enzyme responsible for solubilizing P from soil *i.e.* acid phosphatase, increased with application of FYM in comparison with chemical fertilizer and INM (Table 14.8).

Table 14.8: Effect of Nutrient Source on Soil Biological Properties

Nutrient Source	*Microbial Population Count (Colony formation unit/g soil)*				*Soil Enzymatic Activity*	
	Bacteria ($\times 10^7$)	*Fungi ($\times 10^5$)*	*Actinomycetes ($\times 10^4$)*	*Trichoderma spp. ($\times 10^3$)*	*Dehydrogenase Activity (µg TPF/g soil/24 h)*	*Acid Phosphatase Activity (µg PNP/g soil/h)*
Control	1.10	1.09	6.85	1.14	67.5	994
FYM	3.92	5.18	27.15	4.96	148.1	1410
NPK	1.68	1.15	11.32	1.23	76.7	1031
INM	2.62	4.09	18.28	3.11	117.7	1238
LSD ($P = 0.05$)	0.50	0.59	2.60	0.48	21.9	197

FYM = Application of 20 t FYM/ha; NPK = Recommended NPK application; INM = 50 per cent NPK + FYM 5 t/ha.

Source: Mahanta *et al.* (2013)

The highest microbial population under FYM_{20} treated plots was due to the presence of easily water soluble C (Mahanta *et al.*, 2013), and N in FYM, which acts as a source of energy for soil organisms, whereas the easily soluble C component was missing in mineral fertilizer and, hence, the microbial population was less in the plots under chemical fertilizer. Soil dehydrogenase activity is a good indicator of overall microbial activity in soil and it can serve as a good indicator of soil condition. Soils under FYM_{20} recorded 14.1 and 9.3 per cent higher SOC than NPK and INM treated plots, respectively. Under acid soil conditions (*i.e.* pH 5.64 in NPK plots), large increase in concentrations of Al^{3+} and Mn^{2+} in soil solution could have influenced enzyme function adversely in plots under NPK (Mahanta *et al.*, 2013), and almost neutral pH (6.85) under FYM_{20} might have favoured.

Industrial agriculture inputs in conventional farming system contribute to green house gas (GHG) emission and add numerous forms of environmental degradation. Hence, sustainability of agricultural systems with low or without emission of GHGs is the priority for agriculture. Thus, to reduce the C emission, to provide higher yield and sustainability, an experiment was conducted for six years with FYM and recommended NPK in garden pea-french bean cropping system in the Indian Himalayas. The amount of SOC under FYM_{20} plots (26.6 Mg C/ha) was increased by 2.6 and 2.1 Mg C/ha as compared to NPK and INM treated plots. The per unit CO_2 equivalent emission of GHGs for production of manure is less than manufacture of mineral fertilizer of N, P and K together (Lal, 2004) and hence the quite low emission for 20 t FYM/ha in comparison with NPK and INM (Table 14.9). It is clearly proved from the above emission factor that application of 20 t FYM/ha is highly climate resilient and sustainable than chemical fertilizer and INM.

Table 14.9: Effects of Organic and Chemical Fertilization on Soil Chemical Properties and Carbon Equivalent Emission from Irrigated Gardenpea-French Bean Cropping System after Six Years

Treatment$^{\alpha}$	*SOC Amount/Stock (Mg/ha)*	*CO_2 Emission Reduction through SOC Stock (Mg/ha)*	*CO_2 Equivalent Emission from Production of Nutrient (kg/ha/yr)*
Control	20.2		–
FYM	26.6	23.3	224
NPK	24.0	13.8	479
INM	24.5	15.7	295
LSD (P = 0.05)	3.1	-	-

α: FYM = Application of 20 t FYM/ha; NPK = Recommended NPK application; INM = 50 per cent NPK + FYM 5 t/ha; *SOC = soil organic carbon.

Source: Mahanta *et al.* (2013)

It was estimated that there was 133 kg more CO_2 emission through fertilizer manufacture with application of recommended NPK fertilizer than soil C sequestration (Mahanta *et al.*, 2013). The CO_2 emission for FYM preparation was very less than sequestration through it. Application of 5.04 t FYM/ha was able to nullify the 62 kg CO_2 emitted during FYM preparation through soil C sequestration. Hence, application of more than 5.04 t FYM/ha provided net positive C sequestration. The highest net positive C sequestration (1585 kg CO_2/ha) and the highest C sequestration (1792 kg CO_2/ha) can be achieved by application of 16.6 and 17.1 t FYM/ha, respectively in gardenpea-french bean cropping system in mid-Himalayas (Table 14.10). The highest net positive C sequestering FYM application rate provided 1443 kg CO_2/ha higher C sequestration than recommended NPK. It also provided 48 per cent higher pod yield of system than recommended NPK.

Table 14.10: FYM Requirement to Achieve C Sequestration as Chemical Fertilizer Application and Maximum C Sequestration through FYM Application under Gardenpea-French Bean Cropping System in Indian Himalayas

FYM Required (t/ha) to Achieve C Emission/ C Sequestration as NPK					*C Sequestration amount (kg CO_2/ha/year) through FYM Application*			
*FYM_{em} = FYM_{seq}	NPK_{seq}	NPK_{em}	C Seq_{max}	Net C Seq_{max}	FYM_{em} = FYM_{seq}	NPK_{seq}	FYM Seq_{max}	FYM Net Seq_{max}
5.04	6.07	28.1	17.1	16.6	62	346	1792	1585

*FYM_{em} = C emission with FYM preparation; FYM_{seq} = C sequestration with FYM application; NPK_{seq} = C sequestration from NPK fertilizer application; NPK_{em} = C emission from NPK fertilizer manufacture; C Seq_{max} = Maximum C sequestration from FYM application; Net C Seq_{max} = Maximum net C sequestration from FYM application

Carbon sequestration potential (CSP), defined as the rate of increase in C sequestration over the initial soil C content. The CSP of FYM_{20} plots was about 0.433 and 0.347 Mg C/ha/year more than NPK and INM plots, respectively. The carbon sequestration potential was substantially higher with organic manure application than chemical fertilizer, owing to carbon rich materials in organic products.

6. Conclusions

Organic farming can meet the objectives of sustainable agriculture. Most of the hill and mountain regions are organic by default. Hence the Indian hill and mountain ecosystem should be promoted for organic certification. The organic area in the hilly states of India is about 0.09 m ha. But the actual area under organic is much more than documented. Nine hill states have drafted organic farming policies and out of which Uttarakhand, Nagaland, Sikkim and Mizoram have declared for 100 per cent organic. Sikkim has already fulfilled this. Organic certification complexity, non-availability of bio-pesticide, low yield during transition period and lack of domestic market with high price premium are some of the major constraints for enhancing organic area. Favourable government policy, development of varieties for organic production system and site specific research for organic package of practices can provide more organic products in hills. It has been proved that organic can provide better soil health, enhance soil microbial diversity and it is more climate resilient than conventional chemical system.

7. Future Strategies

- ☆ Many of the concerns of sustainability focus on environment and nature, animal welfare, product quality and health. Therefore, there is need for research support to develop feasible, sustainable and site specific organic agricultural techniques.
- ☆ The amount of stress (available nutrients, weed pressure, insect-pests and diseases) on the organic crop is expected to be more than input-intensive conventional farming. Therefore, choice of a crop variety is more critical in organic situations than conventional farming, where problems can be solved by application of pesticides or mineral fertilizers.
- ☆ Long-term assertions of organic farming systems in various agro-ecological and agro-economic contexts to study the different parameters essential for sustainable development.
- ☆ There must be development of specific biopesticides for specific pest. The biopesticides must be available in common market to be utilized at right time.
- ☆ Long-term evaluation of organic farming influence on yield and yield sustainability, especially in years with extreme climatic conditions such as droughts or floods, the product quality and shelf-life
- ☆ Influence of organic farming on agro-ecosystem stability and availability and quality of natural resources, especially soil fertility, energy resources, biodiversity and beneficial organisms

REFERENCES

Agricultural Statistics at a glance 2013. Government of India, Ministry of Agriculture, Department of Agriculture and Cooperation, Directorate of Economics and Statistics.

Anonymous 2011. Census of India 2011. Office of the Registrar General and CensusCommissioner, India, Ministry of Home Affairs, 2/A, Mansingh Road, New Delhi - 110011.

Anonymous 2012. Basic Animal Husbandry Statistics – 2012. AHS Series-13. Government of India, Ministry of Agriculture, Department of Animal Husbandry, Dairying and Fisheries, Krishi bhawan, New Delhi.

Araya H and Edwards S. 2006. The Tigray Experience: A success story in sustainable agriculture. Third World Network: Environment and Development Series No. 4.

Bello WB. 2008. Problems and Prospect of Organic Farming in Developing Countries. Ethiopian Journal of Environmental Studies and Management, 1(1): 36-43.

Bhattacharyya P and Chakraborty G. 2005. Current Status of Organic Farming in India and other Countries. Indian Journal of Fertilizers, 1: 111-123.

Castillo X and Joergensen RG. 2001. Impact of ecological and conventional arable management systems on chemical and biological soil quality indices in Nicaragua. Soil Biology and Biochemistry, 33: 1591-1597.

Chanda TK, Sati K, Robertson C and Arora C. 2013. Fertiliser statistics 2012-13. The Fertiliser Association of India, FAI House, 10, Shaheed Jit Singh Marg, New Delhi - 110 067.

Chhonkar PK and Dwivedi BS. 2004. Organic farming and its implications on India's food security. Fertilizer News, 49(11): 15-38.

Das, M., 2012. NE can be organic food hub. The Times of India. Jul 31, 2012.

Escobar MEO and Hue NV. 2007. Current developments in organic farming. Recent Research Development in Soil Science, 2: 29-62.

FAO 1999. Guidelines for the production, processing, labeling and marketing of organically produced foods. Joint FAO/WHO Food Standards Program Codex Alimentarius Comission, Rome, CAC/GL 32, p. 49.

Fließbach A, Oberholzer HR, Gunst L and Mader P. 2007. Soil organic matter and biological soil quality indicators after 21 years of organic and conventional farming. Agriculture, Ecosystems and Environment,118: 273-284.

Garcia-Ruiz R, Ochoa V, Hinojosa MB and Carreira JA. 2008. Suitability of enzymatic activities for the monitoring of soil quality improvement in organic agricultural systems. Soil Biology and Biochemistry, 40: 2137-2145.

Gopinath KA, Saha S, Mina BL, Pande H, Kundu S and Gupta HS. 2008. Influence of organic amendments on growth, yield and quality of wheat and on soil properties during transition to organic production. Nutrient Cycling in Agroecosystems, 82: 51-60.

Gopinath KA, Saha S and Mina BL. 2011. Effects of organic amendments on productivity and profitability of bell pepper-french bean-garden pea system and on soil properties during transition to organic production. Communications in Soil Science and Plant Analysis, 42: 2572-2585.

Gopinath KA, Saha S, Mina BL, Pande H, Srivastva AK and Gupta HS. 2009. Bell pepper yield and soil properties during conversion from conventional to organic production in Indian Himalayas. Scientia Horticulturae, 122: 339-345.

Ibeawuchi II, Iwuanyanwu UP, Nze EO, Olejeme OC and Ihejirika GO. 2015. Mulches and organic manures as renewable energy sources for sustainable farming. Journal of Natural Sciences Research, 5(2): 139-147.

Kaur AD and Kaur G. 2014. Organic Farming for Sustainable Agriculture: Global and Indian Perspective. Global Journal For Research Analysis, 3(2): 57-58.

Lal R. 2004. Soil carbon sequestration in India. Climatic Change, 65: 277-296.

MacRae RJ, Hill BS, Mehuys GR and Henning J. 1990. Farmscale agronomic and economic conversion from conventional to sustainable agriculture. Advances in Agronomy, 43: 155-198.

Mäder P, Fließbach A, Dubois D, Gunst L, Fried P and Niggli U. 2002. Soil fertility and biodiversity in organic farming. Science, 296: 1694-1697.

Mahanta D, Bhattacharyya R, Gopinath KA, Tuti MD, Jeevanandan K, Chandrashekara C, Arunkumar R, Mina BL, Pandey BM, Mishra PK, Bisht JK, Srivastva AK and Bhatt JC. 2013. Influence of farmyard manure application and mineral fertilization onyield sustainability, carbon sequestration potential and soil propertyof gardenpea–french bean cropping system in the Indian Himalayas. Scientia Horticulturae, 164: 414-427.

Mahanta D, Bhattacharyya R, Sahoo DC, Tuti MD, Gopinath KA, Arunkumar R, Mina BL, Pandey BM, Bisht JK, Srivastva AK and Bhatt JC. 2015. Influence of farmyard manure application and mineral fertilization on yield sustainability, carbon sequestration potential and soil property of gardenpea–french bean cropping system in the Indian Himalayas. Journal of Plant Nutrition, **doi:** 10.1080/01904167.2015.1009096.

Martini EA, Buyer JS, Bryant DC, Hartz TK and Denison RF. 2004. Yield increases during the organic transition: improving soil quality or increasing experience. Field Crops Research, 86: 255-266.

Mitra SK. 2008. Organic tropical and subtropical fruit production in india prospects and challenges. ISHS Acta Horticulturae 975: IV International Symposium on Tropical and Subtropical Fruits.

Mittal S, Tripathi G and Sethi D. 2008. Development strategy for the hill districts of Uttarakhand. Working Paper No. 217. Indian council for research on international economic relations. admin.indiaenvironmentportal.org.in/files/Working_Paper_217.pdf.

NAAS 2005. Organic Farming: Approaches and possibilities in the context of Indian Agriculture. Policy paper 30. National Academy of Agricultural Sciences, India, February 2005.

Narayanan S. 2005. Organic farming in India: Relevance, problems and constraints. Occasional Paper - 38. Department of Economic Analysis and Research, National Bank for Agriculture and Rural Development, Mumbai.

Pandey J and Singh A. 2012. Opportunities and constraints in organic farming: an Indian perspective. Journal of Scientific Research, 56: 47-72.

Parrot N and Marsden T. 2002. The real Green Revolution: Organic and Agroecological farming in the London: Green peace Environment Trust. pp. 1-6.

Patel D, Das A, Kumar M, Munda GC, Ngachan SV, Ramkrushna GI, Layek J, Pongla N, Buragohain J and Somireddy U. 2015. Continuous application of organic amendments enhances soil health, produce quality and system productivity of vegetable-based cropping systems in subtropical Eastern Himalayas. Experimental Agriculture, 51: 85-106.

Pielou EC. 1966. The measurements of diversity in different types of biological collections. Journal of Theoretical Biology, 13: 131-144.

Prasad K and Gill MS. 2009. Developments and strategies perspective for organic farming in India. Indian Journal of Agronomy, 54(2): 186-192.

Prasad R. 2008. Organic farming. Modern concept of Agriculture. http: //nsdl. niscair.res.in/jspui/bitstream/123456789/670/1/Revised per cent 20Organic _farming.pdf.

Ramesh P, Panwar NR, Singh AB, Ramana S, Yadav SK, Shrivastava R and Subba Rao A. 2010. Status of organic farming in India. Current Science, 98: 1090-1194.

Roychowdhury R, Banerjee U, Sofkova S and Tah J. 2013. Organic farming for crop improvement and sustainable agriculture in the era of climate change. OnLine Journal of Biological Sciences, 13(2): 50-65.

Saha S, Appireddy GK, Kundu S and Gupta HS. 2007a. Comparative efficiency of three organic manures at varying rates of its application to baby corn. Archives of Agronomy and Soil Science, 53: 507-517.

Saha S, Gopinath KA, Mina BL, Kundu S, Bhattacharaya R and Gupta HS. 2010. Expression of soil chemical and biological behavior on nutritional quality of aromatic rice as influenced by organic and mineral fertilization. Communications in Soil Science and Plant Analysis, 41: 1816-1831.

Saha S, Mina BL, Gopinath KA, Kundu S and Gupta HS. 2008a. Relative changes in phosphatase activities as influenced by source and application rate of organic composts in field crops. Bioresource Technology, 99: 1750-1757.

Saha S, Pandey AK, Gopinath KA, Bhattacharyya R, Kundu S and Gupta HS. 2007b. Nutritional quality of organic rice grown on organic7 composts. Agronomy for Sustainable Development, 27: 223-229.

Singh DK, Raverkar KP, Dhyani VC and Charturvedi S. 2012. Organic Farming: Status and Research Experiences at Pantnagar. Research Bulletin No.184, Directorate of Experiment Station, G.B. Pant University of Agriculture and Technology, Pantnagar-263 145, Uttarakhand, India. p. 41.

Singh S and Meena VS. 2013. Promoting organic farming- an analysis of status and issues of Uttarakhand Organic Commodity Board. International Journal of Research in Commerce, Economics and Management, 3(03): 127-130.

Subrahmanyeswari B and Chander M. 2007. Registered organic farmers in Uttarakhand state of India: A profile. In: Proceedings of International Conference on organic agriculture and food security, 55–56. Rome, Italy: FAO.

Vidal J. 2004. Eco sounding. The Guardian, Wednesday 7 April 2004.

Willer H and Lernoud J. 2014. The World of Organic Agriculture 2014: Summary. The World of Organic Agriculture. Statistics and Emerging Trends 2014. 15 edition. (Eds.) FiBL and IFOAM, Frick and Bonn.

Willer H and Lernoud J. 2015. The World of Organic Agriculture 2015: The World of Organic Agriculture: Summary. Statistics and Emerging Trends 2015. FiBL and IFOAM, Frick and Bonn.

Wright DL, Marois JJ and Katsvairo TW. 2013. Transitioning from conventional to organic farming using conservation tillage in Florida. IFAS Extension (SS-AGR-11), University of Florida. http: //edis.ifas.ufl.edu.

Yadav AK. 2011. Organic Agriculture. National Centre of Organic Farming, Department of Agriculture and Cooperation, Ministry of Agriculture, Govt of India, CGO-II, Kamla Nehru Nagar, Ghaziabad, 201 001, Uttar Pradesh, India.

Yawalkar KS, Agarwal JP and Bokde S. 2002. Manures and Fertilisers. Ninth Revised Edition. Agri-Horticultural Publishing House, Nagpur - 440010, India.

Chapter 15

Organic Farming in Coastal Ecosystems of India: Present Status and Future Prospects

B. Gangaiah, T. Subramani, K.R. Kiran, A. Velmurugan, Sachidananda Swain and T.P. Swarnam

ICAR-Central Island Agricultural Research Institute, Port Blair – 744 101, Andaman and Nicobar Islands
E-mail: bandlagangiah1167@gmail.com

1. INTRODUCTION

Coastal areas (more apt term than coastal zone in absence of regulation and management) are defined/described as the interface or transition areas between land and sea, including large inland lakes. Estuaries, lagoons, mangroves, backwaters, salt marshes, rocky coasts, sandy stretches and coral reefs are part of this area. They are diverse in function, form and dynamic. Unlike watersheds, there are no exact natural boundaries to unambiguously delineate coastal areas. India is bestowed with vast coastal region on 7517 km length starting from Gujarat in the West coast and ending in the East coast of West Bengal (5423 km long spread in 9 states and 2 union territories in 69 districts). In addition, the coastal areas of Islands (Andaman and Nicobar Islands and Lakshadweep Islands in 4 districts spread on 2094 km long coasts) too fall under this category. The Exclusive Economic Zone (EEZ) extending from coastal line to 200 nautical miles into the sea also forms part of coastal areas. These areas are hubs of many economic activities that includes agriculture. Some of the activities that are exclusive to these regions include sea ports, marine fishing, beaches, tourism *etc.*

The landmass of coastal areas/ecology is spread on 3 Agro-Ecological Zones (AEZs) out of the total 20 AEZs of India. The three AEZs *i.e.* Eastern coastal plains (AEZ 18); Western Ghats and coastal plains (AEZ 19) and Islands regions (AEZ

20) are spread over 20.4 m ha geographical area (6.5 per cent of countries total) of which 11.87 m ha (58.2 per cent) is gross cultivated area (Sehgal *et al.*, 1992). As per FAO (2001) *Coastal artisanal fishing mixed farming systems* that is one of the 8 farming systems of developing world is practised in these areas. In South Asia (Bangladesh, Maldives, India) this farming system was adopted by ~18 m agriculturists on 2.5 m ha of cultivated land (0.8 m ha is under irrigation) supporting 45 m population.

Artisanal fishing entails fishing with primitive gears that is still prevalent in Islands regions and villages along the coastal regions of India. However, commercial fishing gears have become common in cities around sea ports. Commercial aquaculture has also emerged by converting paddy fields into fish ponds. Mixed farming means crops, animals, fish *etc.* The coastal areas of different states (union territories) of India are depicted in Figure 15.1. (Source: Centre for Coastal Zone Management and Coastal Shelter Belt: http://iomenvis.nic.in/index2.aspx?slid=3680 and sublinkid=259 and langid=1 and mid=1) and details of districts are given in the Table 15.1.

High precipitation of AEZ 19 and 20 (2000-3200 mm) facilitates year round cropping (with length of growing period >210 days) of plantation crops in multi-

Figure 15.1: Coastal Ecosystem of India.

Table 15.1: Coastal States and Districts of India

State/Union Territory (No. of districts)	*District*
Gujarat (9)	Porbandar, Navsari, Bharuch, Amreli, Valsad, Kachchh, Anand Bhavnagar and Surat
Daman and Diu (2)	Daman and Diu
Maharashtra (5)	Sindhudurg, Ratnagiri, Raigarh, Greater Bombay and Thana
Karnataka (3)	Udupi, Uttar Kannada and Dakshin Kannada
Goa (2)	South Goa and North Goa
Kerala (10)	Kasaragod, Kottayam, Alappuzha, Kannur, Kollam, Kozhikoda, Ernakulam, Thrissur, Thiruvananthapuram and Malappuram
Tamil Nadu (11)	Thiruvarur, Ramanathapuram, Pudukottal, Nagapattinam, Thoothukudi, Kanyakumari, Cuddalore, Thruneivell, Villupuram, Kancheepuram and Chennai
Pondycherry (3)	Mahe, Karaikal, Yanam
Andhra Pradesh (9)	Srikakulam, Vizhianagaram, Vishakapathnam, East Godavari, West Godavari, Krishna, Guntur, Prakasam and Nellore
Odhisa (5)	Jagatsinghpur, Bhadrak, Kendrapara, Baleshwar and Ganjam
West Bengal (4)	Hara, South 24 Parganas, North 24 Parganas and East Midnapore
Andaman and Nicobar Islands (3)	North and Middle Andaman, South Andaman, Car Nicobar
Lakshadweep (1)	Kavaratti

storey cropping mode with annual crops like rice, vegetables *etc.* The AEZ 18 with lesser rainfall (1200-1900 mm) has shorter growing period (60-210 days) and annual crops (rice, fruits, and vegetables) has predominance over the plantation crops. The major dams constructed across the rivers in east coast make it a food grain producer zone. To support the year round cropping, fertilizers are used [AEZ 20; 14.3 kg/ha NPK < AEZ 19: 80.3 kg/ha; AEZ 18: 144.7 kg/ha as against national average of 89.8 kg/ha in 2004-05 (FAO, 2005)]. The low use of fertilizers in AEZ 20 and 19 are ascribed to physical isolation of the region from mainland and plantation based farming systems (with high recycling of residues) respectively in contrast to the cereal based farming systems of AEZ 18. It shows that island regions are ideal for organic farming and West coast also has scope for moving towards organic farming as compared to east coast. The allied agricultural (livestock, fishing, forestry) activities are finely interwoven with crops cultivation could aid in promoting organic farming. The coastal districts have vast livestock population (16.4, 36.9 and 129.5 million bovines, sheep & goat and poultry birds). Marine fishing being exclusive to these regions of the country and inland capture and culture fisheries too are important in rivers, lakes *etc.* Sea weed culture adds more avenues for promoting organic farming. Tourism, the most employment generating activity of coasts adds to more consumer base to the organic products. The anticipated climate changes and associated sea level rise are threats to this ecosystem that calls for ecological farming that aims at protecting the coastal regions. The recent floods in Kerala indicates the impacts of climate change (varied monsoons *i.e.* down pours) that got transformed into a disaster owing to human induced development activities in ecologically sensitive coastal regions shows the need to have ecologically permitted activities including agriculture. Earthquakes and associated Tsunamis are also experienced by these regions and the recent Tsunami incidence being on 26th December, 2004 has left immense damage in islands and Tamil Nadu. The coastal salinity and the associated low productivity of crops add to the woes of the region. In this scenario, ecologically benign livelihood avocations are highly pertinent for coastal areas that in agriculture include organic farming.

2. Status of Organic Farming in the Region

As per the organic farm products produced in and traded by the country during 2015-16 (Table 15.2), plantation crops especially tea, spices-condiments, medicinal, aromatic and herbal plants, tuber crops *etc.* are pertinent crops of coastal region that can be targeted for more organized production and trading on commercial scale. The vast forest resources in the region also provide scope for wild collections exploitation as organic produce. Milk and fish are the other important allied agricultural activities can also be produced and traded as organic produce. The coastal saline soils with low productivity can be put to organic farming effectively. The research investigations on organic farming in the region are reviewed below:

Table 15.2: Crops of Coastal Areas Produced under Organic Farming in India

Commodity	*Certified Organic Production (tonnes)*	
	Total	*Export*
Plantation crops	37656.6	5403 (tea)
Spices and condiments	22484.6	3085.5
Medicinal, Aromatic, Herbal plants	34819.2	2242.0
Tuber crops	1471.7	-

2.1 Organic Farming Studies in Plantation Crops

Studies on plantation crops have indicated that about 30-50 per cent of the total produce is available as biomass for recycling (Nampoothiri, 2001). At CPCRI, Kasargod, Kerala (Table 15.3), Subramanian *et al.* (2005) have found that coconut generates 11.67 t/ha recyclable biomass (leaves, spathe, bunch waste and husk *etc.*) under no fertilizer application, that got almost doubled (20-24 t/ha) under high density multispecies cropping system (HDMSCS). The recyclable biomass generation under areca nut was only 50 per cent as that of coconut *i.e.*, 5-6 t/ha in sole stand and 8-11 t/ha under HDMSCS (Hussain *et al.*, 2008). In Andaman and Nicobar Islands, the organic biomass (litter) generated in a year by coconut–clove–*Gliricidia* and coconut –nutmeg – *Gliricidia* plantation *i.e.* 14.9 and 14.4 t/ha contained 190-13-129 and 183-12-122 kg N-P-K nutrient (Pandey and Singh, 2010) whose recycling could meet 81–100 per cent N, 9–10 per cent P and 30–35 per cent K demands of these systems. The residues of each crop having differential decomposition rates needed 15 (*Glyricidia*) to 1093 days (coconut husk) to decompose fully. It means that there will be huge deficit of P and K nutrients (more so in ginger and turmeric) in organic mode of production that needs to be augmented through various means like vermicomposting and bioferilizers *etc.*

Table 15.3. Total Annual Biomass Available for Recycling from a ha of Coconut Based HDMSCS under different Fertilizer Levels

Fertilizer Level (Per cent recommended)	*Recyclable Biomass (t/ha) from Crop*				
	Coconut	*Clove*	*Banana*	*Pineapple*	*Total*
100	15.80	0.666	1.295	0.435	18.196
66.7	16.46	0.676	0.962	0.399	18.497
33.3	14.11	0.619	0.927	0.387	16.043
25	12.50	0.524	0.738	0.351	14.133
20	11.65	0.392	0.575	0.263	12.832
0	11.60	0.349	0.503	0.215	12.667

To hasten the decomposition the plantation crops residues/wastes like coconut leaves, vermicompost/vermiwash production technology was developed by CPCRI, Kasargod and their quality is given in Tables 15.4 and 15.5. The biomass produced from a hectare homestead farms when converted to vermicompsot could yield 35 - 60 kg N, 3.5 - 6.0 kg P and 6.5 - 12.0 kg K (Krishnakumar *et al.*, 2007). Use of the vermicompost (2 kg/plant) + vermiwash (2 litres/plant) so produced along with biofertilizers (NP) + *in situ* green manuring with cowpea + husk incorporation (at the time of planting and once in two years in between rows) + coconut leaf mulching in noni + coconut resulted in not only higher noni fruit yield (6.16 t/ha) but also increased the coconut yields by 25.5 per cent over 4 years as compared to sole coconut (Maheshwarappa *et al.*, 2017). In vanilla (*Vanilla planifolia*) + coconut cropping system, cow dung slurry (6 t/ha in 2 splits in 1:10 dilution) proved promising for vanilla yields (1.87 kg/tree) as compared to vermicompost, vermiwash, bio-fertilizers and their combinations. There has been significantly higher build up of fungal population and P-solubilisers with slurry application (Maheshwarappa *et al.*, 2016). Similarly better yield performance of black pepper and banana (Njalipoovan) in fully organic treatment under coconut plantations as compared to inorganic treatment were reported (Maheswarappa *et al.*, 2013). The slowly decomposing coir pith was composted and its application @12.5 t/ha in saline soils at Ramanathapuram, Tamil Nadu was found to improve soil physico-chemical properties besides improving finger millet productivity (Rangaraj *et al.*, 2007). Improved productivity of napier bajra hybrid fodder (Subramanian *et al.*, 2007) and pine apple (Subramanian *et al.*, 2009) with coir pith compost application under coconut gardens were reported. Incorporation of husk/coir pith in coconut + pineapple system was found to increase the soil fertility parameters as compared to control treatment (CPCRI, 2009). Coconut husk by combining with poultry manure and vermicompost was converted into a cheap liming material (Swarnam and Velmurugan, 2013) with good acidic soils reclaiming potential (Figure 15.2).

Table 15.4: Properties of Coconut Leaf Vermicompost

Parameter	*Value*
C:N ratio	9.95 – 17.0
Total C (per cent)	35-37
Organic C (per cent)	17-20
Humic acid (per cent)	10 -13
IAA (ppm)	0.52-1.15
GA (ppm)	0.23 – 1.61
Phenols (ppm)	10 - 14
WHC (per cent)	116-150
Total N (per cent)	1.8 – 2.1
Ca (ppm)	19,500 – 20,413
K (ppm)	1600 – 4013
Mg (ppm)	4290 – 4679
P (ppm)	2100 – 3043
Na (ppm)	1411 – 1525
S (ppm)	2915 – 3041
pH	6.2-7.9

Table 15.5: Characteristics of Coconut Leaf Vermiwash

Parameter	*Concentration (ppm)*
N	2.8
P	10.2
K	205
Ca	37.9
Mg	6.5
Fe	Traces
Cu	Traces
Zn	0.07
Mn	0.17
Biochemical constituents (μg/ml)	
Total sugars	61.6-111.2
Reducing sugars	41.9-88.4
Free amino acids	20.8-32.3
Proteins	615-890
IAA	0.52-1.15
Gibberellic acid (GA)	0.23-1.61
Total Phenols	10.2-14.8
pH	7.6-8.9

Parameter	Concentration (ppm)
Population of microorganisms (cfu/ml)	
Bacteria	$2x10^3$
Fungi	85
Actinomycetes	$9x10^3$
Phosphate solubilizers	$25x10^2$
N_2 fixers	15
Fluorescent *Pseudomonads*	$8x10^2$

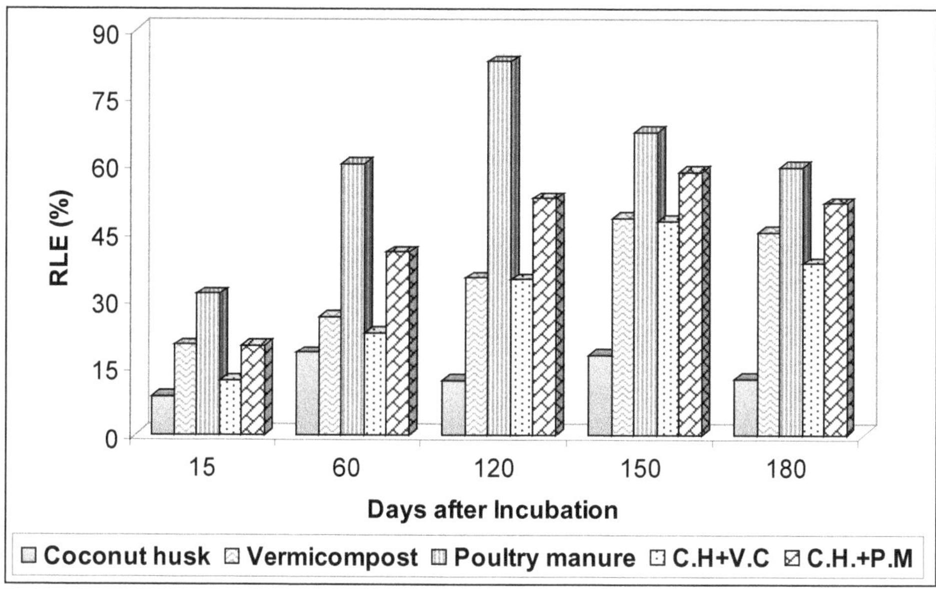

Figure 15.2: Relative Liming Efficiency (RLE) of different Organic Amendments on Acid Soil.

Studies of CPCRI regional station Vittal (Karnataka) on areca nut based intercropping systems (Sujatha *et al.*, 2011) with various short duration medicinal and aromatic plants under organic mode of production have established the compatibility of all crops with areca nut, except *Nilgirianthus ciliatus* that reduced the areca nut productivity by 32.6 per cent as compared to sole arecanut. However, all the intercrops enhanced system productivity (Table 15.6) and thus better economics.

The isolation of 2 efficient phosphate solubilising microorganisms *viz.*, *Pseudomonas* sp. and *Aspergillus niger* from the rhizosphere of coconut and cocoa with inorganic P solubilising capacity of 49.0 per cent and 49.7 per cent (Nair and Rao, 1977) have opened up avenues for identifying more such organisms and related studies. Fungal populations (AM fungi, *Glomus* spp., *Gigaspora* spp., and *Acaulospora* spp.) were found to be of maximum diversity in coconut based HDMSCS with banana and black pepper as intercrops in comparison to others (Ambili *et al.*, 2012). The encouraged microbial and enzymatic activities (Table 15.7) and these microbial populations were found to be more (Table 15.8) under organic mode of production (Thomas, 2010).

Table 15.6: Yields (kg/ha) of Medicinal and Aromatic Plants (MAPs) as Intercrops of Arecanut

Crop	*MAPs*	*Areca nut*	*System***
Vetiveria zizaniodes (Vetiver)	1006	2515	3231
Asparagus racemosus (Shatavari)	10666	2835	4359
Piper longum (Long pepper)	231	2718	2990
Bacopa monnieri (Brahmi)	2070	3586	4325
*Nilgirianthus ciliates** (Kurinji)	8104	1884	3313
Catharanhus roseus (Periwinkle)*	2711	3440	4144
Aloe vera (Aloe)	15490	3081	3552
Cymbopogon flexuous (lemon grass)	8460	3121	4338
Cymbopogon martini (palmarosa)	3249	2678	3164
Ocimum basilicum (basil)	8130	3311	3708
Pogostemon cablin (patchouli)	9082	3362	4225
Artemisia pallens (davana)	5248	3595	4224
Sole areca nut	-	2795	2795
LSD (0.05)	-	756	553

* Leaf + root yield ** Areca nut equivalent yield.

Table 15.7: Microbial and Enzyme Activities in Coconut Mono and Mixed Cropping

Parameter	*Values*
Microbial biomass (μg C/g soil)	
Monocrop	151.06
Mixed farming	181.98
Phosphatase (μg p-nitrophenol/g soil/h)	
Monocrop	31.16
Mixed farming	41.23
Dehydrogenase (μg formazan/g soil/h)	
Monocrop	5.02
Mixed farming	9.36
Bacteria $x10^6$	
Monocrop	2.27
Mixed farming	8.03
Fungi $x10^3$	
Monocrop	2.27
Mixed farming	9.33
Actinomycetes $x10^5$	
Monocrop	3.67
Mixed farming	9.50

Table 15.8: Microbial Activity in Coconut System under Various Nutrient Management Practices

Treatment	*Bacteria (10^5cfu/g soil)*	*Fungi (10^3cfu/g soil)*	*Actinomycetes (10^5cfu/g soil)*
Mono crop + recommended fertilizer	13.45	6.86	9.62
Mixed farming + 50 per cent organic +50 per cent inorganic	18.22	6.6	7.83
Mixed farming +100 per cent organic	23.17	18.16	11.33
Mixed farming + 100 per cent inorganic	11.76	7.12	7.00

2.2 Tuber and Rhizomatous Crops

Studies on standardization of package of practices for organic cultivation of elephant foot yam (EFY) under coconut plantations by ICAR-CIARI, Port Blair, Andaman and Nicobar Islands during 2012-13 indicated 300 g corm size planted at 90 cm x 90 cm as ideal. The field demonstrations (26) during 2012-17 on organic cultivation of "Gajendra" cultivar of EFY under coconut palms showed an EFY yield ranging from 15- 21.5 t/ha. Suja *et al.* (2006) observed not only higher yields of but also better quality of produce as evident from increased starch, crude protein contents and reduced oxalates in EFY produced in organic mode (Table 15.9). An increase in nut yield of 8.71 (12.91 per cent) due to intercropping of EFY with coconut was noticed compared to sole coconut cropping (Ghosh *et al.*, 2008). Ginger and turmeric were also found suitable intercrops of coconut plantation with high profits that too without impacting main crop of coconut (Sharma *et al.*, 1996).

Table 15.9: Comparative Performance of Elephant Foot Yam under different Modes of Production

Mode of Production	*Corm Yield (t/ha)*	*Starch (Per cent)*	*Oxalate (Per cent)*	*Crude Protein (Per cent)*
Conventional	54.6	13.63	0.204	1.74
Traditional	51.7	16.62	0.189	1.88
Organic	64.5	16.57	0.172	2.16
Biofertilizers alone	49.2	15.56	0.181	2.03

2.3 Tea and Coffee

Tea and coffee are the most important crops of hilly regions of coastal India with immense demand for organic produce. The success of organic coffee cultivation by Tribal's of Araku valley in Eastern Ghats region of Visakhapatnam district, Andhra Pradesh in lieu of their traditional shifting cultivation in forests in a cooperative mode have demonstrated its potential to transform their lives and economies. Over 7000 tonnes of organic coffee is produced by them during 2013-14 and is traded under brand name of *Araku*. Such development programmes are being taken up by Odhisa government too in Koraput district, and the same may be adopted in coffee growing coastal districts of Kerala (Kozikode and Mallapuram). Organic farming studies in tea have showed a decline in yield over years (Figure 15.3) and have 65.3 per cent higher cost of production *i.e.* 152.98 Rs/kg (Figure 15.4) when compared to its inorganic farming (92.53 Rs/kg). These cost escalations in organic tea production were ascribed to increases in variable and fixed costs (FAO, 2014). However, the premium price of organic produce from elite sections of the society may still favour its organic cultivation.

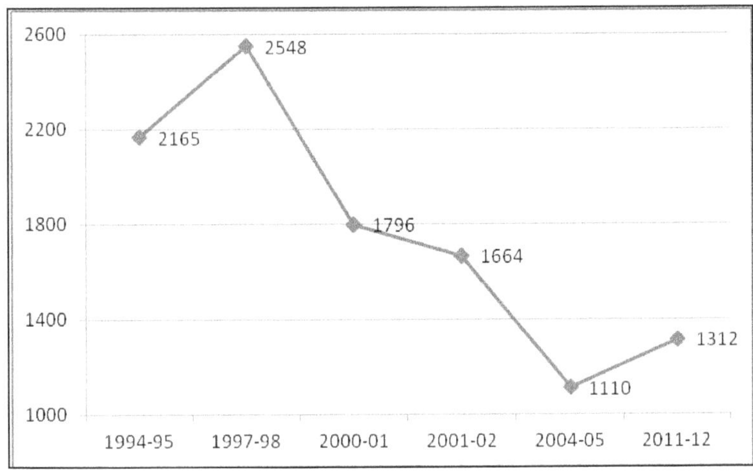

Figure 15.3: Yield (kg/ha) Trend of Organic Tea Garden in South India.

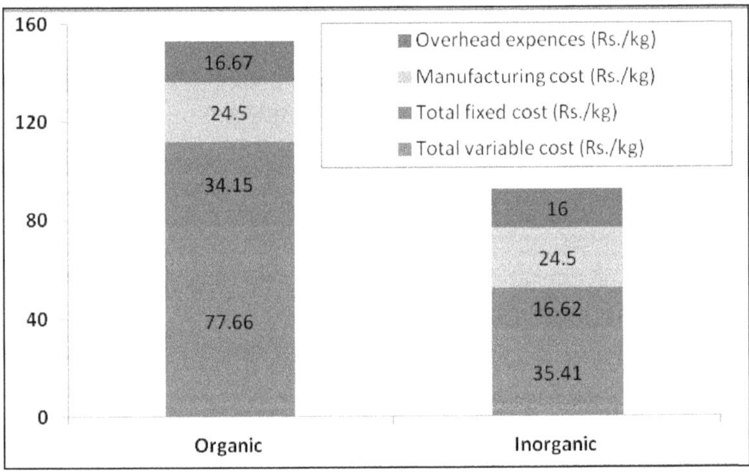

Figure 15.4: Cost of Production (Rs/kg) of Organic and Inorganic Tea.

2.4 Organic Livestock Farming and their Promotional Role in Organic Farming

Livestock production is not highly intensive as is the case of developed countries especially in dairying (Wolde and Tamir, 2016). Backyard poultry farming could also be explored as organic poultry production. The small holder's low input, crop residue and fodder based milk production system that contributes to 70 per cent of total milk production of the country can easily be made organic. However, these organic milk production systems may face unique nutrient management challenges especially for phosphorus whose deficiencies were reported in top soils layers (Loes and Ogaard, 2001). Therefore milk fever or hypocalcaemia or grass tetani is more prevalent in organic farming than in conventional farming systems (Patra, 2007). Inability to meet phosphorous demand by various means in organic farming from basically P deficient soil, can become a potential threat to the fertility of soil in organic dairy farms.

The vast quantity of manure and urine produced in the livestock production systems can be recycled in organic farming. Studies have indicated that entire maize fodder crop nutrient demands could be met from animal waste (dung and urine washings) recycling from a dairy unit (Gangaiah, 2004). Cow (especially desi cows) is the basic unit of organic farms from which Panchagavya and other products used in organic farms are produced. Studies on organic eggplant production (Swarnam *et al.*, 2016) indicated the enhanced growth and yield with foliar spray of panchagavya. Better performance of egg plant was ascribed to increases in the chlorophyll concentration of leaves and the associated promotion of photosynthesis. Further, organic egg plants so produced were found to have more phenolic content (15.7 per cent), carotenoids (24.8 per cent), DPPH activity (51 per cent) and ascorbic acid (1.1 per cent) as compared to the control.

2.5 Exploitation of Oceans for Promoting Organic Farming

Sea weeds (Marine macro algae), the non-flowering plants commonly found in intertidal as well as in the sub-tidal regions of the sea (Domettila *et al.*, 2013) constitute one of the commercially important renewable marine living resources (Rajasree and Gayathri, 2014). Seaweed farming offers array of benefits to the coastal communities and entrepreneurs. Seaweed aquaculture is also seen as sound strategy for coastal developing nations to contribute to climate change mitigation (Duarte *et al.*, 2017). Seaweeds besides their utility as source of polysaccharides known as phycocolloids such as agar, carrageenan and align also contributes to livestock feed additives and fertilizers. The use of sea weeds as soil amendment and fertilizer by coastal communities is well known (Rebours, 2014) and thus could aid in promoting organic farming. India annually produces 3,01,646 tonnes marine algae (Chennubhotla *et al.*, 2013). Scientific experiments have shown that sea weed product can improve and enhance various aspects of plant growth, development (Crouch and Staden, 2008) yield and quality (Zodape *et al.*, 2008) owing to the plant hormones (mainly cytokinins) and trace nutrients present in the extracts (Verkleij, 2012). Few field studies on sea weeds role was done in India like Singh *et al.* (2015) and Layek *et al.* (2015) which established the utility of *Kappaphycus alvarezii* and *Gracilaria edulis* foliar spray in enhancing the maize productivity. The cultivation technology of the seaweeds, namely, Eucheuma (*Kapppaphycus*) and *Gracilaria*, figuring in top 5 cultivated seaweed species in the world has been standardized by CSIR-CSMCRI, Bhavnagar for growing in Indian coastal waters. The institute has also developed the technology to extract phyco-colloids (carageenan, agar/agarose) and biostimulant from these sea weeds. The biostimulant developed has been tested in 20 states of India at 43 centres in collaboration with SAUs, ICAR Institutes and farmers with very encouraging results. The results of this multi-crop multi-locational trials demonstrated an improvement of 10-37 per cent in yields of the crop. The biostimulants of these two seaweed species are given in Table 15.10.

Table 15.10: Composition of *K. alvarezii* and *G. edulis* Sap

Constituent	*Amount in mgL^{-4}*	
	Kappaphyeus alvarezii Sap	*Gracilaria edulis Sap*
Indole 3-acetic acid (IAA)	27	8.7
Zeatin	20	3.1
Gibberellin (GA_3)	24	ND
Choline	57	36
Glycine betaine	79	63
Betain aldehyde	Present	Present
N^+	198	1952
K^+	33654	682
Ca^{2+}	321	352
Mg^{2+}	1112	311
Zn^{2+}	4.7	0.63
Mn^{2+}	2.1	33
Fe^{2+}	86	13
Cr^{3+}	32	0.20
Cu^{2+}	0.65	0.04
Ni^{3+}	3.5	0.21
P^{3+}	17	ND

Source: Layek *et al.* (2015).

Fish wastes from both marine and fresh water generated in huge quantity in coastal regions could become important resources for promoting organic farming (supply of nutrients and also plant protection compounds). It is estimated that the catching and processing of fish generates > 2 million tonnes of waste in the form of viscera, fins, scales and bones (Mahendrakar, 2000). These aquatic waste could be utilized through three most common methods *i.e.* i) manufacture of fish meal/ oil, ii) production of silage and iii) manufacture of organic fertilizer. The later way could aid in furthering the cause of organic farming in the coastal regions. Studies at Centre for Ocean Research, Sathyabama University, Chennai, Tamil Nadu, India have indicated that marine trash fish (MTF) by chopping, treating with molasses and its bacterial degradation (*Bacillus subtilis*) could become promising organic liquid fertilizer for tomato. The fatty acids present in such fertilizer were also found to enrich organic carbon (OC) content of soil (Aranganathan and Radhika Rajasree, 2016).

2.6 Organic Aquaculture

The marine fish catch is naturally produced food that is devoid of any chemical input application except pollution load. The freshwater aquaculture in farms is mostly in inorganic mode of production. The fish coming from the ponds, lakes, natural streams, and rivers is receiving chemical inputs from runoff waters. There is emerging market for organic fish production world over. To capitalise on the market, during 2007 organic aquaculture activity was started in India in Andhra Pradesh and Kerala states through India Organic Aquaculture Project (IOAP) by Marine Products Export Development Agency (MPEDA) with the technical consultancy from Blueyou of Swiss Import Promotion Programme (SIPPO), Switzerland with Naturland of Germany as the certifying agency and INDOCERT in Kerala as the inspection body (Dube and Chanu, 2012). The project has also successfully developed certified stakeholders in organic aquaculture such as certified organic scampi hatchery (M/s Rosen Fisheries, Thrissur), certified organic feed mill (M/s Waterbase Limited, Nellore) and certified organic seafood processor (M/s Jagadeesh Marine Exports, Bhimavaram). Under the IOAP, scampi or giant freshwater prawn (*Macrobrachium rosenbergii*) farms with a water spread area of 20 ha in Kuttanad, Kerala owned by four individual farmers have been stocked with 3,40,000 organic scampi seeds. Organic farming operations have been successful and the world's first organic scampi harvest has taken place during 2008 at Kuttanad, Kerala with a production of about 1.5 t from each farm (Dube and Chanu, 2012). Further studies on organic aquaculture in Kuttanad area under IOAP scheme resulted in evolving rotational cropping of rice and freshwater prawns with 20 per cent higher revenues (Nair *et al.*, 2014). In Andhra Pradesh, organic scampi farming was carried out by the farmers of two societies in West Godavari District with the support of National Centre for Sustainable Aquaculture (NaCSA) and IOAP, MPEDA. A total of 27 farmers from Sri Venkateswara Aqua Farmers Welfare Society, Matsyapuri and Sri Sainadha Aqua Farmers Welfare Society, Velivela were involved in the project covering 31 ha area. Both the societies were cluster certified by Naturland for organic scampi farming which is the first of its kind in aquaculture. A total of 12.6 tons of organic scampi was produced during 2009 with an average production of 407 kg/ha and profit of Rs. 55,495/ha from these societies (Anonymous, 2010). An experiment on the efficacy of organic carp farming found that feed conversion ratio, specific growth rate and total production of Indian major carps (IMCs) were significantly higher in organic culture system than conventional and normal culture systems. Water quality parameters were found to present in the desirable range in organic culture system which indicates that carp farming in organic culture system will improve the fish production and also maintain good water quality (Beg *et al.*, 2016).

2.6.1 Integrated Mangrove Fishery Farming System (IMFFS) for Mangroves

Mangroves that are the protective covers of coastal ecosystems are increasingly threatened by erosion, mining, commercial shrimp farming, sea level rise *etc.* They are spread on 380 km along the coasts of India on 0.3555 m ha. Majority of this is on East coast (82 per cent) and rest 18 per cent on West coast (Ramakrishnan, 2008). Similarly, there are estuaries that are spread over 1.25 m ha along the coast of all maritime states including Andaman and Nicobar Islands (52 per cent back waters in Odhisa and West Bengal, 19.7 per cent in Kerala, rest in other 6 states and 1 UT). Integrated Mangrove Fishery Farming System (IMFFS) has been evolved to utilize mangroves (that holds true for estuaries) effectively. The basic idea of IMFFS is to convert saline areas that cannot be used for aquaculture or other livelihoods into productive land once more. In this system, ponds are constructed in degraded mangrove regions and suitable halophytes or mangrove species are planted on the bunds and brackish water fish are introduced in the pond (Figure 15.5). Ravindran *et al.* (2007) found that *Suadera maritime* and *Sesuvium portulacastrum* promising halophytes for saline soils. Grown up halophyte vegetation will continue to provide mangrove ecosystem services (protection from cyclone, sea water surge) while the fish is frequently harvested leading to sustainable income generation. These systems are tested at Rameshwaram (Tamil Nadu) and Sorlagondi, Nalli and

Figure 15.5a: IMFFS.

Figure 15.5b: Sea Bass Harvest from IMFFS.

***Source*: IUCN and MSSRF.**

Basavanipalem Villages of Krishna Mangrove Wetlands area of Andhra Pradesh by M. S. Swaminathan Research Foundation (MSSRF) in collaboration with The International Union for Conservation of Nature (IUCN). Such interventions are also highly relevant and required for coastal ecosystems and are organic by default. In Rann of Kutch and other places of Gujarat suitable interventions like sea weed culture/salt production from sub-soil brine needs adoption.

Salt production (22.1 m t during 2012-13) is most promising and exclusive economic activity of the coastal regions with domestic and export market. Gujarat with 75 per cent of countries salt production followed by Tamil Nadu (11 per cent) and other states (AP, WB, *etc.*) are major players of salt trade. Though sea water is an inexhaustible source of salt, its production along the coast is limited by weather and soil conditions. Marine salt production that is going on along the coast of Gujarat (Jamnagar, Mithapur,Jhakhar, Chira, Bhavnagar, Rajula, Dahej, Gandhidham, Kandla, Maliya, Lavanpur) can further be intensified and sub-soil brine based salt production is important in Rann of Kutch (Kharaghoda, Dhrangadhra; Santalpur). Scope also lies for salt production along coasts in Maharashtra (Bhandup, Bhayandar, Palghar), Tamil Nadu (Tuticorin, Vedaranyam, Covelong), Andhra Pradesh (Chinnaganjam, Iskapalli, Krishnapatnam, Kakinada and Naupada), Orissa (Ganjam, Sumadi) and West Bengal (Contai). Agro-chemicals production from such salts for use in organic farming needs attention.

3. Future Prospects

The continued and ever increasing demand for organic products especially spices, tuber crops and other plantation crops *etc.* that are the principal crops of coastal regions calls for its capitalization. The vast waste generated from plantation crops through vemicompsting technology utilizing animal manures could meet the nutrient demands of the crops. However, there is need to balance phosphorus and potassium and other deficient nutrients through use of permitted organic fertilizers like rock phosphate and sulphate of potash. Bio-fertilizers including foliar spray of sea weeds extracts may help in this matter. Fish wastes could be utilized in organic farming and there is need to protect the resource base especially in saline coasts through suitable interventions. Islands regions could be immediately declared organic with suitable certification mechanisms of produce. In West coast regions and in tribal inhabited coastal areas, organic farming can be adopted with suitable conversion approach. There is need to go for value addition and organized marketing after thorough certification. The ecologically sensitive coastal regions could be used sustainably through organic farming.

REFERENCES

Ambili K, Thomas GV, Indu P, Gopal M and Gupta A. 2012. Distribution of arbuscular mycorrhizae associated with coconut and arecanut based cropping systems. Agricultural Research, 1(4): 338-345.

Anonymous. 2010. An update on organic scampi aquaculture in Andhra Pradesh. Aquaculture Asia Magazine, XV(1): January-March, 2010, pp. 14-17.

Aranganathan L and Radhika Rajasree SR. 2016. Bioconversion of marine trash fish (MTF) to organic liquid fertilizer for effective solid waste management and its efficacy on Tomato growth. Management of Environmental Quality: An International Journal, 27(1): 93-103.

Beg MM, Mandal B, Moulick S, Mukherjee CK and Mal BC. 2016. Efficacy of organic fish farming for sustainability. International Journal of Advanced Biological Research, 6(2): 180-187.

ChennubhotlaVSK,UmamaheswaraRao M and Rao KS. 2013. Commercial importance of marine macro algae. Seaweed Research and Utilization, 35(1 and 2): 118-128.

CPCRI (Central Plantation Crops Research Institute). 2009. Annual Report 2008-2009. Central Plantation Crops Research Institute, Kasaragod, India, 145p.

Crouch IJ and Staden JV. 1993. Commercial seaweed products as biostimulants in horticulture. Journal of Home and Consumer Horticulture, 1(1): 19-76.

Domettila C, Brintha T, Sukumaran S and Jeeva S. 2013. Diversity and distribution of seaweeds in the Muttom coastal waters, South-West coast of India. Biodiversity Journal, 4: 105-110.

Dube K and Chanu TI. 2012. Organic aquaculture: way to sustainable production. In: *Advances in Fish Research* (Goswami, UC, Ed.), Narendra Publishing House, New Delhi, pp. 219-229.

FAO. 2001. Farming Systems and Poverty: Improving Farmers Livelihoods in a Changing World. (Malcolm Hall, Principal Editor), A Joint publication of FAO and World Bank, Rome, Italy.

FAO. 2014. Report of FAO working group (India, China, Bangladesh, Japan and Sri Lanka) on organic tea, Intercessional Meeting of the Intergovernmental Group on Tea held at Rome during 5-6 May 2014.

Gangaiah B. 2004. Effect of animal urine application on growth and yield of fodder maize (Zea mays). Indian Journal of Agricultural Sciences, 74(12): 678-679.

Ghosh DK, Hore JK and Bandopadhay A. 2008. Standardization of spacing and size of planting material of elephant foot-yam grown as intercrop in coconut plantation. Indian Journal of Horticulture, 65(1): 44-47.

Hussain M, Ray AK, Maheswarappa HP, Krishnakumar V, Ravi Bhat, Subramanian P and George V. Thomas. 2008. Recycling of organic biomass from arecanut based high density multi-species cropping system models under Assam condition. Journal of Plantation Crops, 36(1): 53-57.

Krishnakumar V, Reddy DVS, Thamban C and Sairam CV. 2007. Restructuring of homestead farms for sustainable income and employment opportunities. Journal of Plantation Crops, 35(3): 146-151.

Layek J, Das A, Ramkrushna GI, Trivedi K, Yesuraj D, Chandramohan M, Kubavat D, Agarwal PK and Ghosh A. 2015. Seaweed sap: a sustainable way to improve productivity of maize in North east India. International Journal of Environmental Studies, 72(2): 308-315.

Loes AK and Ogaard AF. 2001. Long-term changes in extractable soil phosphorus (P) in organic dairy farming systems. Plant and Soil, 23: 321-332.

Mahedrakar NS. 2000. Aqua feeds and meat quality of cultured fish. Biotech consort India Ltd, New Delhi. pp.26-30.

Maheswarappa HP, Subramanian P and Geetha Kumari A. 2017. Influence of organic nutrient management practices for noni (*Morinda citrifolia*) when grown as an intercrop in coconut garden. Indian Journal of Agricultural Sciences, 87(9): 1155-1157.

Maheswarappa HP, Krishnakumar V, Alka Gupta and Geetha Kumari A. 2016. Influence of organic sources of nutrients on vanilla (*Vanilla planifolia*) as an intercrop in coconut garden. Journal of Plantation Crops, 44(2): 85-89.

Maheswarappa HP, Dhanapal R, Subramanian P and Palaniswami C. 2013. Evaluation of coconut based high density multi-species cropping system under organic and integrated nutrient management. Journal of Plantation Crops, 41(2): 130-135.

Nair CM, Salin KR, Joseph J, Aneesh B, Geethalakshmi V and New MB. 2014. Organic rice-prawn farming yields 20 per cent higher revenues. Agronomy for Sustainable Development, 34(3): 569-581.

Nampoothiri KUK. 2001. Organic farming – Its relevance to plantation crops. Journal of Plantation Crops, 29(1): 1-9.

Pandey CB and Singh, RK. 2010. Organic black pepper (*Piper nigrum*) cultivation on *Gliricidia* standards in coconut (*Cocos nucifera*) plantations in South Andaman: organic matter production and recycling of nutrients. Indian Journal of Agricultural Sciences, 80 (11): 1–10.

Patra AK. 2007. Nutritional management in organic livestock farming for improved ruminant health and production – an overview. Livestock Research for Rural Development,19: 75-98.

Rajasree R and Gayathri, S. 2014. Women enterprising in seaweed farming with special reference fisherwomen widows in Kanyakumari district Tamil Nadu, India. Journal of Coastal Development, 17: 12-15.

Ramakrishnan, Korakandy. 2008. Fisheries Development in India: The Political Economy of Unsustainable. Gyan Publishing House, 400 p.

Rangaraj, Somasundaram E, Mohamed Amanullah M, Thirumurugan V, Ramesh S. and Ravi S. 2007. Effect of Agro-Industrial Wastes on Soil Properties and Yield of Irrigated Finger Millet (*Eleusine coracana L. Gaertn*) in Coastal Soil. Research Journal of Agriculture and Biological Sciences, 3(3): 153-156.

Ravindran KC, Venkatesan K, Balakrishnan V, Chellappan KP and Balasubramanian T. 2007. Restoration of saline land by halophytes in Indian soils. Soil Biology and Biochemistry, 39: 2661-2664.

Rebours. 2014. Seaweed – a resource for organic farming. Bioforsk FOKUS, 9(2): 107.

Sehgal J, Mandal DK, Mandal C and Vadivelu S. 1992. Agro-ecological zones of India. Second Edition. Nagpur, India. Technical Bulletin, No. 24. NBSS and LUP (ICAR). 130 p.

Sharma R, Prasad S, Mohan NK and Medhi G. 1996. Economic feasibility of growing some root and tuber crops under intercropping system in coconut garden. The Horticulture Journal, 9(2): 167-170.

Singh S, Singh MK, Pal SK, Thakur R, Zodape ST and Ghosh A. 2015. Use of seaweed sap for sustainable productivity of maize. The Bioscan, 10(3): 1349-1355.

Subramanian P, Dhanapal R, Palaniswami C and Joseph S. 2007. Feasibility studies on growing hybrid bajra napier fodder grass as intercrop in coconut under coastal sandy soil. Journal of Plantation Crops, 35(1): 19-22.

Subramanian P, Dhanapal R, Palaniswami C, Maheswarappa HP, Gupta A and Thomas GV. 2009. Coastal sandy soil management for higher coconut productivity. Technical Bulletin No. 50, Central Plantation Crops Research Institute, Kasaragod, 20p.

Subramanian P, Srinivasa Reddy DV, Palaniswami C, Gopalasundaram P and Upadhyay AK. 2005. Studies on nutrient export and extent of nutrient recycling in coconut based high density multispecies cropping. CORD, 21(1): 20-27.

Suja G, Nayar TVR, Potty VP and Sundaresan S. 2006. Organic farming a viable strategy for high yield and quality tuber crop production. Indian Horticulture, 51(6): 4-5.

Sujatha S, Ravi Bhat, Kanna C and Balasimha D. 2011. Impact of intercropping of medicinal and aromatic plants with organic farming approach on resource use efficiency in areca nut (*Areca catechu* L.) plantation in India. Industrial Crops and Products, 33: 78-83.

Swarnam TP and Velmuruan A. 2014. Evaluation of coconut wastes as an amendment to acidic soils in low-input agricultural system. Communications in Soil Science and Plant Analysis, 45: 1–12.

Swarnam TP, Velmurugan A, Jaisankar I and Nutan Roy. 2016. Effect of foliar application of panchagavya on yield and quality characteristics of Eggplant (*Solanum melongena L*). Advances in Life Sciences 5(7): 2636-2639.

Thomas GV. 2010. Technological advances in organic farming of plantation crops In: *Organic Horticulture- Principles, Practices and Technologies* (Singh HP and George V Thomas, Eds.), Westville Publishing House, New Delhi. pp. 440,

Verkleij, F N 1992. Seaweed extracts in agriculture and horticulture: a review. Biological Agriculture and Horticulture, 8(4): 309-324.

Wolde DT and Tamir B. 2016. Organic Livestock Farming and the Scenario in the Developing Countries: Opportunities and Challenges. Global Veterinaria, 16: 399-412.

Zodape ST, Kawarkhe VJ, Patolia JS and Warade AD. 2008. Effect of liquid seaweed fertilizer on yield and quality of okra. Journal of Scientific and Industrial Research, 67: 1115-1117.

Chapter 16

Relevance of Agroforestry Systems in Organic Farming

G. Venkatesh, K.A. Gopinath, V. Visha Kumari and K. Adilakshmi

ICAR-Central Research Institute for Dryland Agriculture, Hyderabad – 500 059, Telangana
E-mail: mamootii73@gmail.com

1. INTRODUCTION

Organic farming offers an alternative that can eliminate many of the environmental issues of conventional agriculture. Agroforestry is a soil fertility enhancement land use system in which woody perennials (trees, shrubs *etc.*) are grown in association with herbaceous plants (crops and pasture) or livestock in spatial arrangement, in a rotation or both. There are both ecological and economical interaction between trees and others components of the system (Lundgren and Raintree, 1982). Hence, practising and adopting agroforestry systems in conjunction with organic farmning may make organic farming more affordable system in India.

The steady decrease in soil fertility due to many drivers is a serious constraint for sustainable agriculture in India. Topsoil erosion is the most detrimental form of soil degradation and is likely to be aggravated by long-term removal of surface litter and crop residues. The shortage of mineral fertilizers and poor performance of current agricultural policies have directed discussions on food security towards sustainable farming practices (Muchena *et al.*, 2005; Kiptot and Franzel, 2012). Agroforestry in conjuction with organic farming has potential to improve soil fertility. This is mainly based on the increase of soil organic matter and biological nitrogen fixation by leguminous trees. Trees on farms also facilitate tighter nutrient cycling than mono- culture systems, and enrich the soil with nutrients and organic matter (Lehmann *et al.*, 1998), while improving soil structural properties. Hence, through water tapping and prevention of nutrient leaching (Bayala *et al.*, 2008), trees help recover nutrients, conserve soil moisture and improve soil organic matter (Duguma and Hager, 2011). The potential of agroforestry components to reduce the yield gap varies depending on the biophysical and human context. There are a number of successful agroforestry technologies, such as trees that improve soil, fast-growing trees for fuel wood, indigenous fruit trees to provide added nutrition and income, and trees that can provide medicinal plant products (Molua, 2005). In practice, there is a need to differentiate between simple agroforestry systems (such as alley cropping, intercropping and hedgerow systems) and complex agroforestry systems that function like natural forest ecosystems but are integrated into agricultural management systems (Oke and Odebiyi, 2007). The interest of investigating agroforestry integeration is gaining importance in a changing climate, comes from the potential of agroforestry practices to produce assets for farmers, combined with opportunities for climate change mitigation and potential to promote sustainable production that enhances agroecosystem diversity and resilience.

2. Agroforestry Practices in India

The agroforestry systems practised in India include trees on farms, community forestry and a variety of local tree based practices. In India, the practice of growing scattered trees on farmlands is quite old and has not changed much over centuries; these trees are multipurpose, used for shade, fodder, fuel wood, fruit, vegetables and medicinal uses. Trees like *Eucalyptus* and *Populus* are also grown in agricultural fields or on field bunds often on farm boundaries in Punjab and Haryana. Shifting cultivation in the Northeast India and Taungya cultivation in Kerala, West Bengal, and Uttar Pradesh and to a limited extent in Tamil Nadu, Andhra Pradesh, Orissa, Karnataka, as well as in the Northeast hill regions are examples of traditional Indian agroforestry systems. Besides, home gardens, wood lots, large cardamom plantations of Eastern Himalayas and plantations elsewhere and Alder based agriculture in Northeast India are other kinds of agroforestry systems (Dhyani *et al.*, 2009).

On the basis of nature of components, Dhyani *et al.* (2009) have reported the common twenty agroforestry systems being practised in different agroecological regions of India and these are:

1. Agri-silviculture (trees + crops)
2. Boundary plantation (trees on boundary + crops)
3. Block plantation (block of trees + block of crops)
4. Energy plantation (trees + crops during initial years)
5. Alley cropping (hedges + crops)
6. Agri-horticulture (fruit trees + crops)
7. Agri-silvi-horticulture (trees + fruit trees+crops)
8. Agri-silvipasture (trees+crops + pasture or animals)
9. Silvi-olericulture (trees + vegetables)
10. Horti-pasture (fruit trees + pasture or animals)
11. Horti-olericulture (fruit trees + vegetables)
12. Silvi-pasture (trees+pasture/animals)

13. Forage forestry (forage trees + pasture)
14. Shelter-belts (trees + crops)
15. Wind-breaks (trees + crops)
16. Live fence (shrubs and under- trees on boundary)
17. Silvi or Horti-sericulture (trees or fruit trees + sericulture)
18. Horti-apiculture (fruit trees + honeybee)
19. Aqua-forestry (trees + fishes)
20. Homestead (multiple combinations of trees, fruit trees, vegetable *etc.*).

The prominent agroforestry systems from different regions of the country are enumerated and presented in Table 16.1.

Table 16.1: Predominant Agroforestry Systems

Agroclimatic Zone	*Agroforestry Systems*	*Tree Component*	*Crop/Grass*
Western Himalayas	Silivipasture (RF)	*Grewia optiva*	*Setaria* spp.
		Morus alba	*Setaria* spp.
	Agrihorticulture	*Malus pumila*	Millets, Wheat
	Agrihorticulture	*Prunus persica*	Maize, Soyabean
Eastern Himalayas	Agrisilviculture	*Anthocephalus cadamba*	Rice
	Agrihorticulture	*Alnus nepalensis*	Large Cardamon/Coffee
	Silivipasture	Bamboos, *Parkia roxburghii, Morus alba*	-
	Silivipasture		Napier
Lower Gangetic plains	Agrisilviculture (Irri)	*Eucalyptus, Albizia lebbeck*	Rice
	Agrihorticulture (Irri)	Mango/Banana/Litchi	Wheat, paddy,Maize
	Silivipasture	*Morus alba, Albizia lebbeck*	*Dicanthium, Pennisetum*
Middle Gangetic plains	Agrisilviculture (Irri)	*Populus deltoides*	Sugarcane-Wheat
	Agrisilviculture (Irri)	*Eucalyptus* spp.	Rice-Wheat
	Agrisilviculture	*Dalbergia sissoo*	Sesamum
	Agrihorticulture (Irri)	Mango/Citrus	Rice-Wheat
	Silivipasture	*Albizia lebbeck*	*Chrysopogan, Dicanthium*
Trans Gangetic plains	Agrihorticulture (Irri)	*Emblica officinalis*	Black gram/Green gram
	Agrisilviculture	*Azadirachta indica*	Black gram –Wheat/ Mustard
	Silivipasture	*Bauhinia variegate, Albizia lebbeck*	*Cenchrus, Pennisetum*
Upper Gangetic plains	Agrisilviculture (Irri)	*Populus deltoides*	Wheat, Bajira, fodder
	Agrisilviculture (Irri)	*Eucalyptus* spp.	Rice-Wheat
	Silivipasture	*Bauhinia variegate, Albizia lebbeck*	*Chrysopogan*, Poa

Agroclimatic Zone	*Agroforestry Systems*	*Tree Component*	*Crop/Grass*
Eastern Plateau and Hills	Agrisilviculture	*Gmelina arborea*	Rice, Linseed
	Agrisilviculture	*Acacia nilotica*	Rice
	Silivipasture	*Acacia mangium*, Bamboos	-
	Silivipasture	*Leucaena leuocephala*	*Chrysopogan, Pennisetum, Dicanthium*
Central Plateau and Hills	Agrihorticulture (Irri)	*Psidium gujava*	Bengal Gram/Groundnut
	Agrihorticulture (RF)	*Emblica officinalis*	Black gram/Green gram
	Agrisilviculture	*Acacia nilotica/Leucaena leuocephala/Azadirachta indica/Albizia lebbeck*	Soyabean, Black gram-Mustard/Wheat
	Silivipasture (RF and degraded lands)	*Albizia amara, Leucaena leuocephala, Dichrostycus cinerea*	*Chrysopogan, Stylosanthes hamata, Stylosanthes scabra*
	TBOs (RF)	*Jatropha curcas*	-
Western Plateau and Hills	Agrihortisilviculture	*Tectona grandis, Achrus zapota*	Rice, Maize
	Agrihorticulture	*Areca catechu*	Black pepper, cardamom
	Silivipasture	*Prosopis juliflora, Ailanthus excelsa*	-
	Silivipasture	*Acacia mangium, Albizia amara*	*Cenhrus*
Southern Plateau and Hills	Agrisilviculture (RF)	*Eucalyptus* spp., *Casuarina equisetifolia, Ailanthus excelsa*	Cotton, Groundnut
	Agrisilviculture (Irri)	*Eucalyptus tereticornis, Melia dubia*	Chilli
	Silivipasture (RF)	*Leucaena leuocephala, Acacia leucopholea*	-
		Eucalyptus spp.	-
	Agrihorticulture	*Tamarindus indica*	Chilli
	TBOs	*Pongamia pinnata*	-
East coast plains and Hills	Agrisilviculture (RF)	*Ailanthus excelsa, Acacia leucopholea*	Cow pea
	Silivipasture	*Casuarina equisetifolia, Leucaena leuocephala*	-
	TBOs	*Pongamia pinnata*	-
	Silivipasture	*Artocarpus* spp.	*Chrysopogan*, Napier, *Cenchrus*
West coast plains and Hills	Agrisilviculture (RF)	*Acacia auriculiformis*	Black pepper
	Agrihorticulture (RF)	*Artocarpus heterophyllus*	Black pepper
	Agrisilviculture (RF)	*Acacia auriculiformis*	Ricw

Agroclimatic Zone	Agroforestry Systems	Tree Component	Crop/Grass
	Agrihorticulture	*Cocos nucifera/Areca catechu*	Rice
	Agrisilviculture	*Casuarina equisetifolia*	Rice
	Silivipasture	*Hardwickia binnata, Albizia lebbeck*	*Cenchrus*
Gujarat coast plains and Hills	Agrisilviculture	*Azadirachta indica, Ailanthus excelsa*	Cow pea, Green gram
	Silviculture	*Prosopis juliflora, Acacia nilotica*	-
	Silivipasture	*Leucaena leuocephala*	*Cenchrus, Setaria* spp
Western Dry Region	Agrisilviculture	*Prosopis cineraria, lecomella undulate, Acacia nilotica, Azadirachta indica*	Pearl millet
	TBOs	*Jatropha curcas*	-
	Silviculture	*Albizia lebbeck, Hardwickia binnata*	*Cenchrus*
All Islands	Agrihorticulture	*Cocos nucifera*	Rice
	Silivipasture	*Bauhinia* spp, *Erythrina indica, Leucaena leuocephala*	*Cenchrus, Pennisetum*

Source: Dhyani *et al.* (2009).

3. Agroforestry as a Changeover Solution

An important question that arises in the discussion on agroforestry in the context of organic farming is: What is the reason for incorporating multipurpose trees into agriculture? One reason is that when multipurpose trees and crops complement each other functionally and structurally, the performance of at least one component is improved. An example is bacteria symbiotic on the roots of the trees, fix nitrogen from the atmosphere, and this nitrogen then becomes available to the annual plants, either through leaf litter fall, or through sloughing off of roots. Trees also may bring up nutrients like potassium from soil horizons below the reach of the annual plants roots, and make it available through litter fall to the annual plants. Other benefits of trees include prevention of soil erosion and increasing the availability of phosphorus. Phosphorus is often bound in insoluble compounds, but organic acids leached from leaf litter of trees renders this phosphorus into a form available to plants (Jordan, 1995). Alleycropping is a type of agroforestry (Nair, 1993). In the type of alleycropping that has been developed for the tropics, leguminous trees or shrubs are planted in hedgerows that border strips or 'alleys' in which an economic crop such as vegetables or grains are grown (Nair, 1993). Alley cropping benefits the economic crop when the branches or roots of the hedgerow species are pruned. As the leaf litter and roots decompose, nutrients, especially nitrogen, contained in them are released to the soil and become available to the crop plants inside the alley. The system can benefit farmers where fertilizers are not available, or as in the case of organic agriculture, commercial fertilizers are not permitted.

3.1 Solving Problems of Organic Agriculture

Applying mulch to control weeds in the soil is a common method in organic farming. The layer of organic matter inhibits the germination of weed seeds in the mineral soil below. If the seeds do germinate, the mulch prevents light from reaching the seedling, and forms a barrier that the seedling must penetrate to survive. Mulch can result from a winter cover crop that is mowed or flattened, or from mulch such as straw brought in from other parts of the farm (Gliessman, 1998). While living mulches can protect the soil (Alley *et al.*, 1999), they are not practical when the economic crops are vegetables or row crops, because of competition. Mulch must be imported, but green manure that has been harvested usually does not last throughout the entire growing season. Once the mulch has decomposed, weeds will appear unless more mulch is supplied. However, if the crops are growing in an alleycropping system, it is relatively easy to apply new mulch. Prunings fall directly from the hedgerow into the alley. Many leguminous trees and shrubs continue putting on new growth throughout the season and can be pruned several times, thereby providing a continuous mulch cover to suppress weeds, and making an important addition to the nutrient supply for the economic crop. In a study of root dynamics of hedgerow species, Peter and Lehmann (2000) found that pruning of above-ground growth can result in the increased dieback of roots, resulting in decreased competition with crop species, and increased nutrient availability as the dead roots decompose. Alleycropping can also help lessen insect herbivory. There are several organically acceptable approaches to insect control that are facilitated by the hedge species in an alleycropping system. One is to increase the habitat for predators and parasites on the pest species. Pests have a more difficult time locating, remaining on, and reproducing on their preferred hosts when these plants are spatially dispersed and masked by confusing visual and chemical stimuli presented by a system such as alleycropping (Altieri, 1995).

3.2 Organic Inputs from Agroforestry Systems

Several studies highlight the potential for multipurpose trees to modify soil chemistry, but the use of trees as sources of organic inputs (*e.g.* prunings/green manure and compost) with the intention of enhancing soil quality and productivity is yet another unexplored area. For organic inputs from agroforestry species to have a significant impact on soil and crops (annual or bi-annual), they firstly, need to have sufficient biomass and nutrients to meet crop requirements (Palm, 1995). This depends on numerous factors including species used, provenance, density of trees, climate, soil fertility, season, age of leaves or plant, and frequency of pruning (Palm, 1995). Secondly, the nutrient release during decomposition of the materials should be synchronized with crop demand (Myers *et al.*, 1994).

The quality (of plant materials) is a function of the initial concentrations and relative proportions of nutrients (*e.g.* N, P), lignin (LG), polyphenols (PL), and to some extent also soluble carbohydrates and cellulose (Palm and Rowland, 1997). Materials with high initial N and low lignin and polyphenols concentrations are presumed to decompose and release nutrients fast and are therefore considered to be of 'high quality. Those with the opposite characteristics are considered low quality; to decompose slowly and often show net immobilization of nutrients (Young, 1997).

In addition to litter quality, management factors (*e.g.* method and timing of application, loading rates and quantities of prunings added) variably affect decomposition of prunings and uptake of released nutrients by crops (Kumar and Goh, 2000). The timing of prunings application is another critical factor that regulates the synchrony of nutrient release with crop demands (Myers *et al.*, 1994). High quality prunings are thought to release nutrients too quickly, *i.e.* before crop demand for nutrients is high, while low quality pruning release nutrients too slowly or even immobilize nutrients during the cropping season. Mixing high and low quality plant material is suggested as one of the options to minimize nutrient losses and to achieve synchrony with crop demand for nutrients (Handayanto *et al.*, 1997).

The use of locally available plant materials from agroforestry species as farm inputs has a number of benefits. The application of organic materials can be more advantageous than costly fertilizers by affecting many biochemical properties controlling nutrient cycling (Nziguheba *et al.*, 2000). They contribute to increased microbial biomass (Wu *et al.*, 2000), SOM build-up, greater efficiency of fertilizer use (Palm *et al.*, 2001), improved soil physical properties (Dudal, 2002) and the supply of other macro/micro nutrients(Nziguheba *et al.*, 2002) especially those not present in commonly used fertilizers.

Organic inputs are also claimed to play a role in reducing P-adsorption (Nziguheba *et al.*, 1998) and soil acidification (Wong *et al.*, 1995). However organic inputs, even if applied at reasonable rates, could sustain crop yields only at limited levels (Vanlauwe *et al.*, 2001), with large amounts of biomass required to elicit moderate yield increase. Insufficiency of nutrients (especially P), the diversity and complexity of input chemical compositions and the difficulty of predicting patterns of nutrient mineralization to synchronize with crop demands all limit yield increase (Dudal, 2002); while immobilisation of nutrients (*e.g.* N) when low quality organic inputs are used may actually decrease yields. Therefore, given the limitations associated with sole use of either organic or inorganic inputs, the judicious use of locally available organic resources from trees combined with mineral fertilizers may be an optimal strategy for smallholder organic farmers (Giller, 2002).

3.3 Ecosystem Services from Agroforestry Systems

The integration of trees, agricultural crops, and/or animals into an agroforestry system has the potential to enhance soil fertility, reduce erosion, improve water quality, enhance biodiversity, increase aesthetics, and sequester carbon (Nair *et al.*, 2009) (Table 16.2). It has been well recognized that these services and benefits provided by agroforestry practices occur over a range of spatial and temporal scales (Table 16.3). Generally, ecosystem services are grouped into three categories in the context of organic farming: (1) provisioning services (2) regulating services and (3) supporting services (Tallis and Kareiva, 2005).

Table 16.2: Examples of Ecosystems Services from different Agroforestry Systems in India

Region	*Challenge*	*Changes Observed Due to Agroforestry*	*Authors*
Himalayas (Kurukshetra)	Improvement of sodic soils	Increase in microbial biomass,tree biomass and soil carbon; enhanced nitrogen availability	Kaur *et al.* (2002)
Himalayas	Restoration of Abandoned agricultural sites	Biomass accumulation (3.9 t/ha in agroforests as compared to 1.1 t/ha in degraded forests); improvement in soil physico-chemical characteristics; carbon sequestration	Kumar (2011)
Western Himalayas	Reducing soil and water loss in agroecosystems in steep slopes	Contour tree-rows (hedgerows) reduced runoff and soil loss by 40 per cent and 48 per cent respectively (In comparision to 347 mm runoff, 39 Mg/ha soil loss per year under 1000 mm rainfall conditions)	Narain *et al.* (1997)
Sikkim Himalaya	Enhancing the litter production and soil nutrient dynamics	Nitrogen-fixing trees increased N and P cycling through increased production of litter and influenced greater release of N and P; nitrogen fixing species helped in maintenance of soil organic matter, with higher N mineralization rates in agroforestry systems	Sharma *et al.* (1996a) Sharma *et al.* (1996b)
Indo-Gangetic plains (Uttar Pradesh)	Biomass production and nutrient dynamics in nutrient deficient and toxic soils	Biomass production (49 t/ha/decade)	Singh (1998)
Himalayas (Meghalaya)	Enhancing tree survival and crop yield	Crop yield did not decrease in proximity to *Albizzia* trees	Dhyani and Tripathi (1998)
Western India (Karnal)	Improvement of soil fertility of moderately alkaline soils	Microbial biomass C which was low in rice berseem crop (96.14 g/g soil) increased in soils under tree plantation (109.12 g/g soil); soil carbon increased by 11-52 per cent due to integration of trees and crops.	Kaur *et al.* (2000)
Western India (Rajasthan)	Compatibility of trees and crops	Density of 417 trees/ha was found ideal for cropping with pulses	Gupta *et al.* (1998)
Central India (Raipur)	Biomass production in N and P-stressed soils	*Azadirachta indica* trees were found to produce biomass in depleted soils.	Puri and Swamy (2001)
Central India	Soil improvement	Decline in proportion of soil sand particles; increase in soil organic C, N, P and mineral N	Pandey *et al.* (2000)
Southern India (Kerala)	Growing commercial crops and trees	Ginger in interspaces of *Ailanthus triphysa* (2500 trees/ha) helps in getting better rhizome development of ginger, compared to solo cropping	Kumar *et al.*(2001)

Table 16.3: Spatial Scales of Various Ecosystems Services Provided by Agroforestry Systems

Ecosystem Services	*Spatial Scale*		
	Farm/Local	*Landscape/Regional*	*Global*
Net primary production			
Pest control			
Pollination/seed dispersal			
Soil Enrichment			
Soil stabilization/erosion			
Control			
Clean water			
Flood mitigation			
Clean air			
Carbon sequestration			
Biodiversity			
Aesthetics/cultural			

Source: Izac (2003) and Kremen (2005).

3.4 Provisioning Services of Trees

Provisioning services are the products obtained from ecosystems,including genetic resources, food, energy, fibre and fresh water.

3.4.1 Tress as Food, Fuel and Fodder Source

The small and marginal farmers in India have long been practicing agroforestry – to meet their food, fodder and fuel requirements (Kumar, 2006). Apart from ensuring food production, such systems also would enhance economic returns to the growers. Diversified production from agroforestry systems is a form of risk avoidance, which is of special relevance in the context of the current agricultural crises in many countries in Asia are experiencing. The potential of agroforestry to provide alternate sources of income and employment to the rural poor also has been highlighted (Puri and Nair, 2004; Samra *et al.*, 2005). The diverse products (fruits, vegetables, spices *etc.*), which are available year-round in systems such as homegardens not only contribute to food security during the "lean" seasons but also ensure food diversity (Kumar and Nair, 2004). They are also sources of mineral nutrients and vitamins for improving household nutritional security especially for 'at-risk populations' (*e.g.*, women and children) (Table 16.4).

Growing of nutritious fruit and other food crops in agroforestry based organic farming has many links with improving the health and nutrition of the rural poor people. Fruit plants are the most valuable component of agri-horticulture systems and serve as a rich source of minerals, vitamins and important alkaloids. It can play important role in child nutrition.

Table 16.4: Nutritious Agroforestry Fruit Trees for Integration in Organic Farming

Common Name	*Botanical Name*	*Source of Nutrients*
Cashew nut	*Annacardium occidentale*	Rich in protein and fibre
Almond	*Prunus dulsis*	Rich in Fe, Cu and vitamin-B1
Ber	*Ziziphus mauritiana*	Rich in vitamin-C, Na and K
Aonla	*Emblica officinalis*	Rich in vitamin-C, Na and K
Jackfruit	*Artocarpus heterophylus*	Rich in vitamin-B, C and K
Mango	*Mangifera indica*	Excellent source of vitamin-A and flavonoids
Pappaya	*Carica papaya*	Excellent source of vitamin- C and flavonoids
Sapota	*Acharas sapota*	Rich in dietary fibre
Bael	*Aegle marmelos*	Rich in vitamin- C, K, Fe and flavonoids

Source: Behera and Pradhan (2014).

Agroforestry practices can provide significant amounts of fuel wood for small and medium farmers. Studies have also shown that trees grown in contour strips, bunds, interpresed and rotational woodlots and can produce large quantities of fuel wood. Agroforestry plantations are probably the only option to meet a major share of the wood demand. Production of fuel wood exemplify the significance and potential of agroforestry systems in meeting local firewood demands. Biofuel production from species such as *Jatropha curcas* can also be integrated with agroforestry practices including contour planting, live fences and hedges (Sileshi *et al.*, 2007).

3.5 Regulation Services of Trees

3.5.1 Provide Erosion Control and Soil Conservation

Agroforestry systems are a traditional form of land management and soil conservation measure that has been practiced in various parts of the world. Agroforestry practices provide vegetative cover which reduces the impact of rain drop and provide protection to the soil, enhance soil productivity and contribute towards sustainable land management. Trees protect the soil surface *via.*, two canopies: the litter layer and the leaf canopy, thereby decreasing runoff and erosion losses, dampening temperature and moisture fluctuations and in most cases, maintaining or improving soil physical properties. (Hulugalle and Ndi, 1993). The beneficial effects of trees protecting the soil surface depend on the spatial and temporal coverage of the tree component. Moreover, tree roots can loosen the topsoil by radial growth, and improve porosity in the subsoil when roots decompose. The perennial nature of tree root systems provides a dependable source of carbon substrate for microorganisms in the rhizosphere which results in improved soil structure (Tisdall and Oades, 1982). These two processes, surface soil protection and root penetration, take place continually in agroforestry systems instead of temporarily, as in agricultural systems (Sanchez *et al.*, 1997).

Trees integerated in oragnic farming helps in stabilizing the conservation structures or contribute to the productive use of the land (Solanki and Ram Newaj, 1999). In a study at Regional Station of CTRI, Kandukur, comparison was made of three systems namely pasture, agroforestry and sole tobacco. The results indicated that the time taken to generate runoff was 25 minutes in pastures, after the initiation of rainfall, followed by agroforestry (tree + crops) (22 min) and the least was in sole tobacco (5 min). The delay in generation of runoff has positive influence as it allows more water to infiltrate in to the soil. Peak flow rate of runoff was lowered in pastures by 74 per cent (5.5 to 2 mm/hr) and 31 per cent in agroforestry system when compared to sole tobacco. Earlier studies at research farm of CRIDA, also reported lower amount of runoff with agrihorticultural (4.9 per cent) and agrisilvicultural (3.8 per cent) systems compared to arable cropping (10.6 per cent) (Das *et al.*, 1993).

3.5.2 Pest and Disease Suppression in Tree Crop Interface

Pest suppression is a perennial challenge to farmers, and it is a very important ecosystem service. In the modern agriculture, reduction of agro-biodiversity and continuous monoculture of crops with minimal rotation has a tendency to deplete the soil and for crop pests to become endemic (Geist, 1999). The shortening of fallow periods may increase the intensity of serious pests including witch weeds (*Striga* spp) (Sileshi *et al.*, 2006). There is growing evidence showing that some agroforestry practices can drastically reduce serious pests of maize such as termites (Sileshi *et al.*, 2005) and weeds (Sileshi *et al.*, 2006). Agroforestry increases plant diversity and structural complexity, with implications on pest population dynamics. Few examples of a perennial system that exhibits a rich range of natural enemy pest control is the coffee agroforestry system, where there is a wide variety of spatial and temporal diversity determined by the shade trees planted within the cropping system. Greater natural enemy presence has been observed in the more diverse and shaded agroforestry systems (Table 16.5).

Table 16.5: Potential of Agroforestry Systems in Pest Suppression

Type of Diversification	*Nature of Diversification*	*Benefit*	*Examples*
Agroforestry	Growing crops and trees together; spatial and temporal diversity	Pest suppression	Greater shade diversity increased bird natural enemy abundance for larval control on crop plant (Perfecto *et al.*, 2004)
		Pest suppression	Coffee berry borer control increased with greater ant diversity and abundance in shade systems (Armbrecht and Gallego, 2007)
		Pest suppression	Complex landscapes that have areas of woodland and hedgerows interspersed within fields had higher rates of larval parasitism (Marino and Landis, 1996)

The insect pests and associated diseases attack is reduced through creating barriers to moving pathogens. Agroforestry has diverse components which act as biological barriers for the insect-pest activities. Some agroforestry multipurpose tree species has insecticidal properties and used as bio-pesticides. Agroforestry helps in biodiversity conservation, improving wildlife habitat and harbouring birds and beneficial insects which feed on crop pests.

3.5.3 Trees for Enhancing Microclimate to Offset Climate Variability

Trees in agroforestry systems can modify microclimatic conditions by providing shade and windbreak (Jose *et al.*, 2004). The trees bring about a whole complex of environmental changes, affecting not just available light but also air temperature, humidity, soil temperature, soil moisture content, wind movement, pest and disease complexes. These factors impact crops, and the effect can be beneficial to a wide array of crops (Sileshi *et al.*, 2007). Wind speed reductions can extend to 30 times the height of tree belts on the leeward side (Tamang *et al.*, 2010). The resultant decline in wind erosion effects can have multiple benefits for crops including increased growth rate and quality, protection from windblown soil, moisture management and soil protection. Furthermore, higher air and soil temperatures in the lee of a shelterbelt can extend the growing season, with earlier germination and improved growth at the start of the season (Brandle *et al.*, 2004).

Farmers in India have traditionally managed and exploited the shady environment under MPT's. The cultivation of crops under canopies of trees grown scattered in farm lands, homegardens, bunds, hedge row and other agriculture systems are the most notable of traditional agroforestry practices in India. The tree litter and canopy have been documented to influence the microclimate in terms of improved rainfall infiltration, soil structure and microfauna, reduced evapotranspiration and temperature extremes, and increased relative humidity (Saka *et al.*, 1994). Woody perennial based production systems, such as agroforestry, have the potential to sequester large quantities of CO_2 and thereby partially offset the global warming process (FAO, 2004).

In drought-prone environments, such as Rajasthan, as a risk aversion and coping strategy, farmers maintain agroforestry systems to avoid long-term vulnerability by keeping trees as an insurance against drought, insect pest outbreaks and other threats, instead of a yield-maximizing strategy aiming at short-term monetary benefits (Rathore, 2004). During the drought deep root systems of trees are able to explore a larger soil volume for water and nutrients, which will help during droughts. Furthermore, increased soil porosity, reduced runoff and increased soil cover lead to increased water infiltration and retention in the soil profile which can reduce moisture stress during low rainfall years (Singh and Pandey, 2011).

3.5.4 Trees Augment Carbon Storage

Carbon sequestration is defined as the capture and secure storage of carbon that would otherwise be emitted to or remain in the atmosphere (FAO, 2000). The role of trees as an important means to capture and store atmospheric CO_2 in vegetation, soils and biomass products are widely acknowledged (Malhi *et al.*, 2008). Carbon sequestration involves the removal and storage of carbon from the atmosphere in carbon sinks (such as oceans, vegetation, or soils) through physical or biological processes. The incorporation of trees or shrubs in agroforestry systems can increase the amount of carbon sequestered compared to a monoculture field of crop plants or pasture (Sharrow and Ismail, 2004; Kirby and Potvin, 2007). The potential of agroforestry systems to sequester carbon varies depending upon the type of the system, species composition, age of component species, geographic location, environmental factors, and management practices. In addition to the significant amount of carbon stored in aboveground biomass *i.e.*, the assimilation of carbon into the plant matter, agroforestry systems can also store carbon belowground. The amount sequestered in each part differs greatly depending on a number of factors, like the agroclimatic region, the type of system, site quality, previous land use, *etc.* The above ground carbon sequestration rates in some major agroforestry systems around the world are highly variable, ranging from 0.29-15.21 Mg/ha/yr (Nair *et al.*, 2009).

Substantial carbon is sequestered in below ground tree parts, part of which is added to the soil every year thus contributing to soil carbon. The tree based systems also add substantial quantities of litter to the soil every year, which is added to the soil and contributes towards the soil carbon. Samra and Charan (2000) observed increase in soil organic carbon status of surface soil from 0.39 per cent to 0.52 per cent under *Acacia nilotica* + *Saccharum munja* and from 0.44 per cent to 0.55 per cent under *Acacia nilotica*+ *Eulaliopsis binata* after five years. Enhanced soil organic carbon coupled with improved nutrient status in soils under tree systems helps in better crop establishment and early seedling vigour that helps the crop plants to tolerate moisture deficits and drought better. Carbon sequestration potential of different agroforestry systems for different parts India is given in Table 16.6.

Table 16.6: Total C Storage Under agroforestry Systems in different Regions of India

Region	*Agroforestry System*	*Total C Storage (t C/ha)*
Semi-arid region	Silvi-pastoral system (age 5 years)	
	Acacia nilotica+ natural grass	9.5-17.0
	Dalbergia sissoo+ pasture	17.2
North western India	Silvopastoral system (age 6 years)	
	Acacia/Dalbergia/Prosopis + *Desmostacya*	6.8-18.5
	Acacia/Dalbergia/Prosopis + *Sporobolus*	1.5-12.3
Central India	Agri-silviculture system (age 8 years)	
	Gmelina arborea	24.1-31.1
Arid region	Agri-silviculture system (age 8 years)	26.0
North-western Himalayas	Agri-horti-pastoral system	1.15

Source: Prajapat *et al.* (2014).

3.6 Supporting Services of Trees

Supporting ecosystem services are those that are necessary for the production of all other ecosystem services. Some examples include biomass production, production of atmospheric oxygen, nutrient cycling, water regulation and biodiversity conservation.

3.6.1 Biomass Transfer

Biomass transfer is one of the nutrient source input in organic farming. This technology involves ex-situ production of biomass away from the cropping land in designated areas, such as hedges around or within the farm. The trees that stand pruning, have high rates of organic matter production, can be combined in agroforestry systems with agricultural crops with added advantage of yielding products such as fruits, fiber, fodder, timber and fuelwood. The leafy biomass of trees is frequently cut from hedges or uncultivated areas and can be incorporated into crop fields as a source of nutrients in organic farming before and/or after the crop is planted. In this technology, interaction between the components is only through the nutrient release from applied mulch and uptake by crop. Leguminous trees like *Leuceana*, *Glyricidia*, *Sesbania* and *Calliandra* have also been widely used as a source of nutrients, with positive results on crop yields (Dhyani *et al.*, 2009). Biomass transfer offer opportunities on increasing production due to increased nutrients. Woody species grown in alleys/hedges/bunds outside the cultivated fields, therefore, may be able to transform less available inorganic forms of phosphorus into more available organic forms, as well as supply significant quantities of N and K, when their leaves are incorporated into the soil as biomass transfers (Sanchez *et al.*, 1997). These nutrients become organic inputs to the organic farming when the added tree biomass is decomposed in the soil.

3.6.2 Trees as Nutrient Source

Trees can provide nutrient inputs to crops in agroforestry systems by capturing nutrients from atmospheric deposition, biological nitrogen fixation (BNF), and deep in the subsoil, and storing them in their biomass. (Sanchez *et al.*, 1997). The incorporation of multipurpose trees along with crops that are able to biologically fix nitrogen is fairly common in tropical agroforestry systems. Ecological interactions between trees and crops in agroforestry system are beneficial because: leguminous trees have a beneficial effect on soil fertility through nitrogen fixation, greater organic matter production, and recycling of nutrients, a combination of annual crops and trees increases biomass production because differences in rooting depth enable uptake of more water and nutrients (Young, 1986). Non N-fixing trees can also enhance soil physical, chemical and biological properties by adding significant amount of above and belowground organic matter and releasing and recycling nutrients in agroforestry systems. Land use systems such as agroforestry, agro-horticultural, agro-pastoral, and agro- silvipasture are more effective for soil organic matter restoration (Manna *et al.*, 2003).

3.6.3 Maintenance of Soil Fertility under Tree Crops

The role of agroforestry in enhancing and maintaining long-term soil productivity and sustainability has been well documented. This is a consequence of the permanent soil cover and perennial root systems of the trees, which provide continuous soil protection, a favorable environment for soil biological processes and more efficient nutrient cycling than systems based on annual crops. Significant improvement in soil biological activity has been reported under different tree based agroforestry systems and found that fluxes of C, N and P through microbial biomass were also significantly higher in *Prosopis cineraria* based land use system followed by *Dalbergia sissoo*, *Acacia leucophloea* and *Acacia nilotica* in comparison to a no-tree control in Rajasthan (Yadav *et al.*, 2011).

3.6.4 Stimulate Nutrient Recycling from the Subsoil

Trees in agroecosystems can enhance soil productivity through biological nitrogen fixation, efficient nutrient cycling, and capturing and recycling of nutrients from deeper layers of soils. In annual cropping systems, nutrient recycling from the subsoil can be increased by the integration of deep-rooting trees. The uptake of nutrients by tree roots at depths where crop roots are not present can be considered an additional nutrient input in agroforestry systems. Such nutrients become an input upon being transferred to the topsoil *via.*, tree litter decomposition. Tree roots frequently extend beyond the rooting depth of crops (Sanchez *et al.*, 1997). If the subsoil contains substantial amounts of nutrients (such as leached nitrate), is sufficiently moist to allow nutrient transport and uptake, and the trees have sufficient amounts of fine roots in the respective depths, nutrient uptake from the subsoil is likely to occur (van Noordwijk *et al.*, 1996). A favorable effect of deep nutrient uptake is most likely to occur in the case of leguminous 'service trees', whose nutrient-rich leaf and branch biomass is regularly returned to the soil as prunings, and the net nutrient accumulation in the trees is small. For timber and fruit trees, which are not regularly pruned, low competitiveness is particularly important.

3.6.5 Water Quality Regulation

Trees with deep rooting systems in agroforestry systems can also improve ground water quantity and quality by serving as a "safety net" whereby excess nutrients that have been leached below the rooting zone of agronomic crops are taken up by tree roots. These nutrients are then recycled back into the system through root turnover and litterfall, increasing the nutrient use efficiency of the system (van Noordwijk *et al.*, 1996). Trees also have a longer growing season than most agronomic crops, which increases nutrient use efficiency in an agroforestry system by capturing nutrients before and after the cropping season. Integrating trees in organic farming can help to clean runoff water by reducing the velocity of runoff, thereby promoting infiltration, sediment deposition, and nutrient retention (Jose, 2009).

3.6.6 Trees Enhance Functional Biodiversity

Agroforestry plays five major roles in conserving biodiversity: (1) agroforestry provides habitat for species that can tolerate a certain level of disturbance; (2) agroforestry helps preserve germplasm of sensitive species; (3) agroforestry helps reduce the rates of conversion of natural habitat by providing a more productive, sustainable alternative to traditional agricultural systems t; (4) agroforestry provides connectivity by creating corridors between habitat remnants which may support the integrity of these remnants and the conservationof area-sensitive floral and faunal species; and (5) agroforestry helps conserve biological diversity by providing other ecosystem services such as erosion control and water recharge, thereby preventing the degradation and loss of surrounding habitat (Harvey *et al.*, 2007). Recent investigations involving biodiversity and crop productivity data for smallholder tropical agroforests elsewhere suggest that moderate shade, adequate labor, and input level can be combined with a complex habitat structure to provide high biodiversity as well as high agricultural yields and thus supporting both conservation and food security (Clough *et al.*, 2011).

3.6.7 Pollination

Agricultural intensification may affect functionally important pollinator species *viz.*, honey bees and other native bee communities disproportionately (Kremen *et al.*, 2002). Restoring pollination services in areas of greatest agricultural intensity would require both reducing insecticide use and restoring native or surrogate vegetation to provide nesting habitat and floral resources for bees when they are not using crops, thus agroforestrey can play a major role in balancing the pollinators population. Agroforestry can improve beekeeping (Chilufya and Tengnäs, 1996) because some trees used in agroforestry produce nectar and pollen and improved bee hives can be introduced in the fruit orchards, woodlots or fodder banks.

4. Challenges in Adoption of Agroforestry in Organic Farming

In spite of significant impact of agroforestry on India's economy, it has not become as popular with the organic farmers as it should have been. There are certain inherent limitations with this land use system. Some of these constraints are listed below:

- ☆ Small and marginal farmers in general, are reluctant to grow trees on their farm land owing to long gestation period in getting returns from trees, particularly in poor sites where return are even slower and lesser.
- ☆ Proven agroforestry technologies are available only for limited situations/ locations/regions.
- ☆ Lack of quality seed/planting material sources of promising agroforestry tree species.
- ☆ Inadequate harvesting and processing techniques for tree based products.
- ☆ Lack of protocols of value addition.
- ☆ Rigid legal laws restricting harvesting, transporting and sale of trees products.
- ☆ Lack of assured financial and policy support for popularizing agroforestry.
- ☆ Lack of proper transfer of technology, trained manpower, infrastructure and funds.

5. Moving Forward-Policy and Research Needs

For setting the agroforestry agenda in association with organic farming for future research, the following area and topics are to be considered:

- ☆ Quantifying and valuing the environmental services of agroforestry in organic farming
- ☆ Developing crop varieties and cultivators that can perform reasonably well under conditions of reduced supply of growth factors (light, water and nutrients)

- Impact of agroforestry on climate change mitigation and adaptation in organic farming
- System management for exploiting production and protection values of multi purpose trees.
- Policy issues for encouraging agroforestry adaptation in organic farming
- Mitigation of climate change by agroforestry system, enhancing productivity and carbon sequestration.
- Diversification of agroforestry species of high market value and quality.
- Development of innovative techniques to promote agroforestry based value added products
- Agroforestry models need to be developed for small, medium, large land holders under different agro-climate zones.

6. Conclusion

One of the most important reasons for adopting agroforestry systems in regions where commercial fertizers are expensive or unavaibale is the abilty of such systems to recover, recycle or efficiently utilize nutrients. In the past, the paucity of hard evidence has hindered the the relevance of integerating agroforestry with organic farming and its acceptance by practitioners, farmers and policy makers. However, in an era of environmental consciousness and ecological sustainability, the role of woody perennials in conjuction with annuals as an environmentally benign and ecologically sustainable alternative to traditional farming offers a number of ecosystem services. Agroforestry systems offers proven strategies for carbon sequestration, soil health improvement, peat suppression, biodiversity conservation and water regulation for not only the farmers, but for society at large. This realization should help promote agroforestry and its role as an integral part of organic farming systems in India for long-term sustainability.

REFERENCES

Alley JL, Garrett HE, McGraw RL, Dwyer JP and Blance CA. 1999. Forage legumes as living mulches for trees in agroforestry systems. Agroforestry Systems, 44: 281-291.

Altieri MA. 1995. Integrated pest management. In: *Agroecology: The Science of Sustainable Agriculture,* (Altieri MA, Ed.), Westview Press. Boulder, Colorado. pp 267-281.

Armbrecht I and Gallego MC. 2007. Testing ant predation on the coffee berry borer in shaded and sun coffee plantations in Colombia. Entomologia Experimentalis et Applicata, 124: 261-267.

Bayala J, Heng LK, Noordwijk MV and Ouedraogo SJ. 2008. Hydraulic redistribution study in two native tree species of agroforestry parklands of West African dry savanna. Acta Oecolo. doi: 10.1016/j.actao.06.2008010.

Behera MK and Pradhan TR. 2014. Harnessing the benefits of agroforestry for food and nutrition security. Van Sangyan, 10: 7-13.

Brandle JR, Hodges L and Zhou XH. 2004. Windbreaks in North American agricultural systems. Agroforestry Systems, 2004. 61: 65-78.

Chilufya H and Tengnas B. 1996. Agroforestry extension manual for northern Zambia. Regional Soil Conservation Unit (RSU), Technical Handbook Series, 11: p124.

Clough Y, Barkmann J, Juhrbandt J, Kessler M, Wanger TC, Anshary A, Buchori D, Cicuzza D, Darras K, Putra DD, Erasmi S, Pitopang R, Schmidt C, Schulze CH, Seidel D, Steffan-Dewenter I, Stenchly K, Vidal S, Weist M, Wielgoss AC and Tscharntke T. 2011. Combining high biodiversity with high yields in tropical agroforests, Proceedings of the National Academy of Sciences, 108(20): 8311-8316.

Das SK, Sharma S, Sharma KL, Saharan N, Nimbole NN and Reddy YVR. 1993. Land use options on semi-arid alfisols. American Journal of Alternate Agriculture, 8(1): 34-39

Dhyani SK and Tripathi RS. 1998. Tree growth and crop yield under agrisilvicultural practices in North-East India, Agroforestry Systems, 44: 1-12.

Dhyani SK, Kareemulla K, Ajit and Handa AK. 2009. Agroforestry potential and scope for development across agro-climatic zones in India. Indian Journal of Forestry, 32: 181-190.

Dudal R. 2002. Forty years of soil fertility work in sub-Saharan Africa. In: *Integrated Plant Nutrient Management in Sub-Saharan Africa: From concept to practice*, (Vanlauwe B, Diels J, Sanginga N and Merckx R, Eds.), CAB International, Wallingford, UK. pp 7-21.

Duguma LA and Hager H. 2011. Farmers' assessment of the social and ecological values of land uses in Central Highland Ethiopia. Environment and Management, 47: 969-982.

FAO. 2000. Carbon sequestration options under the clean development mechanism to address land degradation. FAO, Rome. Available from http: //www.fao.org/forestry/15528-0534f06d08a9c3cbcd73deefd8d06c674.pdf.

FAO. 2004. Assessing carbon stocks and modelling: Win– Win scenarios of carbon sequestration through land-use changes. Food and Agriculture Organization of the UN, Rome. p 156

Geist JH. 1999. Soil mining and societal responses: The case of tobacco in eastern miombo highlands. In: *Coping with Changing Environments*, (Lohnert B and Geist H, Eds.), Ashgate Publishing Ltd, Aldershot. pp 119-148.

Giller KE. 2002. Targeting management of organic resources and mineral fertilizers: can we match scientists' fantasies with farmers' realities? In: *Integrated plant nutrient management in Sub-Saharan Africa: From concept to practice*, (Vanlauwe B, Diels J, Sanginga N and Merckx R, Eds.), CAB International, Wallingford, UK. pp. 155-171.

Gliessman SR. 1998. Agroecology: Ecological processes in sustainable agriculture. Ann Arbor Press. Chelsea, Michigan, p 357.

Gupta GN, Singh G and Kachwaha GR. 1998. Performance of *Prosopis cineraria* and associated crops under varying spacing regimes in the arid zone of India, Agroforestry Systems, 40: 149-157.

Handayanto E, Giller KE and Cadisch G. 1997. Regulating N release from legume tree prunings by mixing residues of different quality. Soil Biology and Biochemistry, 29: 1417-1426.

Harvey CA and Gonza´lez Villalobos JA. 2007. Agroforestry systems conserve species-rich but modified assemblages of tropical birds and bats. Biodiversity and Conservation, 16: 2257-2292.

Hulugalle NR and Ndi JN. 1993 Effects of no-tillage and alley cropping on soil properties and crop yields in a Typic Kandiudult of Southern Cameroon. Agroforestry Systems, 22: 207-220.

Izac AMN. 2003. Economic aspects of soil fertility management and agroforestry practices. In: *Trees Crops and Soil Fertility: Concepts and Research Methods*, (Schroth G and Sinclair F, Eds.), CAB International, Wallingford, UK. p 464.

Jordan CF. 1995. Conservation: Replacing quantity with quality as a goal for management. Wiley, New York, p 340.

Jose S, Gillespie AR and Pallardy SG. 2004. Interspecific interactions in temperate agroforestry. Agroforestry Systems, 61: 237-255.

Jose S. 2009. Agroforestry for ecosystem services and environmental benefits: an overview. Agroforestry Systems, 76: 1-10.

Karlen DL, Mausbach MJ, Doran JW, Cline RG, Harris RF and Schuman GE. 1997. Soil quality: A concept, definition, and framework for evaluation. Soil Science Society of America Journal, 61: 4-10.

Kaur B, Gupta SR and Singh G. 2000. Soil carbon, microbial activity and nitrogen availability in agroforestry systems on moderately alkaline soils in Northern India, Applied Soil Ecology, 15: 283-294.

Kaur B, Gupta SR and Singh G. 2002. Carbon storage and nitrogen cycling in silvopastoral systems on a sodic in North-western India, Agroforestry Systems, 54: 21-29

Kiptot E and Franzel S. 2012. Gender and agroforestry in Africa: a review of women's participation. Agroforestry Systems, 84: 35-58.

Kirby KR and Potvin C. 2007. Variation in carbon storage among tree species: implications for the management of a smallscale carbon sink project. Forest Ecology and Management 246: 208-221.

Kremen C, Williams NM and Thorp RW. 2002. Crop pollination from native bees at risk from agricultural intensification. Proceedings of the National Academy of Science, USA. 99: 16812-16816.

Kremen C. 2005. Managing ecosystem services: what do we need to know about their ecology? Ecol Lett 8: 468-479

Kumar BM and Nair PKR. 2004. The enigma of tropical homegardens. Agroforestry Systems, 61: 135-152.

Kumar BM, Thomas J, Fisher R F. 2001. *Ailanthus triphysa* at different density and fertilizer levels in Kerala, India: tree growth, light transmittance and understorey ginger yield, Agroforestry Systems, 52: 133-144.

Kumar BM. 2006. Agroforestry: the new old paradigm for Asian food security. Journal of Tropical Agriculture, 44 (1-2): 1-14.

Kumar BM. 2011. Species richness and aboveground carbon stocks in the homegardens of Central Kerala, India, Agriculture, Ecosystems and Environment, 140(3-4): 430-440.

Kumar K and Goh KM. 2000. Crop residues and management practices: effects on soil quality, soil nitrogen dynamics, crop yield, and nitrogen recovery. Advances in Agronomy, 68: 197-319.

Lehmann J, Peter I, Steglich C, Gebauer G, Huwe B and Zech W. 1998. Below-ground interactions in dryland agroforestry. Forest Ecology and Management, 111: 157-169.

Lundgren BO and Raintree JB. 1982. Sustained agroforestry. In: Agricultural research for development: potentials and challenges in Asia (Nestel B. Ed.), ISNAR. Hague. Netherlands. 37-49

Malhi Y, Roberts JT and Betts RA. 2008. Climate change, deforestation, and the fate of the Amazon. Science, 319: 169-172.

Manna MC, Ghoshand PK and Acharya CL. 2003. Sustainable crop production through management of soil organic carbon in semiarid and tropical India. Journal of Sustainable Agriculture, 21: 87-116

Marino PC and Landis DA. 1996. Effect of landscape structure on parasitoid diversity and parasitism in agroecosystems. Ecological Applications, 6: 276-284.

Molua EL. 2005. The economics of tropical agroforestry systems: the case of agroforestry farms in Cameroon. Forest Policy and Econnomy, 7: 199-211.

Muchena FN, Onduru DD, Gachini GN and Jager DA. 2005. Turning the tides of soil degradation in Africa: capturing the reality and exploring opportunities. Land Use Policy, 22: 23-31.

Myers RJK, Palm CA, Cuevas E, Gunatilleke IUN and Brossard M. 1994. The synchronisation of nutrient mineralisation and plant nutrient demand. In: *The Biological Management of Tropical Soil Fertility*, (Woomer PL and Swift MJ, Eds.), John Wiley and Sons, New York. pp 81-116.

Nair PKR, Kumar BM and Nair VD. 2009. Agroforestry as a strategy for carbon sequestration. Journal of Plant Nutrition and Soil Science, 172: 10-23.

Nair PKR. 1993. An introduction to agroforestry. Kluwer Academic Publishers. Dordrecht. p 499

Narain P, Singh RK, Sindhwal NS and Joshie P. 1997. Agroforestry for soil and water conservation in the Western Himalayan Valley Region of India 1. Runoff, soil and nutrient losses. Agroforestry Systems, 39: 175-189.

Nziguheba G, Merckx R, Palm CA and Mutuo P. 2002. Combining *Tithonia diversifolia* and fertilizers for maize production in a phosphorus deficient soil in Kenya. Agroforestry Systems, 55: 165-174.

Nziguheba G, Merckx R, Palm CA and Rao MR. 2000. Organic residues affect phosphorus availability and maize yields in a Nitisol of Western Kenya. Biology and Fertility of Soils 32: 328-339.

Nziguheba G, Palm CA, Buresh RJ and Smithson PC. 1998. Soil phosphorus fractions and adsorption as affected by organic and inorganic sources. Plant and Soil, 198: 159-168.

Oke DO and Odebiyi KA. 2007. Traditional cocoa-based agroforestry and forest species conservation in Ondo State, Nigeria. Agriculture, Ecosystem and Environment,122: 305-311.

Palm CA and Rowland AP. 1997. A minimum dataset for characterization of plant quality for decomposition. In: *Driven by Nature: Plant Litter Quality and Decomposition,* (Cadisch G and Giller KE, Eds.), CAB International, Wallingford, UK. pp 379-392.

Palm CA, Giller KE, Mafongoya PL and Swift MJ. 2001. Management of organic matter in the tropics: translating theory into practice. Nutrient Cycling in Agroecosystems, 61: 63-75.

Palm CA. 1995. Contribution of agroforestry trees to nutrient requirements of intercropped plants. Agroforestry Systems, 30: 105-124.

Pandey CB, Singh AK and Sharma DK. 2000. Soil properties under *Acacia nilotica* trees in a traditional agroforestry system in Central India, Agroforestry Systems, 49: 53-61.

Perfecto I, Vandermeer JH, Bautista GL, NunÞez GI, Greenberg R, Bichier P and Langridge S. 2004. Greater predation in shaded coffee farms: The role of resident Neotropical birds. Ecology, 85: 2677-2681.

Peter I and Lehmann J. 2000. Pruning effects on root distribution and nutrient dynamics in a *Acacia* hedgerow planting in Northern Kenya. Agroforestry Systems, 50: 59-75.

Prajapat K, Choudary GL and Jadhav TA. 2014. A Sustainable way to sequester carbon and mitigate green house gases emission. Indian Farming, 64(3): 22-24.

Puri S and Nair PKR. 2004. Agroforestry research for development in India: 25 years of experiences of a national program. Agroforestry systems, 61: 437-452.

Puri S and Swamy SL. 2001. Growth and biomass production in *Azadirachta indica* seedlings in response to nutrients (N and P) and moisture stress. Agroforestry Systems, 51: 57-68.

Rathore JS. 2004. Drought and household coping strategies: A case of Rajasthan, Indian Journal of Agricultural Economics, 59(4): 689-708.

Saka AR, Bunderson WT, Itimu OA, Phombeya HSK and Mbekeani Y. 1994. The effects of *Acacia albida* on soils and maize grain yields under smallholder farm conditions in Malawi. Forest Ecology and Management, 64: 217-230.

Samra JS and Charan SS. 2000. Silvipasture systems for soil, water and nutrient conservation on degraded lands of Shivalik foot hills (subtropical Northern India). Indian Journal of Soil Conservation, 28(1): 35-42

Samra JS, Kareemulla K, Marwaha PS and Gena HC. 2005. Agroforestry and livelihood promotion by cooperatives. National Research Centre for Agroforestry, Jhansi, India, p 104.

Sanchez PA, Buresh RJ and Leakey BRR.1997. Trees, soils, and food security. Philosophical Transactions of the Royal Society, London. 352: 949-961.

Sharma R, Sharma E and Purohit AN. 1996a. Cardamom, mandarin and nitrogen-fixing trees in agroforestry systems in India's Himalayan region. I. Litterfall and decomposition, Agroforestry Systems, 35: 239-253.

Sharma R, Sharma E, Purohit AN. 1996b. Cardamom, mandarin and nitrogen-fixing trees in agroforestry systems in India's Himalayan region. II. Soil nutrient dynamics, Agroforestry Systems, 35: 255-268.

Sileshi G, Akinnifesi FK, Ajayi OC, Chakeredza S, Kaonga M and Matakala PW. 2007. Contributions of agroforestry to ecosystem services in the miombo eco-region of Eastern and Southern Africa. African Journal of Environmental Science and Technology, 1(4): 068-080.

Sileshi G, Kuntashula E and Mafongoya PL. 2006. Legume improved fallows reduce weed problems in maize in Eastern Zambia. Zambian Journal of Agricultural Science, 8: 6-12.

Sileshi G, Mafongoya PL, Kwesiga F and Nkunika P. 2005. Termite damage to maize grown in agroforestry systems, traditional fallows and monoculture on nitrogen-limited soils in Eastern Zambia. Agricultural and Forest Entomology, 7: 61-69.

Singh B. 1998. Biomass production and nutrient dynamics in three clones of *Populus deltoides* planted on Indogangetic plains, Plant and Soil, 203: 15-26.

Singh VS and Pandey DN. 2011. Multifunctional Agroforestry Systems in India: Science-Based Policy Options. Climate Change and CDM Cell, Rajasthan State Pollution Control Board (RSPCB) Occasional Paper No. 4/2011. pp 33.

Solanki and Ram Newaj. 1999. Agroforestry, an alternate land use system for dryland agriculture. In: *Fifty Years of Dryland Agricultural Research in India* (Singh HP, Ramakrishna YS, Sharma KL and Venkateswarlu B, Eds.). CRIDA, Hyderabad. pp 463-473

Tallis H and Kareiva P. 2005. Ecosystem services. Current Biology, 15: 746-748.

Tamang B, Andreu MG and Rockwood DL. 2010. Microclimate patterns on the leeside of single-row tree windbreaks during different weather conditions in Florida farms: Implications for improved crop production. Agroforestry Systems, 79(1): 111-122

Tisdall J M and Oades JM. 1982. Organic matter and water-stable aggregates in soils. Journal of Soil Science, 33: 141-163.

van Noordwijk M, Lawson G, Soumare´ A, Groot JJR and Hairiah K. 1996. Root distribution of trees and crops: competition and/or complementarity. In: *Tree–Crop Interactions: A Physiological Approach*, (Ong CK and Huxley P, Eds.), CAB International, Wallingford, UK. pp 319-364

Vanlauwe B, Aihou K, Aman S, Iwuafor ENO, Tossah BK, Diels J, Sanginga N, Lyasse O, Merckx R and Deckers J. 2001. Maize yield as affected by organic inputs and urea in the West African moist savannah. Agronomy Journal, 93, 1191-1199.

Wong MTF, Akyeampong E, Nortcliff S, Rao MR and Swift RS. 1995. Initial responses of maize and beans to decreased concentrations of monomeric inorganic aluminium with application of manure or tree prunings to an Oxisol in Burundi. Plant and Soil, 171: 275-282.

Wu J, He ZL, Wei WX, O'Donnell AG and Syers JK. 2000. Quantifying microbial biomass phosphorus in acid soils. Biology and Fertility of Soils, 32: 500-507.

Yadav RS, Yadav BL, Chhipa BR, Dhyani SK and Ram M. 2011. Soil biological properties under different tree based traditional agroforestry systems in a semi-arid region of Rajasthan, India. Agroforestry Systems, 81(3): 195-202

Young A. 1986. The potential of agroforestry for soil conservation!. Erosion control. ICRAF Working Paper 42, Nairobi. p 68.

Young A. 1997. Agroforestry for soil management. CAB Int., Wallingford, UK.

Chapter 17

Organic Livestock Production: Present Status and Future Prospects in India

Mahesh Chander

ICAR-Indian Veterinary Research Institute,
Izatnagar – 243 122, Uttar Pradesh
E-mail: drmahesh.chander@gmail.com

1. INTRODUCTION

India is one among the top ten countries which have made significant progress in organic agricultural production and trade. This progress, however, is limited to high value commercial crops like cotton, spices, tea, basmati rice, herbs and honey exported from India, while organic animal husbandry has not been able to progress much. This is mainly due to limited export prospects of organic livestock products as also the difficulties in compliance of standards for organic livestock production which is knowledge and skill intensive system. Considering the growing interest among domestic consumers in good quality, safe products of animal origin, prospects of organic animal products appears to be bright in India. There are some sporadic efforts across the country, wherein, many milk producers are selling organic milk and Ghee on price premiums to quality conscious consumers. The majority population of India consumes products of animal origin like milk, milk products and meat and it 1 per cent of this population chooses to switch to organic animal products consumption in next 5 years, it would have enormous implications for the livestock sector of India in terms of infrastructure, capacity building, marketing *etc.* Organic animal husbandry standards calls for improved animal welfare, controlled disease conditions, limited/restricted use of allopathic medicines, non use of medicated feeds and hormones, comfortable housing, grazing opportunities *etc.* These requirements make it mandatory on the part of producers to improve their capacities for organic animal husbandry. In order to orient and train farmers, the trainers *viz* SMS of KVKs, veterinarians and extension officers of State Departments of Animal Husbandry, faculties of SAUs need to be trained on principles, practices and standards of organic animal husbandry. The current government has announced several schemes which if implemented would give much needed push to organic livestock production in India.

Organic agriculture is rapidly growing around the world. It is also well recognized that the organic animal husbandry has not grown as faster as cereal crops, fruits, nuts, spices, tea, coffee and cotton. While organic farming is rapidly gaining ground in developing countries, the research and development (R and D) activities in organic animal husbandry is confined to EU and a few other developed countries like USA, Canada and Australia. There are opportunities as well as challenges in organic livestock production in developing countries which need to be addressed. The organic livestock development opportunities in developing countries in Asia, Africa and Latin America can be enhanced with more scientific research in organic livestock production under local conditions and strengthening institutional support (Chander *et al.*, 2011, Nalubwama *et al.*, 2011, Rahmann and Godinho 2012, Chander *et al.*, 2012).

2. Organic Livestock Farming

Organic animal husbandry has been defined as a system of livestock production that promotes the use of organic and biodegradable inputs from the ecosystem deliberately avoiding the use of synthetic inputs such as drugs, feed additives and genetically engineered breeding inputs, while ensuring the welfare of animals (Chander *et al.*, 2011; Chander *et al.*, 2013; Chander and Subrahmanyeswari, 2013). There are four principles of organic farming *viz.*, principle of ecology, principle of health, principle of fairness, and principle of care, which organic systems must always take into consideration. In order to achieve the animal welfare, environmental protection, resource-use sustainability and other objectives, certain key principles are adhered to under organic livestock production systems. Subrahamanyeswari and Chander (2008) found that majority of the livestock production practices of the farmers in Uttarakhand were in line with what the organic standards recommend, and thus, were compatible to organic systems. Since, majority of the animal husbandry practices followed by the farmers were favourable to or closer to the recommended organic livestock production standards which is a clear indication of becoming organically certified by effective interventions. In this context, public and private organizations have a major role to make livestock farmers more compatible with the organic standards and also act as an effective solution for the changing climatic conditions.

According to the International Federation of Organic Agriculture Movements (IFOAM), the organic animal husbandry has multiple objectives, as given below:

- ✰ To raise animals in a system that takes into consideration the wider issues of environmental pollution, human health on consumption of animal products allowing them to meet their basic behavioral needs and reduce stress.
- ✰ Diversify in keeping as many types of livestock on the holding as each furnishes different nutrients at the household level. For example, special

attention should be given to rabbits and poultry as income generated from this enterprise goes directly to the disadvantaged segments of the population *e.g.* women and children. Their nitrogen rich manure is used to increase vegetable production in the kitchen gardens, thus, improving the family diet. Other like donkeys those are useful in transport thus, reducing the consumption of non-renewable sources of energy *e.g.* petroleum based fossil fuels should also receive special attention.

- ✰ Exploit the natural behavior of animals in their production systems to reduce stress *e.g.* chicken like perching at night and perching rails should be provided for this purpose. They should also be raised in deep litter system that allows them to scratch for ants and worms and dust bathe. Dark secluded nest should be provided as they like laying in dark secluded places. Goats being browsers in nature like having their forage suspended high enough so that they can attain an upright posture. Pigs have rooting tendency, for which water and mud facilitate their natural rooting behavior.
- ✰ Use of low external input which lessen the cost of production and allow for a sustainable system of production since most materials can be recycled in the farm and also locally available.
- ✰ Bridging of nutrients gap in soil, crops and animals *i.e.* animals feed on crops and cultivated crops by-products. The animals' waste in the form of farmyard manure is composted and taken back to the soil to replenish the lost soil nutrients through cultivation. This ensures the completion of nutrient cycle in the ecosystem.

In order to achieve the animal welfare, environmental protection, resource-use sustainability and other objectives, certain key principles are adhered to under organic livestock production systems, which include:

- ✰ Management of livestock as land-based systems so that stock numbers are related to the carrying capacity of the land and not inflated by reliance on 'purchased' hectares from outside the farm system, thus, avoiding the potential for nutrient concentration, excess manure production and pollution. As such, landless animal husbandry prevalent in India is not ideally suitable for organic livestock farming; unless, the landless livestock keepers go for land leasing.
- ✰ Reliance on on-farm- or locally-derived renewable resources, such as biologically-fixed atmospheric nitrogen and home-grown livestock feeds, thereby reducing the need for non-renewable resources as direct inputs or for transport;
- ✰ Reliance on feed sources produced organically, which are suited to the animal's evolutionary adaptations (including restrictions on use of animal proteins) and which minimize competition for food suitable for human consumption;
- ✰ Maintenance of health through preventive management and good husbandry in preference to preventive treatment, thereby reducing the

potential for the development of resistance to therapeutic medicines as well as contamination of workers, food products and the environment;

- Use of housing systems which allow natural behaviour patterns to be followed and which give high priority to animal welfare considerations, with the emphasis on free-range systems for poultry;
- Use of breeds and rearing systems suited to the production systems employed, in terms of disease resistance, productivity, hardiness, and suitability for ranging.

Animal health and well-being through better living conditions, improved welfare measures and good feeding practices are ensured through a set of standards and the maintenance of written records by organic livestock farmers. Better management practices and prevention of illness are emphasized over treatment. Thus, the primary characteristics of organic livestock production systems are: well-defined standards and practices which can be verified, greater attention to animal welfare, no routine use of growth promoters, animal offal, prophylactic antibiotics or any other additives, at least 80 per cent of the animal feed grown according to organic standards, without the use of artificial fertilisers or pesticides on crops or grass.

3. Strengths

Integrated crop-livestock farming system predominant in India with well diversified livestock population in terms of species and breeds is ideal for organic livestock production. Besides, limited external input use including for animal production and maximum on- farm reliance brings it further closer to organic systems. The livestock production being largely extensive or semi-intensive, animal welfare too is not much compromised compared to factory type of animal production common in Western developed nations. The Indigenous Technical Knowledge (ITK) and *ayurvedic* medicines for health care are effective substitute for allopathic medicines, giving India an edge over western countries in the matters of organic livestock production. The concerned agencies under Government of India are actively pursuing the development of Indian National standards for organic livestock and poultry production, to bring it under regulation- a welcome move which might boost organic livestock production.

4. Weakness

4.1 Feed and Fodder

The inadequate supply of required organic feed and fodder may be a limiting factor while promoting organic livestock farming, since under organic livestock systems, animals are expected to be fed species specific organic diet in sufficient quantities. Besides, the feed and fodder requirement has to be met on farm or locally and it has to be grown following organic crop production methods. The fodder cultivation area in India has remained more or less static for many years and it is concentrated mostly in irrigated areas or so called green revolution belt of the country. Looking at the deficit in green and dry fodder in country, massive efforts are needed to ensure feed and fodder to livestock in required quantity, while considering organic animal husbandry.

4.2 Sanitary Conditions

Prevention of diseases is paramount in organic systems, so that the medicine interventions like antibiotics *etc.* are minimized to the extent possible. To minimize diseases, sanitation is important, for which the efforts are needed on massive scale to improve hygiene and sanitary conditions especially at production, processing and packaging stages. Looking at the prevailing conditions at production sites, processing units dealing with, India needs to do a lot so as to be eligible for organic livestock producing country.

4.3 Existence of Diseases

Among others, the prevalence of Foot and Mouth Disease (FMD) in various parts of India is one limiting factor for export of livestock products, so its control is number one priority for India. The Disease Free Zones (DFZs) may be created, where; organic livestock production may be encouraged.

4.4 Traceability

Unlike in Western countries, milk and meat is sourced from numerous small farmers in India making the traceability a difficult option. Nevertheless, appreciably, Indian government has introduced a web-enabled application- Tracenet system for organic products being exported from India. Considering the logistic problems, small farms, farmers' educational levels, how far traceability mechanism will be feasible makes it a bit skeptical in case of animal products.

4.5 Small Farms

In India, livestock production is mainstay of landless and small scale farmers. However, the landless animal husbandry is not allowed under the organic systems, unless they go for land leasing to raise livestock. Contract farming may be a potential solution where many small farmers may contract out their farms to companies, which may produce organic food products on consolidated holdings with required expertise and resources.

4.6 Lack of Knowledge, Training and Certification Facilities

Easily accessible information in local languages, locally available training and certification facilities at an affordable cost to small farmers is not available in many parts of the country, restricting Indian farmers to switch over to organic production especially when there is weak domestic market and current poor prospects for exports in case of livestock products.

5. Opportunities

It is expensive for intensive livestock producers to convert to organic production, but converting extensive, pasture-based systems could become economically more attractive, if price premiums could be captured for organic meat and livestock products (Scialabba and Hattam, 2002). India may follow experiences of developing countries like Argentina, Brazil and Namibia which could export organic livestock products. India exports certified organic honey, which may be extended initially to small ruminants, for organic textile/garments including the materials like hides, leather and wool. The Indigenous Technical Knowledge (ITK) of farmers may provide effective option for veterinary care through proper validation, as also the negligible use of agro-chemicals especially in drylands and hilly regions, makes favourable environment for organic livestock production. Grass based extensive production systems prevalent in parts of India have good potential for conversion into organic animal husbandry. Moreover, Indian livestock breeds being less susceptible to diseases and stress, need less allopathic medicines/antibiotics. With rising literacy and the consumers' awareness and concern about animal welfare issues and health foods, domestic consumption of organic foods including of animal origin is likely to get a boost.

The organic agricultural products including of livestock origin are gaining increasing popularity. The farmers can cash upon this growing interest in eco-friendly, animal welfare oriented, safe, nutritious and tastier meat products (as perceived by consumers of organic products). The eggs and meat obtained from such venture can be promoted as specialty item to restaurants; hotels and ethnic food jaunts fetching higher returns, better when local/*deshi* birds are raised, which can better perform in free range system. Poultry can utilize the grazing lands/plantation areas (Rubber, coffee, coconut *etc.*) by feeding on earth worms, small insects, green grass *etc.*, while fertilizing the land with manure.

The free range poultry systems or pastured poultry is a sustainable agriculture technique that calls for the raising of laying chickens, meat chickens (broilers), and/or turkeys on pasture, as opposed to indoor confinement, humane treatment, the perceived health benefits of pastured poultry, in addition to superior texture and flavor, are causing an increase in demand for such products, which are believed to be having medicinal value, rich in antioxidants and least in chemical, medicinal or hormonal residues. Therefore, the growing interest in organic farming and meat and eggs drawn from free range systems might offer an attractive option in the form of market premiums for livestock farmers to venture into organic production.

The growing consumer interest in good quality food products in India signals the need for developing domestic market for local consumption of organic foods. With rising literacy, income and awareness on food quality generated by the mass media like print, radio and TV, people are increasingly becoming quality conscious. Also, they are increasingly showing their willingness to pay for good quality products. For example, people readily pay extra money for unadulterated milk, which is not necessarily organic milk *per se.* This trend indicates that there is good potential for organic livestock products for local consumption. The enterprising farmers are now ready to experiment on new ideas on production and marketing, wherein organic livestock products like milk, meat, poultry and fish ideally fit. Just like marketing of FMCG and other industrial products market segmentation can be done by the farmers by supplying products to different categories of consumers with varying prices. The growing interest in eating out especially by visiting ethnic food jaunts, looking out for something unique, local and something which is natural and healthy while being environmentally safe offers hope for the production and supply of organic livestock products for domestic consumers. The domestic market development is the key for the development of organic animal husbandry and poultry farming in India. The growing market for organic cereals, vegetables, fruits, spices, pulses in Indian metros can be successfully extended to organic livestock and poultry products too.

6. Threats

The international trade in organic livestock products from the developing world is considered a risky business due to poor sanitary conditions, existence of diseases, traceability problems as also the self sufficiency in importing countries, which might discourage producers in India too. But rich segments among the Indians might offer market niche for organic livestock products, which can be tapped.

Under organic livestock production systems, it is expected that- organic meat, poultry and egg products come from farms that have been inspected to verify that they meet rigorous standards which mandate the use of organic feed, prohibit the use of antibiotics, give animals access to outdoor, fresh air and sunlight. The production methods are selected based on criteria that meet all health regulations, work in harmony with the environment, build biological diversity and foster healthy soil and growing conditions. After the production, animals are marketed that were raised without use of toxic persistent pesticides, antibiotics and paraciticides (Borell and Sorensen, 2004). Animal health, well being, better living conditions, welfare measures, feeding practices are to be ensured through a set of standards and maintenance of written records by the organic livestock farmers. Better management practices and prevention are emphasized over treatment. Thus, the primary characteristics of organic livestock production system are: a defined standard; greater attention to animal welfare; no routine use of growth promoters, animal offal or any other additives; at least 80 per cent of feed grown according to organic standards, without the use of artificial fertilizers or pesticides on the crops or grass. To be precise, organic meat, milk and eggs means that are produced, harvested, preserved and processed as per organic standards.

Livestock plays an important role in relation to the general principles of organic agriculture, supporting biological cycles within the farming system and diversifying production (Hermansen, 2003). To be successful, organic agriculture must integrate plant and livestock production to the extent possible to optimize nutrient use and recycling. Consumers buy meat and milk products to experience direct qualities, such as taste, nutritive value, and food safety, but also to benefit from indirect qualities, which are linked to the production process. They expect animals to be treated with compassion and to a high level regarding their welfare, and that production has been carried out in an environmentally friendly manner (Sorensen and Jakobsen, 2005).

The concept of organic livestock production gained momentum recently in developed countries because of concerns for animal welfare, chemical residues, incidence of Bovine Spongiform Encephalopathy (BSE), Foot and Mouth Disease (FMD), Bird Flu, Swine Flu, Genetically Modified Food (GMF) and some bacterial diseases. The affluent sections in developing countries too are influenced by these concerns and this stratum of society may increasingly look for organic livestock products fuelling the demand of organic food products in developing countries too. Moreover, the organic farming is considered to be more sustainable than the input intensive conventional systems of crop and livestock production. Therefore, whether one likes it or not, organic farming is rapidly expanding world over. There may be critics especially in developing countries where chronic food shortages are still common but organic is a growing reality, it's happening everywhere with the support of Non-Governmental Organizations (NGOs), private sector and also the government agencies. The Food and Agriculture Organization (FAO) of United Nations too is helping the growth of organic farming by assisting the developing countries through its programmes. It indicates the acceptance of organic farming as a credible system worth paying attention.

Many constraints and opportunities as regards to organic livestock production in developing countries in general, and India in particular, have been discussed by the authors (Chander *et al.*, 2007, Chander *et al.*, 2012). Those which need further attention towards continuous development in this sector are:

- Organic animal husbandry is land based activity, but livestock are kept by many landless livestock keepers in India. So a good number of livestock farmers are not eligible for organic livestock farming.
- Even when farmers own land, the number of animals to be maintained per hectare are far too less, considering 80 per cent holdings in India are <1 ha and per farmer land ownership is going down further year after year due to division of land in the expanding families. The sustainability of organic livestock production at the given stocking rate is difficult to achieve, at least in developing countries like India.
- Many developing countries in Asia and Africa are not yet free from infectious diseases like Foot and Mouth Disease (FMD), which restricts trade from these countries. The reduced opportunity for export discourages livestock producers to go organic.
- Small farmers find it difficult to comply with traceability requirements, which are strictly followed under organic management and essential for export commodities.
- Sanitary conditions at production sites and processing units need improvement.
- The local demand for organic livestock products *per se* is still very low, though the quality consciousness is on the rise among the consumers. The domestic market for organic livestock products needs to be developed. Grazing for minimum 4hrs is required for ruminants. But grazing land is shrinking due to reducing community land and also change in land use pattern.
- Natural sources of essential amino acids (Methionin, for instance) are not available good enough to meet the requirements of livestock particularly swine and poultry.

- ☆ Green fodder supply is insufficient to meet the requirement of the livestock. Animal survive on poor quality roughages.
- ☆ Housing conditions are often improper, increasing risk of zoonotic diseases.
- ☆ Research and development investment in the area of organic animal husbandry is nearly nil.

7. Organic Livestock Production in Developing Countries: Some Experiences

Organic livestock production is mostly confined to a few developed countries, where the demand too is increasing for such products. Many developing countries including India have impressive livestock strength and other favourable factors to their advantage yet these countries find difficult to export their livestock products, since international trade in livestock from the developing world is a risky business as far as organic livestock products are concerned (FAO, 2002; Harris *et al.*, 2003). An exporter must have an assured certified supply chain in order to successfully enter international markets. For instance, the need to have a completely organic supply chain could present a problem for export of organic meat from these countries. Large-scale commercial farms usually undertake most organic livestock production for export; whereas, livestock sector in many developing countries especially in Asia and Africa is largely dominated by the small scale producers with little risk bearing ability and resourcelessness. Moreover, the self-sufficiency of organic livestock products in EU may lead to reduced import demand, thus, constraining the growth of organic livestock sector in these countries.

The tropical countries like India will have to make sustained efforts, more than what is being already done in case of other agro products to make its presence felt in organic livestock production. One way could be to develop organic livestock sector initially for domestic consumption so as to move gradually to organic livestock production for export. It has been reported that the local livestock production practices especially in drylands and mountainous regions of India are very close to organic production practices and compliant to organic standards to a greater extent (Chander and Mukherjee, 2005; Chander and Wanapat, 2006; Chander *et al.*, 2007; Chander and Subrahmanyeswari, 2007; Chander *et al.*, 2008). Such areas may be targeted to develop as hubs of organic animal husbandry, since; it will be comparatively advantageous over intensive production. The success stories from other developing countries may help Indian producers to switch over to organic livestock production. For instance, among Latin American countries, Argentina is in forefront of organic production, where, alongside the vegetables and fruit market, there is growing interest in animal products. Here, organic milk production increased in 2003 and although numbers in the organic beef herd declined slightly in 2003, exports of organic beef increased from 50 tonnes to 270 tonnes, the majority going to the UK (SENESA, 2003). The FAO too recognized that among developing countries, the largest and most advanced organic livestock sectors are reported in Argentina and Brazil. According to the Brazilian Beef Association, there are approximately 2, 10,000 animals being farmed organically in the country. If 2 ha pasture/animal is required, that there are at least 420,000 ha of pasture under organic beef production (Euclides, 2004). Elsewhere, only limited data are available; however, it is evident that interest in organic livestock products is on the rise in response to not only strong demand for organic products in national markets and export markets, but also to the potential it can offer for maintaining soil fertility (FAO, 2002). The developing countries can move forward with simultaneous development of export and domestic markets starting with focusing on the domestic consumers of milk and meat products.

8. Standards for Organic Livestock and Poultry Production

8.1 Organic Management Plan

During the registration of the farm by the accredited Certification Body, the producer has to present an organic management plan which requires to be verified during the inspection. This plan shall be updated annually.

8.2 Choice of Breeds/Strains

The choice of livestock and poultry, breeds, strains and breeding methods shall be consistent with the principles of organic farming, taking into account, in particular, the following:

- ✰ Their adaptation to the local climatic and socio-economic conditions;
- ✰ Their vitality and resistance to diseases

8.3 Sources/Origin

1. Animals must have been born or hatched from production units complying with these guidelines, or must be the offspring of parents raised under the conditions set down in these guidelines.
2. Transfer of livestock and poultry between organic and non-organic units shall not be permitted. The accredited Certification Body shall ensure that brought-in livestock and poultry from other units comply with these Guidelines.
3. Livestock and poultry raised on non-organic production units shall be converted into organic unit as per these Guidelines.
4. When a producer demonstrates to the satisfaction of the accredited Certification Body that the organic source, breed and required management are not available, the accredited Certification Body may allow such livestock and poultry under the following circumstances:
   - ✰ When the producer is establishing an organic livestock and poultry operation for the first time;
   - ✰ when a farmer wants to change the livestock and poultry breed/strain or when new livestock and poultry specialization is developed;

- ☆ For the renewal of a herd, *e.g.*, high mortality of animals caused by catastrophic circumstances;
- ☆ When the farmer wishes to introduce breeding males into the farm. In such cases the young animals that are introduced to the organic farm shall be as young as possible, preferably as soon as they are weaned.

8.4 Record Keeping and Animal Identification

The animals shall bear unique identification numbers in accordance with these standards. The producer or the veterinarian in-charge shall maintain detailed and up-to-date records as set out in

8.5 Housing and Management

1. Livestock and poultry shall be maintained under natural conditions as far as possible. This shall include utilizing natural breeding methods, housing and management conditions to minimize stress, health management system to prevent diseases, with the ultimate aim to progressively limit use of chemical allopathic veterinary drugs (including antibiotics and hormones), reduce feeding of animals with products of animal origin (*e.g.* meat meal, blood meal), and ensure animal comfort and welfare.
2. The housing and day-to-day management of the animal, maintenance of sanitation, hygiene and environment shall be planned to suit the specific behavioral needs of the livestock
3. Sufficient space to ensure free movement and opportunity to express normal patterns of behavior;
4. The animals should not be tied unless required for specific reasons, such as, at the time of milking or for some medical procedures;
5. Where the livestock and poultry normal behavior demands group living, animals shall not be kept in isolation, but shall have company of like kind;
6. As far as possible two different kinds of animals shall not be kept together, unless for specific purposes, such as, free range poultry birds in cow/ buffalo shed for scavenging on ticks and other insects;
7. The housing system shall ensure prevention of abnormal behavior, injury and disease;
8. Appropriate facilities to cover emergencies such as the fire, the breakdown of essential mechanical services and the disruption of supplies shall be available.
9. Housing for Livestock and Poultry shall not be mandatory in areas where appropriate climatic conditions exist to enable animals to live outdoors without compromising their comfort, health and welfare. Conditions shall be inspected and permitted by the accredited Certification Body on producer and location- to- location basis.
10. Housing conditions shall meet the biological and behavioral needs of the livestock and poultry by providing easy access to feeding and watering;

11. Insulation, heating, cooling and ventilation of the building to ensure that air circulation, dust level, temperature, relative air humidity and gas concentrations are kept within limits which are not harmful to the livestock and poultry;
12. Plentiful natural ventilation and light to enter;
13. Appropriate fencing not harmful to the animals
14. Confinement shall be permitted under the following conditions:
15. Inclement weather to protect animals from injury;
16. Ensure health safety or welfare;
17. Protect plant, soil and water quality;
18. The stocking density shall provide comfort and well-being of the livestock and poultry with regard to the species, the breed and the age; behavioural needs with respect to the size of the group and the sex of the livestock and poultry; sufficient space to stand naturally, lie down easily, turn round, groom themselves, and assume all natural postures and movements such as stretching, lying and rumination and wing flapping in case of birds.
19. The sanitation and hygiene in the livestock and poultry farm shall be as per the national/provincial standards and must follow the standard operating protocols to keep the house, pens, equipment and utensils clean and free from microbial contamination.
20. Free-range, open-air exercise areas, or open-air runs should, if necessary, provide sufficient protection against rain, wind, sun and extreme temperatures, depending on the local weather conditions and the breed concerned.
21 .The outdoor stocking density of livestock and poultry kept on pasture, grassland, or other natural or semi-natural habitats shall be low enough to prevent degradation of the soil and over-grazing of vegetation.

8.6 Mammals

1. All mammals shall have access to open-air exercise or resting area, paddock or run which may be partially covered or shall have space for protection from rains are excess if in the open area. The animals must be able to use those areas whenever the physiological condition of the animal or the weather conditions and the state of the ground permit.
2. The accredited certification body shall grant exceptions for the access of males or bulls to open areas to avoid mixing with female animals for controlled breeding. The other animals may also not have access open-air exercise area or run during the winter period or the final fattening phase.
3. Livestock shed shall have properly laid and smooth floor, although not slippery. The floor shall not be entirely of slatted or grid construction.
4. The housing standards shall be in accordance with the standards laid down in the national or provincial guidelines and shall aim at providing comfortable, clean and dry laying/rest area of sufficient size, consisting of

a solid construction. Wherever possible, straw bedding shall be provided.

5. The calves of different age groups must be housed separately and never in the adult animal shed. Tethering of livestock and poultry is prohibited.
6. Pigs must be kept in groups, except in the last stages of pregnancy and during the suckling period. Piglets may not be kept on flat decks or in piglet cages. Exercise areas must permit dunging and rooting by the animals.
7. The keeping of rabbits in cages shall not be permitted.

8.7 Poultry

1. Poultry for organic products shall be reared in open-range conditions and shall have free access to open-air run whenever the weather conditions permit.
2. Housing of poultry in cages shall not be permitted.
3. Water fowl/duck shall have access to a stream, pond or lake whenever the weather conditions permit.
4. Poultry house floor shall be of solid construction covered with litter material such as straw, wood shavings, sand or turf. In case of layers, the floor area must be large enough to permit dropping collection. Perches/ higher sleeping areas of a size and number commensurate with the species and size of the group and of the birds and exit/entry holes of an adequate size must be provided.
5. In the case of laying hens, manipulation of day length may be permitted through the use of artificial lights. The maximum day length shall be 16 hrs with minimum of 8 hrs of natural light.
6. The producer shall follow the all-in all-out system of rearing and shall avoid mixing of different age group of birds, species and breeds. Mixed farming of poultry and pigs shall not be permitted. Between each batch, the house shall be emptied, and runs shall be left to allow the vegetation to grow (Code of Practice for Poultry Housing shall be as per IS 2732:1985).
7. Each Poultry house shall not contain more than: 4800 chickens,3000 laying hens, 5200 guinea fowl, 4000 female ducks and 3200 male ducks, 2500 turkeys. viii. Poultry shall have access to an open area for at-least one third of their life. ix. Livestock and poultry not reared in accordance with these provisions may also be maintained on the production unit provided that they are segregated and reared separately from the organically maintained animals under the rules x. All animals shall have access to proper ventilation and open-air runs. The accredited Certification Body may permit exceptions in case of the physiological, inclement weather conditions or under certain 'traditional' farming systems that restrict access of animals to paddock. But in these cases, it should be ensured that the welfare of the animals is not compromised.
8. Stocking rates for livestock and poultry should be appropriate for the region in question taking into consideration, general climatic conditions, fodder

production capacity, stock health, nutrient balance, and environmental impact.

9. Conversion Period for Animal Production

1. The establishment of organic animal husbandry requires an interim period, the conversion period. The conversion period shall be for three years. The conversion period can be reduced up to one year in the following cases.
2. Open-air runs and exercise areas used by non-herbivore species;
3. For dairy herds converted for the first time and for bovine, ovine, pig and caprine coming from extensive husbandry system during an implementation period;
4. Simultaneous conversion of livestock and poultry and land used for raising feed/fodder within the same unit should be a preferred approach. In such cases wherein the existing livestock and poultry and their offspring are fed mainly with products from the unit, the conversion period for the livestock and poultry, pasture and/or land used for animal feed, may be reduced to two (2) years.
5. The conversion period shall be determined by the accredited Certification Body and the conversion period shall be accounted from the day of registration.
6. In cases, where the land and livestock and poultry conversion to organic status is not simultaneous and the land alone has reached organic status and the livestock and poultry from a non-organic source is introduced, these must be reared according to these guidelines for at least the following compliance periods before their products are to be sold as organic:

9.1 Bovine Including Buffalo

1. Meat products: Twelve (12) months and at least 3/4th of their life span is spent in the organic management system.
2. Calves for meat production: Six (6) months when brought in as soon as they are weaned and less than six (6) months old.
3. Milk products: Ninety (90) days during the implementation period established by the competent authority, after that six months.

9.2 Ovine and Caprine (Sheep and Goat)

1. Meat products: Six (6) months;
2. Milk products: Ninety (90) days during the implementation period established by the competent authority, after that, six (6) months

9.3 Pig

Meat products: Six (6) months.

9.4 Poultry

1. Meat products: from the second day to the entire life span as determined by the accredited Certification Body;
2. Eggs: Six (6) weeks

10. Feed

1. Livestock and poultry farms shall provide maximum diet from feedstuffs (*including 'in conversion' feedstuff*) produced as organic as per the requirements of these guidelines. Agricultural processed residues of organic origin, such as from grain fermentation, fruit processing, vegetable processing, *etc.*, shall be permitted for purpose of feeding, provided that the overall feeding practices satisfy the daily energy and nutrient requirements of the concerned animals.
2. The agriculture land committed to cultivation of feed/fodder crops intended to be used as feed for livestock and poultry shall be organically grown.
3. During the operations, the products shall maintain their organic status provided that livestock and poultry are fed with at least 85 per cent for ruminants and 80 per cent for non-ruminants calculated on a dry matter basis, feed obtained from organic sources that have been produced in compliance with these guidelines.
4. Notwithstanding the above guidelines, where a producer can demonstrate to the satisfaction of the accredited Certification Body that feedstuffs satisfying the requirement outlined above are not available, due to, for example, unforeseen severe natural or man-made events or extreme climatic weather conditions, drought, crop failure, *etc.*, permission shall be granted to allow a restricted percentage (5 per cent) of feedstuffs not produced according to these guidelines to be fed for a limited time, provided that it does not contain genetically engineered/modified organisms or products thereof. The accredited certification body shall set both the maximum percentage of non-organic feed allowed and any conditions relating to this derogation.
5. Specific livestock and poultry rations shall take into account
6. The need of young animals for natural feed, such as, feeding of maternal milk, milk from other mammal or milk replacer of organic origin that has maximum similarity with maternal milk, provided that it does not contain any genetically modified ingredient, antibiotics, hormone, *etc.*
7. That in herbivores, substantial proportion of the dry matter and energy in the daily rations should consist of roughage, fresh or dried fodder, or silage; need for inclusion of cereals in the fattening phase of poultry; livestock and poultry must have ample, free access to water appropriate to maintain full health and productivity.

8. Due to reasons of animal welfare, health and productivity, if supplements are to be added, it shall be permitted on advice of a qualified veterinarian. The permitted list of such supplements, feed materials (probiotics, and biologicals, immunolgicals and procuring aids *etc.*) and processing aids that comply the prescribed criteria.

11. General Criteria for Feedstuff and Nutrition

Substances permitted as should significantly satisfy feeding requirements of the livestock and poultry fulfilling the physiological, behavioral and welfare needs of the concerned species; and such substances should not contain genetically engineered/modified organisms and products thereof; and are non-synthetic and are primarily of plant, mineral or animal origin.

11.1 Specific Criteria for Feedstuffs and Nutritional Elements

1. The feedstuffs should not be prepared by using chemical solvents and chemical treatment. All the ingredients of the feed including supplements, fed to organic animals should be from organic sources. In case of shortage of these substances, or in exceptional circumstances, well-defined analogic substances may also be used.
2. Feedstuffs of animal origin, with the exception of milk and milk products, fish, other marine animals and products derived thereof shall not be used. The feeding of mammalian material to ruminants is not permitted with the exception of milk and milk products;
3. Synthetic nitrogen or non-protein nitrogen compounds shall not be used.

11.2 Specific Criteria for Additives and Processing Aids

1. The supplements should be from natural sources and in compliance with the list published by the competent authority (Bureau of Indian Standards-BIS).
2. Feed processing aiding supplements like binders, anti-caking agents, emulsifiers, stabilizers, thickeners, surfactants, coagulants if used should be from natural sources.
3. Antioxidants: only from natural sources shall be permitted
4. Preservatives: only natural acids are allowed;
5. Colouring agents (including pigments), flavors, odor masking agents and appetite stimulants: only natural sources are allowed
6. Probiotics, enzymes and microorganisms are allowed; but should not be from genetically modified sources.
7. Any synthetic chemicals, such as, antibiotics, coccidiostat, medicine, growth promoters or any other substance supplemented for purpose to stimulate growth or production shall not be fed to the organic livestock and poultry.

8. Silage additives, additives for enriching crop residues and processing aids may not be derived from genetically engineered/modified organisms or products thereof, and may be comprised of only: Sea salt; Coarse rock salt; Yeasts; Enzymes; Whey; Sugar; or sugar products such as molasses, jaggery, grain bran; Honey; Lactic, acetic, formic and propionic bacteria, or their natural acid product when the weather conditions or the fodder harvesting conditions could be perceived as a constraint to adequate fermentation provided that it is approved by the competent authority.

12. Health Care

The organic livestock and poultry, in general, should follow the basic principles of preventive health and productivity management wherein the focus would be on preventing diseases, detecting underlying fertility and production problems and its correction primarily on correcting management, nutrition and sanitation. The health care shall be based on the following broad principles:

- The choice of appropriate breeds or strains of animals that can acclimatize, adapt to environment
- The setting up of the animal husbandry practices should be appropriate to the requirements of each species and should focus on encouraging strong resistance to disease and prevention of infections
- The use of good quality organic feed, together with regular exercise and access to fodder/roughages, and/or open-air runs, so as to have positive effects on natural immunological defense of the animal; appropriate stocking density of livestock and poultry so as to avoid overcrowding and spread of infections or competition to feeding
- The farm should have an established system of detection of sub-clinical, sick or injured animals and if, so detected, must be treated immediately.
- The use of veterinary medicinal products in organic farming shall comply with the certain principles required by the law.
- The use of allopathic veterinary drugs or antibiotics or drugs derived from genetically modified source for preventative treatments and for enhancing productivity or fertility is prohibited.
- Hormonal treatment may only be used for therapeutic reasons and under veterinary supervision.
- Growth stimulants agents or substances used for the purpose of stimulating growth or production shall not be permitted.

13. Breeding and Management

The major focus of livestock and poultry management shall be to provide care, comfort, and respect to the animals and ensure their welfare in the farming system. Livestock and poultry breeding methods shall be in accordance with and in compliance with the principles of organic farming and shall take into account:

- The breeds and strains most suited to local conditions;
- The preference for reproduction through natural methods, although artificial insemination may be used
- Embryo transfer techniques and the use of hormonal reproductive treatment shall not be used unless prescribed therapeutic directed towards correcting the physiological problem;
- That breeding techniques employing genetic engineering shall not be used.

Mutilation, such as, tail docking, cutting of teeth, trimming of beaks and dehorning are not permitted. In exceptional cases, some of these may be authorized by the accredited Certification Body for reasons of safety (*e.g. dehorning in young animals*) or if they are intended to improve the health and welfare of the livestock and poultry. Such surgical procedures shall be carried out by a registered veterinarian at the most appropriate age; and any suffering to and pain shall be reduced to a minimum. Wherever possible, anesthetic and analgesics shall be used. Physical castration is allowed only in order to maintain the quality of products and traditional production practices (*meat-type pigs, bullocks, capons, etc.*).

14. Manure and Urine Excreta Management

Manure and urine excreta collection and management practices in the organic livestock and poultry farm are a critical component. The collection, handling and disposal of the dung and urine from shed, paddock, open run or grazing areas shall be implemented in a manner that:

- Minimizes soil and water degradation;
- Does not significantly contribute to contamination of water by nitrates, phosphates, and pathogenic bacteria;
- Optimizes recycling of nutrients; and
- Does not include burning or any practice inconsistent with organic practices.

All manure storage and handling facilities, including composting facilities shall be designed, constructed and operated to prevent contamination of ground and/or surface water and shall be in accordance with the national standards established for the purpose. Manure application rates shall be at levels that do not contribute to ground and/or surface water contamination. The accredited Certification Body shall establish maximum application rates for manure or stocking densities as per local conditions. The timing of application and application methods shall not increase the potential for run-off into ponds, rivers and streams.

15. Transport

During transport, the producer shall prevent stress, injury, hunger, thirst, malnutrition, fear, distress, physical and thermal discomfort, pain, disease during the transport and shall observe the following conditions set in law of the land (For

Livestock transport: IS 14904:2007 and for Poultry transport: IS 5238:2001) as given below:

All necessary arrangement be made in advance to minimize length of the journey and meet the animal's need during the journey;

- Animals must be fit for the intended journey;
- Means of transport as well as the loading and unloading facilities must be designed, constructed, maintained and operated so as to avoid injury and suffering and ensure the safety of the animals;
- Personnel that handle animals must be trained and competent as appropriate for this purpose and must carry out their tasks without using violence or any other method likely to cause unnecessary fear, injury or suffering;
- Transport must carry out without delay to the place of destination and the welfare conditions of the animals must be regularly checked and appropriately maintained;
- Sufficient floor area, height and other spacing requirements must be provided for the animals, appropriate to their size and intended journey; and
- Water, feed and rest must be offered to the animals at suitable intervals and should be appropriate in quality and quantity to their species, size and age.

Efforts should be made to avoid or reduce stress of any kind. The use of electric stimulation or allopathic tranquilizers shall not be permitted during loading and unloading of animals.

16. Slaugher of Animals

The slaughter of livestock and poultry shall be undertaken in a manner, which minimizes stress and suffering, and shall be in accordance with the national rules framed for the purpose.

- Livestock (IS 1982:1971): The by-products shall be from animals subjected to proper ante-mortem and post-mortem inspection. The handling, storage and transport of slaughter-house by-products shall follow IS 8895:1978
- Poultry (IS 7049:1973): For handling, processing, quality evaluation and storage of poultry, following shall be maintained. Approved products for cleaning and disinfection of the buildings and installations are listed.

The slaughter, evisceration and packing of poultry should be conducted in such a manner as will result in hygienic processing, proper inspection and preservation for the production of clean and wholesome poultry and poultry products. Separate rooms should be provided for: Live poultry receiving and holding; Washing and disinfection of coops; slaughter and bleeding; feather removal; evisceration, chilling and packing; inedible products room. Water Supply: The quality of water should satisfy the requirements of potable water. Particular attention should be given to ventilation. Illumination should be sufficiently strong, properly situated and should not cause glare. Personnel should wear special working clothes of washable material. Proper training shall be given regarding hygiene, frequent hand washing, disinfection *etc.* Ante-mortem and Post-mortem inspection of poultry shall be done in accordance with IS 6559:1972 Activities such as stunning, bleeding, scalding, plucking, feet removal, evisceration and chilling, draining, grading *etc.* shall be done in accordance with IS 7049: 1973.

The minimum age for slaughter shall be: 81 days for chickens;150 days for capons;140 days for male turkeys;100 days for female turkeys;94 days for guinea fowls, 70 days for female ducks And 84 days for male ducks.

17. Animal Data Reording

The producer shall maintain the health, breeding and production records of animals maintained in the farm. Each animal will be identified with a unique identification number with ear tags as specified by International Committee on Animal Records (ICAR), as adopted by India. The data of the animal should be recorded preferably in digital format so that data retrieval is convenient. The animal data should be accessible to the veterinarian and the inspection Authority.

17.1 Animal Identification

In case of cattle/buffalo, sheep and goat each animal will be applied identification device as prescribed by ICAR and could be two-way plastic ear tag (laser-printed numeric and in bar code) or three-way (additional RFID micro-chip) identification number of the animal. In case of birds, flock may be identified. The identification devices of the out-going animals shall not be recycled and used on other animals.

The above standards are continuously revised/updated by the regulatory authorities to keep pace with international requirements as also the local situations. So, it is important that the latest version of standards should be referred from such agencies. For example, in case of India, the guidelines developed under National Programme for Organic Production (NPOP) available at the Agricultural and Processed Food Products Export Development Authority (APEDA) should be referred.

18. Conclusion

Despite of the favorable situation existing like traditional animal husbandry, Indigenous Technical Knowledge, limited or no antibiotic use, limited chemical fertilizers application, less dependence on market for inputs, there several limitations which restrict the growth of organic animal husbandry in India especially the stocking density, feed and fodder scarcity, sanitation, infectious disease prevalence *etc.* May be the increasing interest in this underdeveloped organic sector by *inter alia* FAO and IFOAM alongside the efforts by the public sector would help develop organic animal husbandry in developing countries. The recent initiative, International Animal Husbandry Alliance (IAHA) by IFOAM may help in this direction.

The Research and development agencies in India as also international organizations should augment funding for research and development efforts to develop organic animal husbandry sector to improve the availability of high quality, safe, organic animal products for the consumers. The Government of India is promoting organic agriculture on priority; especially the current government has announced several measures to encourage organic farming including indigenous breeds and indigenous health systems. The new schemes launched during XII plan especially National Mission on Sustainable Agriculture (NMSA) and Paramparagat Krishi Vikas Youjana (PKVY) have potential to give push to organic livestock farming in India.

REFERENCES

Borell von E and Sorensen J T. 2004. Organic livestock production in Europe: aims, rules and trends with special emphasis on animal health and welfare. Livestock Production Science 90: 3-9.

Chander Mahesh, Kumar Sanjay, Rathore R S, Mukherjee Reena, Kondaiah N and Pandey H N. 2007. Organic-vis-à-vis conventional livestock production potential in India. In Proc. International Conference on organic agriculture and food security, FAO, Rome, Italy, 3-5 May, 48-49.

Chander M, Subrahmanyeswari B, Mukherjee Reena and S Kumar. 2011. Organic livestock production: an emerging opportunity with new challenges to producers in tropical countries. Rev.Sci.Tech.Off. Int. Epiz, 30 (3), 969-983.

Chander Mahesh, Ramswarup Singh Rathore, Reena Mukherjee, Shyamal Kumar Mondal and Sanjay Kumar. 2012. Road Map for organic animal husbandry development in India. In: Gerold Rahmann and Denise Godinho (Eds), Tackling the future Challenges of Organic animal Husbandry, Proc. 2nd Organic Animal Husbandry Conference, Hamburg, Trenthorst, 12-14 September, pp.59-62.

Chander, Mahesh.Organic Animal Husbandry development. 2012. A Road map for Asian countries. Proceedings of Symposium1: Enhancing Livestock Organic Farming, K. Boonyanuwat and S. Phetdikhai(eds), Proceeding of the 15th AAAP Animal Science Congress, 26-30 November, 2012, Thailand.http: // breeding.dld.go.th/biodiversity/images/stories/sirichai/program per cent 20symposium per cent 201.pdf

Chander, Mahesh. 2014. IFOAM IAHA Newsletter. http: //www.ifoam.org/sites/default/files/iaha_newsletter_no_3.pdf

Chander Mahesh and Reena Mukherjee.2005. Organic Animal husbandry: Concept, Status and Possibilities in India – A review. Indian Journal of Animal Sciences 75 (12): 1460-1469.

Chander, Mahesh and Metha Wanapat.2006. Networking for Organic Livestock Production Development in Asian Countries: A suggested paradigm. In: Proc. 1st IFOAM International Conference on Animals in Organic Production, pp. 80-87, 23-25 August, St Paul, Minnesota, USA, 297 p.

Chander, Mahesh, B Subrahmanyeswari and Reena Mukherjee.2013. Organic Animal Husbandry. In: Handbook of Animal Husbandry, pp 365-382. New Delhi, Directorate of Knowledge Management in Agriculture (DKMA), ICAR, p 1549.

Chander, Mahesh and B Subrahmanyeswari.2013 Organic Livestock Farming. Directorate of Knowledge Management in Agriculture, ICAR, New Delhi, 293 p.

Euclides Filho K. (2004). Supply chain approach to sustainable beef production from a Brazilian perspective. Livestock Production Science, 90, 53-61 (Special issue: Trends and Developments in organic livestock farming-papers presented at a session of the IX WCAP congress in Porto Alegre, Brazil on 26-31 October, 2003).

FAO. 2006. Livestock's long shadow: Environmental issues and options. Food and Agriculture Organisation, Rome.

FAO. 2012. http: //www.fao.org/ag/ca/1a.html

FAO. 2002.-Market developments for organic meat and dairy products: Implications for developing countries. Committee on Commodity problems, Intergovernmental group on meat and dairy products, Nineteenth session, 27-29 August, Food and Agriculture Organization of UN, Rome, 28 p.

FAO. 2007. In: Papers submitted, International Conference on Organic Agriculture and Food Security, FAO, Rome, Italy, 3-5 May, 142 p.

Hermansen J E. 2003.Organic livestock production systems and appropriate development in relation to public expectation. Livestock Production Science 80: 3-15.

Harris P J C, Browne A W, Barrett H R and Gandiya F. 2003.The organic livestock trade from developing countries: Poverty, policy and market issues (Final technical report, Programme of Advisory Support Services for Rural Livelihoods, Department for International Development, UK) School of Science and the Environment, Coventry University, UK.

Nalubwama S M, Mugisha A and Vaarst M 2011. Organic livestock production in Uganda: potentials, challenges and prospects, 3(4): 749-57.

Rahmann Gerold and Denise Godinho (Eds). 2012. Tackling the future Challenges of Organic animal Husbandry, Proc. 2nd Organic Animal Husbandry Conference, Hamburg, Trenthorst, 12-14 September, 481 p.

Servicio Nacional de Sanidad y Calidad Agroalimentaria (SENASA) 2003. Situation of organic production in Argentina in 2003. SENASA, Buenos Aires.

Sorensen Jan Tind and Jakonsen Kirsten.2005. Product quality and livestock systems. Livestock Production Science 94(1-2): 1.

Subrahmanyeswari B. and Chander, M.2006. Demonstration units as a strategic research and extension plan to diffuse organic livestock farming in India. In: Proc. 1st IFOAM International Conference on Animals in Organic Production, pp. 282-286, 23-25 August, St Paul, Minnesota, USA, 297 p.

Subrahmanyeswari B. 2007. Knowledge Attitude and Practices of Organic farmers with special reference to livestock farming: An exploratory study in Uttarakhand state of India. Ph.D. Thesis, Indian Veterinary Research Institute (Deemed University), Izatnagar.

Subrahamanyeswari, B. and Mahesh Chander. 2008. Compatibility of animal husbandry practices of registered organic farmers with organic animal husbandry standards (OAHS): an assessment in Uttarakhand. Indian Journal of Animal Sciences, 78(3): 322-327.

Subrahmayeswari B and Chander M. 2008. Livestock production practices of registered organic farmers in Uttarakhand state of India pp. 64-67. In: Proc. Second Scientific Conference of the International Society of Organic Agriculture Research (ISOFAR), held at the 16th IFOAM Organic World Congress, 18-20 June, Modena, Italy, 863 p.

Chapter 18

Integrated Farming Systems for Sustainable Organic Farming

B.L. Manjunath

ICAR-Indian Institute of Horticultural Research,
Hessaraghatta, Bangalore – 560 089, Karnataka
E-mail: manjunath.bl@icar.gov.in

1. Relevance of Integrated Farming Systems in India

The average size of farm holding in India has declined over time to 1.15 ha and out of a total 138.3 million farm holdings in India, 85 per cent are less than two hectares. The average farm size of a marginal holding is 0.39 ha while that of a small holding is 1.42 ha. These two categories of farmers together operate 7,11,52,000 ha out of the gross operational area of 15,95,92,000 ha in India (Agricultural census data base 2010-11). This unique Indian situation of small fragmented holdings and lack of capital investments is not suitable for single commodity farming being practiced in developed countries.

The soil, water, climate, flora and fauna constitute the basic natural resources and the national treasure of any country. The progressive fragmentation of land holdings, degrading natural resource base and emerging concerns of climate change are escalating pressure on land and water. The concept of Integrated Farming Systems (IFS) is more feasible for these small farm holdings as it serves as a tool for linking allied agri- enterprises with the crop production besides offering scope for environment safety and conservation of Agro-biodiversity (Kathiresan, 2009). Further, crop diversification through IFS is intended to give a wider choice in the production of a variety of crops in a given area so as to expand production related activities on various crops. The crop diversification may result in enhanced profitability, reduce pest, spread out labour more uniformly, different planting and harvesting times can reduce risks from weather and new crops can be renewable resources of high value products (Reddy and Suresh, 2009).

Further, integrated farming system is an approach wherein risk in dealing with single component can be through effective resource recycling. In this, a judicious mix of one or more enterprises along with cropping has a complimentary effect through effective reutilization of wastes and crop residues and encompasses additional source of income to the farmers (Korikanthimath and Manjunath, 2009). This approach achieves fuller utilisation of available resources to realise maximum profits and also to stabilise returns. Further, it examines the full range of farm activities closely related to one another by the common use of farmer's land, labour, capital and management factors. IFS activity is focused round a few selected inter-dependent, inter-related and inter-linking production systems, based on crops, animals and related subsidiary professions (Kuruvilla Verghese and Thomas Mathew, 2009). On this basis, IFS models have been suggested by several workers for the development of small and marginal farms across the country (Behera and Mahapatra, 1999; Kathiresan, 2009; Singh *et al.*, 2006).

Development of ecologically stable and environmentally sound integrated farming system models to replace or improve the traditional farming systems would maximize the productivity of the farming systems. It is clear that because of the diversity of the ecological settings, socio-economic condition and history of research systems, no single farming system model would be adequate for all the situations (Merrillsands, 1986). Information on the understandings of the linkages and complementarities of different farm enterprises and the means of resource recycling of one enterprise to others are very meager. Although systems approach to nutrient management is rational, proven research findings are few to adopt. With recent thrust on integrated nutrient management and sustainability of crop productivity, there is an urgent need to exploit the possible means of nutrient management to farming systems as a whole.

Organic farming is basically a farm activity of biological nature. Organic farming introduces a change in the farming system which aims at maximum production with optimal utilization of resources. The farm wastes are better recycled to maintain soil fertility, soil health and bio-chemical activities. Further pest and disease management is attained by means of crop selection, rotation, water management, tillage *etc.* organic farming encourages a balanced host/predator relationship through augmentation of beneficial insect population (Chhonkar, 2003; Dwivedi Vandana 2005; GOI, 2001).

Adoption of organic farming to maneuver the ill effects of chemical farming and to achieve sustainability is the need of the hour. Increasing consciousness about conservation of environment as well as human health hazards associated with agrochemicals and consumer preference to safe and hazard free food are the major factors that lead to growing interest in organic agriculture. Use of organic manures enhance the bulk density of soil and there by reduces resistance to penetration by plant roots (lomte and Pendke, 2004).

For smallholder systems numerous adaptation and mitigation options exist through crop–livestock systems by distinguishing risk management, diversification and sustainable intensification strategies. Despite the potential solutions, smallholders face major constraints at various scales, including small farm sizes, the lack of response to the proposed measures and the multi-functionality of the livestock herd. Hence, there is a need for integrated, system-oriented impact assessments and a realistic consideration of the adoption constraints in smallholder systems (Katrein Descheemaeker *et al.*, 2016).

2. Components of Integrated Farming System

IFS will normally have combinations of three types of enterprises:

2.1 Primary Enterprises

Primary enterprise is normally the mainstay of the farm. An IFS farm may have one or more primary enterprises. They produce the main saleable/utilizable product of the farm and provide the major and sustained source of farm income. By-products of primary enterprises can support complementary enterprises by being their main inputs. Normally the primary enterprise is the highest income earning enterprise on the farm, which the farmer has necessary knowledge and skills, enough infrastructure and adequate labour availability.

Some of the examples of primary enterprises of IFS that are commonly seen in Karnataka are:

- ☆ Crop production: Paddy; paddy – sugarcane; dryland crops like maize, ragi *etc.*
- ☆ Seed production: Field crops; vegetables, flowers *etc.*
- ☆ Horticultural crops: Coconut; arecanut; grapes; mango; pomegranate; banana, vegetable crops; flower crops *etc.*
- ☆ Plantation crops: Coffee; rubber; tea: cashew *etc.*
- ☆ Sericulture: Mulberry cultivation and silk worm rearing,
- ☆ Animal husbandry: Dairy; poultry; sheep rearing; goat rearing *etc.*

2.2 Complementary Enterprises

These enterprises basically help in effective utilization of the by-products of primary and other enterprises. They normally have mutually beneficial relations with primary enterprises. Their by-products are valuable inputs for other enterprises.

Some of the examples of complementary enterprises are;

- ☆ Dairy, poultry, based on by-products of crop production
- ☆ Mushroom production based on paddy straw
- ☆ Food processing – based on fruits, vegetables, cereals, pulses, *etc.*
- ☆ Apiculture – based on the flowers from field and horticultural crops
- ☆ Compost preparation – based on farm wastes
- ☆ Vermicomposting based on farm wastes
- ☆ Biogas production

2.3 Supplementary Enterprises

These enterprises need not necessarily depend on other enterprises on the farm, but will be taken up as additional activities to make effective use of manpower and to supplement the farm income. Some of the examples of supplementary enterprises are; back yard poultry, sheep rearing, pig rearing, kitchen gardening, skill based enterprises like carpentry, smithy, *etc.* and non agriculture based enterprises.

3. Recycling of Organic Manual Resources: The Key in Sustenance

Integrated Farming system approach envisages the integration of agroforestry, horticulture, dairy, sheep and goat rearing, fishery, poultry, pigeon, biogas, mushroom, sericulture and by-product utilization of crops with the main goal of increasing the income and standard of living of small and marginal farmers. Integrated systems are about bringing crops and livestock into an interactive relationship with the expectation that together, as opposed to alone, they will generate positive effects on outcomes of interest, such as profitability, overall productivity, and conservation of non-renewable resources. It is, however, much more than this. The "system" includes the environment, soil characteristics, landscape positions, genetics, and ecology of plant and animals. It involves management practices, goals and lifestyles of humans, social constraints, economic opportunities, marketing strategies and externalities including energy supplies and costs and impacts of farm policies. Systems also reflect natural resources available and the impact on their use, wildlife issues, target and non-target plant and animal species, micro-organisms, and indeed all of the definable and indefinable factors that ultimately interact to result in an outcome that is never constant. (Allen *et al.*, 2008)

The effective recycling of farm resources is possible by adoption of farming system research. Crop by-product is utilized as fodder for animals, and animal by-product *i.e.* milk, and dung may be utilized for increasing income and soil fertility, respectively.

In Punjab, Haryana and Western Utter Pradesh, burning of wheat and rice residues is common practice, which increased the concentration of greenhouse gases in atmosphere, in addition huge amount of nutrients are lost. Such situation could be avoided by introduction of some more enterprises like animal husbandry on the farm. Rice straw may be used as animal feed. This farm residues serve as basal diet for dairy animals with little supplementation of concentrates maintains high levels of production. Thus effective recycling of agriculture by products/crop residues for feeding of farm animals and in turn use of animal by-products like dung and urine recycled via biogas or fish pond will not only maximise the farm income but also will maintain a clean environment.

One of the most important aspects of any animal-based agricultural operation is having an effective waste management plan which reaps the benefits and helps reduce the risks associated with the use and disposal of animal wastes. If sheep and goat and/or lambs spend any part of the year in barns, stalls, pens, loafing areas or feeding areas, you will need to deal with manure from those areas. Manure is not just the urine and feces from livestock, but also the bedding, runoff, spilled feed, and anything else mixed with it. A complete manure management system involves collection, storage (temporary or long-term) and ultimate disposal or utilization.

Manure production varies with breed and feeding levels. The amount of bedding to be handled with the manure depends on the housing system selected. Manure contains valuable nutrients like nitrogen (N), phosphorus (P) and potassium (K). In addition to the three major elements, manure also contains essential micro-nutrients (boron, calcium, copper, iron, magnesium, manganese, molybdenum, sulfur, and zinc. Manure nutrients come from the feed that the animals have eaten.

Anywhere from 75 to 90 per cent of the plant nutrients fed to animals are excreted in their manure, so it should be no surprise that the stuff is an excellent fertilizer. The level of protein and inorganic salts in feed (sodium, calcium, potassium, magnesium, phosphate, and chloride) will be reflected in the characteristics of the manure. The amount of nutrients depends on the type of animal and the way the manure is handled. Manure that contains a lot of bedding will contain fewer nutrients than pure manure.The nutrients in manure are a mixture of inorganic and organic forms. The inorganic forms are similar to those in commercial fertilizer. They dissolve in water and plants can use them right away. The organic forms come from the remains of plant tissue, cells, and bacteria that are in the manure. They are slow-release nutrients that the plants cannot use right away. They become available to the plants as the manure decays. The organic matter in manure is also valuable because it makes soil easier to manage, less likely to erode, and more likely to absorb water. Regular manure application lowers soil pH. The acidifying effect of manure is less than that of inorganic fertilizers.

Storage of livestock wastes and wastewater involves accumulating manure in an environmentally sound manner until they can be applied to land or otherwise utilized. Manure storage facilities allow farmers to spread manure when the conditions are right for nutrient use by crops. The ideal storage site for solid manure is a roofed shed with an impermeable floor (*e.g.* concrete). Dry manure can be stored in solid form in stockpiles; however, the piles should be covered. Obviously, manure storage structures or sites should be located to minimize odour nuisance to neighbors.

The manure from different animal species has different physical and biological characteristics which must be taken into account when deciding on which recycling method to use. The manure from goats, sheep and rabbits is not offensive, but it is not a suitable substrate in a simple biodigester system. The pellet-like nature of the manure causes it to float on the surface of the liquid inside the biodigester, and scum formation becomes a problem. This kind of manure is better recycled through earthworms. It is convenient to house sheep, goats and rabbits on raised slatted floors. It is then a simple procedure to add some earthworms to the manure that has fallen through the floors. Any feed residues can be combined with the manure. Irrigating the earthworm beds with the effluent from the biodigester will speed up the process of decomposition and growth of the worms. The residue from this process (vermicompost) can be removed at intervals and used directly as manure.

The average nutrient composition of a manure depends on the livestock species, their age, feeding material, management *etc.* However, the mean nutrient content is given in the Table 18.1.

Table 18.1: The Mean Nutrient Content of Recycled Manurial Resource from the Integrated Systems

Manure	*Nitrogen (N per cent)*	*Phos-phorus (P_2O_5 per cent)*	*Potassium (K_2O per cent)*	*Calcium (Ca per cent)*	*Magne-sium (Mg per cent)*	*Organic Matter (per cent)*	*Moisture Content (per cent)*
Fresh manure							
Cattle	0.5	0.3	0.5	0.3	0.1	16.7	81.3
Sheep	0.9	0.5	0.8	0.2	0.3	30.7	64.8
Poultry	0.9	0.5	0.8	0.4	02	30.7	64.8
Horse	0.5	0.3	0.6	0.3	0.12	7.0	68.8
Swine	0.6	0.5	0.4	0.2	0.03	15.5	77.6
Treated dried manure							
Cattle	2.0	1.5	2.2	2.9	0.7	69.9	7.9
Sheep	1.9	1.4	2.9	3.3	0.8	53.9	11.4
Poultry	4.5	2.7	1.4	2.9	0.6	58.6	9.2

Source: Prahlad *et al.* (2012).

4. Sustainable Agriculture through Integrated Farming Systems

Sustainable agriculture is a way of raising food that is healthy for consumers and animals, does not harm the environment, is humane for workers, respects animals, provides a fair wage to the farmer, and supports and enhances rural communities. The sustainability is assessed quantitatively through Sustainable Yield Index (SYI) which denotes the minimum guaranteed yield as a percent to the maximum observed yield with high probability as per Singh *et al.* (1990) given by

$$SYI = \frac{Y - \sigma}{Y_{max}}$$

Where Y is the estimated average yield of a treatment over years, σ is its estimated standard deviation and Y_{max} is the observed maximum yield in the experiment. Sankar and Reddy (2005) inferred that highly sustainable practices are with a sustainability index of 0.96. Further, SYI is an approach to evaluate the minimum yield likely to be achieved in relation to changes in soil organic carbon and available NPK status of soil.

It is very essential that the agricultural activities will lead to sustainable quality of life as they support the majority of the population. Field crops occupy a major area of the cultivated lands and as such following sustainable practices are very essential for continued food production from the shrinking resource base.

4.1 Sustainable Farming Systems

Majority of farming activities remain subsistence in nature rather than commercial. Any farm activity, other than crop production, requiring considerable capital· investment is not within the capabilities of such farmers. However, such improved capital requiring enterprises like dairy, poultry, fishery *etc.* have been taken up on commercial scale in and around urban areas where there is no problem for marketing the products. Small and marginal farmers are still adopting low capital requiring traditional farming systems to meet their domestic needs. The intensity of farming systems, generally, increases with increasing size of the farm holdings and family members available for work. More common farming systems of small and marginal farmers are indicated below:

Table 18.2: The IFS Components Suggested for different Land Holdings in India

Sl.No.	*Land Holding (ha)*	*Components Suggested*
1	Less than 1. 0	Crops with a pair of draught animals and 5 to 10 chicks
2	1.0 to 2.0	Crops with a pair of draught animals, one milch cow/buffalo, one or two sheep/goats and 10 to 15 chicks
3	2.0 to 3.0	Crops with two pairs of draught animals along with two milch cows/ buffalos, 2 or 3 sheep/goats and 10 to 20 chicks
4	3.0 to 5.0	Crops including sericulture with 2 to 3 pairs of draught animals along with 3 or 4 milk cows/buffalos and 10 to 25 chicks.

The above are only examples to give an idea of the existing farming systems. Actual combinations and the sizes of enterprises vary greatly depending on farming situations. From the above, it is evident that the small sizes of farming systems, though intensive, are aimed at production stability and subsistence. Farmers are well aware of the role of improved breeds and related production technology for overall farm productivity. Major problem limiting switching over to improved breeds is the necessity for additional inputs and greater care to maintain the improved breeds. With the improved breeds and latest production technology, big landlords and industrialists have made spectacular progress in dairy, poultry and fisheries as commercial enterprises but not as components in farming systems. Several small and marginal farmers are now maintaining one or two dairy cattle as the private dairy industries are collecting milk even from remote areas. This indicate that there is scope for the farmers in remote rural areas to go for integrated farming systems involving livestock, if transport and marketing problems are solved.

On-station and on-farm research in different rainfed regions of the country has resulted in identification of a number of sustainable and profitable IFS models which can be adopted by the farmers. The main advantages of the farming systems approach involve: (i) *In situ* recycling of organic residues including farm wastes generated at the farm to reduce the dependency on external inputs, (ii) reduction in cost of production through effective use of byproducts and enhanced input use efficiency, (iii) minimizing risk associated with dependence on a single farming enterprise, (iv) conservation of natural resources, and (v) food security, employment generation, enhanced system productivity and income/unit area. In the IFS approach, several components like crops, livestock, horticulture, poultry and off-farm enterprises get positively integrated at the household level with a view to optimize income and resource use. Farming systems are highly location specific depending on the natural resources available in a given agro-climatic situation. The success of an IFS and its sustainability depends entirely on strengths and weaknesses of the internal factors like natural resources (soil and water, farm size, availability of recyclable residues, labour and external factors such as market prices, infrastructure and the level of post-harvest processing). Soil health and emerging nutrient deficiencies are important issues in any successful IFS, which need to be addressed.

The virtues of IFS include the less dependency of chemical fertilizers/external sources of nutrients and mainly depend of self sustaining nutrient recycling. However, the adequacy of nutrient supply depends on the number and type of components in IFS. However, for growing cereal components of IFS, external nutrient application is needed. Therefore, managing nutrient needs of annual, perennial and fodder systems through integrated nutrient management (INM) are key component in IFS. Cover crops, legume intercropping, green leaf manuring like *Glyricidia* plantations on farm bunds, biogas plant linked with vermicomposting, crop residue recycling, organic mulching, *etc.* are important forms of nutrient management. Soils of India are deficient in organic carbon, which is critical for most of soil processes such as nutrient transformations and availability, and soil physical, chemical and biological health. IFS offer a good platform to provide regular organic matter addition so that soils in IFS are healthy and sustainable. Whenever cereal residue is incorporated, some external quantity of N fertilizer needs to be applied for avoiding any form of N immobilization. Phosphorus additions are needed depending upon crops. Potassium requirements are almost met through organic waste recycling. Based on long term manurial experiments in dry ecosystems in India, under predominant production systems such as sorghum (*Sorghum bicolor*), finger millet (*Eleucine coracana*), pearl millet (*Pennisetum glaucum*), maize (*Zea mays*), upland rice (*Oryza sativa*), groundnut (*Arachis hypogeae*), soybean (*Glycine max*), cotton (*Gossypium* spp.), food legumes *etc.*, integrated use of locally available organic resources along with need based chemical fertilizers contributed in the improved crop productivity and stability of yields. With some of the above nutrient management practices, micronutrient deficiencies are also limited in IFS modules (Srinivasarao and Vittal, 2007).

REFERENCES

Allen VG, Brown CP, Segarra E, Green CJ, Wheeler TA, Acosta-Martinez V and Zobeck TM. 2008. In search of sustainable agricultural systems for the Llano Estacado of the US Southern High Plains. Agriculture Ecosystems and Environment, 124(1-2): 3-12.

Prahlad, Sreedhara JN and Biradar SA. 2012. Sheep and goat rearing in IFS. In: *Lecture notes of ICAR sponsored winter school on integrated farming systems* (Desai BK *et al.*, Eds.), University of Agricultural Sciences, Raichur, Karnataka.

Behera UK and Mahapatra IC. 1999. Income and employment generation for small and marginal farmers through integrated farming systems. Indian Journal of Agronomy, 44(3): 431-439.

Chhonkar PK. 2003. Organic farming: Science and belief. Journal of the Indian Society of Soil Science, 51(4): 365-377.

Dwivedi Vandana. 2005. Organic farming: Policy initiatives, Paper presented at the *National Seminar on National Policy on Promoting Organic Farming*, 10-11 March, 2005. pp.58-61.

GOI 2001. The report of the working group on organic and biodynamic farming, Planning Commission, Government of India. pp. 1-25.

Lomte MH and Pendke MS. 2004. Organic farming inventory. Proceedings of the State level Seminar on Present and Future Prospects of Organic Farming, Marathwada Agricultural University, Parbhani, pp. 1-133.

Kathiresan RM. 2009. Integrated farm management for linking environment. Indian Journal of Agronomy, 54 (1): 9-14.

Katrien Descheemaeker, Simon J. Oosting, sebine Homann-Kee Tui, Patricia Masikati, Gatien N. Falconnier and ken E.Giller, 2016. Climate change adaptation and mitigation in small holder crop-livestaock systems in sub-Saharan Africa: a call for integrated assessments, *Reg. Environ. Change*, DOI 10.1007/s10113-016-0957-8 published online on 02 April, 2016.

Korikanthimath VS and Manjunath BL. 2009. Integrated farming systems for sustainability in agricultural production. Indian Journal of Agronomy, 54(2): 140-148.

MerrillSands D. 1986. Farming systems research: clarification of terms and concepts. Experimental Agriculture, 22: 89-104.

Singh YV, Jha KP and Radhakant Sahoo. 2006. Integrated farming system model for a two-hectare lowland farmer of alluvial zone. Indian Farming, 56(2): 12-14.

Srinivasarao Ch and Vittal KPR. 2007. Emerging nutrient deficiencies in different soil types under rainfed production systems of India. Indian Journal of Fertilizers, 3: 37-46.

Chapter 19

Organic Processing of Meat and Meat Products in India

Prabhat Kumar Pankaj[1]*, *D.B.V. Ramana*[1], *S. Girish Patil*[2], *M. Muthukumar*[2] *and Baswa Reddy*[2]

[1,2]*ICAR-Central Research Institute for Dryland Agriculture, Hyderabad – 500 059, Telangana*

[2]*ICAR-NRC on Meat, Hyderabad – 500 092, Telangana*

**E-mail: dr.prabhatkumarpankaj@gmail.com*

1. INTRODUCTION

Food and Agricultural Organization (FAO)/World Health Organization (WHO) Codex Alimentarius Commission defines organic farming as "a unique production system which promotes and enhances agro-ecosystem health, including biodiversity, socio-economic biological cycles and soil biological activity, and this is accomplished by using on-farm agronomic, biological and mechanical methods in exclusion of all synthetic off-farm inputs". Organic system of production aims to promote sustainable production system which is environmental friendly and assures highest standards of welfare to animals. Demand for organic products is steadily increasing across the world due to increasing health consciousness among consumers. Scientific studies have also shown that organic meat has very low total lipid content and possess poly unsaturated fatty acid/saturated fatty acid and n-6/n-3 indices within the recommended values for the human diet. Keeping in mind the low input practices followed by Indian farmers especially in rural and tribal areas it is assumed that India holds tremendous potential in organic food production. This has created a business opportunity for producers too.

A lot of confusion exists among consumers as to what "organic" means. Marketers have extracted almost every word that connotes organic – *natural, naturally raised, free range, hormone-free* and *additive-free* – each means far from being "certified organic". Organic meat comes from an animal that has not been fed anything grown with toxic or synthetic fertilizers, pesticides, herbicides, fungicides or fumigants; has not been given any kind of growth hormone, antibiotic or genetically engineered product; has been conceived by organically raised animals; and has been butchered and processed following organic regulations.

In India, organic production practices are guided by National Program for Organic Production (NPOP) guidelines of Agricultural and processed Food Products Export Development Authority (APEDA). Production of organic meat and meat products is challenging as it involves compliance to organic standards at every step starting from production of organic feed, rearing of meat animals, slaughter, value addition, packaging and labeling. Often each of these steps is undertaken by different sets of people at different locations. Hence, organic meat production requires brining about a complete change in the production and processing patterns of meat and meat products. So far production of organic meat and meat products is very limited in India. This chapter provides brief information about different aspects of organic meat processing including method of production and marketing opportunities.

2. Organic Meat Production in India

There is no inherent nutritional advantage in the organic foods over non-organics, but it is a type of production system which makes organic food "healthier" than non-organics. Western countries have more stringent laws for use of pesticides as compared to conventional farming in India. India is one of the largest users of pesticide in Asia and also one of the largest manufactures. So, organic food/meat production has a very special importance in India.

Organic cultivation is particularly suitable for a country like India with a huge population of small farmers who still use traditional methods of farming with few agricultural inputs. It is estimated that 65 per cent of the country's cropped area is organic by default, as the small farmers have no choice but to farm without chemical fertilizers and pesticides as they cannot afford. This default status coupled with India's inherent advantages, such as, its varied agro-climatic regions, local self-sustaining agri-systems, sizeable number of progressive farmers and ready availability of inexpensive manpower translate into the potential to cultivate a vast basket of products organically. Similarly, animal husbandry practices followed in tribal and rural areas are near organic as they do not involve much of chemical and synthetic feed additives.

Currently organic meat marketing is limited to supermarkets catering to niche consumer segments in different metro cities. In general, Indian meat industry is an unorganized sector. Although there are few ultra-modern export units they do not have control over production practices of farmers. On the other hand, farmers do not find it worth to follow organic animal rearing as there is no organized setup for its collection and processing. Entrepreneurs interested in organic production are also often deterred by prevalence of diseases and extreme weather conditions. In addition, lack of availability of medications alternative to traditional allopathic drugs is an impediment in growth of organic meat animal production. Extensive research and production of medications acceptable in organic animal husbandry like Ayurvedic, Homeopathic, *etc.* are crucial for development of the sector in India.

3. Regulations for Organic Meat and Meat Products in India

Organic animal husbandry in India is guided by NPOP published by APEDA in 2001. There are several certifying agencies, which provide certification based on the NPOP guidelines. It is necessary to involve certifying agencies from the initial stage of establishment of organic production units to ensure that all necessary guidelines are followed and documented suitably. Only those farms, which have been certified, can put the label of 'organic'.

NPOP provides a complete protocol to be followed for rearing of meat animals. General principle for organic animal husbandry should be governed by the physiological and ethological needs of the farm animals in question. This requires animals to be allowed to conduct their basic behavioural needs and that all management techniques should be concerned for the good health and welfare of the animals. Animals must be provided space for free movement and access to sufficient fresh air and natural daylight. Ample access to fresh water and feed must be provided according to the needs of the animals. No compounds used for construction materials or production equipment shall be used which might detrimentally affect human or animal health. All animals shall have access to open air and/or grazing appropriate to the type of animal and season taking into account their age and condition.

3.1 Conversion Period

The establishment of organic animal husbandry requires an interim period, the conversion period. The whole farm, including livestock, should be converted according to the prescribed periods. Conversion may be accomplished over a period of time. Replacement animal should be brought onto the holding at the start of the production cycle. Animal products may be sold as "product of organic agriculture" only after the farm or relevant part of it has been under conversion for at least twelve months and provided the organic animal production standards have been met for the appropriate time. The certification programme shall specify the length of time by which the animal production standards shall be met. With regard to dairy and egg production, this period shall not be less than 30 days. Animals present on the farm at the time of conversion may be sold for organic meat if the organic standards have been followed for 12 months. All organic animals should be born and raised in the organic holding. When organic livestock is not available, the certification programme will allow brought-in conventional animals according to the following age limits: 2 day old chickens for meat production, 18 week old hens for egg production, 2 week old for any other poultry, piglets up to six weeks and after weaning and calves up to 4 weeks old which have received colostrum and are fed a diet consisting mainly of full milk.

3.2 Breeds and Breeding

Breeds should be chosen which are adapted to local conditions. Breeding goals should not be at variance with the animal's natural behaviour and should be directed towards good health. Breeding shall not include methods which make the farming system dependent on high technological and capital intensive methods. Artificial insemination is allowed. Embryo transfer techniques are not allowed in organic agriculture.

3.3 Mutilations

Mutilations are not allowed but shall allow the following exceptions: castrations, tail docking of lambs, dehorning, ringing and mulesing (the process of removing folds of skin from the tail area of a sheep, intended to reduce fly strike). Suffering shall be minimized and anaesthetics used where ever appropriate.

3.4 Animal Nutrition

The livestock should be fed 100 per cent organically grown feed of good quality. All feed shall come from the farm itself or be produced within the region. The diet should be balanced according to the nutritional needs of the animals. Products from the organic feed processing industry shall be used. The certification programme shall draw up standards for feed and feed ingredients. The prevailing part (at least more than 50 per cent) of the feed shall come from the farm unit itself or shall be produced in co-operation with other organic farms in the region. The certification programme shall allow exceptions with regard to local conditions under a set of time limit for implementation. For the calculation purpose only, feed produced on the farm unit during the first year of organic management, may be classed as organic. This refers only to feed for animals which are themselves being reared within the farm unit and such feed may not be sold or otherwise marketed as organic.

3.5 Products not Allowed in Farming

Synthetic growth promoters or stimulants, synthetic appetizers, preservatives, except when used as a processing aid *etc.* vitamins, trace elements and supplements can be used from natural origin when available in appropriate quantity and quality. The certification programme shall define conditions for use of vitamins and minerals from synthesized or unnatural sources. All ruminants shall have daily access to organically grown roughages.

3.6 Veterinary Medicine

Management practices should be directed to the wellbeing of animals, achieving maximum resistance against disease and preventing infections. Sick and injured animals shall be given prompt and adequate treatment. Natural medicines and methods, including homeopathy, ayurvedic, unani medicine and acupuncture, shall be emphasized. When illness does occur, the aim should be to find the cause and prevent future outbreaks by changing management practices. Vaccinations shall be used only when diseases are known or expected to be a problem in the region of the farm and where these diseases cannot be controlled by other management techniques.

4. International Standards for Organic Meat Processing

Legislated standards are established at the national level, and vary from country to country. In recent years, many countries have legislated organic production, including the European Union (EU) nations (1990s), Japan (2001), and the US (2002). Non-governmental national and international associations also have their own production standards. In countries where production is regulated, these agencies must be accredited by the government. Since 1993 when EU Council Regulation 2092/91 became effective, organic food production has been strictly regulated in the UK. In India, standards for organic agriculture were announced in May 2001, and the NPOP is administered under the Ministry of Commerce. (http://www.apeda.com/organic/quality.html).

Every country has its own guidelines for production of organic food production. In 2002, the United States Department of Agriculture (USDA) established production standards, under the National Organic Program (NOP), which regulate the commercial use of the term organic, whereas Australia has National Standard for Organic and Bio-dynamic products. With the aim to harmonize organic standards across the world, Food and Agricultural Organization (FAO) has published a set of standards titled 'Guidelines for the production, processing, labelling and marketing of organically produced foods' (GL 32 – 1999, Rev. 1 – 2001). Codex Alimentarius Commission (CAC) has 'Guidelines for the production, processing, labelling and marketing of organically produced foods' (GL 32 – 1999) for guiding organic food production. In India, organic meat production is guided by NPOP guidelines of APEDA which has been recognized by European Commission and Switzerland as equivalent to their country standards. Similarly, USDA has recognized NPOP conformity assessment procedures of accreditation as equivalent to that of US. With these recognitions, Indian organic products duly certified by the accredited Certification Bodies of India are accepted by the importing countries. However, producer aiming to export organic products must essentially know the organic standards of importing countries and follow it accordingly to increase the acceptability of the products.

5. Transport and Slaughter of Livestock for Organic Meat Processing

As per NPOP guidelines, transport and slaughter should minimize stress to the animal. Transport distance and frequency should be minimized. The transport medium should be appropriate for each animal. Animals should be inspected regularly, watered and fed during transport depending on weather conditions and duration of the transport. Stress to the animal shall be minimized, especially taking into consideration: contact (by eye, ear or smell) of each animal with dead animals or animals in the killing process, resting time to release stress and each animal shall be stunned before being bled to death. The equipment used for stunning should be in good working order. Exceptions regarding stunning can be made according to cultural practices. Where animals are bled without prior stunning, this should take place in a calm environment. Throughout the different steps of the process, there shall be a person responsible for the well- being of the animal. Handling during transport and slaughter shall be calm and gentle. The use of electric sticks and such instruments are prohibited. Slaughter and transportation standards that will take into consideration: stress caused to the animal and person in-charge, fitness of the animal, loading and unloading, mixing different groups of animals or animals of different sex, quality and suitability of mode of transport and handling equipment, temperatures and relative humidity, hunger and thirst and specific needs of each animal. No chemical synthesized tranquilizers or stimulants shall be given prior to or during transport. Each animal or group of animals shall be identifiable during all steps. Where the transport is by axle, the journey time to the slaughter house shall not exceed eight hours. In general, animals slaughtered must be certified organic and record trail should be maintained till its marketing to ensure quality.

6. Organic Sanitizing and Disinfecting Agents for Meat Processing Plants and Handlers

Sanitizers and disinfectants are extremely essential for maintaining food safety standards in meat processing plant. In India, no specific list of organic sanitizers is approved under NPOP. Hot water under pressure will considerably help in maintaining hygiene standards. Any compounds which do not adversely affect the quality of meat can be used. In United States, chlorine is allowed as fungicide, disinfectant and sanitizer for organic production. Ozone is also approved and is having comparable disinfecting power as that of chlorine, but it is more costly as it is a highly unstable compound. Apart from the above, Peroxy-acetic acid (PAA) is also allowed in organic production.

List of products authorized for cleaning and disinfection of buildings and installations include potassium and sodium soap, water and steam, milk of lime, lime, quicklime, sodium hypochlorite (*e.g.* as liquid bleach), caustic potash, hydrogen peroxide, natural essences of plants, citric, peracetic acid, formic, lactic, oxalic and acetic acid, alcohol, nitric acid (dairy equipment), phosporic acid (dairy equipment), formaldehyde and sodium carbonate.

7. Ingredients, Additives and Processing Aids for Organic Meat Products

Production and marketing of organically processed value added meat products provides logical completion to effort put for maintaining organic standards in the farm, transportation and slaughter of animals. Major ingredients used for production of value added meat products are meat, fillers, condiments, spices, oil, nitrite, phosphate, salt and water. To get the organic certification, all the ingredients used for production processed meat products must be certified organic. With the growth of organic agriculture across India, availability of organic ingredients of plant origins is rapidly increasing. Consequently, organic flours, spice mix, oil, vegetables, *etc.* are easily available in super markets especially in major cities. Nitrate/nitrite and phosphates are the commonly used chemicals in the production of value added meat products. But for producing organic processed products, these chemicals should not be used. But nitrite is an important component which gives cured color and flavor and inhibits *Clostridium botulinum*. Several vegetable ingredients contain high content of nitrate which needs to be converted to active form nitrite to make it functional in meat system. Addition of lactic acid starter culture can convert nitrate to active form nitrite (Agricultural Utilization Research Institute, http://www.auri.org). During production of organic processed meat products, care should be taken not to allow food to come in contact with chemicals contraindicated under organic system. 100 per cent of the ingredients of agriculture origin shall be certified organic. For the production of enzymes and other micro-biological products the medium shall be composed of organic ingredients. In cases where an ingredient of organic agriculture origin is not available in sufficient quality or quantity, the certification programme may authorize use of non-organic raw materials subject to periodic re-evaluation. Such non-organic raw material shall not be genetically engineered. The same ingredient within one product shall not be derived both from an organic and non-organic origin. Water and salt may be used in organic products. Minerals (including trace elements), vitamins and similar isolated ingredients shall not be used. The certification programme may, grant exceptions where use is legally required or where severe dietary, or nutritional deficiency can be demonstrated.

Preparations of micro-organisms and enzymes commonly used in food processing may be used, with the exception of genetically engineered micro-organisms and their products. The use of additives and processing aids shall be restricted. In cases where an ingredient of organic origin is unavailable in sufficient quality or quantity, the standard-setting organization may authorize use of non-organic raw materials subject to periodic review and re-evaluation. These materials shall not be genetically engineered.

Processing methods should be based on mechanized, physical and biological processes. The vital quality of an organic ingredient shall be maintained throughout each step of its processing. Processing methods shall be chosen to limit the number and quantity of additives and processing aids. The following kinds of processes are approved, like mechanical and physical, biological, smoking, extraction, precipitation and filtration. Extraction shall only take place with water, ethanol, plant and animal oils, vinegar, carbon dioxide, nitrogen or carboxylic acids. These shall be of food grade quality, appropriate for the purpose. Irradiation is not allowed. Filtration substances shall not be made of asbestos nor may they be permeated with substances which may negatively affect the product.

As per USDA processed organic meat products are classified as follows:

- **100 percent organic**: Product must contain 100 percent organically produced ingredients, not counting added water and salt.
- **Organic:** Product must contain at least 95 per cent organic ingredients, not counting added water and salt.
- **Made with organic ingredients**: Product must contain at least 70 per cent organic ingredients, not counting added water and salt.
- **Less than 70 per cent organic**: Products with less than 70 per cent organic ingredients are not allowed to be labeled as organic and are only permitted to list those ingredients that are organic on the information/ingredient panel of the product.

7.1 Natural Sources for Organic Meat Preservatives

A number of different natural-based organic acids offer a significant improvement to food safety. One of the main reasons consumers choose organic foods is to limit their intake of manufactured ingredients. Most ingredients and technologies that serve as antimicrobials-ingredients that can improve the safety by either suppressing, inhibiting, or destroying any pathogenic bacteria-are not able to be used in products labeled 'natural' and 'organic. Fortunately, there are some compounds from natural sources that work as well as standard preservatives such as sodium nitrate and nitrite or sodium erythorbate. The problem is that it can take heavy doses of some natural ingredients to provide equivalent results. Cranberry concentrate is a very effective natural anti-microbial, but if you use enough to control the growth of bacteria, the meat turns cranberry red. The challenge is to ensure that organic meats are both safe and appetizing. Cherry powder combined with celery powder is already being adopted by processors that make organic and natural meats because of how effective these ingredients are in improving meat safety and quality.

8. Packaging and Labeling of Organic Meat

Being a highly perishable product, meat needs to be packaged appropriately to maintain its quality. Packaging plays an important role to protect, contain and promote the food products. While finalizing the packaging strategy for organic food, it is essential to understand the nature of products like their physical form (minced, whole cut, pieces, *etc.*), characteristics (moisture level, pH, fat content *etc.*) and stage of processing (convenience, semi-convenience, ready to eat, *etc.*). Problems of preserving food and transporting it to the final consumer in the best possible condition must be addressed. Basic principle of organic system is sustainability. Hence, packages used for containing organic meat must be safe, sourced in responsible way, environmental friendly and maximize the use of renewable material. Packaging material must not release any chemicals to meat products during storage and transportation.

Certified organic meat labels are applied to animals raised without antibiotics, growth hormones or genetically modified feed. These animals have access to pesticide/insecticide-free pastures, as well as access to fresh air, sunshine, water and exercise. The dams of organically raised animals must also have been feed organic feed for at least the last third of gestation. Feed destined for organically raised livestock must be 100 per cent organic, and it must be free from animal parts of any kind. The feed must not contain plastic pellets for roughage, or contain urea or manure – all allowable practices in conventional agriculture.

Providing information to consumers through labeling is also an important function of packaging. Organic labeling must display clearly the guideline followed for producing the product, certification details and the logo approved by the authorities. Effective labeling containing all the information necessary to consumers in attractive way will enhance the marketability of the products. As per NPOP guidelines, the person or company legally responsible for the production or processing of the product shall be identifiable on the label. Single ingredient products may be labeled as "produce of organic agriculture" or a similar description when all standards requirements have been met. Mixed products where not all ingredients, including additives, are of organic origin may be labelled in that way. The word "organic" may be used on the principal display in statements like "made with organic ingredients" provided there is a clear statement of the proportion of the organic ingredients. An indication that the product is covered by the certification programme may be used, close to the indication of proportion of organic ingredients and where less than 70 per cent of the ingredients are of certified organic origin; the indication that an ingredient is organic may appear in the ingredients list. Such product may not be called "organic". Added water and salt shall not be included in the percentage calculations of organic ingredients. The label for in-conversion products shall be clearly distinguishable from the label for organic products. All raw materials of a multi-ingredient product shall be listed on the product label in order of their weight percentage. It shall be apparent which raw materials are of organic certified origin and which are not. All additives shall be listed with their full name. If herbs and/or spices constitute less than 2 per cent of the total weight of the product, they may be listed as "spices" or "herbs" without stating the percentage. Organic products shall not be labelled as GE (genetic engineering) or GM (genetic modification) free in order to avoid potentially misleading claims about the end product. Any reference to genetic engineering on product labels shall be limited to the production method.

Today, many consumers are concerned about the meat they eat; hence, accurate labelling is important to inform consumer choice. In addition, accurate labelling is important to support fair trade. While regulations enshrined in national and international law underpins mandatory label information, unfortunately, regulations are not sufficient to prevent food fraud. To ensure adherence to regulations, and to enforce punitive measures when needed, robust analytical tests are required.

One restriction in production of organic meat is the use of veterinary drugs and rules dictate how and when such drugs may be used (Anon, 2008). Fluorescence microscopy (Kelly *et al.*, 2006) of cross sectional bone cuts can estimate the number of tetracycline doses administered to pig and chicken, and illegal serial or prophylactic dosing of tetracyclines can help verify whether the animal complies with the specified organic rules. This strategy is not conclusive as absence of tetracyclines is not a guarantee of organically produced meat.

One strategy is to explore the differences in animal fat from organically and conventionally raised animals. A gas chromatographic study of fatty acid methyl esters showed an increase in polyunsaturated fatty acids in organically produced lamb (Angood *et al.*, 2008), pig (Kim *et al.*, 2009), and broiler (Castellini *et al.*, 2002; Husak *et al.*, 2008) as compared to conventionally produced meats. However, fatty acid composition is also dependent on specific dietary intake such as pasture and concentrates (Pérez-Palacios *et al.*, 2009). Again, this strategy is not conclusive as a high level of polyunsaturated fatty acids could result from the specific dietary intake and not in from organically produced feed alone.

Another strategy could be analysis of isotopic composition. Stable ratio mass spectrometry of carbon, nitrogen, and sulphur isotopes succeeded in differentiating between organic and conventional Irish beef (Schmidt *et al.*, 2005). The differences in isotopic composition in organic and conventional beef are partly due to the difference in the feed intake (grass or concentrate) (Schmidt *et al.*, 2005). However, the higher content of ^{15}N in conventional beef compared with organic beef (Bahar *et al.*, 2008) might be a result of the mineral fertilizers applied to the soil where conventionally grown animals are fed (Watzka *et al.*, 2006).

Table 19.1: Analytical Techniques Applicable in Authentication of Organic vs. Conventional Meat and Meat Products

Analytical Techniques	*References*
Fluorescence microscopy	Kelly *et al.* (2006)
Gas chromatography	Angood *et al.* (2008); Castellini *et al.* (2002); Husak *et al.* (2008); Kim *et al.* (2009)
Stable ratio mass spectrometry	Schmidt *et al.* (2005)

Approved additives are allowed for use in manufacture of packaging films for packaging of foodstuffs. However, many of these are restricted for use in packaging of organic foodstuffs. The word restricted here indicates conditions and procedures for use shall be set by the accredited certification programmes. The following are approved additives under restriction, such as, 4,4′-Bis (2-benzoxazolyl) stilbene, 9,9-Bis (methoxymethyl) fluorine, carbonic acid, copper salt, Diethyleneglycol, 2-(4,6-Diphenyl-1,3,5-triazin-2-yl)-5-(hexyloxy) phenol, ethylene di-amine tetraacetic acid, 2-(2-Hydroxy-3,5-di-tert-butyl-phenyl-5-chlorobenzotriazole, 2-Methyl-4-isothiazolin-3-one, Phosphoric acid, trichlorocthylester, polyesters of 1,2 propanediol and/or 1,3-and 1,4 butanediol and/or polypropyleneglycol with adipic acid, also end-capped with acetic acid or fatty acids C10-C18 or n-octanol and/or n-decanol, 1,1,1-trimethylolpropane, 3-hydroxybutanoic acid, 3-hydro xypentanoic acid and its copolymers.

9. Pest and Disease Control for Stored Meat and Products

Meat is a highly perishable food. Cold chain is mandatory for meat and meat products unless they are shelf stable processed products. Packed meat and meat products are frozen after its production till its consumption. Shelf life of products in refrigeration and freezing varies depending on types of products and the level of processing. Well packaged products can be stored at low temperature as per the product requirement and it is allowed under the organic production practices. However, care must be taken that there is no contamination of products during the period of storage of products.

Product integrity should be maintained during any storage and transportation and handling by use of the following precautions: a) Organic products must be protected at all times from co-mingling with non-organic products; and b) Organic products must be protected at all times from contact with materials and substances not permitted for use in organic farming and handling. Where only part of the unit is certified, other product not covered by these guidelines should be stored and handled separately and both types of products should be clearly identified.

Bulk stores for organic product should be separate from conventional product stores and clearly labelled to that effect. Storage areas and transport containers for organic product should be cleaned using methods and materials permitted in organic production. Measures should be taken to prevent possible contamination from any pesticide or other treatment not listed before using a storage area or container that is not dedicated solely to organic products. Besides storage at ambient temperature, the following special conditions of storage are permitted, like controlled atmosphere, cooling, freezing, drying and humidity regulation. Ethylene gas is permitted for ripening.

Any handling and processing of organic products should be optimized to maintain the quality and integrity of the product and directed towards minimizing the development of pests and diseases. Processing and handling of organic products should be done separately in time or place from handling and processing of non organic products. Pollution sources shall be identified and contamination avoided. Flavouring extracts shall be obtained from food (preferably organic) by means of physical processes.

Pests should be avoided by good manufacturing practices. This includes general cleanliness and hygiene. Treatments with pest regulating agents must thus be regarded as the last resort. Recommended treatments are physical barriers, sound, ultra-sound, light, and UV-light, traps (incl. pheromone traps and static bait traps), temperature control, controlled atmosphere and diatomaceous earth. A plan for pest prevention and pest control should be developed. For pest management and control the following measures shall be used in order of priority:

- Preventive methods such as disruption, elimination of habitat and access to facilities
- Mechanical, physical and biological methods
- Pesticidal substances contained in the Appendices of the national standards
- Other substances used in traps
- Irradiation is prohibited.

There shall never be direct or indirect contact between organic products and prohibited substances. (*e.g.*, pesticides). In case of doubt, it shall be ensured that no residues are present in the organic product. Persistent or carcinogenic pesticides and disinfectants are not permitted.

10. Inspection and Certification of Organic Products

Certification helps in ensuring that products and its labeling regarding organic claims are authentic. Certification assures consumers of quality and authenticity of the product. Organic meat production consists of four different processes: production of organic feed, rearing of animals in farm, harvesting of meat in abattoir and processing of meat to value added products. All these entities must follow organic guidelines to get certification of final meat product. Producer interested in organic meat production must first study the relevant guidelines given under

NPOP and understand them. After studying the guidelines producer must align his production practices in line with the organic requirements. All the detailed production practices must be documented properly for inspection by the certifying bodies. Annual production plan must be written and submitted for accredited inspection agencies. Only accredited agencies can issue organic certifications. Inspectors appointed by accredited certifying agencies will visit the applicant's production units for inspecting and monitoring the compliance as per organic standards. Inspectors will visit the units to ascertain the compliance to set standards. After initial inspection, frequent visits are also made by the inspection agencies for monitoring. Usually conversion certificate will be given and after certain specified period final organic certification will be issued. Only certified producers can put 'organic' label on their products.

10.1 Identification of Meat Processing Treatments

10.1.1 Irradiation is not Permitted in Organic Meat Products

Electron spin resonance (ESR) spectroscopy has been found to be useful in detection of irradiated poultry meat (Marchioni *et al.*, 2005a; Marchioni *et al.*, 2005b) while gas chromatography can be used to measure volatile hydrocarbons and 2-alkylcyclobutanones present in irradiated poultry meat (Horvatovich *et al.*, 2000). Another possibility is the Comet assay (single-cell gel electrophoresis) (Ostling and Johanson, 1984; Singh *et al.*, 1988), which is based on electrophoresis of lysed cells embedded in agarose on a microscopic slide. The intensity of the Comet tail relative to the Comet head reflects DNA damage (Collins, 2004). The Comet assay finds application in the study of irradiation induced DNA degradation (Klaude *et al.*, 1996); however, the Comet assay cannot be used as a confirmatory tool as different treatments such as freeze-thaw cycles also cause DNA damage (Park *et al.*, 2000).

10.1.2 Meat Preparation (Baking, Cooking *etc.*)

Production of Maillard compounds (Skog *et al.*, 1998) during heating could perhaps distinguish among different procedures used in boiled, fried, and grilled meat (Ballin, 2010). Maillard products from above certain temperatures, which, for example, make it possible to distinguish between long-term cooking at a low temperature from short term cooking at a high temperature. For instance, acrylamide is formed above 120 °C (Mottram *et al.*, 2002) and could be used as an indicator of products exposed to temperature above 120°C. HPLC (Eerola *et al.*, 2007; Paleologos and Kontominas, 2007) and LC-MS/MS (Granby and Fagt, 2004) can be used to quantify acrylamide.

10.2 Identification of Non-Meat Ingredient Additions

10.2.1 Additives

The vast number of organic compounds used as additives makes it difficult to present a detailed description of each. These organic compounds might be colorants, aromas, or preservatives. Colorants and some reducing chemicals can be used to increase fresh meat appearance. Aromas, such as smoke aroma can fraudulently be used instead of natural smoking of meat. Preservatives can be used to preserve meat and make it appear fresh much longer than normal. The nature of the chemical compound determines the appropriate analytical technique. Since colour, aroma, and preservative relate to organic compounds added to the meat products both HPLC and GC methods may be appropriate.

Another group of additives are enzymes that take part in blood clotting. These can be used as blood-based binding agents and added to meat cuts or minced meat to form portions of desired mass and shape. In the binding process, thrombin cleaves fibrinogen to fibrinopeptides A and B. LC-MS/MS can be used to detect bovine (Grundy *et al.*, 2007) and porcine (Grundy *et al.*, 2008) fibrinopeptides A and B in concentrations down to 5 per cent.

10.2.2 Water

Addition of water is a real problem (Elliot, 2007) and regulations dictate the permitted amount of extraneous water in meat. A standard method to determine extraneous water in meat is to study the water/protein ratio. For instance, added water can be determined from a plot of water/protein ratio against extraneous water in boneless, skinless chicken breast (Anon, 2005). Water content can be determined by ISO 1442 (Anon, 1997), which describes the measure of mass before and after drying of meat, and protein content can be determined by ISO 937 (Anon, 1978), which describes an indirect protein determination method based on Kjeldahl analysis.

11. Marketing of Indian Organic Products: Status, Issues and Prospects

Organic animal husbandry requires most of the feed ingredients to be produced at the farm. Landless animal husbandry is discouraged. Absence of synthetic inputs for growth and disease control, and involvement of only organic ingredients reduces the growth rate of meat animals. Consequently, cost of production of the meat will remain high. This is true even for organic agricultural commodities wherein production level is compromised in the absence of use of fertilizers and pesticides. This leads to higher production costs and consequent higher price of commodities as compared to conventionally grown products. Willingness of consumers to pay higher price for health benefits is an important enabler for acceptability and marketability of organic products. Hence, growth of per capita income and affordability is expected to boost the sales

APEDA does not give any statistics about organic livestock product production or exports. In general, no significant effort is being made to tap the international demand for organic meat. Buffalo meat is the major component of meat exports from India and total buffalo meat exports in India stood at Rs. 26,457 Crores in 2013-14. Most of the buffalo meat is produced from spent buffaloes, which have completed their productive period. Rearing of buffaloes for meat purpose is not a widespread practice. Total worth of export of poultry products was Rs. 565 Crores in 2013-14. Due to short life cycle, there is a good scope for rearing broiler birds under organic system. As there is a huge demand for chicken and its products in domestic market due to high preference among consumers, scope for organic chicken is high compared to other meats. In view of fluctuating price of chicken in Indian market organic chicken can be a good business proposition to entrepreneurs for sustained returns.

Consumers' interest in the authenticity of the foods they purchase is increasing, especially where it concerns more expensive 'value-added' products such as organic foods, fair trade products or products with a protected designation of origin (PDO). The scope for marketing organic food in India is vast and still not yet explored to its full potential. The following points can be taken into consideration while designing the strategies:

1. Literacy rate has gone up and people are more health conscious now. They think twice before buying a product. Likewise, agriculturists are now more literate and are ready to experiment with new generation of crops and improved farming methodologies.
2. Stressful lifestyle and so many diseases around have created a need for healthier and contamination free food.
3. Organic food market is still restricted primarily to metros and other major cities of the country.
4. Awareness about benefits of using organic food is very less.
5. Competitive pricing can really open a new avenue in the eatables and grocery section.

12. Opportunities for Tropical Countries

To remain relevant to the global economy, the developing countries too have to produce what the consumers are demanding globally. Therefore, it is a necessity to focus attention on this system of production. India certainly needs to move forward with its organic farming activities. This is what is rightly happening in India currently in case of high value commercial crops. Some South East Asian countries especially Malaysia and Thialand have already initiated research and development work in the area of organic animal husbandry, which may guide to other developing countries in the region. There is a huge demand for organic agricultural products in international market and domestic market is also slowly picking up due to increasing per capita income and living standards.

There are strong reasons for tropical countries to focus attention on organic livestock production. Some of the encouraging factors are: (i) Demand for organic livestock products is growing, (ii) European countries import significant proportions of organic meat and its products, (iii) Organic products tend to retail at a higher price than their conventionally grown/produced counterparts, mainly because of lower yield and certification costs and consumers are ready to pay a large premium price for organic food, (iv) Some developing countries do trade livestock products to developed countries, (v) Non-food livestock products, like organic textiles/garments, hides, wool offer hope for organic livestock production in tropical countries, (vi) Indigenous Technical Knowledge (ITK) available in developing countries may provide effective substitute for veterinary care, (vii) The use of agro-chemicals is almost nil in major parts of tropical countries offering ideal opportunity of organic livestock production to grow, (viii) The native livestock breeds which are prominent in tropical countries are less susceptible to diseases and stress, need for allopathic medicines is very less, (ix) Grass or forest based extensive livestock production system prevalent in most parts of tropical countries offer a potential for conversion into organic animal husbandry, (x) Growing domestic market for organic products in developing countries may help to boost organic market at country, and regional level.

An overview of the situation offers that organic livestock production *per se* is yet to come up in most of the developing world except Latin American countries. The developing countries in Asia and Africa may draw lessons from these Latin American countries in coming years. Organic aquaculture (shrimp and fish), is emerging in China, Indonesia, Vietnam, Thialand, Malaysia and Myanmar for export and domestic market (Willer and Kilcher, 2011). This signals positively for the future growth of organic animal products.

13. Recent Trends and Future of Organic Meats and Processing

There are several opportunities that may impact the future of organic agriculture, including organic meat and meat processing. Grazing animals in marginal land that is organic through wild collection is promising. It can increase income of livestock producers operating on these lands. Global acceptance of organic products like goat cheeses, meat and fiber is also an important opportunity for the future growth of organic markets. While energy and chemical costs are high, practicing sustainable organic meat production in general has an economic edge. Alternative medicine as a result of prohibition of restricted materials to treat diseases and illness can be promising.

A number of research opportunities are also apparent, the list include: emerging health issues, welfare and production constraints; epidemiological surveillance of key production diseases; breeding studies on disease resistance and commercial traits, nutritional deficiencies in organic systems, livestock breeding, biological control and the use of novel plants and plant extracts, development of animal welfare assessment methods, and development of welfare-friendly production systems.

Recently, ICAR-NRC on Meat in a collaborative project with ICAR-Central Research Institute for Dryland Agriculture (CRIDA) have been awarded with organic certification of sheep for meat purpose. Further ICAR-CRIDA is also providing organically certified fodder (hybrid Napier) in an area of 0.8 ha (Annual Report, 2017-18).

Animal health and welfare, with a greater emphasis on disease control and eradication will likely be the main challenge for organic meat production in the future. In order to accomplish goals of disease prevention, control and eradication, monitoring and evaluating of alternative health products will gain importance. To capitalize on the opportunity of using marginal land for organic meat production, converting hill and upland systems to organic production efficiently will most likely be required in the future. Due to the extensive use of grazing for organic meat production, mineral deficiencies as a result of soil characteristics and pasture management system should be important considerations (Qi *et al.*, 1993). In addition, prevention of fraud and quality assurance for organic goat, sheep, dairy, meat and fiber products will continue to be a concern for consumers.

The interest of consumers in organic products mainly stems from health and environmental considerations. Consumers are concerned about the safety of what they eat and about the use of pesticides, hormones and other veterinary drugs in farming practice. Most of the meats produced in India are organic by default as the animals are grown traditionally with little or no synthetic chemicals. Moreover, the country has a wide range of local animal breeds, those are tolerant to disease. However, there is no active initiative from the public sector extension organizations to promote organic meat production. It is high time to establish better partnership and cooperation among livestock farmers, NGOs, certifiers, marketing people and the programs that will support organic meat production and ultimately it will contribute in improving the health status of population as well as agro-ecosystem. Certainly more research is needed in this field, but, at present, the advertising claims that consumption of organic meat rather than conventional meat can reduce the exposure to environmental hazards.

14. Conclusions

Organic meat production is important for many reasons (less environmental impact, less use of energy, and other factors); however, even if the scientific literature in this field is scarce, there is no clear scientific evidence that organic meat can better protect consumers from chemical contamination than conventional meat, even if this is one of the major forces driving consumers to buy organic products. Organic meat production can improve animal welfare, protect the environment, and sustain rewarding rural lifestyles. Traditional and alternative medicine holds the promise for alternative prevention and treatment of animal diseases. The future of organic goat production is to continue searching for alternatives that are environmentally friendly, human health conscientious and animal considerate. Understanding organic farming from economic, ecological, and animal welfare perspectives will increase the likelihood of success. Organic meat production will have to strive for a more sustainable system than the conventional one, offsetting the increased costs of organic livestock production by higher product prices, and certified organic meat products that are healthier than those conventionally produced. To get certification for meat, feeding of certified organic feed ingredients is necessary. Hence, promoting and popularizing production of organic feed ingredients is prerequisite for development of organic meat sector. Community approach in organic animal husbandry can bring about a major advance in respect of organic meat production.

REFERENCES

Angood KM, Wood JD, Nute GR, Whittington, FM, Hughes, SI, and Sheard, P.R. 2008. A comparison of organic and conventionally-produced lamb purchased from three major UK supermarkets: Price, eating quality and fatty acid composition. Meat Science, 78(3): 176–184.

Annual Report. 2017-18. ICAR-Cental Research Institute for Dryland Agriculture, Hyderabad, 181 p.

Anon. 1978. Meat and meat products—Determination of nitrogen content (Reference method). International Organization for Standardization, 937.

Anon. 1997. Meat and meat products—Determination of moisture content (Reference method). International Organization for Standardization, 1442.

Anon. 2005. Commission Recommendation (EC) 1. of March 2005 concerning a coordinated programme for the official control of foodstuffs. L59. Official Journal of the European Union, 27–39.

Anon. 2008. Commission Regulation (EC) No 889/2008 of 5 September 2008 laying down detailed rules for the implementation of Council Regulation (EC) No 834/2007 on organic production and labelling of organic products with regard to organic production, labelling and control. L250. Official Journal of the European Union, 1–84.

Bahar B, Schmidt O, Moloney AP, Scrimgeour CM, Begle IS., and Monahan FJ. 2008. Seasonal variation in the C, N and S stable isotope composition of retail organic and conventional Irish beef. Food Chemistry, 106(3): 1299–1305.

Ballin NZ. 2010. Authentication of meat and meat products. Meat Science, 86: 577–587.

Castellini C, Mugnai C, and Dal Bosco A. 2002. Effect of organic production system on broiler carcass and meat quality. Meat Science, 60(3): 219–225.

Chander MB. Subrahmanyeswari R. Mukherjee and Kumar S. 2011. Organic livestock production: an emerging opportunity with new challenges for producers in tropical countries. Rev. sci. tech. Off. int. Epiz, 30 (3): 969-983.

Collins AR. 2004. The comet assay for DNA damage and repair: Principles, applications, and limitations. Molecular Biotechnology, 26(3): 249–261.

Eerola S, Hollebekkers K, Hallikainen A and Peltonen K. 2007. Acrylamide levels in Finnish foodstuffs analysed with liquid chromatography tandem mass spectrometry. Molecular Nutrition and Food Research, 51(2): 239–247.

Elliot V. 2007. Chicken fillets 'secretly pumped up with water' to increase weight. www.timesonline.co.uk/tol/life_and_style/food_and_drink/article1980529.ece

Granby K and Fagt S. 2004. Analysis of acrylamide in coffee and dietary exposure to acrylamide from coffee. Analytica Chimica Acta, 520(1–2): 177–182.

Grundy H H, Reece P, Sykes MD, Clough JA, Audsley N and Stones R. 2007. Screening method for the addition of bovine blood-based binding agents to food using liquid chromatography triple quadrupole mass spectrometry. Rapid Communications in Mass Spectrometry, 21(18): 2919–2925.

Grundy HH, Reece P, Sykes MD, Clough JA, Audsley N and Stones R. 2008. Method to screen for the addition of porcine blood-based binding products to foods using liquid chromatography/triple quadrupole mass spectrometry. Rapid Communications in Mass Spectrometry, 22(12): 2006–2008.

Guidelines for the production, processing, labelling and marketing of organically produced foods, GL 32 – 1999, Rev. 1 – 2001, 1- 52.

Horvatovich P, Miesch M, Hasselmann C, and Marchioni E. 2000. Supercritical fluid extraction of hydrocarbons and 2-alkylcyclobutanones for the detection of irradiated foodstuffs. Journal of Chromatography, 897(1-2): 259–268.

Husak R L, Sebranek JG, and Bregendahl K. 2008. A survey of commercially available broilers marketed as organic, free-range, and conventional broilers for cooked meat yields, meat composition, and relative value. Poultry Science, 87(11): 2367–2376.

Kelly M, Tarbin JA, Ashwin H, and Sharman M. 2006. Verification of compliance with organicmeat production standards by detection of permitted and nonpermitted uses of veterinary medicines (tetracycline antibiotics). Journal of Agricultural and Food Chemistry, 54(4): 1523–1529.

Kim DH, Seong PN, Cho SH, Kim JH, Lee JM, Jo C. 2009. Fatty acid composition and meat quality traits of organically reared Korean native black pigs. Livestock Science, 120(1–2): 96–102.

Klaude M, Eriksson, S, Nygren J, and Ahnstrom G. 1996. The comet assay: Mechanisms and technical considerations. Mutation Research/DNA Repair, 363(2): 89–96.

Mahesh Chander and Reena Mukherjee. 2005. Organic animal husbandry: concept, status and possibilities in India-A review. Indian Journal of Animal Science, 75 (12): 1460-14 69.

Marchioni E, Horvatovich N, Charon H, and Kuntz, F. 2005a. Detection of irradiated ingredients included in low quantity in non-irradiated food matrix. 1. Extraction and ESR analysis of bones from mechanically recovered poultry meat. Journal of Agricultural and Food Chemistry, 53(10): 3769–3773.

Marchioni E, Horvatovich N, Charon H and Kuntz F. 2005b. Detection of irradiated ingredients included in low quantity in non-irradiated food matrix. 2. ESR analysis of mechanically recovered poultry meat and TL analysis of spices. Journal of Agricultural and Food Chemistry, 53(10): 3774–3778.

Mottram DS, Wedzicha BL and Dodson AT. 2002. Food chemistry: Acrylamide is formed in the Maillard reaction. Nature, 419(6906): 448–449.

Ostling O and Johanson KJ. 1984. Microelectrophoretic study of radiation-induced DNA damages in individual mammalian cells. Biochemical and Biophysical Research Communications, 123(1): 291–298.

Packaging for organic foods. 2012. Published by International Trade Centre (ITC). Packaging for Organic Foods, Geneva, 1–68.

Paleologos E K and Kontominas MG. 2007. Effect of processing and storage conditions on the generation of acrylamide in precooked breaded chicken products. Journal of Food Protection, 70(2): 466–470.

Park JH, Hyun CK, Jeong SK, Yi M A, Ji ST and Shin HK. 2000. Use of the single cell gel electrophoresis assay (Comet assay) as a technique for monitoring low temperature treated and irradiated muscle tissues. International Journal of Food Science and Technology, 35(6): 555–561.

Pérez-Palacios T, Ruiz J, Tejeda JF and Antequera T. 2009. Subcutaneous and intramuscular lipid traits as tools for classifying Iberian pigs as a function of their feeding background. Meat Science, 81(4): 632–640.

Qi K, Lu CD and Owens FN. 1993. Sulfate supplementation of Angora goats: sulfur metabolism and interactions with zinc, copper and molybdenum. Small Rumin. Res. 11, 209–225.

Schmidt O, Quilter JM, Bahar B, Moloney AP, Scrimgeour CM, Begley IS. 2005. Inferring the origin and dietary history of beef from C, N and S stable isotope ratio analysis. Food Chemistry, 91(3): 545–549.

Singh NP, McCoy MT, Tice RR, and Schneider EL. 1988. A simple technique for quantitation of low levels of DNA damage in individual cells. Experimental Cell Research, 175(1): 184–191.

Skog KI, Johansson MA and Jagerstad MI. 1998. Carcinogenic heterocyclic amines in model systems and cooked foods: a review on formation, occurrence and intake. Food and Chemical Toxicology, 36(9–10): 879–896.

Steven C. Ricke Ellen J. Van Loo, Michael G. Johnson and Corlis A. O'Bryan. 2012. Organic meat production and processing. Published by Willey Blackwel, Oxford, United Kingdom.

Watzka M, Buchgraber K and Wanek W. 2006. Natural 15N abundance of plants and soils under different management practices in a montane grassland. Soil Biology and Biochemistry, 38(7): 1564–1576.

Willer H and Kilcher L. (Eds.). (2011). The World of Organic Agriculture: Statistics and Emerging Trends 2011: Bonn: International Federation of Organic Agriculture Movements (IFOAM); Frick, Switzerland: Research Institute of Organic Agriculture (FiBL).

Yadav AK. 2012. Training Manual on certification and inspection systems in organic farming in India. Published by Ministry of Agriculture, Department of Agriculture and Cooperation, National Centre of Organic Farming, Ghaziabad.

Chapter 20

Climate Change Adaptation and Mitigation Potential of Organic Farming

K.A. Gopinath, G. Venkatesh and V. Visha Kumari

ICAR-Central Research Institute for Dryland Agriculture, Hyderabad – 500 059, Telangana
E-mail: ka.gopinath@icar.gov.in

1. INTROODUCTION

Climate change is one of the potential threats for sustainable development of agriculture coupled with food security. Climate change phenomenon is now a global reality. Although climate change is of extreme concern for all countries, India is one of the most vulnerable countries because of both the magnitude and multitude of effects climate change will have on Indian food security.

Agriculture is a major contributor to emissions of methane (CH_4), nitrous oxide (N_2O), and carbon dioxide (CO_2). India's CO_2 emissions have swiftly caught up with the levels of developed countries; however, per capita emissions remain significantly below those in developed countries. In 2006, the world per capita average levels was 4.4 metric tons, while India and the US emitted 1.31 and 18.99 metric tons per capita, respectively (UN, 2009). According to the Human Development Report, "India currently contributes only 4 percent of global emissions despite representing 18 percent of the world's population, but its share is projected to increase to 12 percent by 2050 in the absence of mitigation policy" (UN, 2008). GHGs attributed to agriculture by the IPCC include emissions from soils, enteric fermentation (GHG emissions from the digestion process of ruminant animals), rice production, biomass burning and manure management (Smith *et al.*, 2007). There are other 'indirect' sources of GHG emissions that are not accounted for by the IPCC under agriculture such as those generated from land-use changes, use of fossil fuels for mechanization, transport and agro-chemical and fertilizer production (IFOAM, 2009).

Climate change and variability are a considerable threat to agricultural communities, particularly in India. This threat includes the likely increase of temperature, extreme weather conditions, increased water stress and drought, and desertification. Seasonal variations in weather events may pose risks to traditional methods of crop production either due to water constraints or surplus of water and erosion. In this regard, soil stability will become crucial to store water in the soil profile, to resist extreme weather events and minimize soil erosion. These changes will bring new challenges to farmers. Farmers need tools to help them adapt to these new conditions. Organic farming is one such option which is reported to have both climate change mitigation and adaptation potential particularly in rainfed agriculture.

2. Potential of Organic Farming to Mitigate Climate Change

There is considerable world-wide support at present in advocating organic agriculture for mitigating climate change (Kotschi and Miiller-Samann, 2004; Niggli, 2007; IFOAM, 2008; Goh, 2011). The potential of organic agriculture in mitigating climate change depends on its ability to reduce emissions of GHGs (nitrous oxide, carbon dioxide and methane), increase soil carbon sequestration, and enhance effects of organic farming practices which favour the above two processes (Goh, 2011).

2.1 Reduction of GHG Emissions

The global warming potential of conventional agriculture is strongly affected by the use of synthetic nitrogen fertilizers and by high nitrogen concentrations in soils. However, organic farming systems avoid the use of synthetic fertilizers, and rely on practices such as green manuring, crop rotation with legumes, efficient recycling of residues and the use of organic manures. In addition, these systems avoid the use of synthetic pesticides and rely on practices such as crop rotations, use of bio-pesticides and increase beneficial insects for pest management. These restrictions on fossil-fuel based fertilizer and pesticide inputs can significantly reduce the overall GHG footprint of organic systems in comparison to conventional production systems (Sreejith and Sherief, 2011). Recent experimental results suggest that organic agriculture can significantly reduce GHG emissions.

2.1.1 Reduction of Nitrous Oxide Emissions

N_2O emissions are the most important source of agricultural emissions accounting for 38 per cent of agricultural GHG emissions (Smith *et al.*, 2007). The IPCC attributes a default value of 1 per cent to applied fertilizer nitrogen as direct N_2O emissions (Eggleston *et al.*, 2006). Similarly, emission factors of up to 3-5 kg N_2O-N per 100 kg N-input have been reported by Crutzen *et al.* (2007). These higher values for global N_2O budget are due to the consideration of both direct and indirect emissions, including also livestock production, NH_3 and NO_3 emissions, nitrogen leakage into rivers and coastal zones, *etc.* (Scialabba and Muller-Lindenlauf, 2010). Nitrous oxide emissions are directly linked to the concentration of available mineral N (ammonium and nitrate) in soils arising from the nitrification and denitrification of available soil and added fertilizer N (Firestone and Davidson, 1989; Wrage *et al.*,

2001). High emissions rates are detected directly after mineral fertilizer additions and are very variable (Bouwman, 1995).

In organic systems, the nitrogen input to soils, and hence the potential nitrous oxide emissions, are reduced. The share of reactive nitrogen that is emitted as N_2O depends on a broad range of soil and weather conditions and management practices, which could partly foil the positive effect of lower nitrogen levels in top soil (Scialabba and Muller-Lindenlauf, 2010). In a study by Petersen *et al.* (2006), lower emission rates for organic compared to conventional farming were found for five European countries. One study found no significant differences between mineral and organic fertilization (Dambreville *et al.*, 2007). In a long-term study in southern Germany, Flessa *et al.* (2002). found reduced nitrous oxide emission rates in the organic farm, although yield-related emissions were not reduced.

On the other hand, comparisons between soils receiving manure versus mineral fertilizers found higher N_2O emissions after manure application compared to mineral fertilizer applications, but not for all soil types (Van Groeningen *et al.*, 2004; Rochette *et al.*, 2008). The higher nitrous oxide emissions after incorporation of manure and plant residues are explained by the high oxygen consumption for decomposition of the organic matter (Flessa and Beese, 2000). As there is high uncertainty in N_2O emission factors, further research is needed.

2.1.2 Reduction of Methane Emissions

The reduction or avoidance of CH_4 emissions is of special importance in global warming from the agricultural sector because two thirds of global CH_4 emissions are of anthropogenic origin, mainly from enteric ruminant fermentation in animals (FAO, 2006). and in paddy rice production (Smith and Conan, 2004). Avoidance of methane emissions of anthropogenic origin and especially of agricultural origin is of particular importance for mitigation. In general, the CH_4 emissions from ruminants and rice production are not significantly different between organic and conventional agriculture. Differences are due largely to the extent and intensity of various farming practices and their improvement used within different forms of agriculture.

Although research on CH_4 emissions in organic and conventional paddy rice production is still in its infancy (Goh, 2011). employing better rice production techniques such as using low CH_4-emitting varieties (Yagi *et al.*, 1997; Aulakh *et al.*, 2001). composted manures with low C/N ratio (Singh *et al.*, 2003). adjusting the timing of organic residue additions (Xu *et al.*, 2000; Cai and Xu, 2004). and using mid-season drainage or avoiding continuous flooding have been shown to reduce CH_4 emissions (Smith and Conan, 2004). Further, as herbicides are not used in organic systems, aquatic weeds tend to be present in organic rice paddies and weeds have an additional decreasing effect on methane emissions (Inubushi *et al.*, 2001). In organic farming systems, cropping depends on nutrient supply from livestock and the combination of cropping and livestock provides an efficient means of mitigating GHG emissions especially CH_4 (Goh, 2011). Efficient and direct recycling of manure and slurry is the best option to reduce GHG emissions as this practice avoids long-distance transport (Niggli, 2007).

Methane and N_2O from manure account for about 7 per cent of the agricultural GHG emissions. Methane emissions predominantly occur in liquid manure systems, while N_2O emissions are higher in solid manure systems and on pastures (Smith *et al.*, 2007). There is a very high variance for both gaseous emissions, depending on composition, coverage, temperature and moisture of the manure. Storing manure in solid form such as composting can suppress CH_4 emissions but may result in more N_2O emissions (Paustian *et al.*, 2004).

2.1.3 Reduction of Carbon Dioxide Emissions

Synthetic external inputs like fertilizers and pesticides are banned in organic farming. The energy used for the chemical synthesis of nitrogen fertilizers, which are totally excluded in organic systems, represent up to 0.4–0.6 Gt CO_2 emissions (EFMA, 2005; Williams *et al.*, 2006; FAOSTAT, 2009). This is as much as 10 per cent of direct global agricultural emissions and around 1 per cent of total anthropogenic GHG emissions. Further, CH_4 and N_2O from biomass burning account for 12 per cent of the agricultural GHG emissions. Additionally, the carbon sequestered in the burned biomass is lost to the atmosphere. In organic agriculture, preparation of land by burning vegetation is restricted to a minimum (IFOAM, 2002). CO_2 emissions are reported to be around 40-60 per cent lower in organic farming systems than conventional systems, largely because they don't use synthetic nitrogen fertilizers which require large amounts of energy in their production and are associated with emissions of the powerful GHG nitrous oxide (Sayre, 2003, BFA 2007).

2.2 Soil Carbon Sequestration

Soil C sequestration is an important strategy and is a win-win option of producing more food per unit area besides mitigation of climate change (Lal, 2004). Although soils of the tropical regions have low C sequestration rate because of high temperatures, adoption of appropriate management practices can lead to higher rates particularly in high rainfall regions (Srinivasarao *et al.*, 2012). Soil carbon sequestration at a global scale is considered the mechanism responsible for the greatest mitigation potential within the agricultural sector, with an estimated 90 per cent contribution to the potential of what is technically feasible (Smith *et al.*, 2007, 2008). However, global soil carbon stocks of agricultural land have decreased historically and continue to decline (Lal, 2004). Thus, improved agronomic practices that could lead to reduced carbon losses or even increased soil carbon storage are highly desired (Gattinger *et al.*, 2012).

Soil carbon sequestration is enhanced through agricultural management practices (such as increased application of organic manures, use of intercrops and green manures, higher shares of perennial grasslands and trees or hedges, *etc.*), which promote greater soil organic matter (and thus soil organic carbon) content and improve soil structure (Niggli *et al.*, 2008; Muller, 2009). There is strong scientific evidence that organic farming generally results in higher soil carbon levels in cultivated soils compared to chemical fertilizer based agriculture. Very rough estimates for the global mitigation potential of organic agriculture amount to 3.5-4.8 Gt CO_2 from carbon sequestration (around 55-80 per cent of total global greenhouse gas emissions from agriculture) and a reduction of N_2O by two-thirds

(Niggli *et al.*, 2008). The potential of carbon sequestration rate by organic farming for European agricultural soils has been estimated at 0–0.5 t C per ha per year (Freibauer *et al.*, 2004). Similarly, several field studies have proved the positive effect of organic farming practices on soil carbon pools (Pimentel *et al.*, 2005; Fliessbach *et al.*, 2007; Kustermann *et al.*, 2008). In the USA, a field trial showed a fivefold higher carbon sequestration in the organic system (1218 kg C per ha per year) in comparison with conventional management (Pimentel *et al.*, 2005). Gattinger *et al.* (2012) subjected datasets from 74 studies from pair-wise comparisons of organic vs. nonorganic farming systems to meta analysis to identify differences in soil organic carbon (SOC). The analysis revealed significant differences and higher values for organically farmed soils of 0.18±0.06 per cent points (mean±95 per cent confidence interval) for SOC concentrations, 3.50±1.08 Mg C/ha for stocks, and 0.45±0.21 Mg C/ha/year for sequestration rates compared with nonorganic management. Leifeld and Fuhrer, (2010) found in their review an average annual increase of the SOC concentration in organic systems by 2.2 per cent, whereas in conventional systems, SOC did not change significantly.

3. Organic Agriculture as a Adaptation Strategy

Adaptation in agriculture is not new. Historically, farmers have developed several methods to adapt to changing climate including aberrant weather. However, the adaptation needs to occur at a much faster rate due to impending climate change. The Intergovernmental Panel on Climate Change (IPPC) defines adaptation to climate change as 'adjustment in natural or human systems in response to actual or expected climatic stimuli or their effects, which moderates harm or exploits beneficial opportunities (IPCC, 2001). Long-term crop yield stability and the ability to buffer yields through climatic adversity are critical factors in agriculture's ability to support society in the future (Lotter *et al.*, 2003). Several researchers have reported that organic farming systems perform superior than their conventional counterparts during climate extremes including drought and excessive rainfall.

Several mechanisms may increase drought tolerance of organic cropping systems. Soil organic matter has positive effects on the water-capturing capacity of the soil. Numerous studies have shown soil organic carbon to be higher in organically managed systems (Reganold, 1995; Clark *et al.*, 1998; Liebig and Doran, 1999; Gopinath *et al.*, 2008, 2011). As a result, organically managed soils have high water holding capacity (Liebig and Doran, 1999; Wells *et al.*, 2000). It was found that water capture in organic plots was twice as high as in conventional plots during torrential rains (Lotter *et al.*, 2003). Similarly, Pimentel *et al.* (2005) reported that the amount of water percolating through the top 36 cm was 15-20 per cent greater in the organic systems of the Rodale farming systems trial compared to conventional systems. The organic soils held 816,000 litres per ha in the upper 15 cm of soil. This water reservoir was likely the reason for higher yields of corn and soybean in dry years. In India, most of the organic cotton farmers stated that the capacity of their soils to absorb and retain water was increased after conversion to organic management (Eyhorn *et al.*, 2009). Many farmers also said that they need less rounds of irrigation and the crops can sustain longer periods of drought. In Central America, farmers using organic and sustainable methods reported substantially lower economic losses, and 90 per cent of the neighbours of the study farms indicated a desire to adopt their neighbours' methods after observing the environmental stress tolerance of the organic/sustainable farms (Holt-Gimenez, 2002). In the 21-year Rodale Farming Systems Trial, in which two organic and a conventional crop rotation were compared, the organic crop systems performed significantly better in 4 out of 5 years of moderate drought. In the severe drought year of 1999, three out of the four crop comparisons resulted in significantly better yields in the organic systems than the conventional (Lotter *et al.*, 2003).

Furthermore, plant water uptake and ability to withstand drought are significantly improved by mycorrhizal associations (Syliva and Williams, 1992). Mycorrhizae have been shown to be more abundant in the roots of crops from organically managed systems relative to those of conventionally managed crops (Eason *et al.*, 1999; Mader *et al.*, 2000). This suggests both a physicochemical and biological basis for the increased drought tolerance of organic cropping systems (Lotter *et al.*, 2003).

The mitigation of runoff, erosion and crop losses as a result of rainfall excess is also improved in organically managed systems (Lotter *et al.*, 2003). Organic management of soils leads to improved soil stability and resistance to water erosion compared to conventionally managed soils, due to higher soil C content and improved soil aggregation (Reganold, 1995; Clark *et al.*, 1998; Liebig and Doran, 1999), permeability (Reganold *et al.*, 2001) and lower bulk density as well as higher resistance to wind erosion (Jaenicke, 1998). Lockeretz *et al.* (1981). reported one-third less erosion from Midwest organic farms than from comparable conventionally managed farms. Many studies have reported that organically managed crops have out yielded conventionally managed crops under flood conditions, due to higher levels of water-stable aggregates in organic soils and associated reduced soil compaction after tillage (Denison, 1996; Lotter *et al.*, 2003). Hence, organic crop management techniques will be a valuable resource in an era of climatic variability, providing soil and crop characteristics that can better buffer environmental extremes (Lotter *et al.*, 2003).

4. Conclusion

Organic agricultural systems have an inherent potential to both mitigate climate change through reduced GHG emissions and higher carbon sequestration in the soil, and adapt to climate change. Farming practices such as organic agriculture that preserve soil fertility and maintain or even increase organic matter in soils are in a good position to maintain productivity in the event of drought, irregular rainfall events with floods, and rising temperatures. Soils in organic agriculture capture and store more water than soils of conventional cultivation. Therefore, organic agriculture is one of the adaptation strategies that can be targeted at improving the livelihoods of rural populations that are especially vulnerable to the adverse effects of climate change and variability.

REFERENCES

Aulakh MS, Wassmann R and Rennenberg H. 2001. Methane emissions from rice fields quantification, mechanisms, role of management, and mitigation options. Advances in Agronomy, 70: 247-251.

Balfour, E.B. 1943. *The Living Soil.* Faber and Faber, London. 223 p.

BFA. 2007. Soil makes the carbon cut – BFA Press Release 13th December 2007, Biological Farmers of Australia, http: //www.bfa.com.au/.

Bouwman AF. 1995. Uncertainties in the global source distribution of nitrous oxide. Journal of Geophysical Research, 100: 2785-2800.

Cai ZC and Xu H. 2004. Options for mitigating CH_4 from rice fields in China. Journal of Geophysical Research, 105: 17231-17232.

Chhonkar PK. 2003. Organic farming: Science and belief. Journal of Indian Society of Soil Science, 51(4): 365-372.

Clark MS, Ferris H, Klonsky K, Lanini WT, vanBruggen AHC and Zalom FG. 1998. Agronomic, economic, and environmental comparison of pest management in conventional and alternative tomato and corn systems in northern California. Agriculture Ecosystems and Environment, 68(1-2): 51-71.

Crutzen PJ, Mosier AR, Smith KA and Winiwarter W. 2007. N_2O release from agro-biofuel production negates global warming reduction by replacing fossil fuels. Atmospheric Chemistry and Physics Discussions, 7: 11191–11205.

Dambreville Ch, Morvan Th and German JC. 2007. N_2O emission in maize crops fertilized with pig slurry, matured pig manure or ammonia nitrate in Brittany. Agriculture, Ecosystems and Environment, 123: 201-210.

Denison RF. 1996. Organic matters. The LTRAS century, 4 (March): Long Term Research on Agricultural Systems (LTRAS). College of Agricultural and Environmental Sciences, University of California Davis.

Eason WR, Scullion J and Scott EP. 1999. Soil parameters and plant responses associated with arbuscular mycorrhizas from contrasting grassland management regimes. Agriculture Ecosystems and Environment, 73(3): 245-255.

EFMA. 2005. Understanding nitrogen and its use in agriculture. European Fertilizer Manufactures Association (EFMA), Brussels, Belgium.

Eggleston S, Buendia L, Miwa K, Ngara T and Tanabe K (Eds). 2006. IPCC Guidelines for National Greenhouse Gas Inventories. Volume 4: Agriculture, Forestry and Other Land Use. Prepared by the National Greenhouse Gas Inventories Programme (IGES), Hayama, Japan.

Eyhorn F, Mader P and Ramakrishnan M. 2009. The impact of organic cotton farming on the livelihoods of smallholders: Evidence from the Maikaal bioRe project in central India. Organic Farming Newsletter 5(1): 19-21.

FAO. 2006. Livestock's Long Shadow-Environmental Issues and Options, Food and Agriculture Organization of the United Nations (FAO), Rome, Italy.

FAOSTAT. 2009. FAO Statistical Database Domain on Fertilizers: ResourceSTAT-Fertilizers. Food and Agriculture Organisation of the United Nations (FAO) Rome, Italy.

Firestone MK and Davidson. 1989. Microbial basis of NO and N_2O production and consumption in soil. In: *Exchange of Trace Gases Between Terrestrial Ecosystems and the Atmosphere* (MO Andreae and DS Schimel, Eds.), Wiley, New York, U.S.A. pp. 2-21.

Flessa H and Beese F. 2000. Laboratory estimates of trace gas emissions following surface application and injection of cattle slurry. Journal of Environmental Quality, 29: 262-268.

Fliessbach A, Oberholzer HR, Gunst L and Mader P. 2007. Soil organic matter and biological soil quality indicators after 21 years of organic and conventional farming. Agriculture, Ecosystems and Environment, 118: 273-284.

Freibauer A, Rounsevell, M.D.A. Smith, P and Verhagen, J. 2004. Carbon sequestration in the agricultural soils of Europe. Geoderma, 122: 1-23.

Gattinger A, Muller A, Haeni M, Skinner C, Fliessbach A, Buchmann N, Mäder P, Stolze M, Smith P, Scialabba N El-Hage and Niggli U. 2012. Enhanced top soil carbon stocks under organic farming. PANS, 109(44): 18226-18231.

Goh KM. 2011. Greater mitigation of climate change by organic than conventional agriculture: A review. Biological Agriculture and Horticulture, 27(2): 205-229.

Gopinath KA, Supradip Saha and Mina BL. 2011. Effects of organic amendments on productivity and profitability of bell pepper-french bean-garden pea system and on soil properties during transition to organic production. Communications in Soil Science and Plant Analysis, 42(21): 2572-2585.

Gopinath KA, Supradip Saha, Mina BL, Kundu S, Harit Pande and Gupta HS. 2008. Influence of organic amendments on growth, yield and quality of wheat and on soil properties during transition to organic production. Nutrient Cycling in Agroecosystems, 82: 51-60.

Holt-Gimenez E. 2002. Measuring farmers' agroecological resistance after Hurricane Mitch in Nicaragua: a case study in participatory, sustainable land management impact monitoring. Agriculture, Ecosystems and Environment, 93: 87-105.

Howard, A. 1943. An Agricultural Testament. The Oxford University Press, Oxford. 228 p.

IFOAM. 2002. Basic Standards for Organic Production and Processing, International Federation of Organic Agricultural Movements, Bonn, Germany.

IFOAM. 2008. FAO Workshop on Organic Agriculture and Climate Change. Workshop at the 6th IFOAM Organic World Congress, Modena, Italy, June 18, 2008.

IFOAM. 2009. The contribution of organic agriculture to climate change mitigation. Available online at http: / /classic.ifoam.org/growing_organic/1_arguments_for_oa/environmental benefits/pdfs/IFOAM-CC-Mitigation-Web.pdf.

Inubushi K, Sugii H, Nishino S and Nishino E. 2001. Effect of aquatic weeds on methane emission from submerged paddy soil. American Journal of Botany, 88(6): 975–979.

IPCC. 2001. Climate change 2001 synthesis report: a summary for policymakers. Wembley, UK, Intergovernmental Panel on Climate Change. Available at www.ipcc.ch/pub/un/syreng/spm.pdf.

Jaenicke EC. 1998. From the Ground Up: Exploring soil quality's contribution to environmental health. Policy Studies Report No. 10. Henry A. Wallace Center for Agricultural and Environmental Policy,

Kotschi, J, Müller-Sämann, K. (2004): The Role of Organic Agriculture in Mitigating Climate Change. International Federation of Organic Agriculture Movements (IFOAM), Bonn. 64 p.

Kustermann B, Kainz M and Hulsbergen KJ. 2008. Modelling carbon cycles and estimation of greenhouse gas emissions from organic and conventional farming systems. Renewable Agriculture and Food Systems, 23: 38-52.

Lal R. 2004. Soil carbon sequestration impacts on global climate change and food security. Science, 304: 1623-1627.

Leifeld J and Fuhrer J. 2010. Organic farming and soil carbon sequestration: What do we really know about the benefits? Ambio, 39: 585-599.

Liebig MA and Doran JW. 1999. Impact of organic production practices on soil quality indicators. Journal of Environmental Quality, 28(5): 1601-1609.

Lockeretz W, Shearer G and Kohl DH. 1981. Organic farming in the corn belt. Science, 211: 540-546.

Lotter D, Seidel R and Liebhardt W. 2003. The performance of organic and conventional cropping systems in an extreme climate year. American Journal of Alternative Agriculture, 18: 146-154.

Mader P, Edenhofer S, Boller T, Wiemken A and Niggli U. 2000. Arbuscular mycorrhizae in a long-term field trial comparing low-input (organic, biological) and high-input (conventional) farming systems in a crop rotation. Biology and Fertility of Soils, 31(2): 150-156.

Mahale, P. and Soree, H. 1999. Cosmovisions in health and agriculture in India In: *Food for Thought: Ancient visions and new experiments of rural people* (B Haverkort and W Hiemstra, Eds.), Books for Change (India), Bangalore, pp. 33-42.

Merrill, M.C. 1983. Eco-agriculture: A review of its history and philosophy. Biological Agriculture and Horticulture, 1: 181-210.

Muller A. 2009. Benefits of organic agriculture as a climate change adaptation and mitigation strategy for developing countries. EfD Discussion Paper 09, 19 p.

Niggli U, Fliessbach A, Hepperly P and Scialabba N. 2008. Low greenhouse gas agriculture: Mitigation and adaptation potential of sustainable farming systems, Rome: FAO.

Niggli U. 2007. *Organic Farming and Climate Change*. Available at www.intracen.org/Organics/documents/Organic_Farming_and_Ciimate_Change.pdf.

Paustian K, Babcock BA, Hatfield J, Lal R, McCarl BA, McLaughlin C, Mosier A, Rice C, Robertson GP, Rosenberg NJ, Rosenzweig C, Schlesinger WH and Zilberman D. 2004. Agricultural mitigation of greenhouse gases: Science and policy options. CAST (Council on Agricultural Science and Technology) Report R141. Ames, IA, USA.

Pereira, W. 1993. *Tending the Earth*. Earthcare Books, Mumbai. 315 p.

Petersen SO, Regina K, Pollinger A, Rigler E, Valli L, Yamulki S, Esala M, Fabbri C, Syviisalo E and Vinther FP. 2005. Nitrous oxide emissions from organic and conventional crop rotations in five European countries. Agriculture, Ecosystems and Environment, 112: 200-206.

Pimentel D, Hepperly P, Hanson J, Douds D and Seidel R. 2005. Environmental, energetic and economic comparison of organic and conventional farming systems. Bioscience, 55: 573-582.

Randhawa, M.S. 1986. *A History of Agriculture in India 1980-1986*, Vol. I-IV, Indian Council of Agricultural Research, New Delhi.

Reganold JP, Glover JD, Andrews PK and Hinman HR. 2001. Sustainability of three apple production systems. Nature, 410(6831): 926-930.

Reganold JP. 1995. Soil quality and profitability of biodynamic and conventional farming systems: A review. American Journal of Alternative Agriculture, 10(1): 36-46.

Rochette P, Angers DA, Chantigny MH, Gagnon B and Bertrand N. 2008. N_2O ûuxes in soils of contrasting textures fertilized with liquid and solid dairy cattle manure. Canadian Journal of Soil Science, 88: 175-187.

Rodale J.I. 1946. *Pay Dirt*. Rodale Press, Emmaus, PA.

Sayre L. 2003. *Organic farming combats global warming – big time*, Rodale Institute, http: //www.rodaleinstitute.org/ob_31.

Scialabba NE and Muller-Lindenlauf M. 2010. Organic agriculture and climate change. Renewable Agriculture and Food Systems, 25(2): 158-169.

Singh SN, Verma A and Tyagi L. 2003. Investigating options for attenuating methane emission from Indian rice fields. Environment International, 29: 547-553.

Smith KA and Conan F. 2004. Impacts of land management on fluxes of trace greenhouse gases. Soil Use and Management, 20: 255-263.

Smith P, Martino D, Cai Z, Gwary D, Janzen H, Kumar P, McCarl B, Ogle S, O'Mara F, Rice C, Scholes B, Sirotenko O, Howden M, McAllister T, Pan G, Romanenkov V, Schneider U, Towprayoon S, Wattenbach M and Smith J. 2008. Greenhouse gas mitigation in agriculture. Philosophical Transactions of the Royal Society of London. Series B, Biological sciences, 363: 789-813.

Smith P, Martino D, Cai Z, Gwary D, Janzen H, Kumar P, McCarl B, Ogle S, O'Mara F, Rice C, Scholes B and Sirotenko O. 2007. Agriculture. In: *Climate Change 2007:*

Mitigation (B Metz, OR Davidson, PR Bosch, R Dave and LA Meyer, Eds.), Contribution of Working Group III to the Fourth Assessment Report of the Intergovernmental Panel on Climate Change. Cambridge University Press, Cambridge, UK. pp 497-450.

Sreejith A and Sherief AK. 2011. *The wanted change against climate change: assessing the role of organic farming as an adaptation strategy. In: Climate Change Adaptation Strategies in Agriculture and Allied Sectors* (GSLHV Prasad Rao, Ed.), Scientific Publishers, New Delhi, India. pp. 193-200.

Srinivasarao Ch, Venkateswarlu B, Lal R, Singh AK, Vittal KPR, Kundu S, Singh SR and Singh SP. 2012. Long-term effects of soil fertility management on carbon sequestration in a rice-lentil cropping system of the Indo-Gangetic plains. Soil Science Society of America Journal, 76(1): 168-178.

Steiner R. 1924. *In* "Agriculture: A Course of Eight Lectures". Rudolph Steiner Press/Biodynamic Agriculture Association, London.

Syliva DM and Williams SE. 1992. Vesicular-arbuscular mycorrhizae and environmental stress. In: *Mycorrhizae in Sustainable Agriculture* (GJ Bethenfalvay and RG Linderman, Eds.), Proceedings of a symposium, 31 October 1991. American Society of Agronomy, Crop Science Society of America, and Soil Science Society of America, Madison Wisconsin, pp. 101-124.

United Nations. 2009. Millennium Development Goals Indicators. Accessed From http: //mdgs.un.org/unsd/mdg/SeriesDetail.aspx?srid=751 and crid=).07

United Nations. 2008. Human Development Report 2007/2008. Fighting Climate Change: Human Solidarity in a Divided World. Accessed from <http: //hdr.undp.org/en/>

USDA. 1980. *Report and recommendations on organic farming*. USDA Study Team on Organic Farming, Washington, D.C.

Van Groeningen JW, Kasper GJ, Velthof GL, van den Pol/vanDasselaar A and Kuikman PJ. 2004. Nitrous oxide emissions from silage maize ûelds under different mineral fertilizer and slurry application. Plant and Soil, 263: 101-111.

Wells AT, Chan KY and Cornish PS. 2000. Comparison of conventional and alternative vegetable farming systems on the properties of a yellow earth in New South Wales. Agriculture Ecosystems and Environment, 80(1-2): 47-60.

Willer H, Lernoud J and Kilcher L. 2013. The World of Organic Agriculture. Statistics and Emerging Trends 2013. FiBL, Frick, and, IFOAM, Bonn. 344p.

Williams AG, Audsley E and Sandars DL. 2006. Determining the environmental burdens and resource use in the production of agricultural and horticultural commodities. Main Report. Defra Research Project IS0205. Cranfield University, Bedford, and the Department for Environment, Food and Rural Affairs (Defra) of the United Kingdom Government, London. Available online at http: //www.silsoe.cranfield.ac.uk, and www.defra.gov.uk.

Wrage NGL, Velthop GL, van Beusichem ML and Oenema O. 2001. Role of nitrifier denitrification in the production of nitrous oxide. Soil Biology and Biochemistry, 33: 1723-1732.

Xu H, Cai ZC, Jia ZJ and Tsuruta H. 2000. Effect of land management in winter crop season on CH_4 emission during the following flooded and rice growing period. Nutrient Cycling in Agroecosystems, 58: 327-332.

Yagi K, Tsuruta H and Minami K. 1997. Possible options for mitigating methane emission from rice cultivation. Nutrient Cycling in Agroecosystems, 49: 213-220.

Chapter 21

Nutritional Quality of Organically Grown Food Crops

V. Visha Kumari and K.A. Gopinath

ICAR-Central Research Institute for Dryland Agriculture,
Hyderabad – 500 059, Telangana
E-mail: v.visha@icar.gov.in

1. INTRODUCTION

The term "organic" to consumer in today's life signifies quality and nutrition. It has rather become a status symbol. The interests and preferences of consumers have gone high. Certified organic produce are those which are produced in accordance with given organic farming standards. The key principles and practices of organic food production try to support and enhance biological cycles within the farming system. This in turn maintain and increase fertility of soils, minimize pollution, avoid the use of synthetic fertilizers and pesticides, maintain genetic diversity, consider the wider social and ecological impact of the food production and processing system, and produce food of high quality in sufficient quantity (IFOAM, 1998). In the consumer's mind, organic produce must be better and healthier than that produced under conventional farming system. This image is also the main motive for consumers who are willing to pay premium prices for purchasing organic food. Organic agriculture can be viewed as an attempt to overcome contamination of food supplies with pesticides, pollution, and radioactive fallout *etc.*, associated with processed food and a chemically-based agriculture. From a scientific point of view, however, it is difficult to provide or substantiate the supposed health benefits, since food quality is composed of various partial aspects and without uniform evaluation standards. Several investigations have clearly shown that the type of fertilizations, contrary to the principle of organic farming, does not significantly affect crop quality. Crop quality is not dependent on the principle difference between inorganic fertilization and organic manuring. Side effects caused by synthetic pesticides and drug feeding are not found in organic farming, which is a positive result. The use of herbicides has been documented to increase cyanide, potassium nitrate, and other toxins in crops. The evaluation of food quality by taking into account the criteria such as appearance and nutritional value exclusively is not satisfying. Even though, a range of factors has been investigated comparing organic and conventional food production systems which majorly includes economics, crop yields, soil properties, soil microbiological activity, insect-pest and disease burdens *etc.*, nutritional comparison is one which has gained momentum. Nevertheless, the link between organic products and their enhanced nutritional/environmental values is far from being fully understood (Maggio *et al.*, 2013).

2. The Nutritional Value of Organic Food

A large number of studies have been reported that attempt to investigate if there is a difference in the nutritional value of organically and conventionally grown food. There is considerable variation in the types of studies and study designs. However, the majority involve one of four main approaches (Bourn and Prescott, 2002).

- ☆ The chemical analysis of organic and conventional foods purchased from retailers
- ☆ The effect of different fertilizer treatments on the nutritional quality of crops
- ☆ The analysis of organic and conventional foods produced on organically and conventionally managed farms
- ☆ The effect of organic and conventional feed/foods on animal and human health (predominantly reproductive health)

However, it is very difficult to have a comparison with the first approach mainly because of the study design, consideration of limited production design and unknown about the origin. Different fertilizer treatments are easy to be conducted, but has not resulted in any clear picture. The third approach is found highly useful because the effects of whole systems of production on nutritional value are essentially being evaluated.

2.1 Macronutrients

There is not much on effect of different production systems on starch/ carbohydrate to have a comparison between organic and conventional produce. But, protein content as influenced by organic farming has received importance. In wheat it was found that organically grown and conventionally grown one have comparable protein content (Shier *et al.*, 1984) but somewhat lower levels of protein than the conventional ones (Woese *et al.*, 1997). Gopinath *et al.* (2008) also reported that in the first year of transition, protein content of wheat grain was higher (85.9 g/kg) for mineral fertilizer treatment, whereas, in the second year, there were no significant differences among the mineral fertilizer treatment and the highest application rate (150 kg N/ha) of three organic amendments (FYM, vermicompost and lantana compost). The grain P and K contents were, however, significantly higher for the treatments involving organic amendments than their mineral fertilizer counterpart in both years. Similarly, Saha *et al.* (2007) reported that the protein content in rice grains was the highest (8.98 per cent) in the inorganic treatment (100:60:40 kg N, P, K/ha) compared to organically grown rice. Another study showed that organic virgin olive oil had a higher oleic acid level (Gutierrez *et al.*, 1999). Similarly dairy products *viz.*, hen eggs (Kouba *et al.*, 2002) and raw cow's milk (Toledo *et al.*, 2002) did not show any noticeable change in protein levels. However, a study conducted in Sweden showed that organically-bred cows had more lean meat than their conventional counterparts (Hansson *et al.*, 2000). More qualitatively, meat from organically-grown cows had more polyunsaturated fatty acids (Pastsshenko *et al.*, 2000).

2.2 Minerals and Vitamins

Worthington (2001) reported a 21 per cent increase in iron content, 29 per cent increase in magnesium and 13.6 and 15 per cent higher phosphorus and nitrate in organic produce of different crops. A detailed review by Lairon (2009) also showed that organic food had 21 and 29 per cent more iron and magnesium than non organic food. Among the vitamins, ascorbic acid (vitamin C) was found higher in many organic fruits and vegetables. Hajslova (2005) reported lower levels of nitrate and higher levels of ascorbic acid and chlorogenic acid in organically grown potatoes. In an organic orchard of yellow plums with soil left as natural meadow, ascorbate, tocopherols, and beta carotene were found highest. However total polyphenols were higher in conventional farm (Lombardi *et al.*, 2004). Higher vitamin E level in organic olive oil was reported by Gutierrez *et al.* (1999). In rice, Saha *et al.* (2007) reported significantly higher iron content of 52.2 µg/g with organic fertilization than inorganic fertilization (42.1 µg Fe/g). However, inorganic fertilization was superior in terms of copper content (4.1 µg Cu/g) compared with organic treatments (3.1–4.0 µg Cu/g). In another study, Saha *et al.* (2010) reported higher P and K contents in organically grown rice compared to conventional rice (Table 21.1). Similarly, iron, Cu, and Mn contents were slightly more in organic treatments, unlike the Zn concentration, which was more in inorganic treatments.

2.3 Other Phytomicronutrients

Phytomicronutrients were treated one among the other micronutrients. However, it has gained importance in last two decades. They include carotenoids, flavonoids and other polyphenols. Flavonoids are good antioxidants (Pietta, 2000) and carotenoids have been found to reduce cancer risk (Karppi *et al.*, 2009). It has been anticipated in a recent review that organic plant foods overall contain double the amount of phenolic compounds (Rembialkowska, 2007). One study reported higher levels of resveratrol in organic wines (Levite *et al.*, 2000). Anthocyanic compound in berries have been reported to reduce neuronal and cognitive brain functions (Zafra-Stone *et al.*, 2007). Because of the importance they play in human health more focus is now being given for phytomicronutients. In a study done by Wang *et al.* (2008), higher percentage of sugar, malic acid, total phenols, total anthocyanin and antioxidant activity were recorded in organically grown blueberries than the one conventionally grown. Pragya *et al.* (2007) compared the quality characteristics and sensory quality in fresh green peas grown by organic, inorganic

Table 21.1: Mineral Contents of Rice Grain grown under different Manure and Fertilizer Treatments

Treatment	*P (per cent)*	*K (per cent)*	*Fe (μg/g)*	*Zn (μg/g)*	*Cu (μg/g)*	*Mn (μg/g)*
Control	0.215a	0.256a	30.17a	14.30a	1.18a	4.52a
FYM 5 t/ha	0.233abc	0.272abc	35.16ab	14.69a	1.48b	7.13bc
FYM 10 t/ha	0.239abc	0.281bc	41.15ab	16.07b	1.82d	9.36de
FYM 15 t/ha	0.246abc	0.302d	43.85bc	16.66b	1.87de	10.19e
FYM 20 t/ha	0.261c	0.313d	53.39c	18.00cd	2.06g	10.38e
Fertilizer eq. FYM 5 t/ha	0.228ab	0.258a	30.42a	17.46c	1.42b	6.53b
Fertilizer eq. FYM 10 t/ha	0.234abc	0.266ab	32.20ab	18.63de	1.64c	7.37bcd
Fertilizer eq. FYM 15 t/ha	0.237abc	0.277bc	35.15ab	19.28e	1.80d	8.40bcde
Fertilizer eq. FYM 20 t/ha	0.249bc	0.286c	37.40ab	19.30e	1.98fg	8.75cde

Means in the same column with different letters are significantly ($P < 0.05$) different.

Source: Saha *et al.* (2010).

and integrated methods and found higher copper and zinc levels as compared to inorganically grown peas and peas grown by integrated method of cultivation.

3. Food Safety

3.1 Pathogenic Microorganisms

Since organic production systems rely on animal manures and vegetable/crop wastes for supplementing nutrient requirement there is always concern about the possible contamination of food products. Organic farming has fascinated the notice of the entire food production sector around the globe since it restores on eco-agricultural principles that consider soil, water and air quality. However, organic produce is more exposed to microbiological contamination than conventional produce, since organic fertilizers often consist of manure, and manure may harbor pathogenic microorganisms such as *Salmonella* spp., *Listeria monocytogenes* and *Escherichia coli* (Johannessen *et al.*, 2004; McMahon and Wilson, 2001). Of all organic foods, vegetables stand out as important sources of food borne illness. Use of manure as such may invite more pathogens, but if composted properly would avoid the contamination of food stuff by pathogenic microorganisms. While reporting for dairy products, we have some contradicting results. A Danish study reported that about 100 per cent of poultry samples were contaminated by *Campylobacter* sp in organic farms, whereas 36–49 per cent of samples in conventional farms were (Heuer *et al.*, 2001). In contrast, in a survey conducted in France in four different regions found comparable levels for total bacteria count or butyric microorganisms in milk produced with the two husbandry systems (Echevarria, 2001). Since we have more of incomparable results, it is really difficult to conclude the supremacy of organic food over conventional one with respect to pathogenic microorganisms.

3.2 Phytochemical Contaminants

No use of chemicals is a clear advantage for organic agriculture. However, the question has repeatedly been raised of the level of contamination of organic food by environmental pollution. In a study of the potential of vegetables to suppress the mutagenecity of various environmental toxins, including benzo pyrene (BaP the main carcinogen in cigarette) organic vegetables were found more active (Ren *et al.*, 2001). Against the chemical 4-nitroquinoline oxide, organic vegetables suppressed 37-93 per cent of mutagenic activity (Olsson *et al.*, 2006). A study performed on vegetables and strawberries in Sweden did not show any contamination of organic ones, while 17–50 per cent of conventional ones contained pesticide residues (Bourn and Prescott, 2002). A survey conducted in Italy in the 2002–2005 period on 3500 samples of food of plant origin also concluded that the vast majority (97.4 per cent) of organic farming products do not contain detectable pesticide residues (Tasiopoulou *et al.*, 2007). It is meaningful mentioning that only some natural extracts are used in organic agriculture for pest and disease control such as neem oil, fly ash, garlic extracts, chilli and ginger extracts *etc.* which are quickly degraded and hence no contamination (Moore *et al.*, 2000). The two major fungicides of mineral origin permitted for used in organic farming for disease control are sulphur and copper salts and their various compounds. However, these are allowed with certain restriction. It is also noteworthy that copper being an inorganic compound, does not breakdown like organic compounds and may accumulate in soil over several years resulting in toxicity.

3.3 Mycotoxins

Mycotoxins are poisonous compounds produced by the secondary metabolism of poisonous fungi (moulds) like *Aspergillus*, *Penicillium* and *Fusarium*, which occur in food products (Kouba, 2003). They have a negative impact on human health, *i.e.* are carcinogenic and disabling to the immune system. Mycotoxin production is mainly dependent on temperature, humidity and other favorable environmental conditions. Recent studies have not shown that organic food is more susceptible to mycotoxin contamination than conventional food (Kouba, 2003; Benbrook, 2006; Lairon, 2009).

4. Conclusions

Organic farming practices largely avoid synthetic fertilizers, pesticides, growth regulators and livestock feed additives. Organic farming systems rely on crop rotations, manures, organic wastes and biological pest controls to maintain soil productivity, supply nutrients to growing plants and control pests. Several studies do reveal some differences in quality between conventionally and organically produced foods. In general, many reviews have concluded that organic plant products contain more dry matter, minerals (Fe, Mg) and Vitamin C; and contain more anti-oxidant micronutrients such as phenols and salicylic acid. Organic animal products contain more polyunsaturated fatty acids. However, data on carbohydrate, protein and vitamin levels are insufficiently documented. Almost all of the of organic food (94-100 per cent) does not contain any pesticide residues. Organic vegetables contain far less nitrates, about 50 per cent less. In addition, there were several trends showing less protein but of a better quality, more nutritionally significant minerals, and lower amounts of some heavy metals in organic crops compared to conventional ones. Organic agriculture has the potential to produce high quality products with some relevant improvements in terms of contents of anti-oxidant phytomicronutrients, nitrate accumulation in vegetables and toxic phytochemical residue levels.

REFERENCES

Benbrook CM. 2006. FAQS on pesticides in milk. Organic Center. Calculated from USDA's Pesticide Data Program. http: //www.organic-center.org/science. pest.php?action =view and report_id=79

Bourn D and Prescott J. 2002. A comparison of the nutritional value, sensory qualities and food safety of organically and conventionally produced foods, Critical Reviews in Food Science and Nutrition, 42, 1–34.

Echevarria L. 2001. Qualité du lait livré par les élevages agrobiologiques de quatre régions françaises, Renc. Rech. Rum. 8, 95.

Gopinath KA, Supradip Saha, Mina BL, Kundu S, Harit Pande and Gupta HS. 2008. Influence of organic amendments on growth, yield and quality of wheat and on soil properties during transition to organic production. Nutrient Cycling in Agroecosystems, 82: 51-60.

Gutierrez F, Arnaud T and Albi MA. 1999. Influence of ecological cultivation on virgin olive oil quality. Journal of the American Oil Chemists' Society, 76: 617-621.

Hajslova J, Schulzova V, Slanina P, Janne K, Hellenäs KE and Andersson C. 2005. Quality of organically and conventionally grown potatoes: four-year study of micronutrients, metals, secondary metabolites, enzymic browning and organoleptic properties. Food Additives and Contaminants, 22(6): 514-34.

Hansson I, Hamilton C, Ekman T and Forslund K. 2000. Carcass quality in certified organic production compared with conventional livestock production. Zoonoses and Public Health, 47(2): 111-120.

Heuer OE, Perdersen K, Andersen JS and Madsen M. 2001. Prevalence and antimicrobial susceptibility of thermophilic Campylobacter in organic and conventional broiler ûocks, Letters in Applied Microbiology, 33: 269-274.

IFOAM. 1998. Basic standards for organic production and processing. IFOAM General Assembly, Argentina.

Johannessen GS, Froseth RB, Solemdal L, Jarp J, Wasteson Y and Rorvik LM. 2004. Influence of bovine manure as fertilizer on the bacteriological quality of organic Iceberg lettuce. Journal of Applied Microbiology, 96(4): 787–794.

Karppi J, Kurl S, Nurmi T, Rissanen TH, Pukkala E and Nyyssonen K. 2009. Serum lycopene and the risk of cancer: the Kuopio Ischaemic Heart Disease Risk Factor (KIHD) study. Annals of Epidemiology, 19: 512-518.

Kouba M. 2003. Quality of organic animal products. Livestock Production Science, 80: 33-40.

Kouba M, Enser M, Whittington FM, Nute GR and Wood JD. 2002. Eûet d'un regime riche en acide linolénique sur les activités d'enzymes lipogéniques, la composition en acides gras et la qualité de la viande chez le porc en croissance, Neuvièmes Journées des Sciences du Muscle et Technologie de la Viande, 15-16.

Lairon D. 2009. Nutritional quality and safety of organic food. A review. Agronomy for Sustainable Development, 30: 33-41.

Levite D, Adrian M and Tamm L. 2000. Preliminary results of resveratrol in wine of organic and conventional vineyards. In: *Proc. of the 6th International Congress on Organic Vinivulture*, Basel. pp. 256-257.

Lombardi-Boccia G, Lucarini M, Lanzi S, Aguzzi A and Cappelloni M. 2004. Nutrients and anti oxidant molecules in yellow plums (*Prunus domestica* L.) from conventional and organic productions: A comparative study. Journal of Agricultural and Food Chemistry, 52: 90-94.

Maggio A, De Pascale S, Paradiso R and Barbieri G. 2013. Quality and nutritional value of vegetables from organic and conventional farming. Scientia Horticulturae, 164: 532-539.

McMahon MAS and Wilson IG. 2001. The occurrence of enteric pathogens and *Aeromonas* species in organic vegetables. International Journal of Food Microbiology, 70(1-2): 155-162.

Moore VK, Zabik ME and Zabick MJ. 2000. Evaluation of conventional and "organic" baby food brands for eight organochlorine and ûve botanical pesticides, Food Chemistry, 71: 443-447.

Olsson ME, Andersson CS, Oredsson S, Berglund RH and Gustavsson KE. 2006. Antioxidant levels and inhibition of cancer cell proliferation *in vitro* by extracts from organically and conventionally cultivated strawberries. Journal of Agricultural and Food Chemistry, 54: 1248-1255.

Pastsshenko V, Matthes HD, Hein T and Holzer Z. 2000. Impact of cattle grazing on meat fatty acid composition in relation to human nutrition. In: *Proceedings 13th IFOAM Scientiûc Conference*, pp. 293-296.

Pietta P. 2000. Flavonoids as Antioxidants. Journal of Natural Products, 63: 1035-1042.

Pragya A, Bhattacharya L, Kulshrestha K and Mahapatra BS. 2007. Effect of organic inorganic and integrated methods of cultivation on quality of fresh green peas. Journal of Eco-friendly Agriculture, 2(1): 20-22.

Rembialkowska E. 2007. Quality of plant products from organic agriculture. Journal of the Science of Food and Agriculture, 87: 2757-2762.

Ren H, Bao H, Endo H and Hayashi T. 2001. Antioxidative and antimicrobial activities and ûavonoid contents of organically cultivated vegetables. Nippon Shokuhin Kagaku Kaishi, 48: 246-252.

Saha S, Gopinath KA, Mina BL, Kundu S, Bhattacharaya R and Gupta HS. 2010. Expression of soil chemical and biological behavior on nutritional quality of aromatic rice as influenced by organic and mineral fertilization. Communications in Soil Science and Plant Analysis, 41(15): 1816-1831.

Saha S, Pandey AK, Gopinath KA, Bhattacharaya R, Kundu S and Gupta HS. 2007. Nutritional quality of organic rice grown on organic composts. Agronomy for Sustainable Development, 27: 223-229.

Shier NW, Kelman J and Dunson JW. 1984. A comparison of crude protein, moisture, ash and crop yield between organic and conventionally grown wheat, Nutrition Reports International, 30: 71-73.

Tasiopoulou S, Chiodini AM, Vellere F and Visentin S. 2007. Results of the monitoring program of pesticide residues in organic food of plant origin in Lombardy (Italy). Journal of Environmental Science and Health B, 42: 835-841.

Toledo P, Andren A and Bjrk L. 2002. Composition of raw milk from sustainable production systems. International Dairy Journal, 12: 75-80.

Wang SY, Chen CT, Sciarappa W, Wang CY and Camp MJ. 2008. Fruit quality, antioxidant capacity, and flavonoid content of organically and conventionally grown blueberries. Journal of Agricultural and Food Chemistry, 56(14): 5788–5794.

Willer H and Julia L. 2016. The World of Organic Agriculture - Statistics and Emerging Trends 2016. Research Institute of Organic Agriculture (FiBL), Frick, and International Federation of Organic Agriculture Movements (IFOAM), Bonn.

Woëse K, Lange D, Boess C and Bögl KW. 1997. A comparison of organically and conventionally grown foods - Results of a review of the relevant literature, Journal of the Science of Food and Agriculture, 74: 281-293.

Worthington V. 2001. Nutritional quality of organic versus conventional fruits, vegetables and grains. Journal of Alternative and Complementary Medicine, 7(2): 161-173.

Zafra-Stone S, Yasmin T, Bagchi M, Chatterjee A, Vinson JA and Bagchi D. 2007. Berry anthocyanins as novel antioxidants in human health and disease prevention. Molecular Nutrition and Food Research, 51*(6):* 675-683.

Chapter 22

Economics and Other Dimensions of Organic Farming in India

C.S. Vaidya

Formerly with H.P. University,
Shimla – 171 005, Himachal Pradesh
E-mail: csvaidya2000@yahoo.com

1. INTRODUCTION

The green revolution technologies favored cereals crops only in limited regions in India and brought many ill effects like water logging, loss of soil fertility, desertification, and salinity *etc.* All these compounded for seeking an alternative sustainable farming system in the form of organic farming or ecological agriculture (Banerjee, 2010). Organic farming systems are perceived as a feasible solution for many problems being currently faced by Indian agriculture (Charyulu and Biswas, 2010). The challenge lies in creating an environment in which organic is treated as a complimentary approach and efforts are focused on harvesting benefits to a large section of the Indian farmers. This holds particularly relevant for rainfed agriculture owing to the fact that input use is low under rainfed agriculture and has low conversion period and smaller yield reduction (Gopinath *et al.*, 2011). Organic agriculture is not new to the farmers in many less developed countries as they have been practicing it traditionally. However, in modern commercial agriculture combined with high financial requirements, organic farming has to be competitive not only in economic terms but also on environmental considerations. The commercialization of agriculture has significantly impacted the environment in a harmful manner. High use of agrochemicals caused deterioration in soil health leading to imbalance in biological equilibrium (Devi, 2012). The first step in this direction has to be the favorable economic situation. The additional and contributing factors are concerns for healthy soil, healthy food and healthy people. The premium product prices available for organic farm products in majority of markets can tilt the scales in its favor (Gopinath *et al.*, 2009).

Although in India, farmers are turning towards organic methods of cultivation, the area currently being used for organic farming and the number of organic farmers clearly shows that certain factors are limiting the organic movement. To begin with, there is a debate regarding comparative cost of production under organic and inorganic farming systems. Simultaneously, there is a widespread apprehension that crop productivity under organic farming declines initially during conversion period which might lead to food and income insecurity. The certification of organic produce requires at least 2-3 years conversion period. This is a period necessary for chemical, physical and biological properties of soil to reach an ecological balance (Gopinath *et al.*, 2011). On the other hand, the protagonists of organic farming argue that there is insignificant decline in productivity. Even if the productivity declines, it is more than offset by the premium prices expected for organic products-coupled with the reduced expenditure on fertilizers, pesticides, *etc.* The productivity which declines initially under organic farming systems is claimed to start increasing and reaches the earlier levels or even more in 3-5 years, provided organic practices are applied properly. There is hardly any authentic evidence for these claims.

2. Case Study

A study was taken up to understand the factors making organic farming important for farmers, to compare the 'cost of cultivation' and 'cost of production' of organic versus inorganic production in different states in India representing diverse agro-ecological zones and to document the benefits other than economic of organic farming. The area selected for this study included apple farming in Himachal Pradesh, rainfed hill food crops in Uttarakhand, rainfed cash and food crops in Rajasthan, cotton and soybean farming in Maharashtra, plantation cash crops in Kerala, irrigated mixed farming in Karnataka and Medicinal and Aromatic Plants and Vegetables in Tamil Nadu. The sample size included 376 farmers of which 199 were certified organic farmers and the rest were inorganic farmers (Table 22.1).

Table 22.1: No. of Farmers Surveyed in Diverse Agro-ecological Zones in India

State	*Crop*	*No. of Farmers*		
		Organic	*Inorganic*	*Total*
Himachal Pradesh	Apple	28	26	54
Karnataka	Plantation crops	26	25	51
Kerala	Spices	23	26	49
Maharashtra	Cotton and soybean	22	18	40
Rajasthan	Cereal and commercial crops	35	30	65
Tamil Nadu	Vegetables and medicinal herbs	32	27	59
Uttarakhand	Traditional food crops	33	25	58
Total		**199**	**177**	**376**

2.1 Importance of Organic Farming - Farmers' Perspective

The state-wise analysis revealed that various factors *viz.*, agro-climatic conditions, socio-economic aspects, mindset of farmers, the level of interaction and relationship by farmers with local service providers influenced growers to opt for organic farming. The decision is ultimately the outcome of interplay of more than one factor and hence the analysis of multiple responses has been carried out and results are furnished in Table 22.2.

Table 22.2: Factors Responsible for Importance of Organic Farming – Farmers' Perspective (Per cent of farmers, Multiple Responses)

Motivating Factor	*Overall*	*State*						
		HP	*KA*	*KL*	*MH*	*TN*	*RJ*	*UK*
Improvement in soil quality and fertility	40	15	85	45	90	25	50	5
Productivity improvomont	50	20	90	50	100	40	60	10
High prices of inorganic inputs	75	90	100	80	100	70	80	20
Higher market access	55	10	75	10	95	50	100	35
Premium product prices	50	5	80	0	80	50	100	40
Health hazards associated with inorganic	70	75	85	75	90	65	90	20

HP: Himachal Pradesh; KA: Karnataka; KL: Kerala; MH: Maharashtra; TN: Tamil Nadu; RJ: Rajasthan; UK: Uttarakhand.

Figures have been rounded off to nearest 5 per cent for convenience.

2.1.1 Improvement in Soil Fertility

Organic farming was preferred by 45 per cent and 50 per cent farmers in Rajasthan and Kerala while to the tune of 85 per cent in Karnataka and 90 per cent in Maharashtra based on the fact that it improves soil fertility and quality. On the other hand, farmers of Himachal Pradesh, Tamil Nadu and Uttarakhand did not consider it as an important factor. Overall level, 40 per cent farmers considered improvement in soil fertility as an important factor for practicing organic farming.

2.1.2 Crop Productivity Improvement

Nearly respectively 50 per cent in Kerala, 60 per cent in Rajasthan, 85 per cent in Karnataka and 90 per cent farmers in Maharashtra attribute scope for improvement in productivity as a reason for practicing organic farming. Overall, 50 per cent farmers preferred organic farming due to improvement in productivity.

2.1.3 Increasing Prices of Inorganic Inputs

A large majority of farmers in all states except for Uttarakhand were concerned about the continuous and heavy increase in prices of inorganic inputs adversely affecting their farm economics and considered organic farming as a viable alternative. This was the opinion of 75 per cent farmers in the surveyed states together, 100 per cent in Karnataka and Maharashtra and 90 per cent in Himachal Pradesh.

2.1.4 Increasing Market Access

Market access for organic products is encouraging majority of farmers in

Karnataka, Maharashtra and Rajasthan to practice organic farming. The percentage of farmers ranged from 95-100 in Kerala and Rajasthan and 55 per cent in the surveyed states together.

2.1.5 Premium Prices for Organic Products

Availability of premium prices especially in metropolitan cities compelled farmers to think in favor of organic farming, except for Himachal Pradesh and Kerala due to the fact that there is hardly any difference in prices of organic and inorganic products. Overall level, 50 per cent farmers expressed this as a motivating factor.

2.1.6 Health Hazards

This is one of the most weighted factor due to which organic farming is considered to be important in all the states. About 70 per cent of the farmers expressed this as a reason. But, only 20 per cent farmers expressed in Uttarakhand and were due to the fact that the states use insignificant quantity of fertilizers and pesticides in the conventional farms.

The analysis points out that the importance of organic agriculture does not emerge out of any single reason. It is a mix of several factors. The study highlights that the market for organic products, the premium product prices and health hazards due to inorganic methods of production were the main factors over the past years. The findings also indicate that the spiraling prices of chemical inputs, was also an important factor. Not only have the prices been escalating, but even the doses required for maintaining the production levels have been increasing over time.

2.2 Economic Analysis of Important Crops

The economics of organic crops is an important concern to the farmers. Favorable economics will certainly be the greatest incentive for wider adoption of organic practices in crop husbandry. Benefits such as human and animal health and environmental implications are definitely the secondary concerns. Under such circumstances, creating favorable economic conditions for organic farming becomes a priority. This section discusses the findings of the cost benefit analysis of various crops grown organically under different farming systems in India. The cost of cultivation and production of different crops includes physical inputs and labour, irrespective of the fact whether these were purchased/hired or were family labour and on farm inputs. The analysis does not take into account indirect costs like the rental value of land, depreciation of farm buildings and implements, interest on fixed and working capital, *etc.* It is based on the assumption that such costs do not vary significantly under the two systems and the farmers are more concerned about the direct costs associated with crop cultivation. The productivity differentials under two situations have been worked out for per unit of land. Similarly, the study does not take into account the crop by-product. It was found during the course of the study that the quantity and prices of crop by-products have insignificant differences and an inclusion of value of by-product in the total returns would have not made any difference in the comparative economics of the two situations. The total cost of cultivation has been used to work out the comparative costs of production. The analysis has been further extended to include the market prices so as to work out the comparative returns under organic and inorganic farming conditions. Overall, it is basically the cost of production and net profits which comprise the main steering forces motivating the farmers to adopt organic farming.

The present analysis has an inherent deficiency that this study was primarily planned with a conventional economic analysis in mind that was crop-centric. However, later on during the data collection, it was found that the organic farming system is not individual crop based but it is rather a mix of crops and encompasses the whole farming system. Either the farmer is organic or not, the main crops always have some intercrops. The intercrops are planned so as to have a symbiotic relationship with the main crop. The associated crops provide nutrition to other crops adding to good plant growth and crop bearing. In addition to this, the aim of intercrops may not be directly economic. They may act as natural insect and pest repellent, reducing expenditure on costly chemical insecticides and pesticides. It contributes to net returns from the crops, even if indirectly. Thus, the economics of organic farming is not about the single crop but costs and returns from a mixture of crops. In order to overcome this difficulty, the costs and returns of individual crops have been worked out by deducting the labour and material costs incurred for intercrop and the returns realized from such intercrops, from total costs and returns of a group of crops. The analysis has been carried out for important crops or crop groups in the selected states; and the results are furnished in Table 22.3.

2.2.1 Himachal Pradesh

Despite having lower cost of cultivation and production under organic situation by 41 and 37 per cent, respectively, the inorganic apple is enjoying better net profits to the tune of 31 per cent. This unique situation is the outcome of the fact that organic apple is about 20-25 per cent heavier and hence a standard box designed for 18-20 kg of inorganic apple had 25-28 kg of organic apple but sold at a price equal to that inorganic apple. The organic apple producers suffered due to absence of any specialized marketing channel catering for organic apple, organic apple being sold as undifferentiated from inorganic. This led to a situation leading to absence of preferential prices in favor of organic apple.

2.2.2 Karnataka

The cost of cultivation and production of five crops *viz.*, plantation and field crops was invariably higher when the crops were grown with inorganic practices. The only exception was paddy, which had lower cost of production under inorganic situation, mainly due to higher productivity; however, organic paddy gave 4 per cent higher net returns. In fact, all the crops under consideration were yielding significantly higher net returns under organic scenario. The highest difference was in case of coconut with 54 per cent higher net returns being for organic ones followed by 27 per cent in banana and 22 per cent in sugarcane. This analysis indicated most favorable environment for organic cultivation of crops.

2.2.3 Kerala

In Kerala, three crops *viz* black pepper, coffee and arecanut were studied. The analysis indicated that both cost of cultivation and cost of production were

Table 22.3: Comparison of Cost of Organic and Inorganic Ways of Crop Cultivation (Cost of cultivation in Rs./ha, cost of production in Rs./kg and net profit in Rs./ha)

Crop	Organic			Inorganic			Difference of Organic Over Inorganic (per cent)		
	Cost of Cultivation (Rs/ha)	Cost of Production (Rs/kg)	Net Profit (Rs/ha)	Cost of Cultivation (Rs/ha)	Cost of Production (Rs/kg)	Net Profit (Rs/ha)	Cost of Cultivation (Rs/ha)	Cost of Production (Rs/kg)	Net Profit (Rs/ha)
Himachal Pradesh									
Apple	13201	1.27	153192	22192	2	200168	-41	-37	-23
Karnataka									
Coconut	24802	3.05	15798	28412	3.67	10288	-13	-17	54
Banana	57200	3.81	152800	67790	4.67	120710	-16	-18	27
Sugarcane	42675	0.45	51600	45820	0.52	42180	-7	-13	22
Potato	26565	1.27	31185	30535	1.45	26165	-13	-12	19
Paddy	23660	4.73	23840	26158	3.74	22842	-10	26	4
Kerala									
Black pepper	28020	26.69	50730	19340	18.42	59410	45	45	-15
Coffee	35520	47.36	13230	30600	40.8	18150	16	16	-27
Areca nut	102757	66.85	97053	99628	64.36	101612	3	4	-4
Maharashtra									
Cotton	17046	17.05	7954	28889	27.25	-9809	-41	-37	N.C.
Soybean	14267	8.91	5093	15631	9.47	2519	-9	-6	102
Rajasthan									
Pearl millet	9978	3.22	8622	13270	4.74	3530	-25	-32	144
Psylium	16395	16.39	8905	16358	16.36	6642	0	0	34
Castor	15890	6.91	19530	17380	7.9	13420	-9	-13	46

Crop	Organic			Inorganic			Difference of Organic Over Inorganic (per cent)		
	Cost of Cultivation (Rs/ha)	*Cost of Production (Rs/kg)*	*Net Profit (Rs/ha)*	*Cost of Cultivation (Rs/ha)*	*Cost of Production (Rs/kg)*	*Net Profit (Rs/ha)*	*Cost of Cultivation (Rs/ha)*	*Cost of Production (Rs/kg)*	*Net Profit (Rs/ha)*
Cumin	22725	36.36	18212	25145	40.55	11404	-10	-10	60
Black mustard	15700	8.72	16250	16470	9.15	13230	-5	-5	23
Brown mustard	10325	6.21	14312	11625	6.47	12607	-11	-4	14
Tamil Nadu									
Medicinal and aromatic plants	53875	5.73	752175	65075	6.86	747835	-17	-16	1
Tea	47700	4.77	12300	44585	4.05	21415	7	18	-43
Potato	63388	2.58	46862	67858	2.42	58142	-7	7	-19
Cabbage	39700	0.61	57800	45120	0.58	51775	-12	5	12
Uttarakhand									
Ragi	56203	12.08	-14353	56575	12.04	-16625	-1	0	-14
Jhingora	55316	14.87	-20906	55075	14.43	- 20740	0	3	1
Wheat	25340	18.77	-11840	27320	18.84	-13545	-7	0	-13
Barley	8420	5.61	-5420	8536	5.69	-5536	-1	-1	-2
Pulses	25920	17.28	8685	24165	16.11	10440	7	7	-17

significantly higher in case of organic black pepper to the tune of 45 per cent whereas these were higher by 16 and about 4 per cent in case of coffee and arecanut. Due to higher cost of production following higher input costs and lack of organic management, the net returns of these crops was lower under organic situation than the conventional one. The highest difference of 27 per cent was in case of coffee followed by 15 per cent in black pepper and 4 per cent in arecanut. Like Himachal Pradesh, the organic products were being sold through inorganic channels as undifferentiated products without premium organic product prices.

2.2.4 Maharashtra

Maharashtra was the state where the adverse impacts of inorganic crop production were highly pronounced. This was most evident in case of cotton where the cost of cultivation and production were not only significantly higher for inorganic crop but the net returns were negative to the tune of Rs. 9809/ha. As a result the difference in net profits for organic and inorganic cotton growers could not be computed mathematically, but the situation is very much clear. This was perhaps the main cause of farmer suicides in this state as the inorganic cotton farmers were not able to repay the debts taken for costly inorganic inputs. This figure for soybean was 102 per cent though the cost of cultivation and production were only 9 and 6 per cent lower, respectively for organic cultivation. This state was observed to be having adequate marketing net work in place for organic products and hence the farmers were able to benefit in the form of highly favorable financial returns.

2.2.5 Rajasthan

The organic farmers of Rajasthan were able to realize high to very high net profits from cultivation of selected 6 crops. The net profit from cultivation of organic pearl millet was about 144 per cent higher whereas the cost of cultivation and production were only lower by 25 and 32 per cent respectively. Despite the fact that the cost of cultivation and production were almost identical for organic and inorganic psylium, the net profit under organic situation was about 34 per cent higher due to positive differentials in productivity and product prices. In the remaining four crops the cost of cultivation with organic method was lower from 5 to 11 per cent and cost of production lower by 4 to 13 per cent. This resulted in 60 per cent higher net profit in case of organic cumin and 46 per cent in case of organic castor. These figures for black and brown mustered were 23 and 14 per cent respectively.

2.2.6 Tamil Nadu

Among four crops *viz.*, medicinal and aromatic plants, tea, potato and cabbage; the cost of organic cultivation was comparatively higher by 7 per cent in case of tea and was lower in all other crops varying from 7 to 17 per cent. On the other hand, the cost production was lower for organic cultivation only in case of medicinal and aromatic plants. The cost of production in this case was lower by 16 per cent whereas it was higher by 18, 7 and 5 per cent in case of tea, potato and cabbage, respectively. As a result of differentials in productivity and product sale prices, the net returns were higher by 1 and 12 per cent in case of medicinal and aromatic plants and cabbage, respectively whereas the organic farmers had to contend with 43 and 19 per cent lower returns in case of tea and potato.

2.2.7 Uttarakhand

The subsistence hill agriculture of Uttarakhand presents a unique scenario farmers are running in net loss but they still continued with the cultivation because of the fact that they did not assign any financial value to free family labour and inputs produced on the farm or gathered free of cost from adjoining common lands. This scenario was present on both organic and inorganic farms. However, the minimal use of purchased chemical inputs being used on inorganic farms and despite being inorganic the stress was on the use of free organic inputs to be supplemented by inorganic inputs in whatever low quantities. This led to scenario where there was not much difference in cost of cultivation and production of two methods of production. The difference varied from 0 to the maximum of 7 per cent. As under both the situations there were no profits, losses have been compared and were found to be lower on organic farms except for jhingora where loss due to cultivation was higher on organic farms but the difference was only 1 per cent. In case of pulses, the farmers were realizing profits and these were higher on inorganic farms by 17 per cent.

2.2.8 Overall Economic Analysis of Organic Cultivation of Crops

The analysis indicated largely a favorable economic environment for organic farming in India. There are positive differentials in the cost of cultivation and production, productivity and market prices in the majority of cases. Farmers in four out of the seven states are better placed as far as organic farming is concerned. Farmers were getting better returns in Karnataka, Maharashtra and Rajasthan. In Uttarakhand, the organic and inorganic farmers are almost equally placed in terms of net profits. It is only in Tamil Nadu and Kerala that organic farmers are not able to generate as high returns as inorganic farmers. Overall, the findings reveal that organic farmers are enjoying higher returns per hectare from the same crops, operating under similar agro-climatic and socio-economic conditions. The benefits may not be yielding monetary returns in the beginning but have far reaching implications which not only enrich the natural resource base of the farmers but also have positive environmental and health implications for the family. Such long-term benefits definitely translate into higher financial returns by lowering the input and labour costs significantly and simultaneously the premium prices for organic production further add to it all.

2.3 Performance of Organic Crops

The comparison of organic crops with inorganic in respect of three factors *viz.*, cost of cultivation and production and profits would indicate its performance. The analysis indicates (Table 22.4) that out of 26 crops studied, 69 per cent had lower cost of cultivation. This fact makes organic farming relevant for marginal and small farmers who find it difficult to invest in costly inputs required under inorganic cultivation methods. The cost of production was lower for half of the crops at overall level with Karnataka and Rajasthan. The net profits were higher for 69 per cent crops under organic farming.

Table 22.4: Performance of Crops grown with Organic Practices (No. of crops)

State	*Total Crops*	*Number of Crops having*		
		Lower Cost of Cultivation	*Lower Cost of Production*	*Higher Profits*
H.P.	1	0	0	0
Karnataka	5	5	4	5
Kerala	3	0	0	0
Maharashtra	2	2	2	2
Rajasthan	6	5	5	6
Tamil Nadu	4	3	1	2
Uttarakhand	5	3	1	3*
Overall	**26 (100)**	**18 (69)**	**13 (50)**	**18 (69)**

Note: *No. of crops having lower losses. Figures in parenthesis are percentages.

This scenario clearly indicated the emerging favorable economic environment for adoption of organic farming. The declining cost of cultivation, which is expected to fall further with development of area specific package of practices, has resulted in higher profits for organic farmers. Same declining trend is emerging in cost of production which combined with premium product prices (in the range of 25 to more than 100 per cent after certification, processing and packaging) has translated into higher per unit profits for organic farmers. However, the price differentials, presently, were not as pronounced especially at farm gate as is evident from Table 22.5.

Table 22.5: Farm Gate Price Differentials – Organic over Inorganic

State	*Crop*	*Unit (per)*	*Organic*	*Inorganic*	*Difference (per cent)*
H.P.	Apple	Box	400	400	0.0
Karnataka	Coconut	Nut	5	5	0.0
	Banana	Qtl	1400	1300	7.7
	Sugarcane	Qtl	100	100	0.0
	Potato	Qtl	275	270	1.9
	Paddy	Qtl	950	700	35.7
Kerala	Black pepper	Qtl	7500	7500	0.0
	Coffee	Qtl	6500	6500	0.0
	Areca nut	Qtl	13000	13000	0.0
Maharashtra	Cotton	Qtl	2800	1800	55.6
	Soybean	Qtl	1210	1100	10.0
Rajasthan	Pearl millet	Qtl	600	600	0.0
	Psylium	Qtl	2530	2300	10.0
	Castor	Qtl	1540	1400	10.0

State	*Crop*	*Unit (per)*	*Organic*	*Inorganic*	*Difference (per cent)*
	Cumin	Qtl	6550	5895	11.1
	Black mustard	Qtl	1775	1650	7.6
	Brown mustard	Qtl	1350	1350	0.0
Tamil Nadu	Medicinal and aromatic plants	Qtl	85.75	85.75	0.0
	Tea	Qtl	600	600	0.0
	Potato	Qtl	450	450	0.0
	Cabbage	Qtl	150	125	20.0
Uttrakhand	Ragi	Qtl	900	850	5.9
	Jhingora	Qtl	925	900	2.8
	Wheat	Qtl	1000	950	5.3
	Barley	Qtl	200	200	0.0
	Pulses	Qtl	2307	2307	0.0

2.4 Price Differentials

Premium product prices are normally available for organic products especially after certification, processing and packaging. In the absence of these factors, the price differentials are quite narrow but pointing towards the possible opportunity in the offing. It was observed that premium prices are usually available when job of marketing is entrusted to NGOs. This indicates a huge gap in the marketing skills of individual organic farmers and is an area for intervention by concerned agencies.

The price premium for organic products was not available for apple crop in Himachal Pradesh and for three crops out of five in Karnataka. Paddy in Karnataka enjoyed a premium of about 36 per cent. The organic farmers of Kerala were also deprived of premium prices. The price differential for organic cotton in Maharashtra was as high as about 56 per cent including the amount of Rs.300 per quintal paid for community development. The marketing of organic cotton in this case was through a NGO. Soybean also enjoyed premium price, 10 per cent higher in comparison to inorganic soybean. Four organic crops of Rajasthan obtained higher prices in the range of about 8 to 11 per cent and for the rest two crops no price differentials existed. The organic cabbage of Tamil Nadu obtained 20 per cent higher prices but there were no premium prices available for medicinal and aromatic plants, tea and coffee; at least at farm gate. Three organic crops of Uttarakhand enjoyed premium prices in the range of about 3 to 6 per cent and two crops were sold at the same prices. This analysis indicates emerging price differentials in favor of organic crop which have every possibility of increasing further and for more crops. Though the fact that there are no negative price differentials is hardly ever appreciated is an important indicator of quality and preference for organic products.

2.5 Benefits other than Economics

The economic sustainability of organic farming described in previous section is only one of the several other benefits of organic which farmers think of. But the farmers' perceptions about other benefits like human or animal health and increased agro-biodiversity also need to be documented. These have not been included in the traditional cost-benefit accounting of crops and farming. The analysis would go topsy-turvy if such costs like the increased medical expenses, higher veterinary expenses, loss of biodiversity *etc.*, are added to cost of cultivation and production. It will be misleading if environmental and health aspects are not taken into consideration while analyzing cost-benefit of organic agriculture, hence, they have been evaluated. In the present analysis such costs and returns are not worked out for adding to cost-benefit analysis, but the idea is to present the farmer's responses in this regard so that such benefits and/or negative outcomes could be documented.

2.5.1 Human Health Benefits

The safe organic food is a boon not only for the health of farming families but for the general consumers as well. About 53 per cent farmers reported positive impact on health as a result of switching over to organic farming. This percentage was highest in Tamil Nadu, may be due to the fact of comparatively recent switch over to organic farming has making immediate and visible impact on the health conditions. In the states like Maharashtra and Karnataka positive health impact has been a routine thing and is not even noticed by the farmers. Lowest percentage has been found in Uttarakhand as they used insignificant quantities of fertilizer and there was no use of other chemicals. As a result, the medical expenses have come down in the range of 45 to 80 per cent in different states besides 51 per cent on an average (Table 22.6). The organic farming is not only building the resistance to common infections but also creating congenial atmosphere where such infections and their careers don't flourish.

Table 22.6: Farmers' Perception about the Positive Health Impact of Organic Farming

State	*Per cent of Farmers Reporting Positive Impact on Health*	*Reduction in Medical Expenses (Per cent)*
Himachal Pradesh	52	65
Karnataka	66	60
Kerala	58	70
Maharashtra	55	80
Tamil Nadu	76	45
Rajasthan	62	50
Uttarakhand	7	10
Overall	53	51

2.5.2 Increased Fodder Supply as a Result of Organic Farming

In addition to ensuring crop productivity, one of the welcome outcomes of organic farming is that on farm fodder supply has increased. This adds to the economic gains by saving on the expenses. The increase in fodder supply after the farmers became organic was there on large number of farms in Karnataka, Kerala and Maharashtra, mainly because of the reason that organic farming movement has been very old in the states and as a result of which the complete benefits of becoming organic have started flowing in. In other states it may take several years before such a healthy scenario develops. Even then 20-45 per cent of the farmers reported increased supply of fodder on their farms. The extent of such increase was maximum in Maharashtra where the on-farm fodder supply increased by 20-30 per cent. In rest of the cases the general trend was 10-20 per cent increase. The higher availability of on-farm fodder, not only reduces the labour requirement in tending the cattle; it also makes savings on expenses for purchased fodder.

2.5.3 Animal Health Benefits

The impact of organic farming on animal health has been due to two counts. First, the increased fodder availability ensures better health but more important it is due to the quality of fodder, obviously organic fodder is chemical free. Although, the percentage of farmers who positively responded to impact of organic farming on animal health, was around 26 per cent.It is due to the failure of farmers to establish the relationship between the two (Table 22.7). The veterinary expenses have come down in the range of 20 to 30 per cent in case of Himachal Pradesh, Tamil Nadu and Rajasthan. This decline was in the range of 40-60 per cent in Kerala and only 0-5 per cent in case of Uttarakhand. The low percentage in case of Uttarakhand was due to the fact that traditionally organic farming has never allowed the veterinary expenses to escalate. But in case of Karnataka and Maharashtra the farmers reported 50-100 per cent reduction.

Table 22.7: Impact of Increased Fodder Availability on Animal Health

State	*Per cent of Farmers Reporting Better Animal Health*	*Savings in Veterinary Expenses (Per cent)*
Himachal Pradesh	15	20-30
Karnataka	45	50-100
Kerala	35	40-60
Maharashtra	50	60-100
Tamil Nadu	20	25-30
Rajasthan	25	25-30
Uttarakhand	5	0-5
Overall	26	-

2.5.4 Resurgence of Medicinal Weeds on Organic Farms

Many weeds having medicinal qualities were becoming extinct under inorganic regime. While changing the production practices to organic, such plants have comeback and are now increasingly found on organic farms. The benefits of these plants are multifarious. The study indicates that larger percentage of organic farmers appreciate the usefulness of on farm medicinal weeds (51 per cent). The percentage of farmers reporting resurgence of medicinal plants on farm was highest in Maharashtra (95 per cent) followed by Karnataka (90 per cent) while lowest percentage in Uttarakhand (Table 22.8).

Table 22.8: Resurgence of Medicinal Plants on Organic Farms

State	*Per cent of Farmers Reporting*
Himachal Pradesh	25
Karnataka	90
Kerala	85
Maharashtra	95
Tamil Nadu	45
Rajasthan	35
Uttaranchal	15
Overall	51

2.5.5 Organic Milk Production

Organic milk is a byproduct, not a primary objective of most of Indian farmers. Neither the farmers nor the consumers are so far much aware of the positive qualities of organic milk. It is because of this fact that the organic milk production has not translated into higher demand or prices (Table 22.9). A small percentage of farmers, ranging from 2 to 12 per cent, 7 per cent at overall level, reported increase in milk production as a result of organic feed and other compatible practices. Higher prices were experienced by meager 1 and 2 per cent farmers in Karnataka and Maharashtra respectively. Demand for organic milk has been reported from Himachal Pradesh, Karnataka, Kerala and Maharashtra only by 2 to 4 per cent farmers only. The organic milk, farmers opined, was more nutritious having better taste. However, none of the respondents has ever tried to create awareness among the consumers about the organic milk.

Table 22.9: Organic Farming and Organic Milk Scenario (Per cent of farmers)

State	*Higher Production*	*Higher Prices*	*Higher Demand*
Himachal Pradesh	6	0	2
Karnataka	11	1	3
Kerala	8	0	2
Maharashtra	12	2	4
Tamil Nadu	5	0	0
Rajasthan	6	0	0
Uttaranchal	2	0	0
Overall	7	0	1

3. Conclusions

The cost benefit analysis indicates favorable economics of organic farming at least in certain crops in India. There are positive differentials in the cost of cultivation and production and market prices for majority of the crops. Farmers in five states are better placed as far as organic farming is concerned. The returns are higher in states like Karnataka, Maharashtra, Rajasthan *etc.* In Karnataka, organic farmers had 4-35 per cent higher returns whereas in Kerala the differentials ranged between 4-37 per cent in favor of inorganic farmers. Such differentials were the highest in Maharashtra. In addition to favorable economics, organic farmers observed positive environmental impacts. They observed an improvement in human and animal health and reported enhanced agro-biodiversity.

4. Future Thrusts

It is important to work out the comparative farm economics rather than limiting the analysis to specific crops. This is more important from farmers' point of view. The future studies should stress on economics of various farm sizes.

REFERENCES

Ananthraj S, Sachithanandan S and Prasanna R. 2009. Organic farming for prosperity. Kisan World, 36(3): 27-29.

Biswajit M. 2006. Boosting organic farming in India: The road ahead. Indian farmers' Digest, 39(6): 25-27.

Banerjee G. 2010. Economics of Banana Plantation under organic and in- organic Farming Systems, www.oldsite.nabard.org/databank/./paperoneconomicsoforganicfarming.pdf

Charyulu DK and Biswas S. 2010. Economics and efficiency of organic farming vis-à-vis conventional farming in India. W.P. No. 2010-04-03, Indian Institute of Management, Ahmedabad, India.

Deshmukh SN. 2010. Organic farming: Principles, prospects and problems. Agrobios, Jodhpur.

Devi M. 2012. Economic dimensions of organic and inorganic apple farming in Himachal Pradesh: A comparative analysis of tribal and non-tribal areas. Ph.d. thesis, Department of Economics, Himachal Pradesh University, Shimla-171 005.

Gopinath KA *et al.*, 2009. Bell pepper yield and soil properties during conversion from conventional to organic production in Indian Himalayas. Scientia Horticulturae, 122: 339-345.

Gopinath KA, Saha S and Mina BL. 2011. Effects of organic amendments on productivity and profitability of bell pepper–french bean–garden pea system and on soil properties during transition to organic production. Communications in Soil Science and Plant Analysis, 42: 2572–2585.

Gopinath KA, Venkateswarlu B, Venkateswarlu S, Srinivasa Rao CH, Balloli SS, Yadav SK and Prasad YG. 2011. Effect of organic management on agronomic and economic performance of sesame and on soil properties. Indian Journal of Dryland Agricultural Research and Development, 26(1): 16-20.

Sahoo BB. 2009. Organic *vs* inorganic farming: Myths and realities. Agriculture Today, 12(12): 40-42.

Chapter 23

Participatory Organic Guarantee System for India

Krishan Chandra and A.K. Yadav

National Centre of Organic Farming,
Ghaziabad – 201 002, Uttar Pradesh
E-mail: akyadav52@icloud.com

1. INTRODUCTION

Third party certification is an essential component to world trade. In India, there are 28 accredited third party organic certification agencies. The inherent expense and paperwork required in a multi level system discourages most small organic producers from being certified at all. This limits local and domestic trade as well as access to organic products and growth of the Organic Movement as a whole.

In an attempt to reduce the inequality of this trend, a number of alternative methods to guarantee the organic integrity of products have been developed for small domestic producers, which are growing rapidly. In 2004, a conference sponsored by MAELA (The Latin American Agroecology Movement) and IFOAM (International Federation of Organic Agriculture Movements) was held in Brazil where in representatives from over 20 countries presented on the "alternative certification systems". This became an eye opener for thousands of small scale producers as they associated themselves with this alternative program; collectively referred to now as *Participatory Guarantee Systems* (PGS). As support has grown in around the world for the idea of PGS IFOAM supported it to ensure that organic producers have access to organic guarantee options that best suited their needs. Thus, for local organic markets, PGS is now regarded as a viable organic guarantee option. As a decision to join the movement through National Project on Organic Farming Government of India launched a participatory organic guarantee system entitled PGS-India during the year 2010 with an aim to reinforce the movement, to take the system closer to lakhs of committed organic farmers without any extra cost and also to ensure the credibility of the system.

2. Participatory Organic Guarantee System

PGSs are quality assurance initiatives that are locally relevant, emphasize the participation of stakeholders, including producers and consumers and operate outside the frame of third party certification. A process that can be formal or informal, whereby people in similar situations (in this case small holder producers) in some way assess the production practices of their peers.

PGS system has number of basic elements which embrace a participatory approach, a shared vision, transparency and trust. Participation is an essential and dynamic part of PGS. Key stakeholders (producers, consumers, retailers and traders and others such as NGOs) are engaged in the initial design and then in the operation of the PGS. In the operation of a PGS, stakeholders (including producers) are involved in decision making and essential decisions about the operation of the PGS. In addition to being involved in the mechanics of the PGS, stakeholders, particularly the producers are engaged in a structured ongoing learning process, which helps them improve what they do. The learning process is usually 'hands-on' and might involve field days or workshops.

3. PGS India Philosophy

The philosophy of NPOF (National Project on Organic Farming) sponsored PGS-India programme stipulates:

- ✰ A farmer empowering approach
- ✰ Based on Farmer group declaration and consumer's trust
- ✰ No intermediaries
- ✰ Group supremacy
- ✰ Entire information to be in public domain
- ✰ Scope for product testing
- ✰ Individual farmer gets certificate with PGS number and farmer sub-code

4. PGS India-Guiding Principles

In harmony with the international trends and IFOAM's PGS Guidelines, PGS India system works on the basis of participatory approach, shared vision, transparency and trust. In addition, it gives PGS movement a National recognition and institutional structure without affecting the spirit of PGS.

4.1 Participation

Participation is an essential and dynamic part of PGS India programme. Key stakeholders (producers, consumers, retailers, traders and others such as NGOs) are engaged in the initial design, operation as well as decision making.

The idea of participation embodies the principle of collective responsibility for ensuring the organic integrity of the PGS. This collective responsibility is reflected through:

- Shared ownership of the PGS
- Stakeholder engagement in the development process
- Understanding of how the system works
- Direct communication between producers, consumers and other stakeholders

Together these help to shape the integrity based approach and trust. An important tool for promoting this trust is having operational processes that are transparent (for all to see). This includes transparency in decision making, easy access to the data base and open farm visits by consumers. Different people and groups have different skills, technical knowledge and access to resources so that; they may play different roles in the development and management of PGS. What holds important is the producers who are directly engaged in the operating model of decision making (who gets certified) and significant stakeholders. They may either be directly engaged or be represented through persons they elect. Ideally consumers are also actively engaged in the PGS but the level of activity may depend on such things as distances to markets, how the products are marketed (directly or via other agents) and the extent to which consumer groups are organized and able to participate. Consumers or retailers in PGS groups are not only buying products but are also engaged in decision making and management. Overall the target is to bridge the gap between producers and consumers.

4.2 Shared Vision

Collective responsibility for implementation and decision making is driven by common shared vision. All the key stakeholders (producers, facilitating agencies, NGOs, social organizations and even the State Governments) support the guiding principles and goals of PGS. This can be achieved initially through their participation and support in the design and then by joining it. This may include commitment in writing through signing an application/document that includes the vision. Each stakeholder organization (or PGS group) can adopt its own vision conforming to the overall vision and standards of PGS India.

4.3 Transparency

Transparency is created by having all stakeholders, including producers and consumers, aware of exactly how the guarantee system works to include the standards, the organic guarantee process (norms) with clearly defined and documented systems and the decisions made. Public access will be ensured for documentation and information about the PGS groups, such as lists of certified producers and details about their farms and non-compliance actions. These will be available through a dedicated National database website. However, it doesn't mean that entire information on National PGS database will be available to everyone in complete.

At the grass roots level, transparency is maintained through the active participation of the producers in the organic guarantee process which can include:

- Information sharing at meetings and workshops
- Participation in internal inspections (peer reviews)
- Involvement in decision making.

4.4 Trust

The integrity based upon which PGS is built is rooted in the idea that producers can be trusted and that the organic guarantee system can be an expression and verification of this trust. The foundation of this trust is built from the idea that the key stakeholders collectively develop their shared vision and then collectively continue to shape and reinforce their vision through the PGS. The ways this trust is reflected may depend entirely on factors that are culturally/socially specific to the PGS group. Further, the idea of 'trust' assumes that the individual producer has a commitment towards protecting nature and consumers' health.

Mechanism for expressing trustworthiness includes:

- Declaration (a producer pledge) via a witnessed signing of a pledge document
- Written collective undertaking by the group to abide by the norms, principles and standards of PGS
- Confirmation of activities, procedures and processes through collective peer appraisals

4.5 Horizontality

PGS India is intended to be non-hierarchical at group level. Through collective responsibility among PGS group over sharing and rotating, engaging producers directly in the peer review of each other's farms and by transparency in decision making process the overall democratic structure will be improved.

5. National Networking

PGS India whilst keeping the spirit of PGS intact; aims to give the whole movement an institutional structure. This is projected to be achieved by networking the groups under common umbrella through various facilitating agencies, regional councils and zonal councils. To make the system entirely transparent and accessible to traders and consumers, complete data will be hosted on a common platform (website). National Centre of Organic Farming shall be the custodian of data and will also have to power to define policies and guidelines, undertake surveillance through field monitoring and quality check the produce. Regional councils and facilitating agencies will facilitate the groups in capacity building, training, knowledge/ technology dissemination and data uploading on the PGS website. However, every stage needs to be ensured for non-interference of agencies including apex body in the working and decision making. Even if surveillance is done and reports are made, the same will be made available to public domain. Further, action to be taken on adverse reports will be left to the group and regional council.

6. PGS-India Structure

To give the entire implementation process an institutional shape, the programme has following functional entities:

- ☆ National Advisory Committee (NAC) at Department of Agriculture and Co-operation
- ☆ NCOF as National secretariat of PGS program and custodian of data
- ☆ Zonal Councils (Regional Centers of Organic Farming)
- ☆ Regional Councils
- ☆ Farmer groups
- ☆ Farmers
- ☆ Surveillance mechanism

National Advisory Committee (NAC) at DAC is the apex policy making body. Although, at present it is headed by DAC/NCOF but in due course of time members will also be choosen from all stakeholders including farmer, trader, retailer and consumer. In the interim period the PGS-NAC comprises of officers from DAC and NCOF. NCOF and Zonal Council (ZC) authorized by NAC are responsible for capacity building, initial screening of the regional councils and surveillance of entire system. Facilitating agency/regional councils are authorized to assist farmer groups on regional/local level. The groups will be provided with a logo and a unique code number on completion of requirements. Farmer groups will facilitate consumer/ trader/retailer visits for verification if required. A group with participation with traders/retailers/consumers will be more trust worthy and preferred.

7. Operational Strategy

- ☆ A National PGS website was launched by NCOF
- ☆ Standards, basic operational manuals, procedures, documents, formats *etc.* are available on website http://ncof.dacnet.nic.in
- ☆ Selection and appointment of National Council and Regional Councils is done by NAC. Capacity building of councils and group leaders to be done by NCOF, national council or by RCs.
- ☆ 10 to 50 farmers belonging to one village or two-three close by villages make a group
- ☆ Group needs to registers on the website on-line directly or through regional council
- ☆ Recognition as PGS group is granted by RC on verification of data.
- ☆ Training and knowledge on PGS operational systems is a prerequisite for recognition of the group
- ☆ Capacity building of group members by peers in the group
- ☆ Up-loading data and peer appraisal forms from time to time

- ☆ Regular meetings and peer inspections on each other
- ☆ At the end of season, collectively decide on status of farmers
- ☆ Upload information on website on crop yield farmer wise
- ☆ Website gives logo with PGS number and farmer sub code
- ☆ Print individual certificate with number
- ☆ Consumer can access details about farmer group with PGS number
- ☆ NCOF and ZC undertake regular surveillance, collect samples being sold or produced under PGS and get them analyzed for residue.
- ☆ Samples analyzed for residue in authorized testing laboratories
- ☆ Residue analysis results hosted on website in public domain
- ☆ Cost of analysis of samples by NCOF/ZC will be borne by Governments through NCOF
- ☆ Results are also linked to concerned group, providing additional trust
- ☆ Defaulter group's ID gets automatically blocked and shifted to suspended groups
- ☆ Regional councils need to act and initiate action against defaulter groups and suggest remedial measures
- ☆ Re-entry of suspended groups is possible only after verification by NCOF, NC or RCs

8. Current Status of PGS-India Program in India

PGS-India program was launched in March, 2011. National Centre of Organic Farming (NCOF) under Ministry of Agriculture was declared as Secretariat of the program. The six regional centers (RCOFs) located at Bangalore, Bhubaneswar, Imphal, Jabalpur, Panchkula and Nagpur were declared as Zonal Councils. Initially to begin PGS-India system, twenty regional councils were approved (Table 23.1). Among 20 RCs, the highest number (11) of RCs are located in southern region under the jurisdiction of southern zonal council/RCOF, Bangalore followed by western zone under RCOF, Nagpur (5) and North-western zone under NCOF, Ghaziabad (2) (Table 23.1). The highest numbers of 7 RCs are present in the state of Karnataka followed by Maharashtra (5) and Tamil Nadu (3) (Figure 23.1). Presently 408 local

PGS- Green PGS- Organic

groups, comprising of 5809 farmers covering an area of 6064.14 ha are operating since last 3 years. These groups are producing approximately 23612.42 tones of different commodities (Table 23.2). Two separate logos are being granted, PGS-India Green for in-conversion products and PGS-India Organic for full certified organic products.

Table 23.1: List of Regional Councils Functional under PGS-India

Sl.No.	*Name of Regional Council*	*State*	*Zonal Council*
1	Participative Watershed and Rural, Development Agency (PRAWARDA) #8-9-270/A-36, Behind Baridshahi Garden, Bidar Karnataka 585401	Karnataka (7)	RCOF, Bangalore (11)
2	Shri Saptgiri Rural Development Society No. 932, 1st Block HRBR Layout, Kalyan Bengaluru Karnataka 560043		
3	Belgaum Integrated Rural, Development Society (BIRDS) KVKNagnaur, BelgaumKarnataka 591319		
4	Association for Promotion of Organic Farming UAS Alumini Building (By the side of Bengaluru Karnataka 24		
5	Social Welfare and Rural Development Society (SWARDS) Raghava Nilaya, GBN Gate Near T.V.V. Tumkur Karnataka		
6	Mysore Green Exports Pvt. Ltd. No. 213/Y, 13th Main Road, 3rd Block, Bengaluru, Karnataka 560010		
7	Janodaya No. 3, 9th Cross, 5th Main, Jayamahal Bengaluru Karnataka 560046		
8	Manarcadu Social Service Society Manarcadu P.O., kottayam Kerala 686019	Kerala (1)	
9	T.V. Srinivasan Centre for Rural, Training Bethalapally, Sipcot II, Hosur Tamil Nadu 635125	Tamil Nadu (3)	
10	Makkal Nala Sangam No.2 Selection Tailors Upstair Back Side, Salem Tamil Nadu 636115		
11	Organic Farming Organization Dhanalakshmi Illam, No. 15, thrid Main Vellore Tamil Nadu 632001		
12	Neem Foundation Village Gondkhairy, Amrawati Road (NH-Nagpur Maharashtra 441501	Maharashtra (4)	RCOF, Nagpur (5)
13	Gramin Krishi Kranti Seva Bhavi Santha Yammewar Complex, Near Market Nanded, Maharashtra 431807		

Sl.No.	Name of Regional Council	State	Zonal Council
14	Centre of Science for Villages Post Box NO. 21, Kumarappapuram, Wardha, Maharashtra 442001		
15	Siddhi Vinayak Group Opp. Govt. Milk Dairy, Amravati		
16	Society for Elimination of Rural Poverty HMDA Hermitage Office Complex, 4th Hyderabad Telangana 500004	Telangana (1)	
17	Paryavaran Sanrakshan Evam Adivasi Vikas Kendra 413/1, Mittal Apartment, South Civil Lines Jabalpur Madhya Pradesh 482001	Madhya Pradesh (1)	RCOF, Jabalpur (1)
18	Society for Organic Agriculture Movement (SOAM) 26, Gayatri Nagar-B, Maharani Farm, Durgapura, Jaipur-302018 (Raj)	Rajasthan (1)	NCOF, Ghaziabad (3)
19	Foundation for Agriculture Resources, Management and Environmental Remediation (FARMER) SJ-14, Shastri Nagar, Ghaziabad, Uttar Pradesh 201002	Uttar Pradesh (2)	
20	Horticulture Produce Management Institute 3/15 Mohna Nagar, Industrial Area, Ghaziabad		

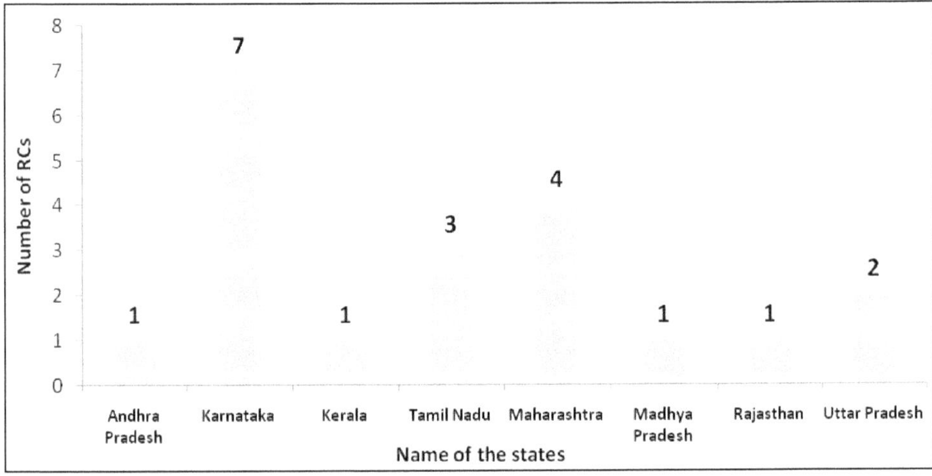

Figure 23.1: Number of Regional Councils Functional at State Level

9. Conclusion

Participatory Organic Guarantee System in India can bring huge number of farmers into a system committed for organic production. Working out in a grass root level with farmers can make the certification process relatively quick. The system also adheres to the strict organic standards. As such PGS offer a needed and

Table 23.2: Status of Regional Councils under PGS-india Programme for the Year 2013-2014

Sl.No.	Name of Regional Council	No. of Groups	No. of Farmers	Area (ha)	Production 2013-14 (tons)
	Andhra Pradesh				
1	Society for Elimination of Rural Poverty	5	100	988	84.2
	Karnataka				
2	Participative Watershed and Rural Development Agency (PRAWARDA)	NA	NA	NA	NA
3	Shri Saptgiri Rural Development Society	10	100	200	35.6
4	Belgaum Integrated Rural Development Society (BIRDS)	7	93	152	11580.0
5	Association for Promotion of Organic Farming	4	63	101	90.0
6	Social Welfare and Rural Development Society (SWARDS)	20	212	52	1850.0
7	Mysore Green Exports Pvt. Ltd.	14	245	405	688.0
8	Janodaya	15	218	108	90.0
	Kerala				
9	Manarcadu Social Service Society	15	1132	1017.88	7695.0
	Tamil Nadu				
10	T.V. Srinivasan Centre for Rural Training	33	373	292	800.0
11	Makkal Nala Sangam	78	1500	607.97	-
12	Organic Farming Organization	22	318	63.9	480.6
	Maharashtra				
13	Neem Foundation Village	129	700	1845.49	--
14	Gramin Krishi Kranti Seva Bhavi Santha	5	31	41	71.6
15	Centre of Science for Villages	-	-	-	-
16	Siddhi Vinayak	-	-	-	-
	Madhya Pradesh				
17	Paryavaran Sanrakshan Evam Adivasi Vikas Kendra	2	84	129.7	147.4
	Rajashtan				
18	Society for Organic Agriculture Movement (SOAM)	46	606	NA	NA
	Uttar Pradesh				
19	Foundation for Agriculture Resources Management and Environmental Remediation (FARMER)	3	34	60.2	NA
20	Horticulture Produce Management Institute,	NA	NA	NA	NA
	Total	**408**	**5809**	**6064.14**	**23612.4**

complimentary system for organic guarantee that builds the organic movement, educate the farmers and consumers and also provide domestic and local market for organic produce.

Chapter 24

Biodynamic Farming

Ravi Koushik

Bhaikaka Krishi Kendra,
Ravipura, Anand – 388 440, Gujarat
E-mail: ravikoushik@yahoo.com

1. INTRODUCTION

Biodynamics can be best thought of as an ecological-spiritual-ethical and self supporting farming system. This specialized organic approach to farming treats every farm as unique and self-sustaining, in balance with soil, plant, animal and cosmic processes. It gives great primacy to the farm as a living entity and places great emphasis on the building of a healthy, biologically rich soil. Such a farm is in turn capable of producing food that is nourishing and vitalizing both to the body and to the mind. Biodynamics is the first conscious approach to a completely holistic and sustainable way of farming.

2. Origin

Biodynamics is a system of farming practices that was developed in the 1920s based on the insights provided by the Austrian philosopher, scientist and social reformer Rudolph Steiner. Steiner, who brought clairvoyant insight into all his work, laid the foundation for anthroposophy - an alternate stream of science that is now recognized as "Spiritual Science". His gift to the world is knowledge from the spiritual world that enables people to recreate and work with natural forces in the fields of education, religion, art and agriculture. Steiner taught that 'Matter is never without Spirit, and Spirit never without Matter. Steiner gave his insights on agriculture through a series of seven lectures to a small group of farmers in Koberwitz in West Poland (at that time in East Germany). This course inspired many people. In response to questions raised by a group of farmers, Steiner was keen to know why their seed did not stay viable for as long and why they were experiencing fertility problems in their animals. Steiner had built up his knowledge from keen observation of plant and animal forms, study of traditional peasant practice, deep spiritual research and study of vedic agriculture. Subsequent practitioners like E. Pfeiffer and Maria Thun carried this seminal work forward.

The word biodynamic means life force. Biodynamic practitioners are concerned with the forces and processes of life that lie behind matter. This is in sharp contrast to today's scientific paradigm, which is focused primarily on matter. Modern science looks for answers by looking at smaller and smaller building blocks of matter. It is a microscopic, reductionist approach which may be thought of as 'knowing more and more of less and less till we know everything of nothing'. Often in this methodology, the very life, which is being explored, is killed in the process of the research.

It is through careful observation that life's secrets are revealed. Steiner noted that there are two complimentary forces that need to be in balance for the growth of a healthy plant. He referred to these as the calcium process and the silica process. These should not be confused with and restricted to just the elemental forms of calcium and silica but are more processes that express themselves in living beings through these elements. Steiner explained about the living processes that work through these elements.

These forces come into play at different times during the growth of plant. When a seed is planted, it is necessary that the calcium forces be encouraged. This force is characterized by a radial growth pattern where the seed sends down roots into the ground and leaves upward. Yield will be determined based on how much it grows during life cycle. As the plant continues to develop, another force begins to exert an influence, the silica force. Later a bud is formed. Steiner refers to this as the cosmic influence. The silica force is related to light and warmth and it is through this process that qualitative aspects of a plant are influenced. For instance, it is this light and warmth that effect flavor, aroma and storage quality.

Modern agriculture has been yield focused and therefore has developed techniques that primarily deal with calcium force. Over emphasis in this direction leads to soft fleshy tissues that are prone to pest and disease attack Hence; chemical agriculture has been prone to all sorts of problems that have led to increased pesticide usage. Even organic agriculture frequently has had the tendency to simply replace chemical inputs with organic ones while leaving the same thinking intact. The calcium forces are similarly over emphasized when fertilizing with un-decomposed chicken manure. Chicken manure grows a plant that is very similar in appearance to ones fertilized with ammonia fertilizers, because the nitrogen in chicken manure is also in the form of ammonia.

Demeter International is the certifying body for Biodynamic farms. The Demeter certification program was established in 1928 and was the first ecological label for organically produced foods. Hence, biodynamic farming can be truly considered the oldest formal form of Organic farming. Biodynamically certified foods fetch a very high premium in the market especially in Europe. Demeter certification forms the highest grade of certification and sets very high and holistic standards for food production. It requires that a farm, in addition to following the biodynamic practices, also pay great attention to biodiversity and ecosystem preservation, soil husbandry, livestock integration, prohibition of genetically engineered organisms and viewing the farm as a living "holistic organism".

3. Core Principles

The core systems of practices in a biodynamic farm can be summarized as follows:

1. The farm uses very few external inputs and is largely self sufficient in terms of its nutrient inputs such as compost
2. The farm returns more to the soil than is removed during the process of growing crops and raising animals
3. The farm is ecologically diverse and has the right balance (*i.e.* ratio) of the animal and plant components – a good balance can be achieved with 2 cows for every acre
4. The farm uses an astronomical calendar to establish a good farming rhythm which in turn provides the most optimum times for all the activities on the farm such as sowing, transplanting, harvesting *etc.*
5. The farm uses a plant positive approach to pest management rather than a pest negative approach. Pests are considered messengers rather than the problem themselves. They point to the real problem.
6. The farm uses the Biodynamic Herbal Preparations BD500 to BD507 to revitalize the soil and farm the air
7. The food produced on the farm is sold locally as far as possible
8. The farm uses open-pollinated seeds

The farm uses all the best practices that a good organic farm would use such as crop rotation, green manuring, companion cropping, mulching, rainwater harvesting *etc.*

4. Origin of Biodynamic Farming in India

Peter Proctor, from New Zealand came to India in the early 1990s and worked relentlessly for over two decades to introduce, adapt and popularize biodynamic farming in India. Initial interest and adoption of biodynamic farming in India, like in many other parts of the world, had largely been export driven and practiced by few farmers. Peter worked with help of organizations such as ICRA and Eco-Agri in Karnataka, Supa-Biotech in Uttarakhand, Kurinji and ICHOR in Tamil Nadu and the Agricultural Extension Offices in Maharashtra. Peter organized hundreds of workshops, travelled relentlessly across the country sharing his knowledge and practice of biodynamic farming. In India, the adoption of biodynamics seemed very natural because the Indian farmer already had a deep connection to the spiritual aspects of farming, a deep reverence for the cow and from ancient times had been conscious of the cosmic connection in farming.

5. Biodynamic Principles and Tecniques

Biodynamics is a way of farming rather than a few dogmatic techniques which are described below.

5.1 Cow Horn Manure (BD 500)

This is a foundation spray in Biodynamics. It is used, in homeopathic doses, as an inoculant to activate the dormant life forces that are already present in the soil. It helps in

- ☆ Humus formation and improvement of crumb structure
- ☆ Increase in soil bacteria like rhizobacteria and phosphate solubilizing bacteria
- ☆ Increase in mycorrhizal fungi and their hyphae
- ☆ Increase in earthworm activity

5.1.1 How to make it

- ☆ Obtain cow horns from an abattoir
- ☆ Collect good quality cow dung from a lactating cow that has been fed only grass for 3 days prior to obtaining the dung
- ☆ Knead the cow dung for 15 minutes to aerate it
- ☆ Fill the cow horns with this dung
- ☆ Plant the horns in a pit in the most fertile part of your farm
- ☆ The horns should have the open face down so that water will not accumulate
- ☆ Cover the horns with soil mixed with compost
- ☆ Cover with a mulch of paddy straw.

5.1.2 When to make it

The planting of the horns should be done at the start of the winter in October.

5.1.3 When is it Ready

- ☆ This special manure is ready in 3 months from the time of planting. The horns are dug out and the manure can be removed by tapping it out of the horn
- ☆ The finished product which is called BD 500 can be stored in a glazed pot surrounded by damp coir pith in a cool place

5.1.4 How to Use it

- ☆ Add 25 g of BD 500 to 15 liters of rain water or water from a natural source
- ☆ Stir biodynamically for an hour

- ☆ Sprinkle the solution over one acre in the form of large droplets using a leafy branch in the late afternoon during a descending moon
- ☆ Sprinkle at least four times a year

5.2 Cow Horn Silica (BD501)

This is a complementary spray to BD 500 and helps to

- ☆ Make the photosynthetic activity more efficient
- ☆ Assist in the mediation of the light forces upon the soil
- ☆ Improve the taste, keeping quality of the crop and increase the resistance to fungus

It is widely used in the vine yards to improve the taste of the wine.

5.2.1 How to make it

- ☆ Collect good quality silica crystals
- ☆ Grind the crystal to a fine powder (grind to a talcum powder consistency)
- ☆ Mix powder with a little water to make a paste
- ☆ Fill the paste in cow horns
- ☆ Allow the paste to dry out and solidify
- ☆ Plant the horn in the soil following the procedure for preparation of BD 500

5.2.2 When to make it

The planting of the horns should be done at the beginning of the summer in April

5.2.3 When is it Ready

The finished product which is called BD 501 is ready in 6 months from the time of planting. It can be stored in a clear glass bottles in a well lit place such as a window sill.

5.2.4 How to Use it

- ☆ Add 1 gram of BD 501 to 15 litres of rain water or water from a natural source
- ☆ Stir biodynamically for an hour
- ☆ Sprinkle the solution over one acre in the form of a fine aerial spray in the early morning during an ascending moon
- ☆ Sprinkle at least four times a year

5.3 Compost Preparations (BD 502-507)

There are six Biodynamic Preparations numbered BD 502 to BD 507 that are called as Compost Preparations because they are typically applied to the soil via the compost. These are made from medicinal herbs and are used to enrich the manures and composts. They enhance all the bacterial, fungal and mineral processes. They are never applied directly to the soil rather they are added while making the Compost or liquid manures so that their effects multiply during composting process. These are used in homeopathic doses. Just 1 gram of each preparation in a compost heap is enough.

- ☆ BD 502 is made by composting the flowers of the herb Yarrow (*Achillea Millifolium*). It helps the processes of potassium, sulphur and trace elements.
- ☆ BD 503 is made by composting the flowers of the herb German Chamomile (*Matricuria chamomilla*). It helps the processes of Ca and N
- ☆ BD 504 is made by composting the leaves of the Himalayan Stinging Nettle (*Urtica parviflora*). It helps processes of Iron and Magnesium.
- ☆ BD 505 is made is made from the bark of the Oak tree. The Himalayan Oak Tree (*Quercus glauca*) has been found suitable in India. It helps processes of Calcium and builds up the plant's immunity to diseases.
- ☆ BD 502 is made by composting the flowers of Dandelion (*Taraxicum officinalis*). It helps processes of Silica (Si).
- ☆ BD 507 is made from the flowers of Valerian (*Valeriana officinalis*). It helps processes of P. This is the only preparation in liquid form.

5.4 Cow Pat Pit (CPP)

CPP is a very effective bio-fertilizer that helps to efficiently get the Biodynamic energies present in the Compost Preparations out onto the farm. It has wide range of colonies of beneficial fungi and bacteria and growth hormones.

5.4.1 How to make it

- ☆ Collect 60 kg of good quality cow dung
- ☆ Make a pit 1 foot deep and line the sides with bricks so that the inner dimensions of the pit are 3 ft.x2 ft. Note that the bottom of the pit should not have any bricks and no cement is used
- ☆ Add 300 g of egg shell powder and 300 gms of basalt powder (or any other rock dust)
- ☆ Knead and fluff the cow dung for 30 minutes
- ☆ Put the raw CPP mixture in the pit and lightly level it
- ☆ Make 5 holes in the mixture about 3 cm deep and add 2 g of the preparations 502-506 separately in each of the holes
- ☆ Add 20 ml of the valerian preparation (BD 507) to ½ a liter of water and potentise it by stirring biodynamically for two and a half minutes. Pour half of this in a hole in the raw CPP mixture. Sprinkle the rest on the mixture and the bricks
- ☆ Cover the mixture with a moist gunny bag. It is important that the moisture of the gunny bag is maintained throughout the time that the CPP is getting ready
- ☆ After four weeks, turn the mixture to aerate it and turn every week thereafter to hasten the fermentation process

5.4.2 When to make it

The CPP can be made anytime of the year. Make it such that you have a year round supply for your farm.

5.4.3 When is it Ready

CPP is ready in 3 months. It can be stored just like that of BD 500. Young CPP (70 to 90 days old) has good colonies of *Bacillus subtillus* and is very effective as a foliar anti-fungal spray. Older CPP has growth hormones such as Giberellic acid and Indole acetic acid and acts as a very good bio-fertilizer

5.4.4 How to Use it

CPP is used at the rate of 1 kg diluted in 40 litres of water per acre. Stir biodynamically for ten minutes before applying it

5.4.5 Uses of CPP

In addition to using CPP as a foliar spray and as a bio-fertilizer for all crops, fruit trees and plantation crops, it can be used as follows

- ☆ In conjunction with BD500 - Add 100 g of CPP while stirring BD500, in the last 15 minutes.
- ☆ To make tree paste - It is added while making tree paste to provide nutrition to the cambium layer and protection against fungus attacks.

- ☆ To dip cuttings, seedlings and saplings before planting – this minimizes the transplant shock and assists with strong root development.
- ☆ CPP is an excellent substitute for Biodynamic Preparations. Knead in 2 kg of CPP while making a new CPP. Use 1 kg of CPP in 100 liters of water for each 5 meters of compost and spread on every carbonaceous layer.
- ☆ To make a seed dressing - CPP can be used to coat the seeds before sowing. This provides both protection and nutrition to the germinating seed. Seed potatoes, treated thus have good protection against blight.

5.5 Biodynamic Compost

Biodynamic compost is made from a diversity of inputs such as:

- ☆ Carbonaceous materials – dry leaves, paddy straw, dried maize stalks, sawdust, sugarcane bagasse *etc.* are rich in carbon
- ☆ Nitrogenous materials – green plant materials (esp. from legume crops like gliricidia, sesbania, cow pea *etc.*), animal manures, bone meal *etc.* The nitrogenous materials help for efficient breakdown of carbonaceous materials. The nitrogen feed the bacteria that break down the carbonaceous material.
- ☆ Beneficial microorganisms - cow dung is the best source for all the beneficial microorganisms that aid in the decomposition process. Previously made compost also can be added to inoculate the compost pile with these microorganisms.
- ☆ Minerals and trace elements – rock dusts such as basalt powder, bore well powder *etc.* are rich in minerals. Rock phosphate is a rich source of phosphorous. Many of these minerals are turned into plant available forms during the composting process by the various solubilizing microorganisms. Sea weed is also an excellent source of many trace elements. A total of about 75 kg will be required for a compost of size 15 m^3.
- ☆ Calcium - hydrated lime is a rich source of Ca. A total of only 250 g will be required for a compost of size 15 m^3.
- ☆ Air - aerobic processes promote beneficial microorganisms and fungi and suppress pathogenic organisms. Hence, it is very essential that there is good air circulation during the composting process. This is achieved by making the compost as a heap (in the form of a windrow) rather than in a pit and by creating an air tunnel with coarse material such as twigs and sticks at the bottom of the pile.
- ☆ Water - proper moisture is essential for the microorganisms that aid in the decomposition to thrive. The carbonaceous materials are moistened before adding to the pile. Also, the cow dung is made into slurry before adding to the pile.
- ☆ Biodynamic Energies – these are introduced via the biodynamic preparations BD502-507. One set is required for a compost of size 15 m^3.

CPP can also be used as a substitute if the preparations are not available or not affordable.

5.5.1 How to Make it

- ☆ The ideal size of a compost pile is 15 m^3 (5 meters length, 2 meters width and 1.5 meters height) constructed in the form of a windrow. The length may be increased depending on the availability of the raw material
- ☆ First create an air tunnel by laying twigs and sticks 10-15 cms on the marked area
- ☆ Put a 10 cm layer of the carbonaceous material
- ☆ Sprinkle the rock dusts
- ☆ Sprinkle the cow dung slurry
- ☆ Put a 15 cm layer of the nitrogenous material
- ☆ Sprinkle a very small quantity of lime
- ☆ Repeat the layers (except for the air tunnel layer) till you achieve the desired height
- ☆ Finally apply a thick plastering of mixture of cow dung and optionally, red soil to prevent loss of moisture and heat
- ☆ The Biodynamic preparations 502-506 are put into cow dung balls separately and the 5 balls placed in the middle of the compost. This can be done when the compost has reached half its height or by making holes in the compost at the very end by means of a crow bar. The BD507 is stirred biodynamically for 10 minutes and added via holes made at the top of the heap.

5.5.2 When is it Ready

The compost pile heats up to a temperature of 65 to 70 °C and stays that way for at least 2-3 weeks. Turn the compost at least once after 6 weeks. This will aerate it and provide a chance to check the moisture content. The compost is ready in 3 to 4 months depending on the ambient temperature.

5.6 Biodynamic Stirring

Biodynamics recognizes and gives great importance to the stirring of different preparations like BD500, BD501, CPP, liquid manures *etc.* before they are applied to the farm. Biodynamic stirring involves the creation of a vortex, rhythmically in the clockwise and anti-clockwise directions. This pulsating action dynamizes the liquid and connects it with the universal creative water rhythms or pulse within the body of the earth. The stirring ensures that the preparations are thoroughly mixed. The oxygen content of the water increases during the creation of the vortex which helps the beneficial microbes to proliferate. It also helps the water to absorb the cosmic forces which then enable it to become a dynamic carrier of the life energy that is present in the various preparations. There are many different ways of stirring biodynamically as described below.

5.6.1 Hand Stirring

This can be used for small farms. Take 13 to 15 litres of liquid to be dynamized in a bucket. Use a stick to create a vortex in one direction. Start slowly at the edge of the bucket and move into the centre, increasing the stirring speed. When a good vortex is formed, change the direction. Change directions every 20 seconds. Total how much time

5.6.2 Verbella Flow Forms

This can be used for large farms when large amounts of water with the biodynamic preparations need to be stirred biodynamically. Flow forms create cascades of water that aid in the reoxygenation of water to restore it to its running stream state.

5.7 Astronomical Planting Calendar

In Biodynamics, an astronomical planting calendar is used to determine the most optimal time for the different farming activities such as sowing, transplanting, harvesting *etc.* This calendar can be downloaded every year from the following website - http://www.organichutbkk.com/resource-gallery-1.html. The planting calendar provides a very good framework around which farming activities can be planned. Plants grown against the backdrop of these rhythms have been consistently found to be better resistant to pests and disease and have produce that is better tasting and longer shelf life.

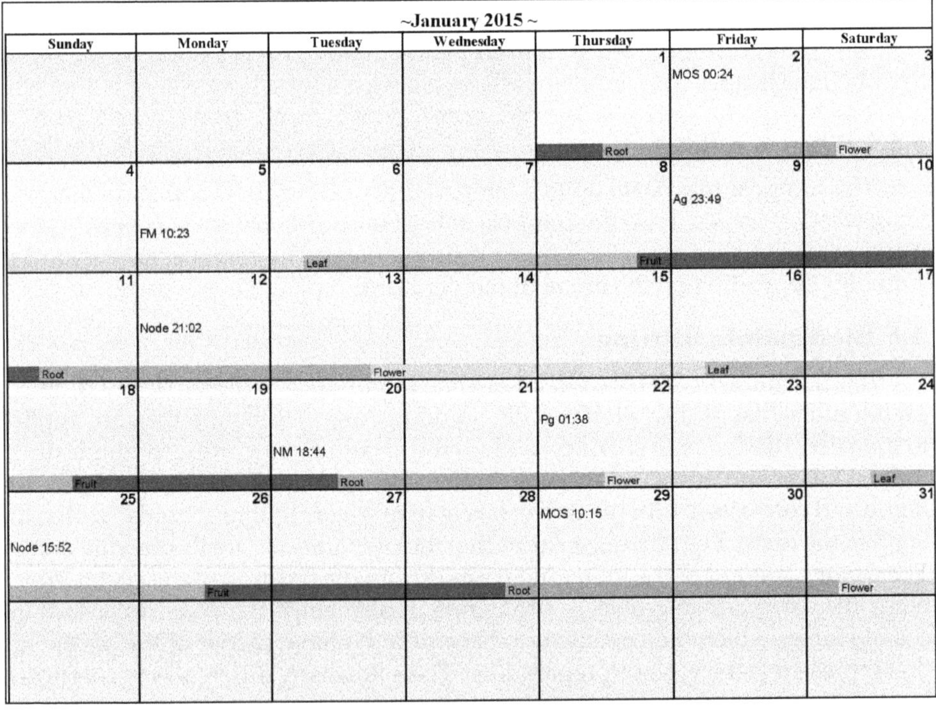

~January 2015~

Sunday	Monday	Tuesday	Wednesday	Thursday	Friday	Saturday
				1 Root	2 MOS 00:24	3 Flower
4	5 FM 10:23	6 Leaf	7	8 Fruit	9 Ag 23:49	10
11 Root	12 Node 21:02	13 Flower	14	15	16 Leaf	17
18 Fruit	19	20 NM 18:44 Root	21	22 Pg 01:38 Flower	23	24 Leaf
25 Node 15:52	26 Fruit	27	28 Root	29 MOS 10:15	30	31 Flower

A Sample Page in the Planting Calendar.

The recommendations are based largely on the indications that Rudolf Steiner gave and on research carried out by farmers and researchers like Lilli Kolisko, Maria Thun and others. It should be noted however that each researcher has found certain (often different) cosmic influences to be the dominant ones for their particular environment. It is therefore important that each farmer experiments with these rhythms, each time observing and documenting the results and finding the pattern that best suits their farm.

It is also important to keep in mind that the planting calendar should never be the most important factor when deciding the timings of the activities. The weather, soil conditions *etc.* will need to override considerations of a cosmic nature.

The different astronomical events (also called as cosmic rhythms) that influence farming activities are detailed below.

5.7.1 New Moon

- In the planting calendar this is marked as 'NM' along with the exact time when it occurs (*e.g.* NM 18:44).
- The New Moon occurs when the Moon is between the Sun and the Earth. The Moon blocks the beneficial influences of the sun and this has a detrimental influence on the Earth.
- Traditional Indian agriculture has recognized the day before New Moon as the No Moon day a day when all agricultural activities are avoided. Biodynamics also recommends that activities such as seed sowing and transplanting be avoided as far as possible on this day.
- The New Moon occurs once every 29.5 days.

5.7.2 Full Moon

- In the planting calendar this is marked as 'FM' along with the exact time when it occurs (*e.g.* FM 10:23).
- The Full Moon occurs when the Moon is opposite the Sun with the Earth in the middle. This is a period of high water content in the sap and foliage. Hence Pests esp. fungi are very active during this time and preventive/ control measures can be planned during this time.
- The Full Moon occurs once every 29.5 days.

5.7.3 Moon Nodes

In the planting calendar this is marked as 'Node' along with the exact time when it occurs.

The moon nodes occur when the moon's orbital path intersects the planetary plane. This is a period of stress on the planet. The influence lasts for 6 hours approx. on either side. Land cultivation, seed sowing and transplanting should be avoided during this time.

The Node occurs once every 13.6 days.

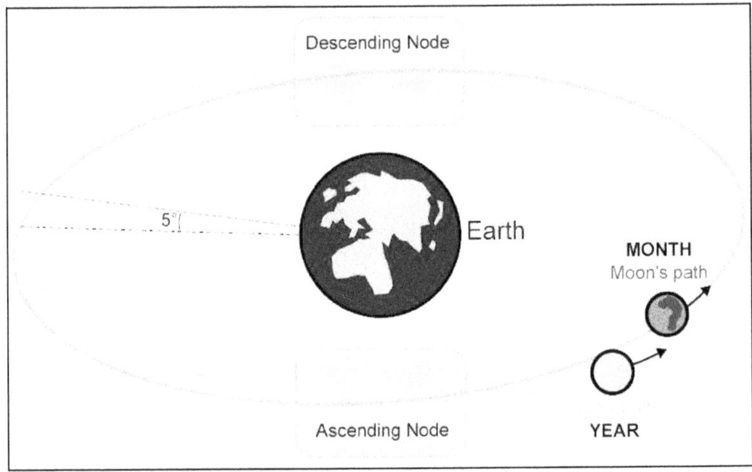

5.7.4 Moon Perigee and Apogee

In the planting calendar Perigee is marked as 'Pg' along with the exact time when it occurs (*e.g.* Pg 01:38).

The Perigee Moon occurs when the Moon is at the closest point to the Earth in its elliptical orbit. There is greater moisture on the earth and hence a tendency towards fungus growth and insect attack especially when the Perigee Moon occurs near a Full Moon. Seed sowing should generally be avoided 12 hours on either side of the Perigee.

The Perigee occurs once every 27.5 days. In the planting calendar, the Apogee is marked as 'Ag' along with the exact time when it occurs (*e.g.* Ag 23:49). The Apogee Moon occurs when the Moon is at the farthest point to the Earth in its elliptical orbit. Apogee seems to be a good time to plant root crops such as potatoes as one gets more potatoes. Other seed sowing should generally be avoided 12 hours on either side of the Apogee. The Apogee occurs once every 27.5 days.

5.7.5 Ascending and Descending Moon

Just as the sun moves southwards (descends) from July to December and moves northwards (ascends) from January to June (referred to as Dakshinayana and Uttarayana, respectively in Indian terminology), the moon also has the same movement. However the Sun takes 1 year to complete this rhythm, while, the moon takes only 27.5 days. The ascending moon can be considered the Moon's spring and summer and the Descending Moon the Moon's autumn and winter. The approx. 14 days when the moon is ascending are marked with a blue background in the calendar and the other 14 days when the moon is descending are marked with a yellow background in the calendar. The ascending moon has very beneficial influence on the above ground activities. The earth breathes out and there is an outpouring of growth activity above the soil surface. This is a good time for activities such as seed sowing, BD501 spraying, harvesting and grafting.

The descending moon has very beneficial influence on the below ground activities. The earth breathes in and draws growth forces back down below the soil surface. This is a good time for activities such as cultivation, BD500 application, pruning and transplanting.

5.7.6 Moon Opposite to Saturn

In the planting calendar, this is marked as 'MOS' along with the exact time when it occurs (*e.g.* MOS 10:15). This is a very special day for Biodynamic Farms when the Moon is exactly opposite the planet Saturn with respect to the Earth. The Moon forces enhance the Calcium forces which are connected with the quantity aspects - growth and propagation of the plant. The Saturn forces enhance the Silica forces which are connected with the quality aspects - substance of the plant in its root, leaf and fruit. The balance of these two important forces produces very strong and healthy plants from seed sown at this time. Hence this is a particularly good day for seed sowing. Other activities that benefit greatly by doing on this day are transplanting and spraying of BD501.The Moon Opposition Saturn occurs once every 27.5 days.

5.7.7 Moon in Zodiac Constellations

The Zodiac comprises of the 12 constellations behind the planetary plane also referred to as the ecliptic path. The moon also moves in front of these constellations. The constellations can be looked upon as fixed markers that divide the ecliptic path into 12 divisions spanning 30 degrees each. The moon's position in this path generates certain favourable conditions to different parts of the plant. This was studied extensively by Maria Thun. Based on her findings, the calendar has Fruit, Root, Flower and Leaf days. The germinating seed receives the influences the greatest, so if it is desired to promote a certain part of the plant, one sows the seeds during the corresponding day in the calendar. For example, peas in a fruit constellation; cabbage in a leaf constellation.

6. Current Status of Biodynamic Farming In India

6.1 Biodynamic Association of India

Biodynamic Association of India (BDAI) is the primary organization to promote and coordinate the biodynamic activities in India.

BDAI coordinates the following activities:

1. Maintaining the BDAI website
2. Publishing of the Biodynamic Calendar. On request, it makes the calendar available for other parts of the world also
3. Memberships
4. Demeter certification in India
5. Intensive workshops on Biodynamics and Preparations making – two such workshops are conducted every year, one in March and the other in September

6. BDAI conferences – These are conducted once in two years
7. Biodynamic seed production

6.2 Demeter Certified Farms in India

The Demeter Office of the BDAI assists Demeter International in the certification of biodynamic farms and products in India. As of June 2015, there are 29 projects in India that are Demeter certified. This includes mixed plantations, fruit orchards, tea and coffee plantations and several NGOs and businesses that run farmers groups. A variety of products like rice, tea, cardamom, cocoa beans, black/green pepper, ginger, turmeric, cloves, nutmeg, mace, cinnamon, cinnamon oil, all spice oil, bay leaf, coconut, coconut milk, coconut oil, dried coconut copra, cashew nuts, coffee, coffee beans, jaggery, amla, mango, mango powder, mango puree *etc.* have been certified.

6.3 Biodynamic Seed Production in India

Getting good quality open pollinated organic and biodynamic seeds is one of the biggest challenges that a biodynamic farmer faces. Under the aegis of the BDAI, a very comprehensive initiative has been launched so that good open pollinated seeds can be produced and shared in a very decentralized and local manner. Under this initiative the know-how and technology is provided to farmers in different regions of India. Biodynamic practices are employed to produce very high quality seeds.

6.4 Formation of OSIB-BDAI

In August/September 2011, Mr. Jayakaran, President of BDAI, being aware of the need for organic seeds for the Indian scenario of organic growers, underwent a five-weeks training program at Sativa Rheinau, Switzerland, guided by Mr. Friedemann Ebner, organic seed breeder at Sativa. Thus, Jayakaran got first-hand experience and exposure to organic seed production. To expose Friedemann to biodynamic practices in India, Jayakaran invited him to participate in the BDAI-organized biodynamic training in February/March 2012. A few days later, in March 2012 in Hyderabad, the BDAI organized an all-India conference on biodynamic research and practices, with additional invitees from abroad, to which Friedemann had been invited as a speaker. Friedemann gave a presentation on the world scenario of the conventional seed industry and the scope for organic seeds as an essential component of the organic movement, enlightening the participants of the need for an initiative for organic seeds in India.

Open Pollinated varieties can be produced successfully in a wide range of agro-climatic conditions; India, being home to all the nine types of agro-climatic conditions, has an advantage over many countries in the world. In November of the same year, BADI called a meeting in Auroville, to discuss the modalities of the program and to formally initiate the organic seed activity within BDAI. In February 2013, BDAI called for a two-day meeting in Bangalore, where BDAI members and other organic enthusiasts were invited to collaborate in the program. The participants were enlightened on scientific methods of seed breeding and multiplication, on processing, cleaning, quality check, packing, storage and on the legal framework in the Indian perspective. The formation of OSIB – the Organic Seeds Initiative of the BDAI – was formally announced there.

6.5 Current Status

Groundwork for Farmers' Training in scientific methods in Biodynamic way for seed multiplication under OSIB started in September 2013 in seven locations across India, namely, in Srinagar (J&K), Anand (Gujarat), Pune (Maharashtra), Tinsukia (Assam), and Ooty, Vinobajipuram and Sevapur (Tamil Nadu). However, success was only achieved in Srinagar, because, there farmers traditionally have been seed growers; in the other locations the program had to deal with very basic lessons to the farmers. The first lot of organic seeds under OSIB was received in September 2014 - more than 42 crops (in temperate vegetables range) exclusively from the Kashmir valley. These seeds are being distributed across India to the organic farmers by receiving requests online and dispatching the seeds directly from the source by courier service. In the 2013/14 season, in Srinagar, 25 farmers were involved in this seed production network, apart from 10 farmers in other locations.

In 2014-15, there is an expansion in range and number of farmers in different locations across India. A few old locations have been discontinued due to lack of interest, and simultaneously some new locations have been considered in Assam, Sikkim, Arunachal Pradesh, Himachal Pradesh, Uttarakhand, Maharashtra, Andhra Pradesh, Karnataka, Tamil Nadu, Chhatisgarh, Odhisa *etc.* where farmers are inviting OSIB to organize training programs in Biodynamic Farming and Scientific method for seed multiplication. The expected range of crops and varieties now is more than 120, including a few cereals, pulses and oilseeds. The total number of all the farmers involved in this network of organic seed multiplication is more than 60.

OSIB is also networking with all other NGOs in India, who are involved in the Organic Farming and Seed actvity *viz.* ICRA (Bangalore), CSA (Hyderabad), Dharamitra (Wardha), SARG Vikas Samiti (Akola) *etc.* for training of their farmers in Biodynamic farming and seed multiplication. Also, OSIB is actively involved in Open Source Seed Network (OSSN) for Apna Beej Program initiated by OFAI, ASHA, CSA and SAEi for generating regional- seasonal specific data on Value for Cultivation and Use (VCU) for existing traditional and improved varieties and formulating policies for free exchange of germplasm with others under Material Transfer Agreement (MTA).

World's two leading organic seeds companies, namely, Sativa Rheinau (Switzerland) and Bingenheimer Saatgut (Germany) have expressed their willingness to extend technical support in breeding for organic and biodynamic varieties, seed quality control, storage and packaging to the OSIB-BDAI.

6.6 Availability of Biodynamic Preparations in India

Most Biodynamic farms, besides making their own Compost and CPP also make their own BD500 and BD501. It is recommended that these preparations be made on the farm itself. Not only are they simple to make but they will have the local microorganisms which are better adapted and conditioned to the farm's micro-eco-system. The other preparations BD502 to BD507 require special herbs that grow only in a cool climate. Typically these are the only inputs on a biodynamic farm that are sourced from outside. The Kodaikanal and Nainital regions have been found to have suitable agro-climatic conditions to grow these herbs. Hence, farms producing the biodynamic preparations 502 to 507 are from these regions. There are three main farms that produce these preparations and make them available to the other farmers.

6.6.1 *Kurinji* Farms

The *Kurinji* Farm is located near Batalagundu, at the base of the Western Ghats just down the hill coming from Kodaikanal. The Kurinji Farms project was started in 1990, and was focused on fruit production and based in Kodaikanal in Tamil Nadu. In a year around 5 to 10 kg of the different biodynamic compost preparations are made. Also around 5,000 cow horns are used for making the BD500. Around 5 kg, on an average, of good quartz crystal stones are used for making BD501. The Kurinji project itself uses nearly half of the biodynamic preparations for its own Demeter project use, and the rest is sold to interested biodynamic practitioners.

6.6.2 ICHOR

ICHOR is a 12 acre biodynamic farm nestled in the Palani hills. This farm specializes in biodynamic cheese making and vegetable production which is in great demand in Kodaikanal and Auroville. ICHOR also produces the biodynamic preparations on this farm.

6.6.3 SUPA Biotech

SUPA Biotech is a pioneer organization in the field of biodynamic agriculture in the country. Thanks to the monumental efforts of this organization and the untiring work of its foot soldiers, biodynamic farming which was limited to few large tea companies in India in the initial years, has been made popular among thousands farmers of the country. SUPA's essential strength lies in technology transfer to farmers especially with collaboration of State governments. The forte of SUPA is to educate farmers on Biodynamic on-farm input production" by providing training and extension support. SUPA is active in six states in India. As a result more than 5,00,000 biodynamic composts and countless CPPs have been made with the help of this organization.

6.7 A Biodynamic Revolution in the Adivasi Belt of Araku Valley

The scale and adoption of biodynamic practices by thousands of farmers can best be seen in the adivasi belt of Araku Valley in Andra Pradesh. Thousands of marginal farmers are growing organic and biodynamic coffee. This is being facilitated by Naandi foundation which has disrupted the NGO space in India by being the first to implement large-scale government programs with efficiency and measurable outcomes. Naandi provides backward and forward linkages to the coffee farmers that include training in sustainable agricultural practices, setting up of a (one of the largest) state-of-art coffee processing unit (CPU) and formation of a Small and Marginal Tribal Farmers Mutually Aided Co-operative Society (SAMTFMACS) with a connect to the international market. This, along with other interventions in health and education has improved the lives of over 12000 farmers significantly. In addition to this, supported by the Livelihoods fund (as part of their carbon offset program), the rebuilding of the forest (3 million trees as of 2015), with a focus on fruit trees has ensured food security for Araku's human and animal populations. Biodynamic practices include application of BD500, BD501, CPP, Compost and the incorporation of the biodynamic planting calendar. The Compost and CPP is made in large quantities as a community effort. The farmers are provided training in bush management, shade lopping, harvesting/picking *etc.* and the cooperative collects the ripe cherries at the doorstep and maintains accurate records of coffee produced by each farmer from the 7 participating mandals. The logistics involved in all this are mind boggling as many of these farms are nestled in far flung remote hills that can be accessed only by a long hard strenuous trek. The CPU provides processes that deliver a consistent quality cup. This in turn has made it possible to build an international brand the 'Araku Emerald'.

6.8 The School of Biodynamic Farming

This school, the first of its kind in India, offers a two year diploma course in biodynamic farming. It is located in the Inba Seva Sangam Commune of Vinobajipuram in Karur District in Tamil Nadu. Its mission is to train rural youth from very poor backgrounds in all aspects of Organic and Biodynamic farming. The school was inaugurated in July 2012 and the first batch of seven students has already graduated from the school. The course teaches practical skills for modern day farming in India with an emphasis on 'Learning by Doing'.

A Model Biodynamic Farm - The *Bhaikaka Krishi Kendra*

The Bhaikaka Krishi Kendra near Anand, Gujarat is one of the best organic and biodynamic farms in the world. All the practices and principles involved in biodynamics in particular and organic farming in general and the benefits that derive from these practices can be seen on this farm. This farm demonstrates that ecologically safe, sustainable and economically viable farming practices without sacrificing productivity or quality.

This 40 acre integrated, holistic organic and biodynamic farm incorporates all aspects of farming like vegetables, cereals, fodder, animals, agro-forestry, water harvesting, orchards, *etc.* The main focus of the farm is the dairy of nearly 100 animals and production of a rich variety of fodder to feed these animals. The farm produces about 300-350 litres of milk every day and has been catering to 85 customers and a 500 bed hospital in the nearby town of Vallabh Vidyanagar for the last 25 years. The farm also produces a variety of fruits like chikoo, mango, custard apple, litchi, papaya, mosambi, lime *etc.* Over 40 different vegetables are grown mostly in the winter months. Carefully designed trenches and ponds ensure that all the rain water is harvested. A variety of oilseeds, cereals and pulses like sun flower, oats, bajra, barley, wheat, linseed *etc.* are also grown. Windbreaker and biomass trees protect the boundary. Bamboo, teak and eucalyptus groves provide a nesting and resting place for many migratory birds. Innovative practices such as rotational pasture grazing, poultry 'as a complement to the dairy and as part of the pasture rotation

Cows in the Pasture (above), and Workshop Trainees on a Exploratory Walk in the Vegetable Garden (below).

cycle' *etc.* are always being tried in an attempt to optimize the farm's synergies - the birds keep the dairy clean of insects such as fly larvae/ticks and produce high protein natural fertilizer for the pasture besides also producing very high quality eggs. The cows in turn indulge the chicken by shearing the grass in the pasture; chickens can't navigate in grasses more than six inches tall.

This farm is prominently featured in the internationally acclaimed documentary 'One Man, One Cow, One Planet' where **Peter Proctor** who is considered the Father of Indian Biodynamic Agriculture says that this is one of the best Biodynamic farms in the world that he has seen. Peter stayed on this farm for a year in 2006 and helped turn this organic farm into a vibrant Biodynamic farm.

A prominent goal of the farm is to disseminate the art and knowledge of sustainable farming to small/marginal farmers and anybody interested in farming.

The farm conducts regular, internationally attended, training workshops in organic and biodynamic farming with the help of renowned experts. The third biennial OFAI (Organic Farming Association of India) conference took place amidst the farm's verdant bamboo groves and hosted over 700 farmers from all over India in 2010. In 2014, the farm hosted the second biennial Biodynamic Conference under the aegis of the BDAI which was attended by participants from all over the world. This farm has become a pilgrimage for farmers and farming enthusiasts.

7. Conclusion

Biodynamic farming has taken very strong roots in India. The Indian farmer instinctively understands the importance of the cow in farming and as such, the Biodynamic practices have found easy acceptance. Also the Indian farmer has always followed the lunar rhythms – some consciously and some enshrined in tradition, rituals and festivals. Unlike in the West, where the Biodynamic movement has largely remained an elitist and a closed movement, Peter's work in India has ensured that Biodynamics can be practiced by the smallest of farmers. With organizations such as Naandi, ICRA, SUPA and others working relentlessly at the grass roots level to hand hold and support the small farmer, the Biodynamic movement has started realizing the scale and reach few other countries can boast of. Naandi is already planning to repeat the Araku success in the Vidharbha region around pomegranates, in Uttar Pradesh around mint and in Punjab around fodder and dairy.

Chapter 25

Organic Farming with Cosmic Energy: Rishi-Krishi, Deshpande Farming Technique

Aryakrishak Mohan Shankar Deshpande

Shri Samarth Agriculture Research Centre,
Ajara, Kolhapur – 416 505, Maharashtra
E-mail: shubhada.m.kamat@gmail.com

1. INTRODUCTION

Krishirdhanya Krishirmerdhya jantunam jivanam Krishihi |
Hinsadidoshayuktoapi muchyateatithi pujanat | |

– ***Krishi Parashara 8***

The broad meaning of this is that, the farmer's life is dependent on the living organisms in the soil he cultivates. During the cultivation, some of these organisms get killed. Since his Dharma (primary duty) is to produce food for the sustenance of the world, he gets absolved of the sinful deed of killing the organisms.

Rishi-Krishi Deshpande technique of farming is a novel but ancient technique of agriculture. The greatest advantage of this technique is that the crops prosper with the help of cosmic energy and the farmers do not have to use any inputs in the form of chemical fertilizers and pesticides. The cosmic energy is the cornerstone of Rishi-Krishi. Rishi-Krishi Deshpande technique and the cosmic energy are inseparably bonded. Further, Angara (likened to holy ash), Amrutpani (meaning nectar water), mulch and seed dressing are the mail pillars of Rishi-Krishi Deshpande technique.

☆ Plants have their own language which they use to communicate with the living organisms in the soil. The diet of each and every tree even of the same species is specific, *i.e.* different from all other trees (the word tree includes all the plants)

- ☆ No organic or inorganic manure is needed as food by plants
- ☆ COSMIC ENERGY is the fundamental cause or source of plant growth
- ☆ Salt accumulation is the result of poor drainage therefore trenching and washing will never be a solution to improve the saline land
- ☆ Nothing has to be purchased from the market to apply in Rishi-Krishi Deshpande farming technique including no earthworm culture, earthworm casting *etc.*
- ☆ Input of cosmic energy in its different forms in its natural usage gives bumper crop in the same season. No gestation period is required.

2. Concept

As opposed to 'do-nothing' method, cultivation as per the Deshpande technique involves hard work and a strict adherence to a scientific discipline. Ahimsak Rishi-Krishi requires all the steps to be followed with the love of a caring mother. This technique, which requires creating an environment for proper nourishment of billions of insects and microbes residing in the farm and use them with care and without harm in growing the crops, is in direct contrast to the 'do-nothing' doctrine followed by the pseudo-organic farmers.

The procedure consisting of Angara, Amrutpani, seed dressing and the mulch works wonders in the field. Normally, even after using purchased earthworm culture, a cultivator has to wait for four to four and half months to have sufficient number of earthworms to be effective. But with Rishi-Krishi, the same result is obtained almost instantaneously and within 8 to 10 days, the earthworms are seen in the soil all over the farm. Further, this is achieved without procuring anything from the market. No earthworm culture and no earthworm castings need be purchased.

How does this happen? The living organisms in the soil which disappear or go into hiding after addition of chemical fertilizers, reappear to do their work after the Angara and Amrutpani treatment. Earthworms are the chief operators of the nature's microbe factory which organizes preparation of suitable food at proper time for the individual crops. Also, the earthworms never work in isolation. The army of working soil organisms proliferates in proximity of earthworms. Not only this, earthworms also ensure the growth of specific organisms needed for food and growth by the individual plants in the field.

3. Principles

The Rishi-Krishi Deshpande Technique of farming is based on three major principles:

- ☆ Trees have a language which they use to communicate with the organisms in the soil. The diet of each and every tree even of the same species is specific, *i.e.* different from all other trees. (the word tree includes all the plants).

- ☆ No Chemical fertilizer or organic manure is needed as food by plants.
- ☆ COSMIC ENERGY is the fundamental cause or source of plant growth.

Out of above, the second observation is unconventional and difficult to believe while the third one is mysterious and hence perplexing.

3.1 Trees Have a Language Which They Use to Communicate With the Organisms in the Soil. The Diet of Each and Every Tree Even of the Same Species is Specific

It is a confirmed and proven practical explanation based on the trials at the fields of a thousand farmers. How does a plant grow? How the water travels from the roots in the ground to the farthest and highest tips of a plant, overcoming the force of gravity? The science answers these questions by negatives and probabilities. But the Indian Vedic texts give definitive answers to these riddles. For example, the texts emphatically state that the plants grow in the moonlight. This is now accepted as scientific truth. This was the reason why I turned to the Vedic texts for enlightenment.

Gamavishyacha butani Dharayamyahamojasa |

Pushnami choushadhihi sarvaha somo bhutwa rasatmakah | |

– *Bhagvad Gita 15/13*

Entering the earth with My energy/support the beings

I nourish all the herbs, becoming the watery moon.

Na vapayetithou rikte kshine somay visheshatah |

– *Krishi Parashar 173*

Do not cause (seeds) to be sown particularly in the period of the warning moon: Acting thus properly one gets an increase of crops.

Chandramasa vav sarvanij yotrishi mahiyante |

Taittiriya Upnishad

Through the moon, indeed, all the luminaries flourish.

The above shlokas which emphasize the fact that the moonlight is essential for the growth of plants, have been known for ages to common men and the scholars.

The distinctive feature of the Deshpande technique is that it will progressively increase the fertility of the soil and will improve the production in all types of soils and for all crops. The technique is not a magic trick or a miracle. It has a firm scientific base which has been perfected by our ancestral Rishis and sages over generations of untiring labour and experiments.

3.2 No Chemical Fertilizer or Organic Manure is Needed as Food by Plants

Do the plants use the various chemicals and manures which we feed them through the soil? This is the crucial questions which need to be addressed. Nitrogen, phosphorus and potassium are regarded as the main fertility constituents of the plants. How these are acquired by the plants? The common belief is that the plants absorb these from the soil and we have to add chemical fertilizers to the soil to avoid deficiency of these elements. This is the current scientific faith or belief. When the scientific belief turns into a blind faith or dogma, it is most difficult to eradicate.

An important constituent of the plants is nitrogen. We are being told, day in and day out – 24 hours a day and 365 days a year, - that since the plants take up this element on a large scale from the soil, the nitrogen in the soil gets exhausted. Therefore, the soil has to be given frequent doses of nitrogenous fertilizers. We believe this since it is being conveyed to us by the authorities holding positions of power in the government and the agricultural institutions. Now the true story is that the plants do not take nitrogen from the soil. They absorb it from the atmosphere. Our saints and sages have long back answered these and similar queries. The modern scientists have got out of the difficulty by designating these as mysteries unexplained.

The Vedic answer is that the cosmic energy is at the root of all these mysteries. Like all forms of energies, the cosmic energy is invisible and can be observed or felt through its effects. One element can be converted into another with the help of the cosmic energy. This energy is available to the primitive micro-organisms in the soil as also to the living cells of the plants. And therefore, the plants have the capability to produce all the food elements needed for their growth, without our feeding them the specific elements.

Dr. C.L. Kervan has recorded several instances of un-explained mysteries of the plant world in his book 'Biological transmutation'. The best two out of these:

1. The wood and bark of oak trees is especially rich in calcium (up to 60 or more percent of the is CaO). Oak trees can grow in schist type of soil which is mainly composed of micaceous minerals *i.e.* aluminum silicates. They also grow in sand which is mainly silica. How do they accumulate calcium? The simple answer is that the silica was converted into calcium by the trees with the help of cosmic energy. Dr.Kervan recorded this as an unexplained mystery since he was not aware of the role played by the cosmic energy.
2. Spanish moss, which belongs to the Tillandisia family of air plants, was grown in copper fibers without any soil. Here again the analysis of the plant showed no copper but 17 per cent iron. How was the copper converted into iron?

This proves, that the fundamental scientific premise, that the elements present in the plants, as revealed by the analyses, are absorbed by the plants in the same form (*e.g.* nitrogen as nitrogen), is totally misconceived. It does not take into account the part being played by the cosmic energy in the process.

The plants produce their food by absorbing the carbon-dioxide from the air and water plus the micro-nutrients from the soil in the presence of sunlight. The plants will grow if this food is digested and energy is needed for this process. This energy is produced by the plants by internal combustion. Now, in this era of instant food, one can provide the instant energy to the plants by feeding the chemical fertilizers after which they show rapid growth. This has led to the erroneous belief that the chemicals give a quick boost to the crops. The actual process which is behind this phenomenon can be traced to an observation by Maharshi Vyas who has stated that a living being prospers at the cost of other lives.

When we feed the chemical fertilizers, due to the poisonous or the toxic nature of these chemicals, billions of bacteria and other life forms in the soil are killed. This releases the life energy which is a part of cosmic energy contained in their bodies. This energy is attracted and absorbed by the cosmic energy of the roots of the living plants. The availability of this readymade instant energy gives a big boost to the growing crop. When the store of the live organisms in the soil is exhausted the soil itself dies *i.e.* becomes saline and unsuitable for cultivation.

The role of organic manures is diametrically opposite to that of chemical fertilizers. All the same they do not directly act as food for the plants. Unlike the chemical fertilizers which kill the soil organisms, the organic manures increase their population. Larger numbers of the microbes bring about in an indirect increase in the fertility of the soil with the consequent increase in the crop yields. Plants take the food elements from the soil in an ion form through 'ion fixing' bacteria. After providing the ion, the bacteria die. So the plants get both the food and strength (energy of the dead bacteria) and they thrive. Further, since the dead organisms are being replaced on a continuous basis in presence of the organic manure, the fertility of the soil is maintained.

My question was how to bring the soil to life? In other words, how to increase the population of the living organisms in the soil? My quest ended at our common but sacred village tree, the banyan. I cannot explain how this occurred to me but after the event, I started to note down the reasons for its importance in our lives.

- ✰ During the rainy season the soil under the banyan tree support a very large number of crawler and farm worm type of earthworms.
- ✰ During the heat of the summer it is fully clothed in dark green foliage. It is every green in our climate, providing a welcome shade to humans, animals and birds.
- ✰ Thousands of small fruit it produces is an important source of food for innumerable species of birds. Their droppings make the soil under the tree, highly fertile and teeming with soil organisms.
- ✰ The aerial roots set out by the tree during the monsoon season contain substantial quantities of hormones and enzymes needed for plant growth. In our villages, sterile animals are fed these roots to induce fertility.

4. Steps Involved in the Rishi-Krishi Deshpande Technique of Farming

4.1 Ahimsak Rishi-Krishi Deshpande Technique: Step I

As per the Deshpande method, the cultivator has to incorporate a minimum of 15 kg of soil from the base of a banyan tree into each acre of farmland he wishes to cultivate. Since the quantity is minuscule, it is called "ANGARA "which is the name of the holy ash from our temples. Use of Angara results in a gradual increase in the population of the living organisms in the soil. Arya Chanakya has provided a very appropriate shloka which proves the same.

Kand bijanam chhedlepo, madhudhruten |
Kandanam Asthibijanam shakudalepah | |

– *Arya Chanakya Arthashastra* 2.24.24

Meaning, if the seed is in stick form, coat it with honey and ghee (clarified butter); if it is in bulb form or with a hard cover, coat it with wet cow dung. If you are a farmer you will realise how difficult and impossible appears the procedure laid down in this shloka. No wonder the mantra was neglected by generations of cultivators. I, on the other hand, spent months trying to unlock the wisdom hidden in the mantra.

4.2 Ahimsak Rishi-Krishi Deshpande Technique: Step II

The second step is the preparation and use of Amrutpani or Nectar water. We all know that Amrut is the heavenly drink which refreshes the gods and has the power to resurrect the dead. In the same manner my Amrutpani invigorates the living soil and converts a dead soil into a living one.

4.2.1. Preparation of Amrutpani (Nectar water)

Materials Needed

200 litres of Amrutpani is needed for one acre of farm land which is prepared as under:

1. 0.25 kg ghee (clarified butter) from a desi (indigenous breed) cow
2. 0.5 kg of honey
3. 10 kg of fresh dung from a desi cow
4. 200 litres of water

4.2.2 Process

Thoroughly mix 0.25 kg of ghee into 10 kg of cow dung. Blend 0.5 kg of honey into this mixture and add 200 litres of water stirring all the time. The mixture thus obtained is Amrutpani.

4.2.3 How to use Amrutpani?

For the crops like Sugarcane, turmeric, ginger *etc.*, it is advised to dip the planting material in Amrutpani before sowing. In the case of crops where the seedlings are transplanted, dipping the roots into Amrutpani before planting is advisable. While watering sugarcane and other crops with canal or well water, mix Amrutpani in the main watering channel stirring all the time. For rainfed or monsoon crops, the seeds need dressing as explained in Step III (next step). Here when the soil is damp it should be drenched with Amrutpani. The drenching needs to be done between the rows and not directly to the plants.

For crops like chilli, tobacco or fruit trees, small amount Amrutpani around the plants can be used. Excess of Amrutpani is always beneficial and will not harm the young plants. The procedure of preparation of Amrutpani mentioned above and its use (mentioned earlier and later) has to be followed with strict discipline.

4.3 Ahimsak Rishi-Krishi Deshpande Technique: Step III

Dressing of seeds for planting: The seeds are living and hence they need dressing for best results after planting. Correct seed dressing is very improtant for success while following the Deshpande system.

- ☆ Dressing of seeds with a hard coat such as rice, wheat, corn, bhendi (okra) *etc.*: Add 1 kg of Angara (soil from the base of a banyan tree) to sufficient Amrutpani to make a thick paste or muck. Mix a small quantity of the paste with the seeds in a sifting pan and keep on rotating the pan till the seeds get a covering of the muck. Dry the seeds in shade, store and use as needed.
- ☆ For seeds with a soft or thin coat such as hemp, cereals, moong, groundnut, *etc.*: The muck as prepared above should be lightly sprinkled and the seeds should be used immediately.

4.4 Ahimsak Rishi-Krishi Deshpande Technique: Step IV

Mulching which means covering the soil in the crop area is an essential part of the Deshpande system. Rishi-Krishi lays great emphasis on covering the soil with a suitable material as mulch. For example, sowing of hemp every four finger or three inch spacing along both sides of the furrows while planting sugarcane. After one and a half month, this should be uprooted and used to cover the ridges. Similarly sugarcane bagasse can be used as a mulch. Perennial plants such as glyricidia and subabul provide excellent mulch for soil. One can plant these along the farm fences. It must be ensured that subabul trees are pruned well before the seed pods ripen as otherwise the flying seeds germinate all over the farm creating a big headache for the farmer.

5. Pest Management for Success of Rishi-Krishi

The chemical fertilizers are not needed as food by the plants. In the same manner the crops do not need the modern chemicals to fight pests and diseases (the term chemicals covers pesticides, fungicides, herbicides *etc.*) This does not mean that the crops will not be attacked by the pests or diseases. But it is not necessary for the farmer to purchase the expensive chemicals to combat these. Most of the organic remedies for control of pests and diseases are available at his farm.

When a farmer follows Rishi-Krishi, the crop is healthy and vigorous. This is achieved by the microbes that produce and supply the right type of food as and when needed by the plants. Therefore, it is unnecessary to use strong chemicals. Some of the remedies followed under the Deshpande technique are:

5.1 Chemical Weapons

The general belief is that the discovery and use of chemical weapons belongs to the modern age is misconceived. The chemicals were born together with the life forms on this earth. In our farm or garden whenever the insects or diseases attack a plant the message is instantaneously communicated to the nearby plants by a specific chemical process. This alerts the defence mechanism of the plants which produce the toxins needed to combat the attack. We, however, go into panic and start spraying chemicals which are not necessary. In a recent study in USA scientists at Rutger University found that when a plant – they used tobacco plants for the study - becomes infected with disease, it launches its own violent attack, cleaving the cell walls of the oncoming disease-laden organisms. At the same time it oozes the chemical methyl salicylate, a volatile form of aspirin which is commonly known as oil of wintergreen. Within hours of the attack the volatile chemical emitted by the plants become air borne and wafts over the neighbouring plants, enabling them to activate their own defences. Plants lack a circulatory system and an immune complex which in animals battles offending virus and bacteria with antibodies. Instead, plants employ anti-microbial proteins as their guardian. When the warning of an attack is received say with the wafting of oil of wintergreen, the plants start producing legions of these proteins as a sort of pre-emptive strike, enhancing resistance to infection.

5.2 Water

The traditional farmer knows that when it rains during blossom time the honeydew is washed off and the trees bear plenty of fruit. So why the farmers are not advised to spray water?. The cultivator is a victim of imperfect science. This can be tried and tested further.

5.3 Cow Urine

I insist use of urine and dung from the DESI (Indian breed) cow. This is based on scientific facts and has no religious significance. Most probably the cow was designated as a sacred animal in our religion for scientific reasons. Only, the desi cow knows the rule of cosmic energy. The cow urine is to be used as cure for many plant health problems. It can be stored in glass bottles. The effectiveness of the urine increases with the length of storage. The older the better. It is beneficial to keep the bottle in sunlight. Complimentary but separate spraying of neem extract together with the cow urine is very effective in cures of many diseases from bunchy top in bananas to downy mildew in grape. The cow urine should be sprayed once a week, during the life of a crop as a preventive measure. We should not wait for the insects or disease to attack. The spray solution should be 150 to 200 ml of the urine in 15 liters of water.

5.4 Neem Extract

Soak 15 to 20 kg of small bunches of neem branches in 150 to 200 litres of water. Allow this to stand in shade for four days. The yellow extract thus obtained should be filtered and diluted with water such that on stirring it would produce a soapy lather. One can also safely use the extract without dilution.

If the neem leaves are not available, take 4 to 5 kg of neem oil cake and use after soaking. Filter and dilute as discussed above. The cow urine and the neem extract are very effective as general purpose preventive sprays for control of pests and diseases for all crops. They should be used once a week separately.

5.5 Buttermilk

Gram (cowpea), chillies, egg plant, tomato and many other crops are subject to frequent attacks by leaf curl disease. The cure for this is spraying sour buttermilk together with neem extracts or cow urine. The butter milk should be such that it tastes mildly sour.

5.6 Vavding Decoction (*Embelia ribes* Decoction)

The Vavding decoction which is prepared by boiling the berries in water is a very common household medicine in Maharashtra. The spraying of this on fruit crops including grapes, papaya, mango, tomato *etc.* gives a healthy shine to the produce. The berries are available at all herbal product shops. Boil 250 g of the berries in 2 litres of water till the solution has been reduced to 1½ litres. Spray once a week after dilution at the rate of 10 cc of the decoction per litre of water.

5.7 Vekhand (Orris Root)

A farmer should always carry a small 2-3 cm piece of orris root in his pocket. This will ensure total protection from poisonous snakes and scorpions. The aborigines, tribal and other forest dwellers sprinkle the Vekhand water around their huts in the evening to keep away the poisonous snakes, and scorpions. Like Vavding, Vekhand is also a common home remedy in our country. The Vekhand decoction is a proven remedy against white grubs which are destroying productive lands along the river banks and sugarcane farms.

Vekhand decoction can be prepared by boiling 250 g of Vekhand pieces in 2 litres of water till the decoction is reduced to 1 ½ litres. Dilute the extract to 200 litres and feed it to the soil through irrigation water.

To get rid of the white grubs from the sugarcane field, feed the extract as above during the first and second week of June. This should be done even when the soil is damp or wet. The decoction should be mixed into the main irrigation channel in small doses. After the watering is completed the Amrutpani must be sprinkled all round the crop with the help of small leafy branches.

5.8 Pusphak Fan

During the hours of early morning before 10 and late afternoon after 4, the plants produce almost 80 per cent of the food. The radiation from the sun contains higher proportion of red colour during these hours. One can take advantage of this phenomenon by using red coloured three bladed fans which should be installed at right angles to the soil surface. Two ball bearings will keep the blades rotating in the breeze and the light all over the farm will have a red hue for the whole day. A further improvement will be fitting small mirrors on the blades. The beam of reflected light from the rotating fan will keep the birds away from the crop.

5.9 Smoke

One pest which is immune to cow urine or neem extract is the white fly, which attacks some vegetable and fruit crops such as cabbage, cauliflower *etc*. One simple weapon to combat this pest is smoke. In the evenings when the wind is from west to east, burn waste material at the west end of the farm to produce smoke. The white fly will leave the field for more hospitable environment. Continue the practice till the crop is free of the pest. While making smoke, use waste part of leather shoes or waste pieces of leather.

6. Impact

All the trails carried out by me using the Rishi-Krishi technique have been 100 per cent successful. It has been used from Nagpur to Goa and from Pune to Ramdurg. Here the soils were of different types and the crops were also different. And everywhere the same tools, namely the Angara and Amrutpani ensured full success. A number of case studies and success stories are documented by Deshpande (2003).

7. Conclusion

The procedure consisting of Angara, Amrutpani, seed dressing and the mulch works wonders in the field. Normally, even after using purchased earthworm culture, a cultivator has to wait for four to four and half months to have sufficient number of earthworms to be effective. But with Rishi-Krishi the same result is obtained almost instantaneously and within 8 to 10 days the earthworms are seen in the soil all over the farm. Further, this is achieved without procuring anything from the market. No earthworm culture and no earthworm castings need be purchased. The living organisms in the soil which disappear or go into hiding after addition of chemical fertilizers reappear to do their work after the Angara Amrutpani treatment. Earthworm is the chief operator of the nature's microbe factory that organizes preparation of suitable food at proper time for the individual crops. The technique involving Angara, Amrutpani, seed dressing and mulch results in improved crop yields not only for good and medium soils but for low grade, salty and non productive soils as well.

REFERENCE

Deshpande MS. 2003. Organic farming wrt cosmic energy, Non-violence Rishi Krishi: Deshpande Farming Technique, published by Mrs. Pushpa Mohan Deshpande, Kheda-Ajra, Kolhapur. 238p.

Chapter 26

Natural Farming

Bharat Mansata

C/o Earthcare Books
10, Middleton Streeet, Kolkata – 700 071, West Bengal
E-mail: bharatmansata@yahoo.com

1. INTRODUCTION

Natural farming is holistic and bio-diverse organic farming in harmony with nature. It is low-intervention, ecological and sustainable. In its purest advanced form, it is a 'do-nothing' way of farming, where nature does everything, or almost everything, and little needs to be done by the farmer. This can best be achieved in a progressive manner with tree crops. On 24th October, 2015, Bhaskar Save – the acclaimed 'Gandhi of Natural Farming' – left this world at the age of 93 years. As he explained, "When a tree sapling planted by a farmer is still young and tender, it needs some attention. But as it matures, it can look after itself, and then it looks after the farmer." Save has inspired and mentored 3 generations of organic farmers. In 1997, Masanobu Fukuoka, the legendary Japanese natural farmer, visited Save's farm. He described it as "the best in the world", even better than his own farm. Indeed, Save's farm is a veritable food forest; and a net supplier of water, energy and fertility to the local eco-system, rather than a net consumer.

Save's way of farming and teachings are rooted in his deep understanding of the symbiotic relationships in nature, which he is ever happy to explain in a simple, down-to-earth idiom to anyone interested. In 2010, the International Federation of Organic Agriculture Movements (IFOAM) honoured Save with the 'One World Award for Lifetime Achievement'.

Bhaskar Save's 14 acre orchard-farm, Kalpavruksha, is located on the Coastal Highway near village Dehri, District Valsad, in southernmost coastal Gujarat, a few km north of the Maharashtra-Gujarat border. The nearest railway station is Umergam on the Mumbai-Ahmedabad route. About 10 acres at Save's farm are under a mixed natural orchard of mainly coconut and chikoo (sapota) with fewer numbers of numerous other tree species. About 2 acres are under seasonal field crops cultivated organically in traditional rotation. Another 2 acres is a nursery for raising coconut saplings that are in great demand. The farm yield is superior to any farm using chemicals. This is true in all aspects of total quantity, nutritional quality, taste, biological diversity, ecological sustainability, water conservation, energy efficiency, and economic profitability. The costs (mainly labour for harvesting) are minimal, and external inputs almost zero.

2. Natural Farming and its Fruit

Natural farming is holistic and bio-diverse organic farming in harmony with nature. It is low-intervention, ecological and sustainable. In its purest advanced form, it is a 'do-nothing' way of farming, where nature does everything, or almost everything, and little needs to be done by the farmer. This can best be achieved in a progressive manner with tree crops. With annual or seasonal field crops, more continuing attention and work by the farmer are needed, but even here, the work and input needed progressively diminishes as the soil regains its health and symbiotic biodiversity is re-integrated.

"*Who* planted the great, ancient forests?" asked Bhaskar Save. "Who tilled the land? Who provided seed, manure, irrigation, or protection from pests? ... In our forests, untended by man, the food trees – like *ber*, *jambul*, mahua, mango, wild fig, wild sapota, tamarind, *etc.* – yield so abundantly in their season, that the branches sag with the weight of the fruit. The annual yield per tree is commonly over a tonne, year after year, carried away by forest dwellers, including man. But the earth around each tree remains whole and undiminished. There is no gaping hole in the ground! If anything, the soil is richer.

"From where do these forest trees – including those on rocky mountains – get their water, their nitrogen, phosphorous, potash? Though stationary, Nature provides their needs right where they stand. But arrogant modern technology, with its blinkered, meddling itch, is blind to this. "Our ancient sages understood Nature's ways far better than most modern day technologists," said Bhaskar Save. He quoted the Upanishads:

"Om Purnamadaha Purnamidam Purnat Purnamudachyate

Purnasya Purnamadaya Purnamewa vashishyate'

Meaning, this creation is whole and complete. From the whole emerge creations, each whole and complete. Take the whole from the whole (respectfully, as many times as you need) the whole yet remains, undiminished, complete!"

"Gandhi believed in *gram swaraj* (or village self-governance)," said Save. "Central to his vision was complete self-reliance at the village level in all the basics needed for a healthy life. He had confidence in the strength of organic farming in this country. Similarly, Vinoba pointed out that industries merely *transform* 'raw materials' sourced from Nature. They empty one bag to fill another, but cannot create anew. Only Nature is truly creative and self-regenerating – through synergy with the fresh daily inflow of the sun's energy.

"There is on earth, a constant inter-play of the six *paribals* (key factors) of Nature, interacting with sunlight. Three are: air, water and soil. Working in tandem with these, are the three orders of life: *vanaspati srushti*, the world of plants; *jeev srushti*, the realm of insects and micro-organisms; and *prani srushti*, the animal kingdom. These six *paribals* maintain a dynamic balance, harmonising Nature's grand symphony!

Man has no right to disrupt any of the *paribals* of Nature. But modern technology, wedded to commerce – rather than compassion – has proved disastrous at all levels. We have despoiled and polluted the soil, water and air. We have wiped out our forests and killed its creatures. And relentlessly, modern farmers spray deadly poisons on their fields, massacring Nature's *jeev srushti,* or micro-organisms and insects – the unpretentious, but tireless little fertility workers that maintain the vital, ventilated quality of the soil, recycling all life-ebbed biomass into nourishment for plants. The noxious chemicals also inevitably poison the water, and Nature's *prani srushti* or animal kingdom, including humans.

Gandhi declared, 'Where there is *soshan,* or oppression, there can be no *poshan,* or nurture!' Vinoba Bhave added, 'Science wedded to compassion can bring about a paradise on earth. But divorced from *ahimsa,* or non-violence, it can only cause a massive conflagration that swallows us in its flames.'

Trying to increase Nature's 'productivity,' is the fundamental blunder that highlights the arrogant ignorance of agricultural scientists. Nature, unspoiled by man, is already most abundant in her yield. When a grain of rice can reproduce a thousand-fold within months, where is the need to increase its productivity! What is required at most is to help ensure the necessary natural conditions for optimal, wholesome yield.

In all the years a student spends for an M. Sc. or Ph.D. in agriculture, the only goal is short-term – and narrowly perceived – economic (rather than nutritional) 'productivity. For this, the farmer is urged to *buy* and *do* a hundred things, greatly increasing his costs. But not a thought is spared to what a farmer must *never* do so that the land remains unharmed for future generations and other creatures.

A quarter century ago, 'Poison in your Food' – a well-researched, lead feature in 'India Today', 15th June, 1989 – starkly exposed that Indians are daily eating food laced with some of the highest amounts of toxic pesticide residues found in the world. In the process, they are exposed to the risk of heart diseases; brain, kidney and liver damage; and cancer. More recently, even the Food Safety and Standards Authority of India, Union Ministry of Agriculture, reported that the toxic pesticides and chemicals contained in the foods we commonly buy are hugely in excess of permissible limits, exposing consumers to unacceptable risk of myriad diseases. Such poisons are even more dangerous for pregnant women, the babies they bear, and young children, as well as the ill and diseased.

3. Differnces between Chemical Farming and Organic Farming

Bhaskar Save listed 18 major points of difference between chemical farming and organic farming in harmony with nature:

1. Chemical farming fragments the web of life; organic farming nurtures its wholeness.
2. Chemical farming depends on fossil oil; organic farming on living soil.
3. Chemical farmers see their land as a dead medium; organic farmers know theirs is teeming with life.
4. Chemical farming pollutes the air, water and soil; organic farming purifies and renews them.
5. Chemical farming uses large quantities of water and depletes aquifers; organic farming requires much less irrigation, and recharges groundwater.
6. Chemical farming is mono-cultural and destroys diversity; organic farming is poly-cultural and nurtures diversity.
7. Chemical farming produces poisoned food; organic farming yields nourishing, poison-free food.
8. Chemical farming has a short history and threatens a dim future; organic farming has a long history and promises a bright future.
9. Chemical farming is an alien, imported technology; organic farming has evolved indigenously.
10. Chemical farming is propagated through schooled, institutional misinformation; organic farming learns from Nature and farmers' experience.
11. Chemical farming benefits traders and industrialists; organic farming benefits the farmer, the environment and society as a whole.
12. Chemical farming robs the self-reliance (and self-respect) of farmers and villages; organic farming restores and strengthens it.
13. Chemical farming progressively leads to bankruptcy and misery; organic farming liberates from debt and woe.
14. Chemical farming is violent and entropic; organic farming is non-violent and synergistic.
15. Chemical farming is a hollow 'green revolution'; organic farming is the true green revolution.
16. Chemical farming is crudely materialistic, with no ideological mooring; organic farming is rooted in spirituality and abiding truth.
17. Chemical farming is suicidal, moving from life to death; organic farming is the road to regeneration.
18. Chemical farming is the vehicle of commerce and oppression; organic farming is the path of culture and co-evolution.

4. Bhaskar Save's Plea for India's Agro-Ecological Resurgence

On 29th July, 2006, Bhaskar Save addressed a detailed 8 page Open Letter (along with six annexures) to M.S. Swaminathan, then chairman of the National Commission on Farmers. This was at a time of an unrelenting wave of farmer suicides in various parts of India, particularly Vidarbha and Andhra Pradesh, but also Punjab, the frontline state of India's 'green revolution', now turned black. Bhaskar Save's Open Letter – widely circulated and translated all over the world – presented a devastating critique of the government's agricultural policies favouring chemical farming, while making an eloquent plea for urgent and fundamental reorientation. Save stated, "I say with conviction that only by mixed organic farming in harmony with Nature, can India sustainably provide abundant wholesome food and meet every basic need of all – to live in health, dignity and peace." Swaminathan wrote back to Save, 'I have long admired your work and am grateful to you for the detailed suggestions… valuable comments and recommendations. We shall take them into consideration in our final report.'

A further independent Open Letter from Bhaskar Save, dated 1st November, 2006, was sent to the Prime Minister. Save asked in his letter, "In this vast nation, does any government agricultural department or university have a single farm run on modern methods, which is a net supplier of water, energy and fertility to the local eco-system, rather than a net consumer? But where there is undisturbed synergy of Nature, this is a reality! By all criteria of ecological audit, my farm has only a positive contribution to the health of the environment. Economically too, I get a manifold higher income than 'modern' farmers."

The success demonstrated by Bhaskar Save in decreasing and eliminating external fertility inputs while achieving high productivity, is thus a model for promoting food security; and his method of tree-cropping – integrating short lifespan, medium lifespan and long lifespan species – has been hailed as potentially revolutionary for wasteland regeneration, while also offering sustainable and rewarding livelihoods to large numbers of people.

5. Natural Abundance at *Kalpavruksha*

About twenty steps inside the gate of Bhaskar Save's farm is a sign that says: 'Co-operation is the fundamental Law of Nature.' – A simple and concise introduction to the philosophy and practice of natural farming. Further, inside the farm are numerous other signs that attract attention with brief, thought-provoking *sutras* or aphorisms. These pithy sayings contain all the distilled wisdom on nature, farming, health, culture and spirituality, Bhaskar Save has gathered over the years, apart from his extraordinary harvest of food!

If you ask this farmer where he learnt his way of natural farming, he might tell you – quite humbly - 'my university is my farm.' His farm has now become a sacred university for many, as every Saturday (Visitors' Day) brings numerous people from all over India, and occasional travellers from distant lands.

Kalpavruksha compels attention, for its high yield easily out-performs any modern farm using chemicals. This is readily visible at all times. The number of coconuts per tree is perhaps the highest in the country. A few of the palms yield over 400 coconuts each year, while the average is closer to 350. The crop of *chikoo* (sapota) – largely planted more than forty- five years ago – is similarly abundant, providing about 300 kg of delicious fruit per tree each year. Also growing in the orchard are numerous bananas, papayas, areca-nuts, and a few trees of date-palm, drumstick, mango, jackfruit, toddy palm, custard apple, *jambul*, guava, pomegranate, lime, pomelo, *mahua*, tamarind, neem, *audumber;* apart from some bamboo and various under-storey shrubs like *kadipatta* (curry leaves), crotons, tulsi; and vines like pepper, betel leaf, passion-fruit, *etc.*

Nawabi Kolam, a tall, delicious and high-yielding native variety of rice, several kinds of pulses, winter wheat and some vegetables and tubers too are grown in seasonal rotation on about two acres of land. These provide enough for this self-sustained farmer's immediate family and occasional guests. In most years, there is some surplus of rice, which is gifted to relatives or friends, who appreciate its superior flavour and quality. Excluding the two acres under field crops and a coconut nursery, the remaining ten acres of orchard have consistently yielded an average food yield of over 15,000 kg per acre per annum! (This has declined in the past 15- 20 years following pollution from progressive industrialization of the area.) In nutritional value, this is many times superior to an equivalent weight of food grown with the intensive use of toxic chemicals, as in Punjab, Haryana and many other parts of India.

The diverse plants in Bhaskar Save's farm co-exist as a mixed, harmonious community of dense vegetation. Rarely can one spot even a small patch of bare soil exposed to the direct impact of the sun, wind or rain. The deeply shaded areas under the *chikoo* trees have a spongy carpet of leaf litter covering the soil, while various weeds spring up wherever some sunlight penetrates. The thick ground cover is an excellent moderator of the soil's micro-climate, which – Bhaskar Save emphasized – is of utmost importance in agriculture. "On a hot summer day, the shade from the plants or the mulch (leaf litter) keeps the surface of the soil cool and slightly damp. During cold winter nights, the ground cover is like a blanket conserving the warmth gained during the day. Humidity too is higher under the canopy of dense vegetation, and evaporation is greatly reduced. Consequently, irrigation needs are very low. The many little insect friends and micro-organisms of the soil thrive under these conditions.

5.1 Nature's Tillers and Fertility Builders

Bhaskar Save stated, 'A farmer who aids the natural regeneration of the earthworms and soil-dwelling organisms on his farm, is firmly back on the road to prosperity.' Earthworms flourish in a dark, moist, aerated soil-habitat, protected from extremes of heat and cold, and having an abundance of biomass. These tireless workers digest organic matter like crumbling leaf litter along with the soil, while churning out in every cycle of 24 hours, one and a half times their weight of rich compost, high in all plant nutrients, and with a bacterial population that is nearly a hundred times more than in the surrounding soil.

The earthworm's burrowing action efficiently tills the land, imparting a porous structure to the soil. This increases its capacity to hold air and moisture, the most important requirements of plant roots. The worm castings too are well aerated and absorbent, while allowing excess water to drain away. They form stable aggregates, whose soil particles hold firmly together, resisting erosion. Various other soil-dwelling creatures – ants, termites, many species of micro-organisms – similarly aid in the physical conditioning of the soil and in the recycling of plant nutrients;

and there are innumerable such helpful creatures in every square foot of a natural farm like Kalpavruksha.

In stark contrast, modern agricultural practices have proved disastrous to the organic life of the soil. Many of the burrowing creatures are killed by the toxic effect of the chemicals used, or crushed under the weight of heavy tractors. The consequent soil compaction, resulting from their death, has reduced soil aeration and the earth's capacity to absorb moisture. By ruining the natural fertility of the soil, we actually create artificial 'needs' for more and more external inputs and unnecessary labour for ourselves, while the results are inferior and more expensive in every way. 'The living soil,' stressed Bhaskar Save, 'is an organic unity, and it is this entire web of life that must be protected and nurtured. Natural Farming is the Way.'

5.2 Weeds as Friends

In nature, every humble creature and plant plays its role in the functioning of the eco-system. Each is an inseparable part of the food chain. The excrement of one species is nutrition for another. In death too, every organism, withered leaf, or dry blade of grass leaves behind its contribution of fertility for bringing forth new life. Consequently, if we truly seek to regain ecological harmony, the very first principle we must learn to follow is, 'Live and let live'. In a country like India, a variety of weeds rapidly cover bare ground with the first showers of the rainy season. When torrential downpours follow as the monsoon progresses, the weeds buffer the hammering force of the raindrops, while their roots bind the soil against erosion. Such soil erosion could otherwise be severe in our tropical conditions, particularly on sloping terrain. Bhaskar Save thus observed – it is our foolish ignorance that we fail to understand how great a blessing the weeds are!

The roots of the weeds also improve aeration in the passages they make in the soil. Moisture absorption and retention are higher. By shading the ground, the weeds moderate the temperature of the earth, reducing evaporation and maintaining suitable conditions for soil organisms. And when the weeds die, the earthworms, ants and decomposer- bacteria that feed on their dead leaves and roots, return their mineral nutrients to the soil to help the next generation of plants, and the long life-span trees. Weeds may additionally perform a variety of specialised functions. As soil conditions change, there is a natural progression of different kinds of weeds that inhabit the earth. Some are excellent pioneers that steadily work to improve the soil where little else yet grows. Some are leguminous, and provide nitrogen. Yet others may function as reproduction inhibitors of the little insects that sit on them, thereby checking the plant damage that some of these creatures might cause.

5.2.1 When Weed Control is Needed and How

While weeds, in general, are friends of a farmer, in certain unnatural conditions, some species may become stubbornly rampant. Such weeds may then be a nuisance if they rapidly overgrow the crops planted by the farmer, blocking off sunlight. However, here too, the weeds help check and heal a more fundamental problem – that of soil erosion or impoverishment. They persistently signal to the farmer that s/he is planting a wrong crop in the given circumstances, or growing it in a wrong way, hurting the earth and her creatures. The only sensible and lasting 'root-cure' to situations of weed rampancy among field crops is to adopt mixed planting and crop rotation, while discontinuing chemicals and deep tillage. Since the problematic weeds will only phase out gradually as the soil regains its health, they may still tend to over-shade the food crops in the interim period of recovery. The way to manage this is to periodically cut the weeds (before they flower), and mulch them at least 3-4 inches thick on the soil under the crops. Without any sunlight falling on the weed seeds in the soil, their fresh germination is effectively checked.

There may thus be some competition between crops and weeds for sunlight, though not for soil nutrients. If the crops emerge taller, their shade will suppress the weeds, which will then be unable to cause any problem. This happens naturally in healthy, living, non-acidic soils. Our ancestors have been farming for many generations. But because their soil was healthy, they never faced any serious problem from weeds, even as recently as a few decades ago. There is a thumb-rule for seed spacing while planting your crops. If your soil is poor/weak, increase the quantum of seeds you plant. In other words, plant closer. By this stratagem, the crops cast shade on the ground more rapidly, retarding the weeds. If, however, your soil is fairly healthy, plant fewer seeds, that is, keep a larger gap between them. When farmers shift back to organic farming, their soil steadily improves in health each year. Correspondingly, crop growth gets better, while weed growth declines. In just 2-3 years, there should be no need for any weeding at all. Until then, the farmer is better advised to cut and mulch the weeds.

The cutting of weed growth above the land surface – without disturbing the roots – and laying it on the earth as 'mulch,' benefits the soil in numerous ways. With mulching, there is less erosion of soil by wind or rain, less compaction, less evaporation, and less need for irrigation. Soil aeration is higher. So is moisture absorption, and insulation from heat and cold. The mulch also supplies food for the earthworms and micro-organisms to provide nutrient-rich compost for the crops. Moreover, since the roots of the weeds are left in the earth, these continue to bind the soil, and aid its organic life in a similar manner as the mulch on the surface. For when the dead roots get weathered, they too serve as food for the soil-dwelling creatures.

5.2.2 Mulching for Weed Control

Mulching is effective in checking the rapid re-emergence of the cut weeds, only if the mulch layer is thick enough to block off sunlight. For example, the weeds cut from a plot of 100 sq. feet will never provide a thick enough layer to fully cover the entire 100 sq. feet. It may be adequate for 25 sq. ft., or perhaps just 10 sq. ft., depending on the density of weed growth. If sunlight penetrates through a layer of mulch that is too thin (less than 3 inches), the weeds may grow back vigorously again. Thus, if 10 sq. feet is the area that can be adequately mulched, at least 3 to 4 inches thick, with the weeds cut from 100 sq. ft., that is what the farmer should stick to, unless additional biomass can be obtained from an external source. The fresh weed growth from the balance unmulched land would again need to be cut and mulched in the selected area. In this manner, the mulch method of shading out weeds can be successful in 4 or 5 stages. The decomposition of the weeds may take several months, but the compost formed will be very helpful to the crop. What was viewed as an enemy, will now serve as friend!. It is also important that the cutting and mulching operation should be done before the weeds have flowered and become pollinated. If the farmer is too late, and the mulch contains pollinated weed seeds, a new generation of the same weeds will re-emerge strongly in the mulched areas.

5.2.3 Weed Control through Over-Shading Plants

The *Dabhro* weed is considered a menace by most farmers. To control it, one needs to plant crops that thickly shade the ground. No matter how often you remove it, the *dabhro* comes up again from its deep reaching roots. You cannot destroy it this way. Rather, you should plant an over-shading crop like banana at 4 ft by 4 ft, or 5 ft by 5 ft. When these have grown a little, provide them a good quantity of dung manure. The leaves that emerge will span out such that the canopies of adjacent plants will touch, thickly shading the ground and thereby suppressing the *dabhdo*, and gradually destroying it.

5.3 Multi-Storey, Multi-Function

Above the ground cover of weeds that constitute the lowest storey of vegetation in the orchard area (where any sunlight penetrates to the ground), there are numerous shrubs like the *'kadipatta'* (or curry leaf, *Murraya koenigii*) and the homely croton that line the pathways through the orchard. The latter plant, of various spotted and striped varieties, is relatively shallow rooted. It serves as a 'water meter', indicating by the drooping of its leaves that the moisture level of the soil is falling!. The shrubs of curry leaf contribute to moderating the population of several species of crop-feeding insects, while also providing an important edible herb widely used in Indian cooking. From this minor crop alone, Bhaskar Save earned an income of at least Rs 2,500/- each month, at zero cost. (Even the harvesting and bundling is done by the purchaser.) Here and there, one might see climbers like the pepper vine or betel leaf in a spiral garland around a *supari* (arecanut) palm, or perhaps a passion fruit vine arching across a clearing. These provide additional bonus yield.

6. The Principles of Farming in Harmony with Nature

"The four fundamental principles of natural farming are quite simple!" declared Bhaskar Save. The first is, 'all living creatures have an equal right to live'. To respect such right, farming must be non-violent. The second principle recognizes that 'everything in Nature is useful and serves a purpose in the web of life'. The third principle is farming is a *dharma*, a sacred path of serving Nature and fellow creatures; it must not degenerate into a pure *dhandha* or money-oriented business. Short-sighted greed to earn more – ignoring Nature's laws – is the root of the ever-mounting problems we face. Fourth is the principle of perennial fertility regeneration. It observes that we humans have a right to only the fruits and seeds of the crops we grow. These constitute 5 to 15 per cent of the plants' biomass yield. The balance 85 per cent to 95 per cent of the biomass, the crop residue, must go back to the soil to renew its fertility, either directly as mulch, or as the manure of farm animals. If this is religiously followed, nothing is needed from outside; the fertility of the land will not decline."

7. Plant Needs – What? How Much?

Dispelling a common misunderstanding, Bhaskar Save clarified that the organic matter we add to the soil is not the 'food' of a plant, at least in any direct sense. Rather, it is food for the innumerable soil-dwelling creatures and micro-organisms, which function ceaselessly to maintain the fertility of the land. And there are more micro-organisms in half a cup of good soil than there are humans on earth!. Through the digestive processes of the soil dwelling creatures, including earthworms, the organic matter added to the soil gets decomposed into a progressively more inorganic or mineral form. The mineral rich excreta of these creatures must then dissolve in moisture, before being absorbed by the roots of plants. More serious yet is the misconception about *how much* of the minerals or water is needed. Bhaskar Save never tired to emphasize that plants are actually *mitahari*, or very small consumers of the nutrients in the soil. Sunlight and air are what they need in abundance, while the moisture requirement of most plants – barring aquatic and semi-aquatic species like mangroves and rice – is best met when the soil is just damp rather than soaked.

Since India has no lack of sunlight, it is the porous, humus-covered soils that absorb and hold more air and moisture, which are the most productive in giving a sustained, high yield of biomass. This is ancient knowledge, though less understood these days. Scientific analysis confirms that approximately 88 per cent of the weight of a plant – or any organic matter – consists of just carbon and oxygen, with roughly equal contributions of about 44 per cent each. Much of these two elements is drawn by the plant from atmospheric carbon dioxide, absorbed through minute pores or stomata in the underside of leaves. Hydrogen, drawn from moisture, is third in the list and contributes about 6 per cent of the plant's weight. The moisture also provides some of the oxygen, as does the air contained in the pores of the soil.

Where the porosity, or internal pore space of the soil is high – as in all good, living soils characterized by a granular, 'crumb structure' – this enables it to hold enormous reserves of both air and moisture. Such a condition – known to local farmers as *waafsa* – where dampness and air (also warmth) are simultaneously present in the soil, is ideal for plant growth. Continuing with the list of the 'building-blocks' required by a plant, nitrogen comes a distant fourth, contributing between 1 to 2 per cent of its weight. This element, abundant in the air, is made available in the soil through the action of billions of rhizobia — the micro-organisms that dwell in the root nodules of leguminous plants. Nitrogen is also supplied when dead organic matter is broken down in the soil, under the action of even larger numbers of decomposer bacteria.

Less than 5 per cent of the weight of a plant originates in the various other mineral nutrients provided by the soil itself. These are elements like phosphorous, potassium, calcium, silicon, magnesium; and a number of trace elements or micronutrients required in very minute quantities, such as iron, copper, zinc, boron, cobalt, manganese, *etc.* The earthworm castings in a mixed natural farm or forest provide an abundant supply of these minerals and trace elements. Myriad other animals, birds, insects and micro-organisms (bacteria, fungi, molds *etc.*) add their contribution in recycling nutrients to the soil. In fact, every creature – in excretion and in death – is an integral part of the continuous fertility cycle of nature. Additionally, deep-rooted trees draw up fresh supplies of minerals dissolved over time from the underlying parent rock or sub-soil. Thus a farmer, who is mindful of the natural, biological processes of fertility regeneration, scarcely needs to bother about the chemical analysis of his soil. The important thing is to religiously return all crop residues and bio- wastes to the earth. Any pronounced 'nutrient deficiency' in the topsoil – often caused by cash-cropping monocultures – then becomes largely corrected in a few years by reverting to mixed cropping. Of course, checking soil erosion and shunning agro-chemicals is also essential.

7. Do Nothing?

While the physical work on a natural farm is much less than in a modern farm, regular mindful attention is a must. Hence the saying: 'The footsteps of a farmer are the best fertilizer to his plants!' In the case of trees, this is especially important in the first few years. Gradually, as they become self-reliant, the work of the farmer is reduced – till ultimately, nothing needs to be done, except harvesting. In the case of coconuts, Bhaskar Save had even dispensed with harvesting. He waited for the coconuts to ripen and fall on their own, and merely collected those fallen on the ground. For growing field crops like rice, wheat, pulses, vegetables, *etc.*, some seasonal attention, year after year, is unavoidable. This is why Bhaskar Save termed his method of growing field crops – organic farming, while a fairly pure form of 'do-nothing natural farming' is only attained in a mature, tree crop system. However, even with field crops, any intervention by the farmer should be kept to the bare minimum, respecting the superior wisdom of nature, and minimizing violence.

8. Conclusion (The Five Concerns of Farming)

Bhaskar Save summarized the key practical aspects of his approach to natural farming with reference to the five major areas of activity that are commonly a preoccupation of farmers all over the world. These are: tillage, fertility inputs, weeding, irrigation, and crop protection. i) Tillage: Tillage in the case of tree-crops is only permissible as a one-time intervention to loosen the soil before planting the saplings or seeds. Post planting, the work of maintaining the porosity and aeration of the soil should be left entirely to the organisms, soil-dwelling creatures and plant roots in the earth. ii) Fertility inputs: The recycling of all crop residues and biomass on the farm is an imperative for ensuring its continued fertility. Where farm-derived biomass is scarce, initial external provision of organic inputs is helpful. However, no chemical fertilizer whatsoever should be used. iii) Weeding: Weeding too should be avoided. It is only if the weeds tend to overgrow the crops, blocking off sunlight, that they may be controlled by cutting and mulching, rather than by uprooting for 'clean cultivation'. Herbicides, of course, should never be used. iv) Irrigation: Irrigation should be conservative, no more than what is required for maintaining the dampness of the soil. Complete vegetative cover – preferably multi-storied – and mulching greatly reduce water needs. v) Crop protection: Crop protection may be left entirely to the natural processes of biological control by naturally occurring predators. Poly-cultures of healthy, organically grown crops in healthy soil have a high resistance to pest attack. Any damage is usually minimal, and self-limiting. At most, some non-chemical measures like the use of neem, diluted *desi* cow urine, *etc.* may be resorted to. But this too is ultimately unnecessary. By thus returning to Nature many of the tasks that were originally hers, a weighty burden slips off the back of the half- broken, modern day farmer. And the land begins to regenerate once more.

Wendell Berry, a perceptive thinker, organic farmer and writer states: "When we change the way we grow our food, we change our food, our values, our society. ... Natural farming is about healing our relationships." Bhaskar Save said, "Children have a birth-right to suckle the sweet wholesome milk from their mother's bosom! But tragically, our modern rapacious way of farming, rampant industrialism and consumerist culture draw on Mother Earth's life-blood and flesh. How then can we hope to receive her continuing nourishment?" He added, "Non-violence, the essential mark of cultural and spiritual evolution, is only possible through natural farming."

REFERENCE

Bharat Mansata, 2012. The Vision of Natural Farming. Earthcare Books, Kolkata. 277p.

Chapter 27

Indigenous Technologies of Organic Agriculture

R.A. Ram and R.K. Pathak

ICAR-Central Institute for Subtropical Horticulture,
Lucknow – 226 101, Uttar Pradesh
E-mail: raram_cish@yahoo.co.in

1. INTRODUCTION

It is pertinent to mention that layer of useful atmosphere surrounding planet earth is very thin. Any change induced in this layer automatically disturbs the nature's energy cycle of nature. Environmental degradation or pollution has acquired global dimensions as on today. Green Revolution came as alternative to enhance production, and country witnessed quantum jump in production of cereals and number of other commodities. But indiscriminate use of agro-chemicals and water over 5-6 decades, have adversely affected soil fertility, crop productivity, produce quality and particularly the environment in many parts of the country. On close observation, it is evident that with this, we have contaminated ultimate sources of energy *i.e. Pancha Mahabhta's* particularly land, water and air owing to our greed. Now world is at cross road, and trying for few new techniques such promotion of genetically engineered seeds, development of complex fertilizers and new molecules for pest and weed management. It is obvious that polluted air, loaded with breakdown products of industrial and city combustions, oil fumes, sulphuric acid and many others are detrimental to soil fertility, water quality and the environment. After closely observing for one and half decade, we are of firm view that instead of sustainable agriculture, these are further going to deteriorate the situation.

Looking back for a while to ancient time, when traditional systems of farming were practiced, country cherished "Golden Era" and social fabric was at the helm of prosperity. In ancient India, farming was based on the principles of "*Jiva Jivasya jivanam/jiwa jiwat jayate-* meaning there by life sustains life and one organism live on another organism. By adopting these principles, there used to be record production in most of the commodities. As per information available, yield of paddy varied from 7-8 tons per hectare in Tamilnadu, (Claude, 2010). Country was known for production and export of high quality spices, cotton and silk fabrics. The system reveals holistic crop management through use of suitable cultivars, use of quality seeds, process of seed treatment, use of suitable soils, multi-storeyed crops with optimum plant population, balanced nutrition, optimum use of water, timely weeding, and protection from disorders. Cow and cow family was integral component of farming system. *Kunnappa jala,* special preparations made after fermenting animal flesh, cow dung, and urine and plant products was the main strength for soil fertility and healthy crop production (Nene, 2005). It is interesting to note that most of the agricultural activities were performed on the basis of planetary movement and even *Mantras* were used for pest management. Farmers were self dependent for their food, fodder, fuel, fibre to the whole community in a given area. India was an organic country. Slowly we lost our heritage and food production reached to such a low level that country was forced to import food grains from other countries.

These adverse effects of intensive agriculture have compelled to think for alternative and sustainable system of agriculture. As a result of dedicated efforts by few individuals, number of organic farming systems *i.e. Biodynamic Farming, Rishi Krishi, Natural Farming, Panchagavya Krishi, Nateuco Farming, Homa Organic Farming, and Jaivik Krishi* have emerged in different parts of the country. In these systems, maximum reliance is placed on self-regulatory agro systems and locally available or farm derived renewable resources (Pathak, 2013). Since soil, water and environment all have been polluted, hence even organic production systems, once effective are not capable of assuring sustainable agriculture. This has created number of myths in mind of policy makers and large number of farmers in acceptance of organic farming.

2. Qualm in Organic Farming

- ☆ Can yield level be maintained in organic farming to feed ever growing population?
- ☆ Are there are enough organic wastes available for conversion of entire area in the country under organic farming?
- ☆ Is organically produced foods are more nutritive as compared foods produced in conventional systems?
- ☆ Can pests, diseases and weeds managed by organic means?
- ☆ Is organic techniques are eco friendly and cost effective?
- ☆ Are there enough markets for organic produce?

For sustainability in agriculture, one has to change the mind set and conceive a farming system approach, which can enhance rhizosphere and biosphere simultaneously for sustainable production.

Study published in British Journal of Nutrition showed significantly lower levels toxic heavy metals in organic produce. Cadmium along with lead and mercury was found to be almost 50 per cent lower in organic crops than conventionally grown one. Concentrations of total nitrogen were 10 per cent, nitrate 30 per cent and nitrite 87 per cent lower in organic food compared to conventional. Professor Leifert added that organic vs non organic debate has rumbled on for decades now but the evidence from this study is overwhelming – that organic foods are rich in antioxidants and lower in pesticides residues. He suggested that this study should be a starting point for promotion of organic cultivation in the world.

On this background, we intended to promote *"Homa Jaivik Krishi"* basically based on harnessing solar and cosmic energy in agriculture. It is also interesting to note that it systematically perused, it is capable of enhancing the other sources of energy *i.e.* soil, water, air and can reset the planet earth energy cycle in active mode which can benefits every one and future generations.

3. Homa Organic Farming

Since sustainability and safe agriculture had been our dream for the couple of years, we conceived the concept of *"Homa Jaivik Krishi"*. We must emphasize that with little of support wherever it has been attempted, it has shown remarkable impact. The concept is discussed as under:

- ✰ Promoting farming system approach;
- ✰ Adopting use of agriculture calendar for various farm activities;
- ✰ Encouraging massive plantation for maintenance of biodiversity and creation of water bodies for conservation of rain water;
- ✰ Integration of legumes in the cropping system;
- ✰ Integration of indigenous cows with hump and use of her products in farming;
- ✰ Enhancing biological quality of soil through higher soil humus to harness and retain the energy;
- ✰ Integration with Homa Organic Farming to heal atmosphere for resolving current crises.

3.1 Biodynamic Agriculture Calendar

Life pattern of all living organisms are interwoven into cosmic rhythms. Human life, as well animal and plant life is strongly dependent on the rhythms of the earth and moon. On the basis of such influences, planting calendar is prepared for a year. The agricultural operations such as compost preparation and use, field preparation, seed sowing/transplanting, harvesting, storage are performed during different timings of the year. Since moon is nearest to the earth, hence agricultural activities in tune with moon's movement have impressive impact on yield and quality of produce. As per results derived at CSK, Palampur, Himachal Pradesh, 10-15 per cent increase in yield, minimization in pest's incidence; better quality have been obtained simply by adhering to agricultural calendar (Sharma *et al.*,

2011). Implication of agriculture calendar is non monetary input, hence need to be researched for more elucidation, tried and shared by fellow farmers/amateur growers after proper observations even in conventional system of farming.

3.2 Habitat Development

Proper habitat development in organic village/area is one of the requirements for conversion of any area in to organic. It is done by encouraging plantation of wide range of plants, trees, bushes, perennials, annuals trees in the area, creation of water bodies for storage of rain water which are helpful in creation of diversity of flora and fauna. Green vegetation helps in capturing of solar energy, release of oxygen and storage of CO_2 as carbon in the wood. They also act as source of biomass; provide congenial conditions for birds, predators, parasites helpful in farming. Plants restrict soil erosion, aerate soil, and provide nutrition to soil biota. A large number of plants have Biological Nitrogen Fixers (BNF) capacity and are potent source for capturing N (78.6 per cent) freely available in the atmosphere. These groups include trees, shrubs, climbers, annuals, biennials and provide fuel, fodder, timber, pulses, grains, oilseeds, medicines *etc.* Hence efforts are required to encourage these as main crop, intercrop crop, green manures, auro green (multi crops) *etc.* It is interesting to record that rhizosphere soil of certain trees are rich in microbial consortia, while different plants parts have pesticidal properties. These are utilized in preparation of composts, bio-enhancers and bio-pesticides. All these are helpful in one or other way in assuring sustainable production without any external aid (Nandish, 2010). As external input only cow has to be added in organic production and role of cow in organic farming has been enumerated as under:

Cow has been identified as *Kamadhenu* (wish fulfilling) and considered mainstay symbol of purification, health, wealth and prosperity. In all organic farming systems prevalent in the country, cow plays a major role and all her product is used in one or other way. It is interesting to record that indigenous breed of cow (*Bos indicus*) with hump and horns have capacity of absorbing cosmic and solar energy. Dung which permeates through her intestine is enriched with microbial consortia, while urine, milk and butter milk are used for preparation of composts, bio- enhancers and bio-pesticides, major strength of organic farming. Cow milk when applied to seedlings of plants, enhances overall growth and development. A consortium of 3 bacterial strains *viz. Bacillus lentimorbus* (B-30486), *B.subtalis* B-30487 (B-30487), a *B.lentimorbus* B-30486 (B-30486) isolated from Sahiwal cow's milk act as bio control agent and plant growth promotion. Strain B-30488 was evaluated for its biological control ability against *Fusarium oxysporum* f.ap. ciceri and *Alternaria solani* for protecting wilt disease in chickpea and early blight disease in tomato (Khan *et al*, 2012). Strain B-30488 mediated induction of dietary antioxidant in vegetables (*Trigonella foenum-graecum, Lactuca sativa, Spinacia oleracea* and *Dacus carota*) and fruit (*Citrus sinensis*) after minimal processing (fresh, boiled, and frozen) was tested by estimating the total phenol content, level of antioxidant enzymes has also been demonstrated (Das *et al.*, 2006; Nautiyal *et al.*, 2006 and 2008). An actinomycete isolated from fresh cow dung, showed antipathogenic potential against *Colletotrichum gloeosporioides* (anthracnose pathogen) and *L. theobromae* (gummosis, stem end rot and die back pathogen) of mango (Garg, 2012). It is interesting to mention that one cow can help 4-5 hectares area to manage as organic farm.

3.3 Bio-enhancers

Bio-enhancers are organic preparations, obtained by active fermentation of animal and plant residues over specific duration. There are number of bio enhancers, name coined by the person who attempted their efficacy and are being used by farmers. Few of these bio-enhancers which are in use are *Amrit Pani, Cow Pat Pit, Bijamrita, Jivamrita, Panchagavya, Vermi wash, Biosol etc.* In fact these are rich source of microbial consortia, macro and micronutrients and plant growth promoting substances including immunity enhancers. All bio-enhancers have phosphorus solublizers and nitrogen fixtures. These are utilized to treat seeds/seedlings, enrich the soil and induce better plant vigour. It is also pertinent to mention that with frequent use of these, one can enhance seed germination/plant vigour, manage pests and ensure better quality of produce. But, one must remember that affectivity of bio enhancer will depend upon organic carbon of the soil managed locally through harnessing solar energy (Pathak and Ram, 2013).

3.4 Homa Organic Farming

A number of studies indicate that due to climate change, nature is out of balance and pollution of entire environment seems to be one of the important causes for maintaining productivity in most of the crops. Analysis by Jennifer Burney, scientist at University of California (Hindustan Times, 6th November, 2014) predicted that air pollution caused wheat yields in densely populated states of India, to be 50 per cent lower than what they could have been in 2010. She also cautioned that up to 90 per cent of the decrease in potential food production seems to be linked to smog, a mix of black carbon and other pollutants.

With more than four decades experience of Five Fold Path Mission, we are of the firm view that the ancient knowledge of *Agnihotra* and *Homa* Therapy still can help us to set things right. Many reports and some preliminary studies in different countries provide a clue that how *Homa* therapy is helpful to bring nature back to harmony. *Homa* farming is based on regular performance of *Agnihotra*. Additionally some other fire is performed daily for a few hours and *Agnihotra* Ash is applied to the soil, plants and water.

Agnihotra is a gift to humanity from most ancient Vedic Sciences of bio energy, medicine, agriculture and climate engineering. It is the process of purification of the atmosphere through the agency of fire, prepared in the copper pyramid tuned to the biorhythm of sunrise/sunset. In fact it is radiation effects of astrological combinations and "*Mantras*" leads to better capture of cosmic energies from Sun and Moon which helps to reset the energy cycle of planet in natural harmony to benefit all livings. Materials used and its impact has been enumerated below.

3.4.1 Materials Used for *Agnihotra*

Pyramid

The word 'pyramid' Greek meaning is fire in the middle preferably made of copper. Copper is universally acknowledged for its excellent conduction of electricity and heat. The pyramid shape is widely experienced to generate and store a special energy field, which possess bacteriostatic properties. Copper act as a good energy conductor and it has anti-microbial properties (Abhang and Patahde, 2017). The inverted pyramid shape of the *Agni kunda* allows controlled generation and multidimensional dissipation of energy (Joshi, 2009; Narang, 2009). Pyramid acts as a generator of unusual energy fields and spreads in its surrounding area. The dimensions of the pyramid are: 14.50 cm x 14.50 cm at the top, 5.25 x 5.25 cm at the bottom and 6.25 cm in height with three steps.

Cow Dung

It is a valuable excretory product, possess antibacterial, anti viral and anti radiation properties. Dung is a well recognized disinfectant, keeps environment free from pollution and does not allow any radiation effect. It contains plenty of Ammonia, Menthol, Ammonia, Phenol and Indol. Thin patties are prepared from the dung and dried in the sun. Dung from cow family can be used but one should take precautions that nothing should be added in the dung during preparation of cakes/patties. *Agnihotra* fire is prepared with patties.

Cow Ghee

Cow ghee is richest source of energy among all organic compounds. It comprises of glycerol, a number of saturated and unsaturated fatty acids, which vaporize easily. It helps in rapid combustion of materials used for performance of *Agnihotra* and other *Yagnays*. When used in *Agnihotra* fire, it acts as a carrier agent for subtle energies in the nearby areas. If possible homemade ghee is preferred.

Rice

Rice is rich source of carbon and universally available. Unbroken (*Akshata*), brown non polished rice is used for performance of *Agnihotra*. Highly polished rice looses nutritional value and in broken rice energy structure is disturbed. Organically produced rice is preferred for *Agnihotra*.

3.4.2 *Agnihotra* Timing

The time component of *Agnihotra* is one of the most important. Light radiated by sun at sunrise and sunset is termed as diffused light in the sky which has great ecological significance. Since sun at the rise or set becomes near to horizon in transverse/oblique position, solar rays cover a longer distance to reach to the earth. The visible rays of higher wavelengths *i.e.* yellow, orange, red and infra red of invisible spectrum reach to the earth and the sky looks yellowish red at the time of sunrise and sunset. These rays are much energetic and penetrating, hence more effective. While those of lower wave length *i.e.* violet, indigo and blue of visible spectrum and probably ultraviolet, X-rays and Y-rays are scattered and lost in the atmosphere. For calculation of timings, software has been developed in Germany to get the exact sunrise and sunset timings for every place on the planet. (www.homatherapie.de; www.homatherapyindia.com; www.agnihotra.com.au).

3.4.3 Preparation of *Agnihotra* Fire

Place a flat piece smeared with *ghee* at bottom of copper pyramid. Arrange pieces which have been coated with *ghee* in pyramid in such a way that air can pass freely. Smear a little *ghee* on the small piece of patty, light it outside and insert in centre of pyramid. Soon it will catch fire. If needed, one can use a hand fan to accelerate the fire. Do not use any other aid and blow through mouth for increasing the fire. Good flame is required during offering.

3.4.4 *Mantras*

Vibrations exist everywhere. Chanting of *Mantras* is creation of vibrations. Where there is vibration, there is also sound and at the time of chanting of mantras vibrations create specific effects thereby exciting results are realized. It has been claimed that it is such energetic that it travels 12 km above in atmosphere, heals and captures cosmic *i.e. Pranic* (subtle) energy spread all around.

3.4.5 Process of *Agnihotra*

Take a few grains of rice in a small copper dish or in the left palm and apply a few drops of *ghee*. Exactly at sunrise utter the first mantra and while chanting *"SWAHA"* offer a few grains of rice with right hand in the flame. Utter second Mantra and after word *SWAHA* repeat the same process. At sunset do the same by chanting evening *Mantras*.

3.4.6 Precautions

If times are missed, it is not *Agnihotra* and healing effect will not be there. After each *Agnihotra* try to spare as many minutes as one can spare for meditation, a minimum till fire extinguishers itself. Just before the next *Agnihotra*, collect the ash and keep it in a glass or earthen container. It can be used for spray on plants or making folk medicines to treat human and animals.

Since atmosphere is very much polluted, to strengthen impact of *Agnihotra*, one should practice two simple *Agnihotra* fires. The person who performs *Agnihotra* for special purpose or any other *Agnihotra* must be a regular practitioner of *Agnihotra* as it creates the basic healing cycle around the physical body of the person. Special fires are Vyahruti *Homa* and *Om tryambkam Homa*. These can be performed at any time of the day or night except the times specified for *Agnihotra* fire. Fire is prepared as in *Agnihotra* (Paranjpe, 2005). There are four *mantras*. After each mantra, few drops of *ghee* is offered in the fire. After the fourth *mantra*, a teaspoon full of *ghee* is added to the fire. These Mantras are:

Vyahruti Homa

Bhooh Swaha Agnaye Idam, Na Mama

Bhuwah Swaha, Wayawe Idam Na Mama

Swa swaha, Sooryaya Idam Na Mama

Bhooh, Bhuwah, Swa Swaha,Prajapataye Idam Na Mama

With these four offerings, we convey our gratefulness to soil, air, sun and creator of universe. After these four mantras offering, one should start chanting *Tryambakam homa* in the same pyramid with mantra given as under:

"Om Tryambakam Yajamahe Sugandhim Pusti Wardhanam

Urwarukamiwa Bandhanan Mrityour Mukshiya Mamritat Swaha"

A drop of *ghee* should be added to the fire after the word *Swaha*.

This *Homa* can be done while establishment of orchard, sowing/transplanting any crop or before the start of any gathering or meeting. It creates a nice atmosphere to the field and for people who participate. This *Homa* can be done by a single person or a family or group in a house or in the field.

3.4.7 *Homa* Resonance Point Technology

Since incidence of diseases and insect-pests in most of the crops have become major problem, hence to cover large areas and get quick remedy in crops/plantations/forests, resonance technology has been developed by few European scientists (Paranjpe, 2005; Berk, 2014). Resonance is a special geometrical configuration of 10 pyramids which have been activated with special *Mantras* and set up to magnify the power and spread over large area.

Establishment of Resonant Point

Ten new pyramids are charged with *Mantras* and placed on the farm in a special configuration by a *Homa* volunteer who is authorized to install resonance points. Also two simple huts are needed, constructed with inexpensive, natural materials, found locally, like wood, adobe bricks, mats, bamboo, stone, cane, *etc.*

They are simply to protect the person performing the *HOMA* healing fires from the sun, rain and to prevent entering of animals like dogs, cats, chickens, *etc.* in the hut.

Activation of Resonance Pyramids

Activation of the ten pyramids is done in the *Agnihotra* hut only once. All pyramids are charged at the same time with fire and *Mantra*. The practice of *Agnihotra* in this hut has to be started at sunset, not at sunrise.

Agnihotra Hut

The main hut or *Agnihotra* hut is where the *Agnihotra* fire is performed daily at sunrise and sunset. It is ideal to build this hut in the centre of the area. Size should be approximately 3 x 4 meters, longer side aligned with the East/West axis. Entrance has to be from the west and one will sit down facing east to do the fires.

Parallel to east wall a hole of approximately 50 cm depth and 30 cm x 30 cm is dug. After activation, one main pyramid is buried, a column of mud is built on top of it, and to an approximate height of 50 cm other activated pyramid is placed on top of it, directly above the buried pyramid. Pyramid on the column acts as a Resonance Pyramid. Two activated pyramids are placed on smaller mud platform in front of the main column. The left one is used for daily *Agnihotra* and right one is for performance of other occasional fires. *Agnihotra* hut is a place of silence, no words, other than *Mantras*, are spoken inside.

Om Tryambakam Hut

This hut is little larger than first one (approximately 4 x 5 m). It is also known as the "HEALING HUT". Sick people can sit for getting benefitted. Entire area where resonance point has been established becomes healing area for living beings. In this hut, two pyramids are placed on small mud platforms, one on the right for OM TRYAMBAKAM HOMA and left for *AGNIHOTRA*. To get maximum effect, it is advisable to perform four hours Om Tryambakam Homa daily and preferably twenty four hours on full and no moon days. It is better to construct this hut near the entrance of the farm so that outsiders can come and go without disturbing privacy that live or work on the farm.

Resonance Pillars

Four pyramids are installed at the boundary of the farm, or at a maximum distance of 500 metres exactly north, south, east and west from the central point of *Agnihotra* hut. At each point mud column of height of 1.20 m is built. One activated pyramid is placed on top of the column. Small wooden boxes, if necessary with locks, can be built on top of each column, to maintain pyramids clean and safe. One resonance point can be used to heal 80 hectares of land as shown in Figure 27.1.

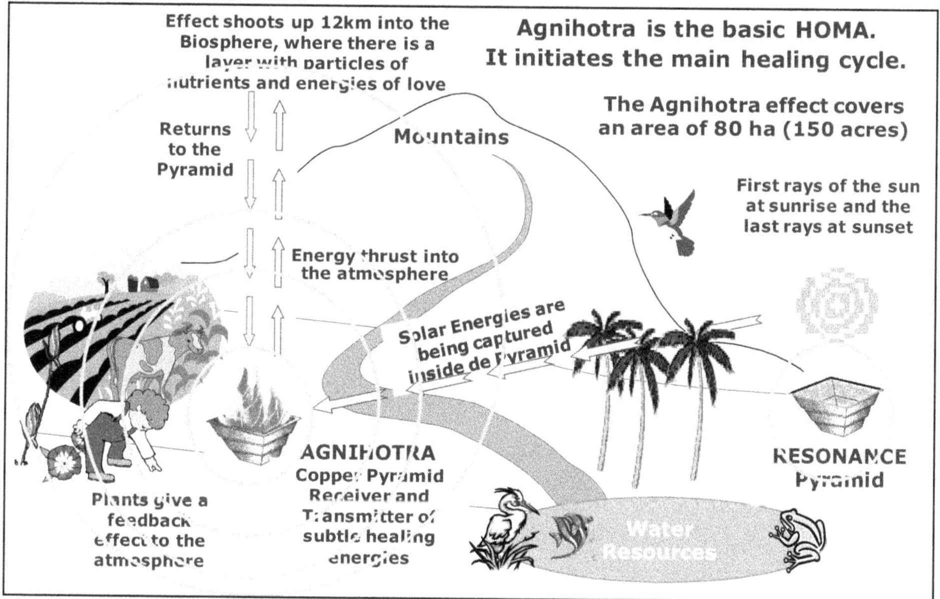

Figure 27.1: Schematic Presentation of Working of Agnihotra Energy Cycle.

Flood of subtle energies at sunrise creates strong purifying effects at all levels wherever it touches the earth. *Agnihotra* amplifies these purifying effects further. This flood of subtle energies carries music along with. Morning *Agnihotra* creates quintessential sound. If fire prepared in the prescribed copper pyramid and mantras are uttered by offering rice mixed with ghee to the fire, then a channel is created through the entire atmosphere and PRANA - life energy, is purified.

Tremendous quantities of energies are gathered around the copper pyramid at the time of *Agnihotra*. Pyramid acts as generator and fire is turbine. A magnetic field is created which neutralizes negative energies and reinforces positive energies. When *Agnihotra* fire is created it is not just energy from the fire, rhythms of mantras generate subtle energies which thrust into the atmosphere by the fire. As result of *Agnihotra* atmosphere, flora and fauna develops capacity to sustain biotic and abiotic stresses. When flame extinguishes subtle energy is stored in resultant ash. This ash is used in agricultural activities, treatment of water and preparation of folk medicines used for curing number of human and animal ailments. Therefore, by regular performance of morning and evening *Agnihotra*, a positive energy pattern is created which benefits every features of life in the given area, where resonance point has been established (Bruce and Haschel, 2009).

3.4.8 Science of *Agnihotra*

Sonic power of sound or vibrations has been acknowledged in science since ancient times. With substantial amplification, these vibrations can penetrate energy spheres at subtle and cosmic levels. *Agnihotra* is tuned to the circadian biorhythm of nature *viz.* sunrise and sunset and is based on Vedic science of bio-energy. Burning of specific substances and uttering of Mantras are synchronized with the timings of sunrise and sunset. Effects of *Agnihotra* mantras are multidimensional and of enormous value. *Agnihotra* fumes have been found to be rich in formaldehyde, ethylene oxide, propylene oxide, and Buta propiolactone which inhibit the growth of pathogenic microbes from the atmosphere and soil. Thereby, it purifies the atmosphere, and strengthens *pranic i.e.* life energy. As the *Prana* and mind are interlinked with each other, it automatically unburdens the pollution effects. In fact there are two basic energy systems in the physical world: heat and sound. In performing *Agnihotra*, both of these energies, *viz.*, heat from *Yajna's* fire and sound of chanting of Vedic Mantras are blended to achieve desired physical, psychological and spiritual benefits who reside in the area (Narang, 2009). In addition, the resultant ash is also a powerful tool full of subtle energy and used for number of purposes, as discussed below.

3.4.9 Use of *Agnihotra* Ash

One should remember that when *Agnihotra* is done in proper way *i.e.* correct timings, pure ingredients, offering is done with correct pronunciation of mantras: resulting ash is full of subtle energy (Sharma *et al.*, 2011.). The resultant ash of *Agnihotra* henceforth will be mentioned as Ash in order to repeat it number of times. In fact Ash is the most important input of *Homa* Organic Farmer. It is effective and beneficial at all stages farming operations. Experiments conducted in Russia have shown that ash was not radioactive even with use of radioactive ingredients. Since the resultant ash is full of subtle energy, it is used for many purposes in agriculture, animal husbandry and human health. Ash enhances moisture holding capacity of the soil. It renders the nutrients and elements already existing in the soluble form and easily available to the plants. Ash has been found to enhance solubility of phosphorus and make it available to plants. Use of Ash has been enumerated in detail by Pathak and Ram (2013).

Credit for introduction of *Agnihotra* and *Homa* outside to India goes to dedicated efforts of Shree Paranjpe. He took message to entire world, that since 1972 (Paranjpe, 2005). His main objectives were to teach ancient technology basically for addressing environmental pollution. Incidentally, *Agnihotra* and *Homa* got public attention when it was successfully demonstrated in agriculture. *Homa* Farming on a larger scale started in South America end of last century. One first breakthrough happened when a fungal disease called Black Sigatoka hit banana plantations and no other known methods were able to control this disease.

Large scale management of Black Sigatoka has been demonstrated by Paranjpe as reported by Garcia (2009). She also claimed that within four months of practicing *Homa*, Black Sigatoka, Yellow Sigatoka, Panama wilt disease and nematodes incidence in banana were eradiated. She also reported that banana plants became healthy; owing to rejuvenation of soil which resulted production of healthy new banana suckers. These reports cover all the main crops grown in that area like banana, coconuts, cocoa, coffee, mangoes, cotton, citrus fruits, grapes, palm oil, and different kinds of vegetables like cucumbers, cabbage, tomatoes, avocados and beans.

Use of ash as dusting on the foliage of tomato has been observed to escape infestation of fruit and shoot borer (Punam *et al.*, 2011). In fact Ash and biosol are two potent bio formulations for cultivation of irrespective crops through *Homa* Farming techniques. These negate the effect of polluting factors while increasing quality production in large number of crops which have been tried. In general, ash is used to protect any kind of seeds from pest infestation during storage, treatment of seed/seedlings while sowing/planting, soil and foliar application for the management of plant vigour, pest and disease incidence, and to improve quality production of any crop. Ash is basic components in preparation of one potent bio enhancer known as Biosol. Ash has been also found effective in minimizing vermi composting period and enhancing efficacy of paste for smearing on tree trunk and main limbs in number of fruit plants. Ash is applied by direct dusting/sprinkling, placed in gunny bag in irrigation source, through drip/sprinklers or drenching on mulches. Brief description of biosol is enumerated below.

Gloria Biosol

A special bio-formulation developed by Gloria from Peru and named as Gloria Biosol (Weir, 2009) is prepared after a series of processes which act as nutrient supplement to the plants. Processes include fermentation of organic substances *i.e.* Ash, fresh cow dung, cow urine, vermi compost along with *Yantram* in a plastic made rigid bio digester specially prepared for the purpose for four weeks. Since there is anaerobic fermentation, provision of valve is made for air dispense. A tap size of 5cm diameter and wide of 25 cm diameter is provided for removal of biosol and a lid at top which can be sealed air tight is designed. The size of bio digester could be from 200 to 1000 litres capacity as per requirement. Once constructed, it can be used for years. Biosol is powerful bio fertilizer for plants with high level of macro and micronutrients. It is rich in enzymes, beneficial microorganisms, phyto-hormones and useful components for plants thereby effective in improvement of soil fertility and crop productivity (Pathak and Ram, 2013).

On the basis of this information, Five Fold Path Mission organized International Scientific Conference on Homa Organic Farming sponsored by Planning Commission, Government of India. Scientists from different Universities and Government institutions shared their results, and discussed how to proceed further (Berk, 2014).

Thus, the fact that *Homa* Organic Farming works has been well established. It sustains production, deals with management of pests and diseases, and also improves the livelihoods of the farmers as cost benefits ratio is very high. In a preliminary study, Selvaraj *et al.* (2009) reported reduction in disease incidence in few crops like powdery mildew in roses, leaf spot and *Fusarium* wilt in carnation and gerbera, leaf spot in cabbage and late blight in potato.

4. Benefits of *Homa* Organic Farming

4.1 Reduction in Pollution

As on today pollution all around the planet earth is the major problem being faced by the humanity. It is because of rampant pollution, the atmosphere, soil, water, atmosphere; all are polluted by metallic, non metallic and gaseous toxicants of different types. It is interesting to mention that "*Homa* Organic Farming" is principally based on "you heal the atmosphere and the healed atmosphere will heal us". When cow dung is burnt along with ghee, its disinfectant effect goes in the atmosphere which further gets resonated in larger area; the atmosphere is disinfected, purified (Jarek, 1999). Few preliminary studies showed improvement in concentration of negative ions after *Agnihotra* is an important indicator of atmospheric pollution. Higher concentration of negative ions in the air is clear indication of less pollution. Besides this, regular practise of *Agnihotra* controls pathogenic bacteria in the air and soil.

4.2 Improvement in Soil Fertility

Maintenance of soil fertility is one of the major constraints for sustainability in agriculture. With conventional agriculture, problem of soil salinity, sodicity, acidity, water logging *etc.* have become major constraints in many parts of the country. These types of soil need to be rejuvenated to keep food production enhancing as per need. *Homa* is an effective tool to rejuvenate these problematic soils. In *Homa* atmosphere, soil enriched with beneficial micro-flora and fauna. This results in enhancement of soil water holding capacity and aeration. It has been further observed that addition of *Agnihotra* ash in the soil improves water soluble phosphate level and nutrients which are readily absorbed by the roots of the plants. Absorption of macro nutrients like nitrogen, phosphorus and potassium improves by active cell transport in *Homa* atmosphere. When *Homa* is performed, an atmosphere is created which becomes conducive to crop production by harnessing nutrients, microorganisms and beneficial insects in the area.

The rhizosphere is the soil-plant root inter phase and in practice consists of the soil adhering to the root besides the loose soil surrounding it. Plant growth promoting rhizo bacteria (PGPR) are potential agents for the biological control of plant pathogens. The importance of rhizosphere in *Homa* Organic Farming for plant growth and development has especially been stressed by Pathak (2013). Because these, *Homa* atmosphere and addition of ash has been found effective in improved nodulation in legumes like soya bean. The activity of plant growth promoting rhizo bacteria (PGPR) is enhanced if *Agnihotra* Ash is added to the soil and the plants are in *Agnihotra* atmosphere.

4.3 Earthworms Activity

Earthworms are an important component of soil fertility as they aerate the soil, degrade organic residue, aeration, degradation of organic resides and improvement in water holding capacity. Conventional agriculture has adversely affected the earthworm's population in the soil. As per common observation, *Homa* atmosphere has been found much conducive for earthworm activities. In *Homa* atmosphere, multiplication of earthworms increases and in turn helps in improvement of soil fertility (Paranjpe, 2005). This statement gets further verified from observations at a *Homa* Farm near Dhule, India that earthworm population was doubled in a month. Normally this population requires 3 to 4 months at organic farm. With Homa technology one can obtain minimum 12 harvests of vermin compost per year. Some *Homa* farms have reported from 17 to 18 harvests in a year. In general, maximum 8 harvests of vermicompost are reported in conventional farming.

4.4 Water

Water is an important natural resource made available in plenty by the Mother Nature. It is one of the critical inputs, which is essentially required for survival of all living beings. In recent years, water, is becoming one of the major constraints for the welfare of human beings. In future, water shortages are likely to be widespread. Indiscriminate over exploitation of ground water resources is causing shrinkage of total resource-base. Ground water quality is controlled by geological factors (arsenic, iron, fluoride *etc.*), overdraft, fertilizers and pesticides use, and saline water intrusion in the coastal regions. In fact pollution of ground water is becoming alarming. In number of districts of the country, concentration of nitrate, fluoride, iron, arsenate in ground water are becoming alarming, systematic research in mitigating water pollution through *Agnihotra* could be of considerable importance. In one of the publication, it is cautioned that due to pollution of water resources on earth, dead zones are being created where nothing survives. Water pollution now takes its toll in rich as well as poorer countries. No water is relatively safe to drink now (Paranjpe, 2005). He advised that if *Agnihotra* is practiced on a large scale, the atmosphere is healed and the water resources will get purified and lead to better absorption of Sun's rays. Several reports show that doing *Agnihotra* and use of *Agnihotra* ash in the well, water quality improves considerably. Non-potable water turns into drinking water – in one case the pH reduced from 9.5 to 7.2, and the salinity from 1150 ppm to 720 ppm (Frits and Lee, 2009; www.agnihotra.com.au). In Austria, one well was closed owing to high arsenic contamination and after performing *Agnihotra* and use of *Agnihotra* ash to the well, water quality was significantly improved.

Dr. Masuru Emoto, Japan did pioneer work on water quality and dedicated his life to research on water and human health. He developed a technology by quick freezing of water and through advance photography it provides wonderful results of crystal formation depending upon water purity. With dedicated research efforts for number of years, he concluded that performance of *Agnihotra* and incorporation of Ash is potential tool to improve water quality. He was of the view that regular performance of *Agnihotra* and use of Ash in water bodies may be helpful in resolving water crises in the country (www.rh4.info).

4.5 Radioactivity

Problem of radio activity is becoming alarming in recent years. The ancient Vedic knowledge states that radioactivity can be neutralized by *Agnihotra* atmosphere and use of Ash. This statement sounds quite extraordinary or even incredible as there is no technique available in modern science to reduce radioactivity. No radioactivity, neither in the milk nor in vegetables was observed at *Homa* farm in Austria after Chernobyl - although on all surrounding farms, every thing was contaminated with radio activity.

4.6 Energy Saving

Availability of clean energy is one of the major issues being faced by every country. As mentioned earlier, *Homa* organic farming is principally based on Harnessing Solar and Cosmic energy which is available in plenty in the country like India without any expenditure. Homa organic Farms could become a source for more production of energy which can be shared by local villagers.

4.7 Management of Colony Collapse Disorder

Honeybees are dying on mass scale, worldwide due to pollution. They are also severely affected to their detriment by applications of chemical pesticides, as well as genetically modified organisms and even mobile phone towers. Scarcity of bees is a major threat to agricultural production (www.homafarming.com). No loss of bee colonies were experienced at *Homa* farms - which is astonishing to compare the situation of bees at conventional areas.

Bees are attracted to *Homa* atmosphere as the quantum of energy they receive from *Agnihotra* fire helps to perform at a greater level of efficiency. When environment is provided to all 45 known pollinators, they can help to increase the yields of several crops. This is experienced with corn, tomatoes, berries and fruit crops (Paranjpe, 1989).

5. Conclusion

In fact, regular performance of *Agnihotra* injects energy into atmosphere to prevent pests and diseases and promote occurrence of natural parasites and predators and improves pollinator's activities. As a result, healthy micro-flora and fauna based ecosystem is created. This helps in development of healthy micro environment/ecosystem and after creation of healthy micro environment, creatures like earthworms and soil microbes thrive better. As per experiences from different parts of world, it is claimed that soil health, pest and disease management, reduction in crop production cycle and rejuvenation of plants can be achieved. These positive experiences led to perform some scientific studies. If adopted in organic atmosphere and organic mindset the technology is capable of assuring higher yield with quality production, improving soil fertility, water quality, prevention and eradication of plagues, diseases of the whole plant kingdom. It nurtures and brings balance in all components of the ecosystem and encourages wide range of biodiversity, which includes animals and humans.

6. Future Thrust Area of Research

It is evident that sustained production in agriculture can be achieved by changing the present systems of farming. It is apparent that ancient techniques have much relevance. Field experiences with large number of farmers in India and other regions are indicators of its importance. For acceptance by large number of farmers, policy makers, systematic research efforts are needed so that mystery of exciting results could be explained on the basis of science.

REFERENCES

Abhang P. and Pathade G. 2017. *Agnihotra* technology in the perspective of modern science: A review. Indian Journal of Traditional Knowledge, 16(3): 454-462.

Anon. 2014. Proceedings of National Academy of Sciences (HT, 6th, November, 2014).

Berk U. 2014. Suggested Experiments with *Agnihotra* and *Homa* Therapy: What has been done and what can be done. Ulrich Berk-, German Association of Homa Therapy Ed.). www.homatherapie.de; dght@homatherapie.de.

Bruce J and Heschel K. 2009. *Homa* Organic Farming. In: Proceeding, Brainstorming Conference on Bringing *Homa* Organic Farming into the mainstream of Indian Agriculture, organized by Five Fold Path Mission. pp. 84-94.

Claude A. 2010. Interview with Kabra, S.G. Pesticides are damaging people, pp. 52-53.

Das Gupta, Khan, S M and C S Nautiyal. 2006. Biological control ability of plant growth promoting *Paenibillus lentimorbus* NRRL B-30488 isolated from milk. Current Microbiology, 55: 902-909.

Frits and Lee R. 2009. Homa Organic Farming in Australia: Effects of *Agnihotra* and Homa Therapy on Water Resources. In: Proceeding. Brainstorming Conference on Bringing Homa Organic Farming into the mainstream of Indian Agriculture, organized by Five Fold Path Mission. pp. 39-42

Garcia I. 2009. Homa Organic Farming and Bananas. In: Proceeding, Brainstorming Conference on Bringing Homa Organic Farming into the mainstream of Indian Agriculture, Five Fold Path Mission. pp. 28-32.

Garg N. 2012. Cow dung as source of bio agents. National Conference on Managing Threatening Diseases of Horticultural, Medicinal, Aromatic and Field Crops in Relation to Changing Climatic Situations and Zonal Meeting, Indian Phytopathological Society, Nov. 3-5, 2012, pp. 85-87.

Jarek B. 1999. Homa farming for the new age, a practical guide to Homa Farming based on ancient science of *Agnihotra*, Fundacja *Agnihotra*, Nad Lasem/Wysoka 151, Jordanow, Poland, pp. 34-785.

Joshi RR. 2009. The Integrated Science of Yagna. Pub. Yug Nirman Yojana Vistar Trust, Gayatri Tapobhumi, Mathura-281003. pp. 1-40.

Khan N, Mishra, A and Nautiyal CS. 2012. *Paenibacillus lentimorbus* B-30488 controls early blight disease in tomato by inducing host resistance associated gene expression and inhibiting *Alternaria solani*. Biological Control, 62: 65-72.

Kishi M and Ladou J. 2001. International pesticide use. Int J Occup Environ Health 7: 259-65.

Masuru M. 2010. Purification of water with *Agnihotra* ash. www.agnihotra.com.au, www.rh4.info.

Nandish BN. 2010. Organic culture. In: The Organic Farming (Claude Ed). Other India Press, Goa, India. 229 p.

Nandish BN. Nandish and legume culture pp: 229 in Claude A. Howard, 2002. An Agricultural Testament by Albert Howard-Other India Press (OIB) Goa-India.

Narang I. 2009. The science of *Agnihotra*, Maharshi Dayanand Charitable Trust, Delhi.

Nautiyal CS, Govindarajan R, Lavania M and Pushpangadan P. 2008. Novel mechanism of moulding natural antioxidants in functional foods: Involvement of plant growth promoting rhizobacteria NRRL B-30488. Journal of Agriculture and Food Chemistry, 56: 4474-4481.

Nautiyal CS, Mehta S and Singh, NB. 2006. Biological control and plant growth promoting *Bacillus* strains from milk.

Nautiyal CS, Mehta S and Singh, HB. 2008. Biological Control and plant growth – promoting Bacillus strains from Milk. Journal of Microbiology and Biotechnology, 16: 184-192.

Nene YL. 2005. Cow and Agriculture, Paper presented in National Seminar on Cow in Agriculture and human health. Agri History Society, Udiaipur, Rajasthan, p 55.

Paranjpe V. 1989. Homa Therapy Our Last Chance, Five Fold Path Mission, Inc, Madison, U.S.A.

Paranjpe V. 1989. Purified with Fire. In: Secrets of the Soil (Tompkins-Bird Ed.), Rupa and Co. New Delhi. pp. 243-254.

Paranjpe V. 2005. Why Homa farming. www.terapiahoma.com

Pathak RK and Ram RA. 2013. Bio-enhancers: A potential tool to improve soil fertility, plant health in organic production of horticultural crops. Progressive Horticulture, 45(2): 23-25.

Pathak RK. 2013. Bio Enhancer: A Potent and Affordable Source in Organic Production of Vegetables in Protected Conditions. National Seminar on Protected Cultivation of Horticultural Crops and Value Addition. Allahabad School of Agriculture, Sam Higginbottom Institute of Agriculture and Technical Sciences. pp. 62-68

Pathak RK. 2013. Homa Jaivik Krishi: A Ray of Hope for Sustainable Production of Fruit Crops. In Proceedings, National Seminar on Tropical and Subtropical Fruits, Aspee College of Horticulture and Forestry, Nausari Agricultural University. pp. 192-2001.

Selvaraj S, Anita NB, Ranchana P and Xavier K. 2009. Studies on effect of *Agnihotra* on the yield and quality of horticulture crops. In: Proceeding, Brainstorming Conference on Bringing Homa Organic Farming into the mainstream of Indian Agriculture, organized by Five Fold Path Mission. pp 56-62.

Sharma Punam, Kumari SK, Rameshwar R and Atul. 2011. Effect of different timings and pyramid on properties of *Agnihotra* ash. In: National Symposium cum Brainstorming Workshop on Organic Agriculture, 19-20 April, 2011, CSH Himachal Pradesh Krishi Vishvadyalaya, Palampur, H.P. 149 p.

Weir ML. 2009. Preparation and application of Gloria Biosol. In: Brainstorming conference on Bringing Homa Organic Farming into the mainstream of Indian Agriculture, Five Fold Path Mission in collaboration with Planning Commission, GOI. pp. 95-96.

Chapter 28

Perpetual Yogic Agriculture Technology

Sunita T. Pandey and Kewalanand

Department of Agronomy, College of Agriculture,
G.B. Pant University of Agriculture and Technology,
Pantnagar – 263 145, Uttarakhand
E-mail: sunitatewari_8@yahoo.co.in

1. INTRODUCTION

Agriculture is practiced through various methods applied in the world. Science has made easy availability of food to feed the people. However, the produce lacks in good taste and nutritional value due to impurities and residual effect of chemicals used in farming. In the absence of purity and vegetarian nature of food, our mental and physical health is becoming poorer by each passing day. In the bygone eras, people enjoyed healthy body and long life due to pure and uncontaminated food. To rediscover these processes and virtues again so that our mother earth has blissful state, peace, prosperity, wealth and pure food-grains in abundance, we have to adopt **Shaswat Yogic Agriculture** which is the call of time to wards that dream. For this we have to shift from chemical-intensive farming to bio-intensive farming which includes rhizospheric engineering, carbon sequestration, residue recycling, identification of efficient strains of bio-agent and plants, their application schedule, bio-remediators, in-situ organic farm waste recycling and natural farming *etc.* A lot of work is being done in these areas worldwide, but one new virgin/unexplored dimension may be explored as another area of research in crop production *i.e.* **use of thought power in organic agriculture**. Yogic agriculture means doing various agriculture operations in a yoga (connection/union) state or personally empowered state of mind. It is achieved through Raja yoga meditation technique of personal empowerment which is an ancient technique of India. It is being revived by the Raja Yoga Education and Research Foundation, a sister organization of the Brahma Kumaris World Spiritual University, united nation recognized non-governmental organization headquartered at Mount Abu, Rajasthan. The organization has been teaching methods of personal empowerment based on Brahma Kumaris Raja Yoga meditation (BKRYM) techniques for the last 80 years.

The acceptance and recognition of thought power is increasing worldwide. One of the important reason in increasing this is the awareness of non monitory input for safe farming and food as well as powerful tool to improve health of all lives. The positive thoughts have been found to change water structure when crystallized, seed germination, growth and development of crop plants as well as the health of people. It is a powerful tool to create above changes even from distance. The hidden power of thought is being practiced by all but mostly people do not know this power and use it in improper manner. This is high time when we all need to practice generating the powerful thoughts for well being of this planet. One way is to practice Brahma Kumaris Rajyoga Meditation (BKRYM). In this powerful thoughts are given in pure state of mind to change the present day agriculture and its dependents.

2. Concept

Researchers of quantum physics suggest that we are electromagnetic beings and our thoughts radiate out from us like a broadcast signal inspecting thing inside and around us. Although, we would like to believe that our thoughts are private, research has shown that thought does affect matter outside of our selves. This gives a whole new meaning to watch what you think. Swami Vivekanand in his methodology of Rajyoga explained that "if there is any power which can change the world, that is power of thought vibration energy" Mahatma Gandhi said "man is the product of his thoughts. What he thinks, he becomes" based on such theories Shaswat Yogic Agriculture Technology (SYAT) is defined as "Shaswat Yogic Agriculture Technology is the application of pure, peaceful and powerful thought vibration energy (*created during BK Rajyoga meditation*) to production system, which increases quality and productivity, along with benefiting environment, society and all forms of life."

SYAT component are:

- ✰ Shaswat – Ever lasting
- ✰ Yoga – Union of metaphysical energy (Soul) and Supreme metaphysical energy (supreme Soul).
- ✰ Yogic – Process that simulates *the union* and is as pure, peaceful and powerful.
- ✰ Agriculture - Food production in sustainable manner.
- ✰ Technology - Knowledge to create positive thought vibration energy and its application

The agricultural practices, as mentioned in Rigveda, Krishiprasara, Manusmriti, Agni Purana and Vriksha Ayurveda, are based on the spirituality which is the time tested and time honoured. These Agriculture modes have been devised to pursue the exclusive belief of "harmony" with nature. Having harmony with nature entails its creatures to comprise reverence and gratefulness for living and non-living objects, and maintain synchronization with natural forces or five elements *viz.*, earth, water, air, fire and sky. This production system is applicable in organic mode. Another component of SYA is The Yogic production system; a product of the positive thoughts vibrated during Raj yoga meditation which clearly depicts the use of natural ingredients on farm organic inputs and also natural/biological/mechanical pest control and plant protection measures with minimal use of permissible natural minerals. Thus, the Yogic agriculture results in Healthy Soils, Healthy Food, Healthy Environment and Healthy Human being and Livestock.

3. Energy in Universe

There are two main sources of energy on this planet Earth: The **electromagnetic energy** received from sun. This energy is consumed by green plants for manufacturing the **fixed energy** in the form of energy rich organic molecules. This fixed energy is consumed by herbivores and carnivores. Upon the death of living cells (Detritus), it is consumed by the detritivores which recycle the organic form of nutrients into inorganic form and deposit in soil and further utilized by plants. This cycle is a continuous process responsible for the viability of the ecosystem and thereby improved natural resources and production system. But due to the greedful activities of man this system is disturbing day by day leading to the degradation of natural resources. To overcome this problem we have to adopt the utilization of basic energy of the Universe *i.e.* the most powerful but unutilized energy known as Metaphysical or Subtle energy. This energy is based on our spirituality. Spiritual growth is our only purpose and reason for being alive on earth. Each individual must learn how to utilize this energy for spiritual growth and constructive purposes. Constructive use (positive use) of this energy raises the level of consciousness of man or raises his vibration rate or frequency. Every individual has a different rate of vibration. All of man's earthly problems are created by his thought projections. What we project from our mind in the form of thoughts, we create and receive. Spiritual growth requires "overcoming all negative thinking" which dissipates "the life force" or vital energy. Everything in the Universe is energy and vibration. Each basic element of the known atomic chart consists of energy at different rates of vibration. There is a frequency or vibration of energy that saturates the Universe. This energy is beneficial to all living cells whether human, plant or animal. Man has the power to utilize this energy "with his mind." Every thought is transmitted by this basic energy of the Universe. This energy might be considered the basic power of the Universe or God. The power behind the thought is "energy" or calls it God Power or Creative Power. The Breath of Life, the Holy Spirit is this energy. Science has proven that by projecting "love", or positive energy, to a plant, the plant will flourish and grow vigorously. If a child is injured, its mother will project love (positive energy) to the child by laying her hands on the painful area. She automatically releases "energy", and the pain is relieved. This metaphysical energy is created during BK Rajyoga meditation which is the base of Yogic Agriculture.

4. The Process

With the Organic/Natural agriculture system the farmer uses the metaphysical energy as an additive and catalyst *i.e.* energy of thought Vibrations created during BK Rajyoga meditation to directly enliven the holistic and specific Laws of Nature responsible for the growth of the lives. SYAT is a sustainable technique in agriculture where the farmers employ the thought vibration energy created during BKRYM as one of the key ingredient in every stage of farming. The Rajyoga Meditation can be practised by farmers as they go through their daily chores in the field. This when combined with organic farming results in significantly lower cost of inputs with improved production of better nutrition content as established by the study across various locations in India. This Technology establishes an emotional relationship with nature as a whole. Plants do react to the positive emotions which is scientifically established by different scientists across the world.

4.1 BK Rajyoga Meditation

Brahma Kumaries World Spiritual University with its headquarters at Mount Abu, Rajasthan is reviving/rejuvenating the ancient Rajyoga meditation technique of Bharat, which is an internal, in-cognitive experience that spreads metaphysical or subtle energy through positive thought vibrations. Anonymous (2009). This subtle energy is very powerful. Rajyoga meditation already has been proven throughout the world to be an eminent and widespread practice to develop inner spirituality. Now the call of time is to understand this spiritual science for future sustainability. Rajyoga meditation creates its own natural cycle that brings internal harmony between the mind, intellect, personal traits (resolves) and with the five elements of nature. In practicing BK Rajyoga meditation, persons experience receiving the bliss and powers of Supreme Energy. They become karma yogis and experience their own potential. They bestow more to Mother Earth and society rather expects more.

The Rajyogi should know his original form and remain in that state, connect his intellect with Supreme Energy - the ocean of Knowledge/Power. They are enriched with pure and powerful thoughts during union with supreme power and when they use this pure and powerful thought vibration energy on any living object, it makes the changes in object as per Rajyogi's thoughts power. The pure and the best thoughts will create all successes. Things considered impossible will also become possible. This is not any magic or clairvoyance but the science of spirituality. Such a performance is possible only through the best, the purest, and the most powerful vibrations of rajyogi's emotions. A Rajyogi must avoid the two extremes of luxury and austerity. He must not fast, nor torture his flesh. one who does so, cannot be a Yogi. One who fasts, keeps awake, sleeps much, works too much, does no work, cannot be a Yogi (*Shrimad Bhagwat Gita, VI, 16*).

It is now a firmly established fact that the state of human mind directly affects the state of health of an individual. When a person maintains continuous remembrance of the divine in his mind, his mind remains peaceful. He experiences divine happiness. He lives a life full of self-control and becomes the embodiment of Perfect Being with command over all the senses. As the remembrance of the Supreme energy was incessant in their mind, their mind, intellect and personality became perfectly pure and the Life Force Energy remains maintained at its peak status throughout their existence. The aura of their pure vibrations continued to reach newer and newer horizons. In this way, Rajyoga is a holistic science which is the source of subtle powers of mind. The power of Rajyoga can do such a great activity which is better and more advanced than science.

4.1.1 Steps

The first step of BK Rajyoga Meditation is to attain the soul-conscious stage. There is no need to shut eyes because Rajyoga inspires us to perform karma yoga. One cannot perform his duties (Karma) with closed eyes. Rajyogi maintains his soul-conscious stage at all times. Following are the three main subtle powers (Energies), which when combine, makes the power known as the I the energy (meta physical energy).

- ✰ Mind- The Thinking power of the soul creating thoughts
- ✰ Intellect- The decision making power of the soul for utilizing/filtering thoughts
- ✰ Sanskar-The impression of what we do according to thought filter trough intellect and which remains in our memory

With the help of these powers, we can visualize the existence of the supreme energy during BK rajyoga meditation and can also make others feel him or his presence. We, as energy, possess the original 07 qualities of being knowledgeful, pure, peaceful, loveful, blissful, joyful, and powerful and are adorned with all divine virtues. The Supreme Energy is the ocean or the source of all these divine virtues and powers. Visualize ourselves as meta physical energy (a point of light) when we union with the Supreme meta physical energy (a point of light), all his divine qualities and powers get into the I (Soul as meta physical energy) and the I (Soul as meta physical energy) feels itself recharged with all those 07 qualities and powers. One needs to remain in this state with full concentration power. At this stage one recharge with the 07 qualities whatever thoughts we create, we can achieve the same.

Soul consciousness is Initially, start with one minute then gradually increase it two, three, four, five and so on. If you practice it with concentration, the thought vibrations start spreading in the atmosphere and everybody, including Mother Nature expresses its impact. Because of our body-consciousness, all creatures, birds, beasts, animals, trees-vegetation and also nature are expressing ill-effects of our evil thoughts. In BK Rajyoga mediation, the soul experiences ultimate bliss and ecstasy. When the soul is in that state, the vibrations that emanate from the soul changes the atmosphere. If one has good thoughts during the meditation, the whole agro ecosystem as well as atmosphere, too, changes accordingly.

During Rajyoga, we should considers our self as I (Soul the meta physical energy), become energy (Soul) conscious and the influence of energy-consciousness on the thoughts is positive. Thus our thought vibration energy (subtle energy) starts affecting in a positive way to all lives to which we visualize. To achieve this goal, one can practice BK Rajyoga meditation in the following manner:

4.1.2 Practice of Empowerment through BK Rajyoga Meditation

- ☆ Initially believe yourself as a soul (metaphysical energy). Feel your original form I, the soul, am full with Divine powered point of light, that is knowledgeful, pure, peaceful, loveful, joyful, blissful and powerful.
- ☆ Now remember Supreme energy (Supreme Soul) in Incorporeal World and visualize that, the rays of knowledge are absorbing within me.
- ☆ Supreme energy is the ocean of purity. I, the soul, am becoming completely pure.
- ☆ The rays of peace are coming to me from the ocean of peace.I, the soul, am going to be very calm and cool and embodiment of peace.
- ☆ The rays of love are falling on me from the ocean of love and I, the energy, am going to be filled with love or have become embodiment of love.
- ☆ The rays of bliss are coming to me from the ocean of bliss and making me feel super-sensuous bliss I, the energy, am becoming blissful.
- ☆ Supreme energy is ocean of joy.the rays of joy are coming to me and I, the energy, am filled with joy.
- ☆ Supreme energy (Supreme Soul) rays of Godly powers are coming to me and I, the energy (Soul), am becoming full with all Godly powers.
- ☆ I, the soul, am recharged with all about 07 original properties of Soul and become powerful Soul, embodiment of Supreme Energy.

In Rajyoga meditation, practice more and more channelization of above mention thoughts is required. Initially practice it for one minute then increase as much as one can. To practice Rajyoga, the time of early morning from4-00 to 5-00 a.m. is best. At this time, the atmosphere of surrounding is very cool and mind is fresh. Fix any cool and clean place of home for practicing meditation so that one will get success easily. When we stabilize ourselves in the form of meta physical energy and remember Supreme energy, visualize the above mentioned scene in front of us through intellect and see the experimented farming crop through the third eye of intellect. Do meditation on it. Through this practice, one will see wonderful changes in the crop. The yog energy works as an additive to organic farming system through recharging energy in all lives associated with farming.

5. Practical Applications

5.1 Meditation on Natural Resources

Visualize that:

- ☆ I, the meta physical energy, am the owner of nature
- ☆ Supreme meta physical energy (a point of light) is looking at me and saying that children you are not ordinary, you are the richest and recharging my 07 properties
- ☆ At this stage I the Soul visualize the object to be strengthen and give positive thoughts
- ☆ Do this practice five times throughout the day, sitting in the farm or from the distance.

5.2 Giving thought Vibration to the Crop in the Morning from 4:00 to 4:45 A.M. (Amritvela)

Initially, feel the self (Soul) as if it were complete with all self respect.experience original form of the soul (point of light). then in angelic form (*farishta*), experience yourself as if you were in front of Divine light, the Supreme Energy in subtle world removing all weaknesses, feel yourself perfect, leaving subtle body a side, be bodiless, feel yourself connected with supreme energy getting recharge with 07 original properties of Soul and give *positive thought vibration energy* to the five elements of nature and feel all the elements, *viz.* the earth, water, air, fire, sky, moon-stars, planets are becoming pure; all microbes, insects, viruses, animals, birds and beasts *etc.* are feeling Godly treasures, emerge the self as seating on the globe being a complete angel. Being a complete and pure angel along with supreme energy, I am reaching in the farm. I am chatting with supreme energy about experimented agriculture, looking at the crops and visualize that supreme energy is giving powers happily and the crop is becoming in its purest form. All the microbes of land, water and air also becoming complete with treasures of supreme energy. They are helping in making the crop/nature healthy for cognization. The rays of purity and powers, peace, strength are spreading on the crop and in that area crop/nature is growing in good state and the grain is also becoming powerful. sometime visualize your experimented farm in front of supreme energy who is giving *subtle energy* to the crop and giving such powers to the crop that any type of harmful insect or disease cannot enter into the crop. Do this experiment everyday at one time at one place by sitting either in the farm or in visualizing it from home.

5.3 Practicing Karmyoga Throughout the Day in Farming

- ☆ While working in the farm, emerge the Farm on the globe of world.the crop is receiving all the powers of supreme energy through me.I, the perfect angel, am working in the farm. Experiment the power of silence specially wherever you are in whatever condition and whatever you have with whomsoever are you.
- ☆ Remaining in this position, concentrate on the action which you are performing.
- ☆ While working in the farm there must be flow of positive and elevated thoughts. If one pay attention on his/her thoughts automatically positive thoughts will ponder to make the stability of elevated thought.
- ☆ To nourish and protect the world is human's duty. How positive and negative thoughts affect deeply to the vegetations is seen and experienced practically in organic farming.
- ☆ Vicious thoughts or emotions reduce the power of vegetation. Not only is this but vegetation also affected by our positive or negative thoughts which inversely affect us through the food.
- ☆ Thus, it prepares a cycle of more joy, salvations, happiness, and peace in the life of human beings.
- ☆ Before sowing any seed, give thought vibration energy (Subtle energy) to that seed, supreme energy powers are entering to the seed and cover the seed with the power of purity.
- ☆ Similarly before broadcasting any type of organic manure into the soil, remember supreme energy of the corporeal world or one (Meta physical energy) go to subtle world and make it powerful and cast it into the land.
- ☆ If you want to sprinkle any resistance, medicine or tonic, first you experiment this method then spray it. While spraying feel that I am spraying along with supreme energy directly.
- ☆ There are so many such methods that one can apply as per choice. Whatever experiment one is going to apply, it is very essential to have faith, trust and determination.

5.4 Mechanism of Action

Thought vibration energy is non-specific in nature and mainly acts in form of Metaphysical energy (Subtle energy) for making seed, soil flora and fauna, water and associated metabolic processes powerful. Thus it can be clubbed with any form of ecological or organic practices. Thought vibration energy affects water and water retains this change. In echo system, thought vibration energy is absorbed by water present in life forms leading to growth and development as per thought created. Since plants as well as our body have water in plenty, it is important to make sure that our thoughts should be positive for healthy nature. During B K Rajyoga meditation, pure and powerful thought vibrations (Subtle energy) are created to heal the object. The living forms absorb this energy through water and respond according to the thoughts created. In all life forms in planet earth, water is a key element for life. The thought vibration energy acts on water to form better crystals leading to better ecosystem, germination, growth and development of Plants. It has been well demonstrated in Psychology that negative thoughts adversely affect health of human beings. Perhaps the effect on water in our bodies is one way this happens.

6. Effect of Thought Vibration Energy and Other Lives

It has been demonstrated by Lebbrecht (1999) that water has an observable sensitivity to external stimuli such as thoughts, prayers. He photographed normal water crystals as well as healed water crystals. The temperature and humidity also affects the formation of crystals.

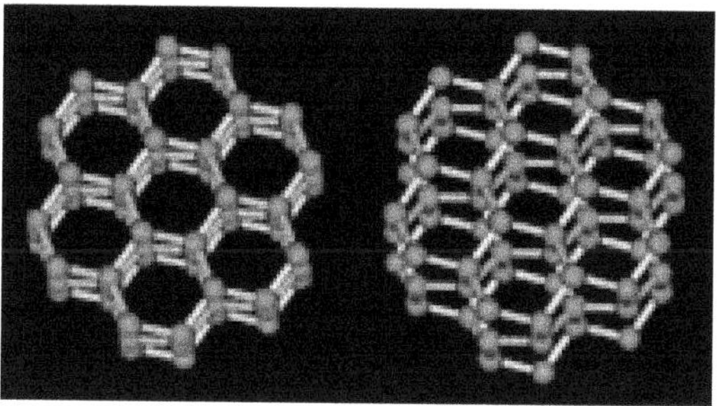

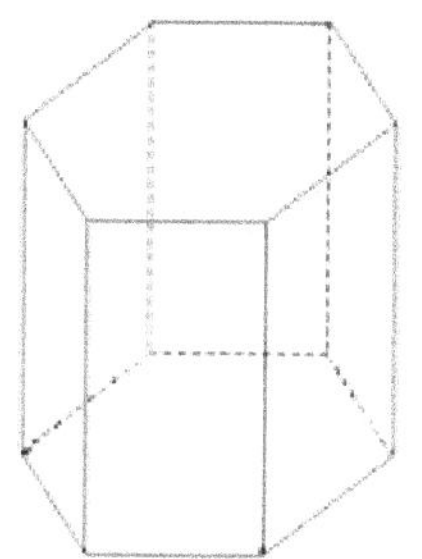

Normal Water Crystals.

Human thoughts affect Ice crystal formation and the structure of crystal is an indicator of water quality. Beautiful and intricate crystals formation indicates high quality water and difficulty in forming crystals indicates low quality water (Emoto Masaru, 2004).

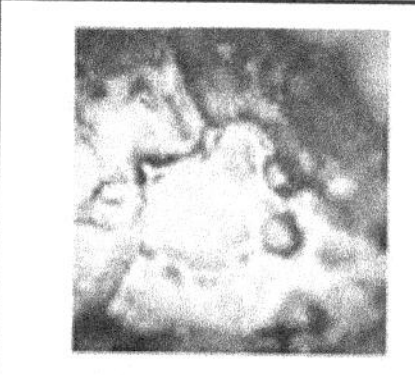

Water Molecule, Before Offering a Prayer

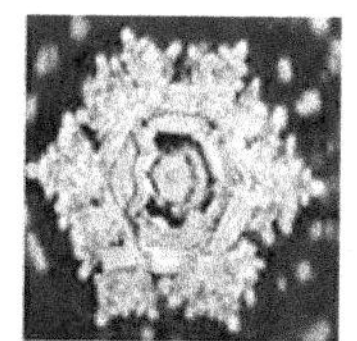

Water Molecule, After Offering a Prayer

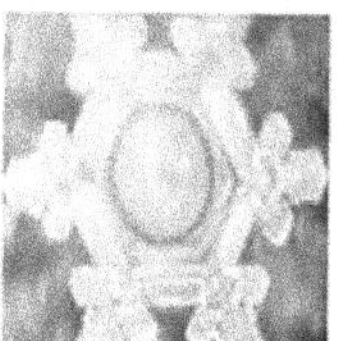

Thank You

Love and Appreciation

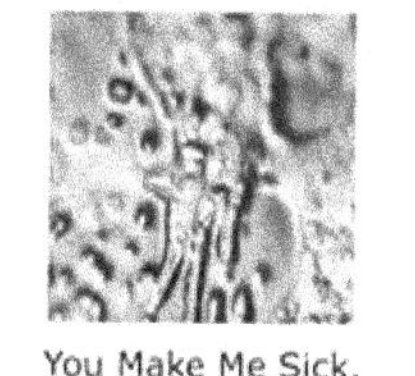

You Make Me Sick, I Will Kill You

Disharmonized thoughts

The well known researcher Sir Jagdishchandra Bose, proved, through his experience that trees and plants too have the same activities as human beings have. He prepared one machine that records the internal movement of vegetation by which he proved that animals, beasts, vegetation also baffles at the death moment. At that time the energy from its inside comes out forcefully. A very exciting body of scientific research has demonstrated the effect of thoughts on water. Water stores thought energy and this energy impacts whoever and whatever use it (Emoto Masaru, 2004). The positive thoughts subjected to water have been found to form beautiful crystals and water exposed to negative thoughts form either no crystal or deformed ones. The research out lined that thought, words and feelings affect the molecular structure of water and thus thought also have an effect on human body and plants as both contain water in plenty. A Canadian Biologist Grad (1963), of MAC Gill University of Montreal, wanted to see if people can actually transmit healing energy to others. To eliminate placebo effects he conducted an experiment on plant growth under controlled conditions. He took two containers having same concentration of salt; one treated by a person known of 'psychic' healing laid his hands on the salt water container and other untreated salt water. The seed soaked in the water treated by intense thoughts grew taller than the untreated batch. The results demonstrated that whatever the 'healer' did to the water was reflected in the health of the plants even though the salt is said to retard germination and growth of plants. This was the effect of positive thoughts. It has been worked out that the opposite would occur with negative thoughts. Taking this into consideration in other study Dr Grad took best source of negative thoughts as disturbed psychic patients. The patients in his experimental group were diagnosed with more psychotic depression and less depressed one. Each patient was given container of plain water and marked them to know which patient held which container and untreated water was used as control. When the water of these containers was used to sprout seeds, all the water samples held by psychiatric patients produced poor growth but the water held by the extremely depressed person suppressed growth since the patients were doing nothing more than being themselves and the water was affected and showed its effect on plant growth. Grad using infrared spectroscopy discovered that water treated by the thought power of healer had minor shifts in its molecular structure and decreased hydrogen bonding between the molecules. This is the same thing observed when water was exposed to magnetic fields. Thus thought we create do affect the water. This fits in nicely with the beautiful crystals due to positive thoughts (Masaru Emoto, 2004). Research demonstrated that thought affects water and water retains this change. The water in turn, affects who so ever uses it. Since our body has about 70 per cent water, it is likely that our negative thoughts adversely affect health, or mind works over matter or energy works over mass. The important lesson to learn from this research is that *thoughts are powerful and affects us. It is important to make sure our thoughts are positive for being healthy.* The intentions have been found to have direct effect on living biological systems and that a seed germination bioassay has the sensitivity to enable detection of effects caused by various applied energetic conditions. This effect was demonstrated by Du Charme, Laurene. J. (2007) by studying the effect of intentional thought in close proximity or at a distance: demonstrating the relation between mind and matter on seed germination. To measure biologic effects of positive and negative intent as a bridge between mind and matter with proximity as a factor, using seed germination as an objective biomarker was evaluated by conducting two experiments, one in close proximity, 10 inches from target and repeated at a distance of 5 miles, utilizing zucchini seeds. The experiment yielded mixed results. The weight of increase of seeds receiving positive intent was significantly greater than the weight increase of seeds receiving negative intent ($p<.001$).However, there was no significant difference in stem length between the positive and negative groups. Stem length increase was unexpectedly greater in the groups of seeds that received intention from a distance when compared groups of seeds that received intention at close proximity ($p<.001$). Intentions both positive and negative together had a statistically significant effect on weight, when compared to control ($p<.001$).

It has been proved that that humans are interconnected with living organisms in significant ways. Backster demonstrated that plants and other living organisms respond to changes in human and gave theory of "primary perception. A polygraph instrument attached to a plant leaf registered a change in electrical resistance when the plant was harmed or even threatened with harm. He argued that plants perceived human intentions. He also reported a finding that human thoughts and emotions caused reactions in plants that could be recorded by a polygraph instrument. When we create positive and negative thoughts, each have a different impact on our surrounding environment. While this idea is not new, not too many have taken the time to scientifically go about proving such ideas -although the field of quantum physics has shed some powerful light on the topic over the years. The Princeton Engineering Anomalies Research (PEAR) Laboratory has also been trying to wrap their heads around the subject and have concluded that the mind does in fact have a subtle capacity to influence the output of devices, Dr. Konstantin Korotkov.

6.1 Effect on Seeds

Classic healing energy experiment was conducted by Barrington, *et al.* (2003). In this experiment, two trays of lettuce seeds were prepared after they had been stressed in some way. These trays were watered using energized water by the healer and the other with ordinary water. Several of these studies indicated that there was greater germination rate and growth of seedlings in the groups that were watered with energized water that had been treated with positive thoughts of the healer. Many healing energy studies have been performed with seeds (Roney-Dougal and Solfvin, 2003) The largest body of work (Grad, 1963) involved studying recovery from injury in barley seeds caused by stress induced by a 1 per cent saline solution comparing bottles of saline solution treated for 15 minutes by a healer to an untreated control. Grad's procedures were carefully randomized and double-blinded with significant differences in four of six experiments. The outcome of this study was a larger number of seeds germinated when watered with saline treated by thought vibration energy held by the healer. These seedlings were taller with a greater total yield per pot compared to the control. Studies have been done on the effects of qi (healing process) on the germination rate of rice seeds (Yiji 1991). Two qi gong masters separately emitted qi from their Laogong point (a point on the palm of the hand) for 30 minutes to different batches of dry rice seeds held in the palm of their hands. "Each procedure was repeated for three different batches of seeds over five periods during the day. The treated seeds were germinated on a wet paper surface in the dark. The percentage of seeds that germinated were generally greater for the qi-treated seeds than that of the control. This was because of the vibration of positive thoughts created by healer with positive power.

Shaswat Yogic Agriculture technology has been applied to seeds through a bonafied research, which has been conducted by Department of Agronomy, College of Agriculture, G.B. Pant University of Agriculture and Technology, Pantnagar, Uttarakhand, to assess the effect of metaphysical energy on seed germination, seedling vigour and dehydrogenase enzymes during storage. Seeds of wheat variety UP-2565 and chickpea variety PG 114 were obtained from the Breeder Seed Production Centre, Pantnagar and treated with Meta physical energy through BK-Raj Yog Meditation (BK-RYM) for 2 hour and 4 hours in wheat and in chickpea. Thereafter, invigorated seeds were stored in poly-lined cloth bag at 10 per cent moisture content (for wheat) and 8 per cent moisture content (for chickpea) under ambient condition from June to October, 2012 up to next planting season. Results revealed that all the seedling vigour parameters decreased as the storage period advanced but in both the crops, metaphysical energy treated seeds showed higher seedling vigour and enzyme activity. Over a period of 6 months of storage, Alpha (α) amylase activity was 2.4 times higher in BK-RYM for 4 hours treated wheat seeds over control seeds. Whereas, in case of chickpea protease activity was 2.0 times higher in BK-RYM treated seeds for 2 hours duration, than that of untreated control seeds. This increase in enzyme activity resulted in enhanced seedling growth parameters and seedling vigour in wheat as well as in chickpea seeds (Omvati and Sunita, 2014).

6.2 Effect on Crops

Research conducted at Karnal (Hisar) at farmers' field indicated that yogic + organic (FYM) treatment enhanced growth and yield than alone organic (FYM) and control treatments and remained at par with chemical fertilizer treatment. This (yogic + organic) treatment also led to boldest grains of wheat than all other treatments. The grain protein content of yogic + organic and Organic treatments was also significantly higher than chemical fertilizers and control treatment (Sunita *et al.*, 2015).

At SDUA&T, Dantiwada, Gujarat the work on groundnut indicated that the pod and haulm yield remained similar in both OFM I (organic farming) and OFM II (organic farming + yogic farming) up to second year, however, in third year the yields increased significantly under OFM II compared to OFM I. Under CIM (chemical farming), both the yields remained significantly more compared to organic module. The data for oil content and yield are available for I and II years only. The results indicated that oil content increased significantly due to OFM II in second year compared to Ist year and oil yield remained similar in both the organic modules. Oil content due to OFM II was similar to CIM in Ist year but significantly more due to OFM II compared to CIM in IInd year. Oil yield was significantly more due to CIM compared to OFM during both the years due to more pod yield. The protein content remained similar in all the modules least being in OFM I in second year of experiment

The observations on beneficial soil microbial population buildup under groundnut-wheat rotation before sowing, at flowering and after harvest of crops at the end of rotation indicated that during *Kharif* (ground nut crop) and *Rabi* (Wheat crop), the highest microbial population (*Rhizobium, Azotobactor, Azospirillum*) and Phosphate Solubilizing Bacteria (PSB) was under OFM II and it was maximum during second year. Chemical input module (CIM) led to least population of beneficial microbial population at all the stages compared to organic modules.This observation suggests that organic farming blended with thought vibration energy (TVE) not only enhances germination, growth and yield of the crops but also leads to better soil health in terms of high soil microbial buildup aiming TVE application to soil microbes. Backester (2003) observed humans thoughts are interconnected with living organisms in significant ways.

Treating the soil culture with high frequency sonic vibrations for 4 minutes showed maximum number of bacteria, actinomycetes and fungi (Stevenson, 1958). The thought energy also has very high frequency of vibration which might have led to enhancement in soil microbial population in this study.

The results obtained from field trials indicate that in future both soil health as well as safe and enhanced food production will be obtained only with organic farming blended with BK Raj yoga meditation energy. In addition, this will build up pure, peaceful and disciplined farming population for transforming agriculture and people.

6.3 Effect of Thought Vibration Energy from Distance

It has been proved that the thoughts send to a distant object creates changes on the object. This type of healing has been practiced on farms and the intended crops grew faster with least diseases. Thus distant healing with thought power can be a strong non monitory input in improving production potential of crops. Under similar growth medium and growing conditions Benor (2001) observed significant changes in germination and growth of maize seeds/seedlings. He conducted an experiment filled three pots of the same size, with soil from the same source and using three batches of maize seeds from the same packet. To test this from distance, the environmental conditions were kept same for all the three pots. He sent positive thoughts to the first, left the middle one alone, and send negative thoughts to the third from distant place. After two weeks visible differences in the growth rates in each pot was observed. This way one tests one's ability to heal from distance. The power of positive thoughts/intentions positively affect the plant growth and development (Alberto, 1977; Mac Donald *et al.*, 1976 and 77; Nicolas, 1977; Russell, 1973 and Saklani, 1989). Under organic farming these effect of distant thought treatments has been found more pronounced. Roney-Dougal and Solfvin (2003) conducted a field trial with lettuce plants on a commercial organic farm. This was a replication of a previous study conducted in 2001.In both experiments, a healer was asked to enhance a group of healthy organic lettuce seeds for greater germination, growth, and better health. Significant results were found in the better health of the treated lettuce seeds. The growth hypothesis was confirmed by significant enhancement in gross weight and net weight for treated seeds, and the health hypothesis was confirmed by the significant reductions in slug and fungal damage in the lettuce that was enhanced by the healer. Overall, the thought vibration based seeds yielded about 10 per cent more during the season than any of the other three groups, suggesting a practical value for the commercial farmers.

7. Awareness Programme

Agriculture and Rural wing of Rajyog Education and Research Foundation is reviving the ancient knowledge of Bharat (Shaswat Yogic Agriculture), with an aim to play a flagship role in the sustainable and ecofriendly agricultural development of the country, "Sustainable Yogic Agriculture Project". Hundreds of farmers in various states of the country have taken benefit by adopting this farming methodology, which is a mix of organic farming, technology that helps in agricultural sustainability and exposure of positive vibrations emanated through BK Rajyoga meditation. Awareness programs have been conducted through seminars at Indian Agricultural Research Institute (IARI), New Delhi; Indian Council of Agricultural Research Complex for Eastern Region (ICAR-RCER), Patna (Bihar); Water and Land Management Institute, Aurangabad (Maharastra); G B Pant University of Agriculture and Technology, Pantnagar (Uttarakhand); Dr. Balasaheb Sawant Konkan Krishi Vidyapeeth, Dapoli (Maharastra); Indira Gandhi Agricultural University, Raipur (Chattisgarh); Charan Singh Agricultural University, Hisar (Haryana); Tamilnadu Agricultural University, Coimbatore (Tamilnadu); Dr. Punjabrao Deshmukh Krishi Vidyapeeth, Akola (Maharastra); and Acharya N G Ranga Agricultural University, Hyderabad (Andhra Pradesh), *etc.*

8. Conclusion

The power of thoughts has been found to play a vital role in transforming agriculture as well as human health. The ill effect of modern farming will be minimized and safe and secured food production will be achieved and the people/ farming population will be more healthy and wealthy. The power of BK rajyoga meditation can make it success as it is a non monitory pure and powerful input for farming system. This kind of technology, even if not very practical in current scenario because of lacking trained healers and mental make up of policy makers/authorities, must be promoted for future or new era likely to come very soon. So research may be planned and conducted in a networking mode in various Universities in coordination with Prajapita Brahma Kumaris Godly University, Mount Abu. This will not only transform agriculture but also facilitate transforming farming families and other human being.

9. Acknowledgements

Authors deeply acknowledge the inspiration of chief and administrative head of PPBKIVV Dadi Janki, valuable advices, and blessings of late Rajyogini B. K. Mohini didi ji, the than Chairperson of Rural Wing of Raj Yoga Education and Research Foundation (RERF), a sister concern of PPBKIVV, Mount Abu, Rajasthan, having an International status of World Spiritual University. We are grateful to Raj yogini B

K Sarla didi ji, Chairperson and Vice-Chairman, B K Raju Bhai ji and B. K. Sumant, Head Coordinator, Mount Abu, Agriculture and Rural Wing of Rajyog Education and Research Foundation, for putting their whole hearted support for bringing awareness about the Yogic Agriculture. The authors are highly obliged to Rajyogi B K Suraj Bhai ji for becoming instrumental to treat the seeds of wheat and chick pea by Yogic (metaphysical) energy of thoughts. Authors record their gratitude to all the Rajayogi farmers of Maharastra and Gujrat, having everlasting touch with this spiritual university and are ever ready on Godly services, a work designed to give benefit to the humankind, for sharing their experiences of applying the powers of Rajayoga Meditation through on farm demonstration in their own farm.

REFERENCES

Alberto B. 1977. Methodology for research on psychokinetic influence over the growth of plants. Psi Communication, 31: 9-30.

Anonymous, 2009. Perpetual Yogik Kheti, Rural Development wing. Rajyog Education and Research Foundation, (PPBKIVV Mount Abu): 64 p.

Backster C. 2003. Primary Perception: Biocommunication with plants, living foods, and human cells. White Rose Millennium Press, ISBN 0-9664354-3-5, Website.

Backster C. 1968. Evidence of a primary perception in plant life. International Journal of Parapsychology, 10(4): 329-348.

Barrington 2003. Field study of an enhancement effect on lettuce seeds: A Replication Study. The Journal of Parapsychology, 67: 102-04.

Benor and Daniel J. 2001. Spiritual healing: Scientific validation of a healing Revolution 2nd Ed., Wholistic Healing Publication PO Box 76 Bellmawr NJ 08099 USA. 316p.

Du C. and Laurene J. 2007. The effect of intentional thought in close proximity or at a distance: Demonstrating the relation between mind and matter on seed germination. Dissertation submitted to the Faculty of Holos University Graduate Seminarary in partial fulfillment of the requirements for the degree of Doctor of theology.

Emoto M. 2004. Healing with water. The Journal of Alternative and Complementary Medicine, 10(1): 19-21.

Grad BA. 1963. Telekinetic effect on plant growth. International Journal of Parapsychology, 6: 473-498.

Hemangi J. 2012. Deployment of natural powers.http: //www.scribd.com/doc/96541571/Yogic-Agri-English

Konstantin Korotkov. 2012. viewing the inner life of plants, sprit of ma'at Vol 3 No. 1.

Lebbrecht K. 1999. A snow crystal primer; The basic fact about snowflakes and snow crystals. *www.its.cantech.edu/-atomic/snow rystals/primer/primer.htm*

MacDonald R, Dakin HS and Hickman JL. 1976, 1977. Preliminary studies with three alleged psychic healers. Research in Parapsychology, 2: 67-71.

Nicholas C. 1977. The effects of loving attention on plant growth. New England Journal of Parapsychology, 1: 19-24.

Omvati and Sunita 2014. Metaphysical Energy (BKRYM): A zero budget technology for seed invigoration. 188-196. In: *Efficiency centric management (ECM) in agriculture.* 197p.

Pandey Suinta, Verma Omvati, *et al.* (2015). Yogic Farming through Brahma Kumaris Rajayog Meditation: An Ancient Technique for Enhancing Crop Performance. Asian Agri History, 19(2): 105-123.

Rondey J, Dougal S, Dougal M and Solfivn. 2003. Field study of an enhancement effect on lettuce seeds: A Replication Study. The Journal of Parapsychology, 67: 279-297.

Russell E. 1973. Report on Radionics: Science of the future, The science which can cure where orthodox medicine fails, Suffolk, England: Neville Spearman. pp. 254-267.

Saklani A. 1989. Psychokinetic effects on plant growth, further studies by Linda Henkel, and John Palmer. Research in Parapsychology, 8: 37- 41.

Stevenson. 1958. The effect of sonic vibration on the bacterial plate count of soil. Plant and soil 10(1): 1-8.

Yiji F. 1991. Spiritual healing: Scientific validation of a healing revolution- A study from rice germination and emitting Qi. Southfield, MI: Vision Publications, 244p.

Chapter 29

The Market for Organic Food in India: Present Status and Future Potential

Nina Osswald

Independent Researcher and Consultant,
Hyderabad, Telangana
E-mail: osswald.nina@gmail.in

1. INTRODUCTION: ORGANIC BOOM – OR BUBBLE?

Over the past ten years, organic products have rapidly become more widely available in India. Since 2010 in particular, the number of organic stores has grown significantly, and other shops and supermarkets also increasingly include organic options in their product range. The media have even claimed that the market is "booming". For a long time, the Indian organic sector was primarily export oriented, and the domestic market has started to grow more significantly only over the past five to ten years. Although official data on the growth of the organic food sector is not available, industry experts estimate it to be growing at around 25 per cent per annum. Some retailers report growth figures of as much as 100-300 per cent over recent years. Nevertheless, organic food still constitutes a small niche of less than 1 per cent of the overall food market. While the growth of domestic demand and sales was initially concentrated in the large metropolitan or Tier I cities, organic food is now becoming more available in Tier II cities and even smaller towns.

Along with the growth of the market, the variety of sales channels and retail formats has also become more differentiated. Among the early pioneers of organic food retailing in the domestic market were small, independent organic and natural food stores. Meanwhile, most organized retailers have included some organic products in their offering. Organic companies have launched the first organized organic retail chains. In addition to retail stores, innovative models such as e-commerce, home delivery and community-supported agriculture are emerging in many places, and several restaurants and catering services now use organic ingredients. The range of sourcing strategies is as diverse as the supply chain models, ranging from production on company-owned farms to associated farmers, producer cooperatives, organic distributors and in some cases even import.

While new organic stores and brands are coming up on a monthly basis, some stores and organic chains that were part of the organic boom have already been forced to close down. To some extent, this is a natural development in a young and dynamic sector, where actors face myriad challenges that range from issues on the production level, lack of adequate infrastructure, insufficient access to finance and training and lack of awareness among customers. It also raises the question whether India has really been experiencing an organic boom, or rather an organic bubble. A differentiated analysis reveals that neither is entirely accurate. While some of the new organic enterprises might prove to be short-lived, the growth in demand is undoubtedly also benefitting many organizations and enterprises with sustainable business strategies. This chapter aims to shed some light on the developments and differentiations in India's fast-growing and highly dynamic organic sector. Findings are largely based on research conducted during 2011 and 2012 in Bangalore, Mumbai and Hyderabad, published previously in the book Organic Food Marketing in Urban Centres of India (Osswald and Menon, 2013), with some updates by the author to account for more recent developments.

2. Organic Production, Certification and Marketing

Estimates on the area under some form of organic cultivation in India vary widely, depending on the source of the data. Official data on certified production is available from the Agricultural and Processed Food Products Export Development Authority (APEDA). However, India lacks an established system for documentation of non-certified organic farming, despite the fact that the bulk of the total domestic production is grown by non-certified farms. These farms practice various forms of sustainable agriculture, such as organic, bio-dynamic and natural farming, non-pesticide management, low external input sustainable farming and traditional farming systems.

The National Programme on Organic Production (NPOP, see Government of India, 2005) was launched by the Ministry of Commerce, Government of India in 2000. The programme established the National Standards of Organic Production (NSOP) and the India Organic label (Figure 1). The primary goal of the NPOP was to facilitate export of organic products. Accordingly, organic certification and accreditation of certifiers came under the authority of the export agency APEDA. Certification and labeling for domestic sales of organic products are currently still voluntary. However, it is recommended that products are only sold as organic if they are certified, and certification and labeling are likely to become mandatory in the future.

At present, only a small part of organic food products sold in the domestic market is certified according to the National Standards of Organic Production. These third-party certified products carry the India Organic label and often also the European and US organic labels. Many small producers are unable to afford the fees for third-party certification or to manage the necessary bureaucratic work.

Therefore they opt for participatory guarantee systems (PGS), a peer-control system for organic farmer groups that ensures organic practices and is more accessible for smallholder producers. For use in the domestic market, products produced under a PGS carry a label developed by the PGS Organic Council India (see Figure 1). In parallel, the Government of India is also working on a PGS project.

Uncertified products are sold in the domestic market under a variety of terms, such as "organically grown", "pesticide-free" or "from natural farming". Without any official guarantee, authenticity of these products is often guaranteed by personal relationships of trust and transparency between growers and retailers, or directly between growers and consumers. The new web portal Eco-Farmers Market (www.ecofarmersmarket.in) distinguishes between three types of organic products: "certified organic" for third-party certified products, "guaranteed organic" for products grown under a PGS system, and "declared organic" for products without official certification or guarantee.

Figure 29.1: The Official India Organic Label for Products Certified According to NSOP and the PGS (Participatory Guarantee System) Organic Label (*Sources*: APEDA; PGS Organic India Council).

As with organic production data, no official statistics are available on the total size of the domestic market. The most reliable estimates are compiled in Table 29.1. For certified organic products, some conclusions can be drawn from the existing data for total production minus export volume. Historically, export sales used to surpass domestic consumption of certified organic products. Only a decade ago the bulk of India's certified organic production was still exported: about 70 per cent according to Carroll. (2005), and as much as 85 per cent of sales according to Garibay and Jyoti (2003). The primary reasons for the historic domination of export sales are strongly export-oriented government policies, better returns for organic products in international markets, and a lack of domestic demand.

The situation has tilted increasingly in favour of domestic sales only recently, since demand in the domestic market has picked up. While exports continue to grow – from 4 per cent of total certified organic production in 2010-11 to 13.3 per cent in 2013-14 – the domestic market is growing even more rapidly. Within a period of only one decade, domestic sales of organic products have grown from a few crore to an estimated INR 1,000 crore (US$ 226 million). Overall sales of products grown under the PGS India Organic system have grown by more than 130 per cent between 2010 and 2013 (Gupta, 2013). For the future, organic companies and retailers, industry experts and market research companies generally express highly positive growth expectations of the domestic organic sector. For instance, a report by (TechSci Research, 2014), expects organic food sales in India to grow at a rate of 25 per cent until 2019. By comparison, exports were reported to be growing at a rate of 7.7 per cent by APEDA in 2013-14. Note that the apparent gap – both exports and domestic sales are growing more rapidly than the total organic production – can be explained by the fact that the bulk of marketable surplus production is sold as conventional. This is largely due to lack of demand and lack of market access for organic products.

Table 29.1: Indian Organic Market Data and Estimates (Various Sources)

	Ministry of Agriculture/NCOF (1)	*APEDA (2)*	*ICCOA*	*Other*
Total organic production	INR 8,000 crore INR (2010-11), of which INR 4,000 crore (US$ 903 million) marketable surplus	1.24 million tonnes (2013-14, certified organic production incl. cotton)	n/a	400,000 tonnes worth INR 4,000 crore (OTA 2010)
Export sales (certified organic, 2013-14)	n/a	165,262 tonnes, incl. 4,985 tonnes organic textiles US$ 403 million incl. US$ 183 million organic textiles 7.73 per cent growth since 2012-13	INR 1,800 crore (US$ 406 million)	n/a
Domestic sales	INR 1,000 crore (2010-11) (US$ 226 million)	n/a	INR 500-800 crore (2012-13) (US$ 90-150 million)	PGS Organic products only: INR 5.4 crore (2012-13) (US$ 1 million) (Gupta, 2013)

(1) Information provided by Dr. Yadav, former Director of National Centre for Organic Farming (NCOF), cited in (Jishnu and Sood, 2012); (2) http://www.apeda.gov.in/apedawebsite/organic/Organic_Products.htm

3. Differentiation of Supply Chains and Retail

Thottathil (2014) has analyzed the organic farming sector in Kerala in great detail and discovered a divide between uncertified organic farmers growing staples like grains, pulses and vegetables on one side, and certified export-oriented farmers growing cash crops like spices and coffee for international markets on the other side. Similarly, the Indian organic farming sector before 2000 had two main strands (Singh, 2009). Export-oriented companies that had their own farm projects or worked with large independent producers on the one hand; and grassroots organizations and cooperatives working with smallholder producers on the other. The latter used to primarily support producers in adopting sustainable farming methods, and only more recently started focusing on processing, branding, distribution and marketing activities. They became the early pioneers in domestic marketing, together with organic specialty retailers who often had their own farms as well as close personal ties with other producers.

The first large, commercial organic companies started selling in the domestic market only since the turn of the century. Some of them were already established in organic exports, while others were newly founded enterprises. Accordingly, most Indian organic companies are relatively young (OTA, 2010). Over the past ten to fifteen years, farmer groups and associated NGOs also increasingly looked at expanding their marketing activities in local and regional markets, especially the bigger cities where demand for organic products is highest.

Along with the growth of domestic sales and consumption of organic food, the market has become more differentiated with regard to supply chain organization and urban retail structures. Table 29.2, shows the most common types or marketing channels that currently exist in urban India, the key characteristics of their supply chain and retail models, and a few examples from different cities. The characteristics of each type are explained in more detail in the following sections.

Table 29.2: Domestic Marketing and Retail Channels for Organic Food

Marketing Channel	*Characteristics*	*Examples*[1]
Organic specialty retail	Organic food stores, health food stores, deli stores, non-food stores, organic bazaars Organic retail chains At least 50 per cent of food product range organic Mix of certified, PGS and uncertified products Mix of local suppliers and large brands, sometimes own farms	Numerous independent organic and natural food stores across India Organic retail chains like Morarka (Down to Earth), Fabindia The Farmers' Market (Mumbai) Good Seeds Organic Bazaar and Dharti Organics (Hyderabad)
Conventional organized retail	Supermarkets, hypermarkets Less than 1 per cent organic sales Organic suppliers: large certified brands	Most major retail chains, some small regional chains and standalone supermarkets
Direct marketing and regional food networks	Community-supported agriculture, regional food networks, farmers' markets 100 per cent of products organic Can be certified, PGS or uncertified Local and regional suppliers, own farms	Adarsh Bio Organics (Gurgaon) Mumbai Organic Farmers and Consumers Association (Mumbai) Gorus (Pune) Sahaja Ahaaram (Hyderabad) The Farmers' Market (Mumbai) Sahaja Samruddha (Bangalore)

Marketing Channel	*Characteristics*	*Examples*[1]
Restaurants and catering	Restaurants, tiffin services, catering, outlets in corporate campuses Usually not 100 per cent organic Mix of certified, PGS and uncertified products Mix of local suppliers and large brands, sometimes own farms	Lumiere (Bangalore) In the Pink (Bangalore) The Green Path (Bangalore) Birdsong Café (Mumbai) Organic Express (Hyderabad, Gurgaon) Hyderabad Goes Green (Hyderabad)
Online retail and delivery services	Online without physical retail infrastructure, only warehouses Orders over phone or online Some are purely organic and natural, some are conventional with organic range Can be certified, PGS or uncertified Mix of local suppliers and large brands	Natural Mantra (Mumbai) Jiyo (Mumbai) I Say Organic (Delhi) Adi Naturals (Bangalore) Jiva Organics (Hyderabad)
General trade	Kirana and general stores, vegetable stalls, traditional markets Very few sell a small range of organic products	Modern Stores (Mumbai)

Examples were selected to illustrate the diversity; no rating or preference is intended by the selection.

The number of stores selling organic products has grown rapidly over just a few years. Until around 2005, very few stores even in the biggest cities sold organic products. The number has now increased to hundreds of stores and online retailers across India, and even thousands if all outlets of organized retail chains are counted. No official, up-to-date record of the exact number of stores currently selling organic food products is available. However, as an indication, a large organic company based in Bangalore reported to be selling certified organic products through a total of 1,000 retail stores across India, of which 350 are in Bangalore, 200 in Chennai and the rest in Hyderabad and other cities.

India's western region is considered by several market studies to be the largest market for organic food products, followed by the southern region (for instance, Rao *et al.*, 2006; TechSci Research, 2014). An empirical research study across three cities in South and West India found that Bangalore is the largest urban organic market in India (Osswald and Menon, 2013). In 2011, the city already had 157 stores selling organic food products (including supermarkets), out of which 23 were organic specialty stores. Hyderabad and Mumbai had a slightly smaller number of stores, and lower overall sales of organic products. Bangalore also had the highest ratio of organic stores per 100,000 inhabitants: 2.12, as compared to 0.83 Mumbai and 1.21 in Hyderabad. Since 2011, the number of stores has continued to grow every year.

While supermarket outlets that carry organic products outnumber organic specialty stores, they usually have a much smaller product range, and therefore contribute only slightly more to overall sales of organic products. An analysis of organic food sales in Bangalore (Figure 29.2) revealed that organized retailers had a share of 34.5 per cent in 2011/12, slightly more than for organic food stores. However, if organic food stores and non-food specialty stores such as Fabindia are taken together, they make up 56 per cent of organic food sales – significantly more than organized retail.

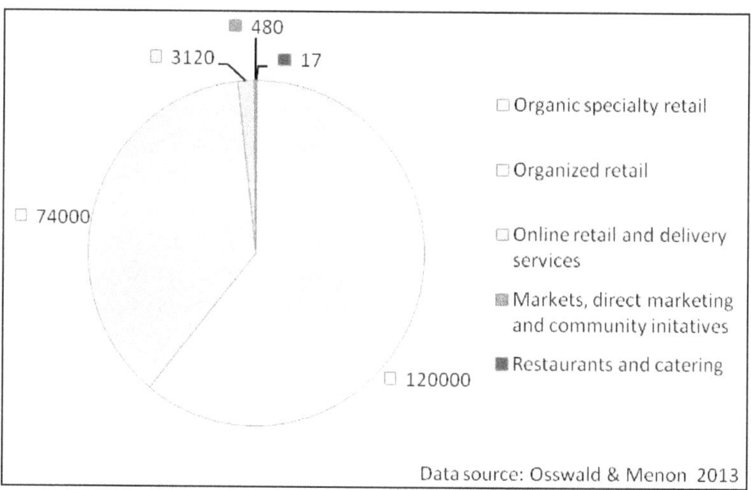

Figure 29.2: Organic Retail Sales by Retail Channel in Bangalore (2011/12, in 1,000 INR).

3.1 Organic Specialty Retail

Most organic specialty stores in India do not sell exclusively organic products. Therefore, the category of organic specialty retail refers to stores and other formats where the food range is comprised to at least 50 per cent of organic products. These products can be certified organic, PGS labeled, or not certified but grown on farms that work with organic or natural farming methods. Many organic stores also sell some non-organic products, because a full product range is not available from organic producers and processors. Some of these products are marketed as "natural", because they do not contain synthetic additives or preservatives and are minimally processed. As product availability is improving across India, the number of stores that offer a fully organic range is also growing.

Small independent organic stores usually source products from a number of suppliers such as local, unbranded and uncertified farmers, organic distributors and third-party certified organic companies. Most organic specialty retailers have personal contacts to farmers, especially for sourcing fresh produce, and some also source products in bulk and package them under their own retail brand. Most organic stores are independent, proprietor-run stores that sell primarily groceries and a small range of non-food items. The average turnover of these small stores in Bangalore was INR 2 lakh per month in 2011 (Osswald and Menon, 2013). More recently, organic retail chains operated by corporate organic companies have begun to emerge across India. For instance, Morarka Organics operates some company-owned stores and also works with franchisees.

In addition to organic stores, organic specialty retail also comprises several other retail formats and channels. Since 2010, organic markets, bazaars or melas were started in several cities. They often include a farmers' market, where growers sell directly to consumers, but they also host a variety of organic retailers, caterers and retailers of other eco-products such as natural cosmetic, home gardening needs, organic textiles, recycled products and crafts items. Some examples are The Farmers' Market in Mumbai which has run since 2010, the Whitefield Farmers' Market and Dakshini Pete in Bangalore, and the Good Seeds Organic Bazaar and Adivaram Angadi in Hyderabad.

Another major subcategory of organic specialty retail are non-food stores. Prominent examples are the India-wide chain Fabindia and the Bangalore-based Mother Earth which sell primarily textiles and crafts items, but also have an organic food section. Most of the food products in these stores are certified organic and some are non-certified or natural. Fabindia started its organic food range in 2004 and currently makes less than 10 per cent of its total turnover with organic products. Products are grown, processed and packaged by other companies but labeled with the Fabindia brand.

3.2 Conventional Organized Retail

The second most important retail channel in terms of organic food sales are supermarkets and hypermarkets. These retail formats have started to come up across India since the turn of the century, and still comprise less than 10 per cent of urban food supply (cf. section 3.6). Before 2010, very few organized retailers sold any organic products. The first organized retailers that started selling organic food products in India were large independent supermarkets and the larger regional and national chains. In Hyderabad for instance, the local up-market supermarket Q-Mart and the South Indian retail chain Spencer's were among the first supermarkets that sold branded organic products before 2010.

Since that time, availability in organized retail has grown rapidly. Many of the larger retail chains that operate India-wide already have an organic product range of varying size, and many smaller local chains and independent stores are also beginning to follow suit. While most retail chains stock organic products only in prime locations – based on customer profile and sales performance– others have plans of making organic products available in all their outlets.

Conventional supermarkets and hypermarkets typically have a smaller organic product range than organic specialty stores, typically less than 100 different products which constitute less than 2 per cent of total food sales. At present, very few organized retailers offer organic fresh produce. In Bangalore, Spencer's was the first supermarket chain that sold organically grown vegetables from farms in conversion in seven selected outlets. Nature's Basket and Namdhari's are two other supermarket chains that sell some organic fresh produce in selected outlets, although the bulk of their fresh produce is not organic.

Organized retailers in India have not yet started to launch any organic retailer brands of their own. Generally, they stock only organic products that are certified and labeled. They preferably source from larger brands with a wide product range and India-wide supply chains. Most organized retailers are not interested in sourcing directly from small farmers because larger suppliers are more able to ensure consistent, timely and year-round supply as well as consistent product quality. Instead, they source mainly through distributors and from large certified organic brands.

3.3 Direct Marketing and Regional Food Networks

As part of the growing organic sector, a number of initiatives have emerged that seek to establish localized supply chains and to bring producers and consumers of organic food closer together. A key element in direct marketing is a short and well-timed supply chain from the farms to the consumers. This saves costs and resources for storage and transport, and it ensures that produce is delivered as fresh as possible. Retail structures are kept to a minimum, and profit-seeking intermediaries are eliminated so that prices can be kept low for consumers while still paying the farmers a better price.

Several direct marketing initiatives emerged out of the work of rural development organizations with smallholder producers. Exclusively organic farmers' markets now exist in all major cities. Often they are integrated into organic bazaars and melas (see section 3.1) where – in addition to organic farmers – different vendors sell organic snacks, packaged organic products, natural cosmetics and other eco-products. Among the pioneers in organic farmers' markets in India are the Thanal Organic Bazaar in Thiruvananthapuram, running since 2003, and the organic bazaars organized by Institute for Integrated Rural Development (IIRD) in Aurangabad since 1999. Several farmers' markets and direct marketing initiatives are supported by NGOs, community initiatives and businesses, for instance the Sahaja Aharam cooperative store in Hyderabad by Centre for Sustainable Agriculture, the Organic Mobile in Hyderabad by Deccan Development Society, the Shyamoli farmers' market in Kolkata by Earthcare Books, and the Navadarshanam direct sales by Carrots vegan restaurant in Bangalore. The Farmers' Market in Mumbai charges fees for all retailer stalls to enable organic farmers to participate free of charge. Some producer companies, for instance Chetna in Hyderabad, and some larger independent farmers sell products directly to consumers through home delivery.

Another innovative model of direct marketing has been developed by community initiatives such as consumer clubs, consumer-producer cooperatives and community-supported agriculture. Even though informal systems of collective farming and direct relationships between growers and consumers have existed for a long time, these community initiatives are a young phenomenon in urban India. Many of them are part of a global movement for localization of production and consumption and for alternative regional food networks. At the core of their work is a conviction that food should be grown organically on small diverse farms, distributed locally, and that farmers deserve a fair share of the price that the consumer pays. In order to share the risks involved in farming, consumer members of community-supported agriculture initiatives help farmers to plan their production according to demand, and they often pre-finance the harvest and give a purchase commitments. These local networks of consumers and producers are based on personal interactions, solidarity and trust. A key overall objective is to build vibrant and sustainable local communities around food and farming.

3.4 Restaurants and Catering

To date, most organic restaurants, caterers and bakeries are not fully organic, but serve food "made with organically grown ingredients, as far as possible." Some of the reasons are the lack of availability of a wide range of organic products, seasonal variations, higher cost and difficulties in sourcing from various scattered suppliers. Bangalore currently has the highest number of restaurants and catering businesses that use organic ingredients, but other cities are also beginning to follow suit.

Most restaurants, bistros and cafés that serve organic food are independent and proprietor-run. A few others are owned by large organic companies or operated by NGOs. Just a small number of restaurants were established before 2008, and there has been a wave of new start-ups since then. There are now several restaurants, cafés, bakeries and caterers serving organic food in all bigger cities in India. For instance, at the time of writing this chapter, at least 13 restaurants, cafés, caterers and bakeries served organic food in Bangalore, 8 in Mumbai, and 8 in Hyderabad.

Event catering and ready-to-eat food delivery businesses have also emerged in the organic sector. Several small, home-based caterers operate on events such as farmers' markets and organic bazaars. Some of them also do home delivery of organic meals and snacks. Generally, food delivery is very popular among the urban upper middle and upper classes. Some organic tiffin services in Mumbai have tied in with the traditional Dabbawala system for deliveries. In other cities, ready-to-eat organic food deliveries are also starting to be available, though not as widely as in Mumbai.

An example of an innovative start-up is Organic Express, operating in Hyderabad and Gurgaon since 2009 and catering mostly to young urban professionals. Despite growing demand, office environments like IT parks and corporate campuses do not offer healthy lunch and snack options. In order to fill this gap, Organic Express offers meals and snacks made with fresh vegetables, millets and unpolished rice, using organically grown ingredients as much as possible.

Some organic restaurants and catering businesses offer specialty foods, such as macrobiotic, vegan or millet-based meals, while others concentrate more on traditional Indian meals and tiffins. Most organic restaurants and caterers rely on a wide range of organic suppliers, from small local farms and NGO initiatives that are not necessarily certified organic to larger organic companies. While some restaurants get deliveries directly from suppliers, others also buy from organic specialty stores. Two organic restaurants in Bangalore, Lumiere and The Green Path, grow part of the ingredients on their own farm and also operate organic stores.

3.5 Online Retail and Delivery Services

Most organic stores already offer home delivery, usually with orders being placed over the phone. In fact, some organic shops make the bulk of their retail sales through home delivery. For instance, one of the pioneers of organic online retail in Bangalore, Adi Naturals, has been successfully using online orders and home delivery alongside a physical retail outlet for several years. Dubdengreen has been following a similar model successfully in Delhi, and has more recently started to offer deliveries in Bangalore as well. A growing number of retailers, from small stores to chains like Fabindia, are now offering an online order option through their websites. Some of the more recent organic startup retailers offer an exclusive delivery service with orders being placed either over the phone or online.

Many customers prefer a combination of home delivery with a retail store which allows them to see products before ordering, have face-to-face contact with the retailer and ask questions about the products and organic production methods. Exclusive delivery, on the other hand, has the advantage of keeping the retailer's cost much lower because they only need a warehouse and no retail space. Successful examples of the latter type are Jiva Organics in Hyderabad and I Say Organic in Delhi. I Say Organic was started only in 2012, and is already reported to have a revenue of Rs. 10 lakh a month (Kumar, 2013).

Usually, organic e-commerce websites that operate regionally put a strong focus on fresh produce. Others that deliver India-wide by courier, like the Mumbai-based Natural Mantra, do not sell fresh produce. Some conventional online retailers have also taken up a range of organic products, though usually no fresh produce. As the online retail space requires the least investment in physical infrastructure, new players continue to enter the market at a rapid pace. An article by (Biswas, 2013) presents several case studies of successful organic online retailers and concludes that the sector has very good growth prospects in the future.

3.6 General Trade

The majority of the urban population still purchases the bulk of their food from traditional retail channels. Traditional retail in India is also known as general trade or unorganized retail. It comprises *kirana* and general stores, convenience stores, street stalls, pavement vendors and pushcart vendors. Traditional retail stores are typically small and proprietor-managed and have a small number of employees. Their supply chains are a mix of local and India-wide suppliers, with products being sourced both from local small and medium enterprises and from distributors selling packaged products from India-wide corporate food processors.

Over the past few decades, the Indian retail sector underwent profound structural changes, which are often referred to as the Indian Retail Revolution (Businessworld, 2011). Since the turn of the century in particular, more and more large malls, supermarkets and hypermarkets started to open, especially in affluent neighbourhoods of the big cities. In 2009, organized retail was still growing at 55 per cent – the highest growth rate of all sectors (IBEF, 2008; Wiggerthale, 2009). Meanwhile, the growth has slowed down somewhat. In 2007, organized retail still had a share of only 4-5 per cent of the total retail sector and 1 per cent of total food purchases. More recently, it was projected to reach 10 per cent of total retail by 2016-17 (Assocham, 2010; Yes Bank, 2012).

Currently, *kirana* and general stores do not usually carry any organic products, and most store managers are not aware of organic food. There are just a few exceptions to this, for instance an old-established general and medical store in Pali

Naka, Mumbai which has been supplied by the Mumbai-based organic company Conscious Foods since 1996, and more recently also took up products by Morarka Organic Foods. Since 2011, several kirana stores in Mumbai included a limited number of branded and certified organic products in their range. However, most of them had little success and discontinued organic products after a while due to limited demand. It would be revealing to investigate whether the low demand was due more to a lack of promotional efforts, or whether customers were not interested in organic food.

4. Challenges and Risks

Even though domestic sales of organic products and the overall number of organic retailers have been growing rapidly in recent years, producers and retailers in the organic food sector face many of challenges. Several new stores and online retail startups were already forced out of business again. Actors in domestic production, distribution and retailing face a range of challenges, weaknesses, threats and risks which are listed in Table 29.3.

Some of the biggest challenges for organic businesses in India lie in the area of supply chain management and logistics. At present, most organic retailers are not able to provide a consistent product availability and quality, especially for fresh produce, due to difficulties in coordinating with producers, seasonal fluctuations of product availability, and logistical challenges such as inadequate transport infrastructure and storage space. Organic products require separate handling to avoid mix-ups and contamination, and the young sector has not yet built sufficient infrastructure for distribution, (cold) storage and processing. Generally, those companies, restaurants, retailers and marketing initiatives who have invested in backward supply chain integration, for instance their own farms or close links with farmers, are better able to ensure a steady supply of products.

Table 29.3: Overview Analysis of India's Organic Food Sector

Strengths	*Weaknesses*
Growing media attention and public awareness of organic (*e.g.* after Satyamev Jayate episode)	No reliable data on uncertified organic and natural production and on domestic market size
Growing awareness of the harmful ecological, economic, social and health effects associated with using chemicals and genetic modification in agriculture	Lack of separate infrastructure, *e.g.* transport, cold storage, distributors
	Lack of publicity, consumer education and awareness on organic production, certification and labels
Double-digit market growth, increasing availability especially in cities and towns and optimistic growth projections	Supply constraints (insufficient number of growers, crop varieties, year-round consistency, quality)
Organic product range growing, including fresh produce, dairy products, eggs and processed products	Insufficient capacities and training opportunities on organic production methods, production planning, quality control and farm-level value addition
Emerging pan-Indian networks in organic movement and industry (*e.g.* ASHA, web portal Eco-Farmers Market, Organic Trade Association of India)	Lack of market information, understanding of the demand situation, market linkages and coordination between producers, processors, distributors and retailers

Strengths	*Weaknesses*
Some support for certification and group certification (Internal Control Systems) available from government and organic companies PGS system strong and recognized by the government Some federal state governments passed organic farming policies and actively promote production	Govt. support for organic mainly for large, commercial and export-oriented projects Unfair competition by continued subsidies and support for chemical farming; prices tied to conventional prices Most R and D done by individuals and NGOs with limited funds; lack of effort from agricultural universities
Opportunities	*Threats*
Increase in purchasing power in the cities Demand for a wider product range, especially fruits, value-added and processed products Sell organic food through traditional retailers to expand customer base Synergies with related social and ecological movements such as biodiversity, seed sovereignty, terrace gardening, climate change, veganism *etc.* PGS as a viable alternative for domestic market Follow the examples of some North-Eastern federal states and set target to go 100 per cent organic Use organic export revenues to subsidize production for domestic market Large uncertified and undocumented production area that can be tapped into for organic production New back-to-the land movement of mid-level professionals showing interest organic farming, returning to ancestral farms	Lack of transparency in the supply chain "Conventionalisation" of the organic industry (*e.g.* highly processed and unhealthy products, centralized supply chains); corporatisation, dominance of a few large companies and price pressure on growers Lack of professionalism and false/exaggerated claims in advertisements by organic companies Organic perceived as expensive, elitist or a fad Price premium attracts investors, food corporations and retail chains without genuine interest in sustainable farming (organic seen mainly as business opportunity) "greenwashing", false claims and attempts to cash in on the sustainable farming and organic food platform by non-organic food corporations; misuse of terms like "natural", "pure", "pesticide-free" *etc.* Certification not accessible for resource-poor smallholders (cost, bureaucratic hurdles) Accusations of certification frauds (against certification agencies, companies and producers), lack of trust in certification and labels among consumers

Limited product availability is one of the major purchasing barriers for consumers who are interested in buying organic food (Rao *et al.*, 2006; Chakrabarti and Baisya, 2007; The Nielsen Company, 2007; Chakrabarti, 2010). The categories that are most highly in demand in the organic field are fruits, vegetables and dairy (Rao *et al.*, 2006; Sally, 2013), and yet availability is lowest for these perishable products. While organic fresh produce has become more widely available over recent years, it still lags behind the demand. One major challenge is quality control on the farm level and training of farmers on practices such as staggered planting, when to harvest crops, and how to handle them post-harvest for minimum losses and damage. The quality of products gets compromised during transport and due to a lack of storing infrastructure such as cold storage, especially with highly perishable products such as green leafy vegetables and fruits.

At present, lack of awareness is one of the biggest obstacles to winning more consumers as buyers of organic food. The majority of the population is still not aware that chemicals used in conventional agriculture are harmful, nor that safer alternatives exist. Studies on awareness levels vary significantly with regard to their findings. Several studies found awareness levels ranging between 18 per cent and 41 per cent of respondents (Garibay and Jyoti, 2003; Lohr and Dittrich, 2007; Menon, Sema and Partap, 2010; Datta, 2010). Other studies that focused on consumers with a higher socio-economic profile discovered awareness levels of 62-80 per cent (Osswald and Dittrich, 2010; Dholakia and Shukul, 2012; Sally, 2013). Awareness and demand for organic products are highest in metropolitan cities.

While overall public awareness of sustainable agriculture and organic food options is still low, it has improved substantially over the past few years. The media are now paying more attention to topics like food adulteration, pesticide residues in food, and the benefits of sustainable agriculture practices. For instance, an episode of the popular television show Satyamev Jayate on "Toxic Food: Poison on Our Plate" in June 2012 reportedly created a surge of interest in organic food. Growing numbers of consumers now turn towards organic food due to health concerns, sometimes following the specific advice of their doctors to eat more millets, whole grains and organic produce. Only a smaller number is aware of the adverse ecological and economic consequences of unsustainable farming practices.

Another major obstacle to winning more consumers for organic products concerns price expectations and willingness to pay. The issue of pricing of organic products in the domestic market is a complex one. From the perspective of farmers, organic farming is more profitable in the long run, especially in rain-fed agriculture. Nevertheless, organic products are more expensive to produce and market due to a number of factors, such as chemical fertilizer subsidies of over INR 60,000 crore per year (Government of India, 2014), lack of comparable support for organic farming and inadequate infrastructure, market information and market access for organic producers. Additionally, organic farming requires more labour and management skills.

At present, organic food costs up to 20 per cent to 30 per cent more on average than conventional food items in the Indian market (TechSci Research, 2013; Osswald and Menon, 2013). At the same time, there are vast differences between different organic suppliers and brands. Some companies brand organic as a premium product and put a correspondingly high price premium on their products. Whether this price difference is passed on to the farmers is not always clear. By contrast, many marketing initiatives and organic companies consciously keep price levels affordable for a broad social spectrum of consumer while still being able to pay producers a good price. Several NGOs and foundations support the marketing efforts of small farmers through externally funded staff, free use of infrastructure, access to capital and volunteer work. Despite the considerable price differences within the organic sector, consumers often perceive organic in general to be more expensive. While some organic companies and experts argue that it is acceptable for organic to be more expensive until it has reached a larger market share, others hold against it that organic is running a high risk of being perceived as elitist and losing its credibility.

While many consumers are willing to pay a higher price for good quality organic products (Kriesemer, Weinberger and Chadha, 2009; Roy *et al.*, 2010; Radhika and Seema, 2012), price is still a significant purchasing barrier (Dholakia and Shukul, 2012; Sally, 2013). Better communication between farmers, retailers and consumers could improve awareness on why organic is more expensive, and promote a better understanding of the real cost of good quality food. Prices for organic produce are often tied to the conventional market prices, which can make it difficult to sustain a viable organic farm if buyers do not follow a fair pricing mechanism. Considering the subsidies for the conventional food industry, organic food might remain inaccessible for the majority of the Indian population for some time to come. One of the major challenges in the organic sector thus lies in ensuring that organic farming is profitable for producers and allows them to earn a decent livelihood, while at the same time keeping organic products affordable.

Related to the issue of pricing is the risk that the organic sector increasingly attracts investors and entrepreneurs who are driven more by a short-term interest in profit than an interest in long-term sustainability of organic farms. In countries with more established organic sectors, a process of commercialization and conventionalization has been observed, with organic producers and brands following more or less the economic model of the conventional food industry (see for instance De Wit and Verhoog, 2007; Best, 2008). This process comprises unsustainable practices such as economies of scale, industrialization and mechanization, monocultures, high levels of processing, centralized supply chains with high food miles, energy-intensive modern retail formats, corporate company structures, competitiveness, price pressure on producers and lack of ownership and participation of both producers and consumers.

Thus, as organic food becomes more popular in India, the challenge of guaranteeing authenticity and credibility of all stakeholders in the supply chain is also growing. Especially with new organic companies that have only loose ties to growers and suppliers, it has become difficult for consumers to judge their credibility. This is true whether a producer or organic company is certified or not, as there have also been occasional rumours of certification frauds.

Organic producers, processors and retailers can communicate product authenticity to consumers either through third-party certification, the PGS system or a close personal relationship of trust. At present, third-party certification is only "recommended" for sale in the domestic market, but not legally binding. Many doubts have been voiced whether this system with its high cost and bureaucratic requirements is appropriate at all for the Indian context where the majority of small-holder producers are not able to afford the certification fees. Internal control systems were introduced to ease the burden on individual farmers by certifying groups. Nevertheless, the high cost does add to the competitive disadvantage vis-à-vis chemically grown food.

Additionally, most consumers in India – including many who buy organic products regularly – are not aware of the Indian organic standards, certification and labels. The organic standards, certification system and India Organic label were developed mainly with export markets in mind, and there has never been a coordinated consumer education campaign for either the India Organic or the PGS label. In their study of consumers in Hyderabad, (Osswald and Dittrich, 2010) found that only 10 per cent had seen the India Organic label before, and 8 per cent the PGS Organic label (See Figure 2). However, even their knowledge about what the labels mean was very incomplete. At the same time, 75 per cent of respondents who were interviewed by Osswald and Dittrich (2010) said they trusted organic labels, or would trust them if they had more information about the meaning, and the certification process.

Despite the limitations of the third-party certification system, consumers need some legal certainty that products sold as organic really are produced strictly according to organic principles. Thus, a stronger focus on transparency along the entire supply chain can ensure the authenticity and credibility of organic products. PGS and trust-based local partnerships could be a viable alternative for quality assurance in regional organic food networks. They can promote transparency in the supply chain and help to make products in the domestic market more affordable (Philip, 2014). If labeling is to become mandatory for the domestic market in future, financial support for the certification process, encouragement of the PGS system and publicity and educational campaigns will be needed to popularize the concept both among producers and consumers.

5. Trends and Future Prospects

Over a period of just a few years, the organic sector in India has developed dynamically, with an impressive number of new organic companies being founded and conventional companies entering the organic market. Organic products have made a full entry into organized retail, and big investors and corporate companies have started taking an interest. Along with the growing number of organic stores and other marketing channels, the available product range has also been expanding. Until recently, only dry provisions such as grains, pulses, spices and jaggery were available on a regular basis. Most organized retailers sell exclusively branded, certified products sourced from distributors or from large organic companies, and very few have ventured into fresh organic produce to date. By contrast, organic specialty retailers have made remarkable progress in improving the availability of organic fresh produce in the cities. Most cities now have organic delivery services as well as plenty of organic stores that sell organic vegetables and fruits on several days a week. Several suppliers now even provide organic dairy products, eggs and chicken meat. The range of processed and value-added products is also being expanded continuously.

Currently organic is still a small niche of less than 1 per cent of the total food market, and only a small part of the population is aware of organic food. However, organic has already become a lot more popular in a relatively short time span, and the growth projections for the coming years are highly positive. For instance, (TechSci Research, 2014), expects an annual growth rate of 25 per cent until 2019. In 2006, a study estimated the overall market potential in the eight largest cities of the country at INR 1,452 crore (US$ 300 million), and the accessible market potential through modern retail at INR 562 crore (US$ 125 million) (Rao *et al.*, 2006). ASSOCHAM estimates that the value of the organic food market will grow to Rs. 6,000 crore (US$ 970 million) by 2015 (Whitehead, 2013).

Table 29.3 summarizes the opportunities and major growth drivers that lie ahead for the Indian organic sector. At present, major constraints are the limited supply situation, inadequate infrastructure and logistical difficulties. More organic producers are needed, and they need to grow more varieties, especially for fruits and vegetables. Expanding the product range further is an opportunity both for organic farmers and processors. Value-added and processed products such as pickles, jams, dehydrated fruits and vegetables, cold-pressed oil, peanut butter, snacks and sweets are a good opportunity for small, home-based businesses and for village-level value addition.

In addition to organic production, market linkages and information flow between growers and buyers is a challenge. One of the success factors for organic companies and marketing initiatives is a strong supply base in the form of own farms, backward supply chain integration or strong relationships with grower groups. These are essential in order to control the range, consistent supply and high quality of produce. In retail, a robust base of reliable suppliers and good rapport with consumers through excellent service quality and friendly, competent staff are essential. A new initiative called Eco-Farmers Market was launched by Alliance for Sustainable and Holistic Agriculture (ASHA) in 2014 to make information flows between organic producers and procurers more smooth, and help producers find new marketing avenues for their organic produce. The web portal is supported by call agents who stay in touch with the producers and continuously update the online information.

It remains to be seen in the future how many of the new actors in the organic arena will be able to sustain themselves, and whether part of the organic boom is in fact a bubble or fad. Empirical evidence indicates that a business strategy based on a high initial investment for branding and retail infrastructure can be difficult to sustain. Several examples of high-investment single-brand organic retail chains such as Organic Garden in Mumbai and Organic Haus in West India proved to be not viable and reduced the initial number of outlets.

At the same time, a more organic growth has been taking place on a smaller scale in many places across India. Small retailers, community-based marketing networks and consumer-producer associations have emerged in several cities with the objective of bringing producers and consumers of food closer together, educating consumers, building trust in organic production methods and assuring authenticity of organic products without high cost or bureaucratic efforts. Instead of – or sometimes in addition to – certification, consumers rely on personal contact and farm visits to ensure organic production methods. Prominent examples are Restore in Chennai, Adi Naturals and Sahaja Samrudha in Bangalore, MOFCA in Mumbai and Sahaja Aharam in Hyderabad. While marketing initiatives and stores that are based on long-standing relationships with grassroots producers and urban consumers may have less impressive growth figures, lower financial returns and are often dependent on external support, they may prove to be more resilient and less dependent on demand fluctuations in the long run. Promoting organic production as part of a strategy for resilient, regionally-oriented sustainable food systems can help sustain livelihoods in agriculture and counter the agrarian crisis, respond to the ecological crisis and climate change, fight malnutrition and achieve food sovereignty, counter the modern-day urban health crisis and build stronger local communities.

A major untapped opportunity in domestic marketing lies in reaching a much broader customer base by introducing organic to mass food distribution channels like *kirana* stores, street vendors and markets. Some of the advantages of general trade for marketing organic products are their decentralized location, broad customer base in terms of socio-economic profile, and close personal contact between retailers and customers. If promoted well, this could be a good and largely untapped opportunity to promote organic staple food products beyond a narrow niche of educated and aware consumers. One of the challenges will be to ensure product authenticity, affordable pricing and good communication of the reasons for any price premium.

6. Conclusion

The domestic market has grown at an impressive rate over the past decade, and has recently begun to outperform export sales. In the metropolitan cities, the bulk of organic food sales takes place in organic specialty retail and conventional retail chains. The spectrum of supply chain models and retail formats is becoming ever more diversified, and comprises several innovative business models such as online retailers, social enterprises and community initiatives.

The main challenges for the future will be to sustain the growth of organic food sales, and to develop the organic sector sustainably and holistically. Organic producers need better post-harvest infrastructure, regional logistics and market linkages so that organic produce can reach urban markets with minimum losses. Regional distribution networks and PGS can be promoted as effective tools for sustainable food supply and quality assurance in the domestic market. Consumers need more information about organic farming, labels and availability of organic products.

REFERENCES

Alvares C. 2009. The Organic Farming Sourcebook, Mapusa: Other India Press and Third World Network.

Assocham. 2010. Organized Retail Share Likely To Surpass 30 per cent By 2013: ASSOCHAM", *Press Release*, 24.06.2010, http: //www.assocham.org/prels/shownews-archive.php?id=2473 (accessed 1.7.2012).

Best H. 2008. Organic agriculture and the conventionalization hypothesis: A case study from West Germany. Agriculture and Human Values, 25/1: 95-106.

Biswas S. 2013. A Click Away, *The Telegraph*, 30.06.2013, http: //www.telegraphindia.com/1130630/jsp/graphiti/17062104.jsp (accessed 30.6.2013).

Businessworld. 2011. The Old Kings And The New, *Businessworld*, 12.12.2011, http: //www.businessworld.in/en/storypage/-/bw/the-old-kings-and-the-new/365798.0/page/0 (accessed 1.7.2012).

Chakrabarti S. 2010. Factors influencing organic food purchase in India: expert survey insights. British Food Journal, 112/8: 902–915.

Chakrabarti S and Baisya RK 2007. Purchase motivations and attitudes of organic food buyers. Decision, 34: 2-22.

Datta P (Ed.). 2010. The Marketing Whitebook 2010-2011: One-stop Guide for Marketers, New Delhi: Businessworld.

Dholakia J and Maneesha S. 2012. Organic food: An assessment of knowledge of momemakers and influencing reasons to buy/not to buy. Journal of Human Ecology, 37(3): 221-227.

Garibay SV and Jyoti K. 2003. Market Opportunities and Challenges for Indian Organic Products", http: //orgprints.org/00002684 (accessed 15.1.2009).

Government of India. 2005. National Programme for Organic Production, New Delhi: Department of Commerce, Ministry of Commerce and Industry, http: // www.apeda.com/organic/ORGANIC_CONTENTS/English_Organic_Sept05. pdf (accessed 10.3.2009).

Gupta A. 2013. Participatory Guarantee System (PGS) and the Small Holder Markets. Paper presented at the IFOAM Asia-Pacific Symposium Entrepreneurship and Innovation in Organic Farming, Bangkok, 2.-4.12.2013.

Hanisch, M and Osswald N. 2012. Community-Supported Agriculture and Consumer Cooperatives in India. Paper presented at the international conference Cooperative Responses to Global Challenges, Berlin, March 21-23, 2012.

IBEF. 2008. Retail: Market and Opportunities. Indian Brand Equity Foundation, http: //www.ibef.org/artdisplay.aspx? cat_id=28 and art_id=19772 (accessed 30.1.2009).

Jishnu L and Sood J. 2012. Organic Universe, Down To Earth, July 16-31: 26–37.

Kriesemer S, Weinberger KK and Chadha ML. 2009. Demand for and awareness of safely produced vegetables in India, The World Vegetable Centre.

Kumar A. 2013. Ashmeet Kapoor's food venture, I Say Organic has a revenue of Rs 10 lakh a month, Economic Times, 10.06.2013.

Lohr K and Christoph D. 2007. Changing food purchasing and consumption habits among urban middle-classes in Hyderabad. Berlin: Humboldt-University, http: //www.sustainable-hyderabad.in (accessed 1.2.2009).

Menon MK, Sema A and Tej Partap. 2010. India Organic Pathway: Strategies and Experiences. In: Organic Agriculture and Agribusiness: Innovation and Fundamentals (Tej Partap and M Saeed, Eds.), Tokyo: Asian Productivity Organisation. pp 75–86, http: //www.apo-tokyo.org/00e-books/AG-22_ OrganicAgriculture/AG-22_OrganicAgriculture.pdf (accessed 31.5.2010).

Osswald N. 2013. Building short supply chains around consumer participation: Community-supported agriculture and consumer cooperatives in India.

Paper presented at the IFOAM Asia-Pacific Symposium Entrepreneurship and Innovation in Organic Farming, Bangkok, 2-4.12.2013.

Osswald N and Dittrich C. 2010. Sustainable food consumption and urban lifestyles: The case of Hyderabad/India. Emerging Megacities Discussion Papers 03/2010, Berlin: Europäischer Hochschulverlag, http: //www.sustainable-hyderabad.de.

Osswald N and Menon MK. 2013. Organic food marketing in urban centres of India, Bangalore: International Competence Centre for Organic Agriculture (ICCOA).

OTA. 2010. India Market and Regulatory Assessment: USDA Emerging Markets Program, Organic Trade Association, www.ota.com/pics/documents/MarketRegulatoryAssessment.pdf (accessed 7.9.2012).

Philip A. 2014. Organic products within reach of common man, thanks to PGS, *The Hindu*, 07.08.2014, http: //www.thehindu.com/todays-paper/tp-national/tp-tamilnadu/organic-products-within-reach-of-common -man-thanks-to-pgs/article6290033.ece (accessed 29.9.2014).

Polasa K, Sudershan RV, Subba Rao GM, Vishnu Vardhana Rao M, Pratima Rao and Sivakumar B. 2006. KABP Study on Food and Drug Safety in India: A Report. Food and Drug Toxicology Research Centre, National Institute of Nutrition, Hyderabad.

Radhika PP Ammani and Seema. 2012. Eating healthy: Consumer perception of organic foods in twin cities. International Journal of Marketing, Financial Services and Management Research, 1(2): 67-72.

Rao K, Raj S, Menon MK and Tej Partap. 2006. The market for organic foods in India: Consumer perceptions and market potential. Findings of a nation-wide survey. International Competence Centre for Organic Agriculture, Bangalore.

Roy D, Birol E, Deffner K and Karandikar B. 2010. Developing country consumers' demand for food safety and quality: Is Mumbai ready for certified and organic fruits?, In: *Choice Experiments in Developing Countries: Implementation, Challenges and Policy Implications* (B Jeff and E Birol, Eds.), Cheltenham/Northampton, MA: Edward Elgar Publishing. pp. 261-277.

Sally M. 2013. Increase in consumption of organic food products: ASSOCHAM survey, Economic Times, 23.05.2013.

Singh S. 2009. Organic produce supply chains in India: Organisation and governance, Ahmedabad, Gujarat.

Sudershan RV, Subba Rao GM, Pratima Rao, Vishnu Vardhana M and Polasa K. 2008. Food safety related perceptions and practices of mothers: A case study in Hyderabad, India. Food Control, 19(5): 506-513.

TechSci Research. 2013. India organic food market forecast and opportunities 2017, Noida: TechSci Research, http: //www.researchandmarkets.com/research/hks2ht/india_organic (accessed 12.9.2014).

The Nielsen Company. 2007. Unavailability and price the major reasons for Indians not purchasing organic products. http: //in.nielsen.com/news/20071203.shtml (accessed 16.1.2009).

Thottathil SE. 2014. India's organic farming revolution: What it means for our global food system. University of Iowa Press, Iowa City.

Whitehead RJ. 2013. Indian farmers should fill home and overseas void by going organic. *FoodNavigator-Asia.com*, 07.10.2013, http: //www.foodnavigator-asia. com/Markets/Indian-farmers-should-fill-home-and-overseas-void-by-going-organic (accessed 12.9.2014).

Wiggerthale M. 2009. *Zur Kasse bitte. Die neue Konsumfreudigkeit und boomende Märkte in Indien: Welche Folgen es haben kann, wenn Supermarktketten nach Liberalisierung des Einzelhandels rasch expandieren*, Berlin: Oxfam, www.oxfam.de/download/ zur_kasse_bitte.pdf (accessed 11.12.2009).

De Wit J and Verhoog H. 2007. Organic values and the conventionalization of organic agriculture. Wageningen Journal of Life Sciences, 54(4): 449-462.

Yes Bank. 2012. FDI in retail: Advantage farmers. Yes Bank Ltd. and ASSOCHAM, http: //www.assocham.org/arb/general/Background-FDI-Retail.pdf (accessed 2.9.2014).

Chapter 30

Policy and Institutional Environment for Promotion of Organic Farming in India

Shilpanjali Deshpande Sarma

The Energy and Resources Institute,
Indian Habitat Centre, Lodhi Road, New Delhi – 110 003
E-mail: shilpas@teri.res.in

1. INTRODUCTION

Traditional and subsistence agriculture practiced in India was assumed mainly in the absence of chemical inputs, informed by the contexts and needs of local agro-ecosystems incorporating indigenous knowledge that had evolved over time. In the Agricultural Testament, published in the 1940s by Albert Howard describes the robustness of Indian farming methods for soil fertility management in contrast to systems in existence at that time in the west. However severe food security concerns led to the advent of the green revolution in India, the ensuing expansion of the industrial mode of agri-production in the country being facilitated by strong institutional support.

The current organic movement in India emerged as a response to the negative environmental and socio-economic externalities, experienced as a part of the green revolution. In addition in large areas of India where agriculture is rainfed, farmers due to their inability to access costly inputs, continue to follow organic by default farming. By the late nineties India was home to farmers practicing organic agriculture with support from a growing number of committed civil society groups. On the other hand in 2001 the recognition of the growing international market as well as the potential for export earnings from organic produce resulted in an institutional framework being developed by Agricultural and Processed Food Products Export Development Authority (APEDA) based under The Ministry of Commerce and Industry (MoCI). In 2004, post introspection on the status of organic agriculture in the country, the National Project for Organic Farming (NPOF) was introduced the Ministry of Agriculture as the flagship program for promoting and supporting organic agriculture particularly as a tool for soil health management.

Despite emphasizing agriculture growth which is sustainable there is no direct reference to organic agriculture in the Agriculture Policy 2000 except for its allusion to promoting use of biomass and organic inputs as part of integrated nutrient or pest management and highlighting availability aspects. However in 2005, the Organic Farming Policy, 2005 was formulated by the Department of Agriculture and Cooperation (DoAC) after the NPOF. Its aim was to promote and develop organic farming for enhancing livelihoods of farmers and strengthening rural economy. The policy document focused on maintenance of soil fertility, identification of appropriate areas and crops for organic agriculture, development of package of practices, establishing model farms, market development, awareness creation and others (MoA, 2005). More recently the NPOF was moved under the National Mission for Sustainable Agriculture (NMSA) following which a new scheme – the Paramparagat Krishi Vigyan Yojna (PKVY) was announced in the 2014-15 union budget.

The mandate for supporting and promoting organic agriculture rests mainly with the MoCI and the Ministry of Agriculture and Farmers' Welfare (MoA and FW). This paper reviews and examines the support provided by the state at the national level for organic agriculture in India through its policies and institutional environment.

2. Promotion of Organic Agriculture through MoCI Led Initiatives

Having formulated NPOP in 2001, MoCI through APEDA has been focused on strengthening the exports market for Indian organic produce. The NPOP was the first comprehensive policy document developed to modernize and streamline the development of organic farming sector. The NPOP came out with its 7th edition in 2014 after a gap of nine years. It now includes regulations for the livestock and aquaculture sectors.

2.1 National Program for Organic Production (NPOP)

The objective of the NPOP is to facilitate the certification of organic products in conformity of Indian standards and that of importing countries besides also enabling the evaluation of certification programs as per established criteria. The NPOP framed the (i) National Standards for Organic Production (NSOP), (ii) accreditation regulations and criteria as well as procedures for inspection and certification agencies (iii) guidelines for certification groups as well as (iv) regulations for granting license for use of certification mark. The standards include those for organic crop production, organic poultry, livestock and products, organic apiculture as well as aquaculture production in addition to organic food processing and handling (including labeling). It also provides an institutional apparatus to facilitate implementation of the NSOP and certification of organic produce through National Accreditation Policy and Programme (DoC, 2014) The standards development was guided by the principles of IFOAM and is in harmony with international standards such as IFOAM and Codex (Bose, undated).

Presently the NPOP is regulated under different acts for domestic and export markets. Initially formulated for the development of exports in the organic sector, NPOP was notified under the Foreign Trade Development and Regulation Act and implemented by MoCI with APEDA leading the initiative. The NPOP standards are recognized by European Union and Switzerland and even the United States Department of Agriculture as equivalent to their standards (Bose, undated).

The National Steering Committee (NSC) is responsible for the implementation and administration of NPOP which has members from MoCI, MoA, APEDA and other boards. The NSC also functions as the National Accreditation Body (NAB) which undertakes accreditation of inspection and certification agencies and also certification programs. The Evaluation Committee (that consists of members similar to NSC but also including Export Inspection Council of India or Export Inspection Agencies) associated with NAB evaluates inspection and certification agencies and submits recommendations to the body. APEDA, one of the members of the committee plays an important role receiving applications, screening them besides coordinating and arranging evaluation visits. The Inspection and certification agencies should have programs in operation for at least a year and must be competent with operational procedures as well as national and international standards (DoC, 2014).

2.2 Agricultural and Processed Food Products Export Development Authority

APEDA, established in 1985, is responsible for the development of scheduled products, their industries and promoting their exports. For organic agriculture APEDA is responsible for the implementation of NPOP, promoting organic exports, establishing equivalency with countries importing Indian organic produce in addition to undertaking training and capacity building of stakeholders. APEDA focuses on reviewing and amending national standards for organic agriculture besides also evaluating, accrediting and monitoring certification bodies. Recently APEDA has also initiated a web based electronic traceability system known as Tracenet for organic traceability (Bose, undated).

2.3 Tracenet

Tracenet is a web based traceability system that enables stakeholders to track organic produce from its source to its export and subsequent import by other countries besides also facilitating certification of produce to be exported (Sood, 2013). This system identifies all operators across the supply chain through obligatory registration of producers, processors, traders by certifying agencies documenting information on the linkages across the organic supply chain as well as on quality assurance. Registered operators must contribute information on their plan for the complete cropping season whereas the certifying agency enters decisions arrived at after field inspections including compliance/non-compliance verdict and the underlining reasons. The operator's farm can be mapped by a Global Positioning system and is verified by certifying agencies at the time of inspection. All this information is not only necessary for the realization of organic certification but also is the basis for tracking organic products in case of future issues of non-compliance emerge. In this way Tracenet not only provides a platform for collating all stakeholders in the organic supply chain (Sood, 2013), it also provides a repository of information on individual organic supply chains in India.

3. Key Programs and Initiatives under MoA and FW for Promoting Organic Agriculture

The MoA and FW led the development of organic agriculture in the country through the NPOF which emphasized the strengthening of input production for organic agriculture. NPOF has now been subsumed under NMSA which besides possessing a theme on Integrated Nutrient Management (INM) and Organic farming as part of the Soil Health component also contains a new scheme known as the Paramparagat Krishi Vikas Yojna (PKVY). A key initiative for promoting research and development on organic agriculture has also been undertaken through the Indian Council for Agricultural Research (ICAR) Network project on Organic Farming. In addition regulations and guidelines for third party certification as well as alternatives like the Participatory Guarantee Systems (PGS) were also developed.

Implementation of the various programs and schemes on organic agriculture rests with the state level line departments of agriculture or horticulture as they are the primary state mechanism for agricultural extension. In case district/block or village level interventions related to organic farming or practices need to be operationalized, it is undertaken by the same personnel that provide advisory or extension services for conventional agriculture At the state level Agriculture Technology Management Agency (ATMA) and Krishi Vigyan Kendras (KVKs) have been known to provide training and other related forms of extension service to organic farmers.

3.1 National Project on Organic Farming

Initiated as a pilot project in 2004, as a part of the X Five Year Plan, the NPOF subsumed the 'National Project on Use and Development of Biofertilizers' to become the flagship program promoting organic agriculture. The existing organizational set up under the bio-fertilizer program were re-invented as the National Centre for Organic Farming (NCOF) and six Regional Centers for Organic Farming (RCOFs) (DoAC, 2010a). With soil health management as the underlining strategy, the scheme was revised in 2010 with these objectives (DoAC, 2010b):

1. Promotion and production of quality biological and organic inputs (biofertilizers and biopesticides) as well as improvement of soil organic carbon
2. Implementation of a quality control regime for biofertilizers as per the Fertilizer Control Order; formulation/upgradation of standards or protocols especially for unregulated biological inputs; the development, maintainance and supply of microbial based bio-inputs for nutrient management and plant protection besides undertaking testing of these biofertilizers for their bio-efficacy was also intended.

3. Technical capacity building of stakeholders through short term certificate courses on organic farming and on farm resource management; regular trainings and refresher courses for quality control personnel; trainers' training on wide ranging aspects for farmers, certification or extension agencies, industries and others. The NCOFs and RCOFs were to lead the development of human resources.
4. Technology development through initiating research for validation of indigenous practices, inputs and technologies for the development of package of practices. Here the implementing agencies were NCOF/RCOFs/ICAR/State Agricultural Universities (SAUs).
5. Building capacity for soil health assessment by way of initiating studies on biological soil health aspects under different farming systems practices on geographic regions. These were to be taken up by NCOF/RCOFs in collaboration with ICAR/SAUs through provision of financial assistance
6. Provision of technical support to certification systems (through formulation of standards, design of implementation protocols, evaluation and surveillance); developing and administering PGS with NCOF as the implementation agency along with interested state government or private agencies/NGOs or unaided. NCOF was also to undertake residue testing of PGS samples.
7. Publication of material on organic agriculture; awareness generation via organizing conferences and trade shows and publicity through print and electronic media. These objectives would be implemented by NCOF/RCOFs/central/state agencies, ICARs, SAUs.
8. To function as a data collection centre and information repository of information on organic agriculture
9. Supporting central and state governments in relation to evaluation and monitoring schemes on organic agriculture.

Given the thrust on inputs generation the *Capital Investment Subsidy Scheme for Vegetable and Fruit Market Waste Compost, and Biofertilizers-Biopesticides Production Units* was devised as part of NPOF with the intention of strengthening the infrastructure for production of organic/biological inputs. The program provides credit linked and back ended capital investment subsidy for establishing compost production units as well as bio-fertilizer and bio-pesticide production units (Table 30.1). Focusing on biofertilizers, biopesticides in addition to fruit and vegetable market waste, the objective of the scheme was to enhance the availability and quality of these inputs to reduce dependence on chemical fertilizers, besides improving soil health and agricultural productivity (DoAC, undated1). This scheme was to be implemented through NCOF in collaboration with National Bank for Agriculture and Rural Development (NABARD).

Some NPOF objectives that existed prior to the revised guidelines of 2010 such as capacity building and training through service providers and other organizations, establishing model organic farms and finance for vermiculture hatcheries have been discontinued as they could be taken up as part of other schemes (DAC,

2010b). Also as per the previous guidelines, NCOF was to function as a central nodal agency coordinating with government agencies as well as other stakeholders (farmers, NGOs *etc.*) besides ensuring convergence of schemes contributing to the development of organic agriculture. These functions were perhaps diluted in the present version.

Table 30.1: Details of the Capital Investment Subsidy Scheme (CISS) for Setting up of Organic Inputs Production Units

Organic Inputs Production Unit	*Financial Assistance*	*Eligible Agencies*
Fruit and Vegetable market waste/agro-waste compost units Financial Assistance	33 per cent of Total Financial Outlay (TFO) or Rs. 60 lakh whichever is less for 100 Ton Per Day (TPD)	Municipalities, APMCs, Public sector/Private sector companies, fertilizer companies or any individual entrepreneurs
Bio-fertilizer/Bio-pesticide production units	25 per cent of TFO or Rs.40 lakh whichever is less for setting up of 200 Ton Per Annum (TPA) capacity production unit.	Public sector/Cooperative/ private sector companies, small agencies/NGOs and Individual entrepreneurs

Source: DoAC (2010b).

3.2 National Mission on Sustainable Agriculture

The National Mission on Sustainable Agriculture was formulated under the National Action Plan for Climate Change for designing and implementing measures to improve resilience of Indian agriculture to climate change impacts. The restructured NMSA program for the XII Plan seeks to improve agricultural productivity through focus on soil and water conservation, water use efficiency, soil health management and rainfed area development (DoAC, 2014a).

The NMSA has a specific component for soil health management under which it merges the NPOF, National Project on Management of Soil Health and Fertility (NPMSH and F) and Soil and Land Use Survey of India. The aim is to promote sustainable practices for location as well as crop specific soil health management such as residue management, organic farming practices through linking soil fertility maps with macro- micro nutrient management, judicious application of fertilizers and minimizing soil erosion. This component has two programmatic interventions for 1. Soil health, which focuses on strengthening soil mapping, testing and reclamation as well as fertilizer quality control, and 2. Integrated Nutrient Management (INM) and organic farming under which the following initiatives are promoted and financially aided (DoAC, 2014a):

- ☆ Establishing compost and organic inputs production facilities
- ☆ Setting up or strengthening quality control laboratories,
- ☆ Adoption of organic villages for undertaking integrated manure management and initiatives for biological nitrogen fixation
- ☆ Adoption of organic farming by cluster approach again under PGS
- ☆ Support for PGS including soil residue analysis and on-line data management respectively

- ☆ Promotion of organic inputs on farms
- ☆ Demonstration of package organic farming practices in farmers field capacity building of stakeholders through various training by NCOF and its regional centres
- ☆ Support for the development of location and cropping system specific package of practices for organic agriculture
- ☆ Establishment of separate departments for organic agriculture research and teaching in state agricultural universities (Table 30.2).

Table 30.2: Details of Assistance Provided under INM and Organic Farming

Sl. No.	*Component*	*Pattern of assistance*	*Implementation Agency*
1	Setting up of mechanized fruit/ vegetable market waste/agro waste compost production unit	100 per cent Assistance to State Govt./ Govt. Agencies upto a maximum limit of Rs. 190 lakh/unit and 33 per cent of cost limited to Rs. 63 lakh/unit for individuals/ private agencies through NABARD as capital investment for 3000 TPA production capacity	NABARD
2	Setting up of state of art liquid/ carrier based Biofertilizer/ Biopesticide units	100 per cent Assistance to State Govt./Govt. Agencies upto a maximum limit of Rs. 160 lakh/unit and 25 per cent of cost limited to Rs. 40 lakh/unit for individuals/private agencies through NABARD as capital investment of 200 TPA production capacity	NABARD
3	Setting up of bio-fertilizer and organic fertilizer testing quality control laboratory (BOQCL) or strengthening of existing laboratory under FCO	Assistance up to maximum limit of Rs. 85 lakh for new laboratory and up to a maximum limit of Rs. 45 lakh for strengthening of existing infrastructure to State Government laboratory under Agriculture or Horticulture Department.	State Government
4	Promotion of organic inputs on farmer's field	50 per cent of cost subject to a limit of Rs. 5000/ha and Rs. 10,000 per beneficiary. Propose to cover 1 million ha area	State government
5	Adoption of organic farming through cluster approach under Participatory Guarantee System (PGS) certification	Rs. 20,000/ha subject to maximum of Rs. 40,000 per beneficiary for 3 year term	State government
6	Support to PGS system for on-line data management and residue analysis	Rs. 200 per farmer subject to maximum of Rs. 5000/group/year restricted to Rs. 1 lakh per Regional Council. Up to Rs. 10, 000/ sample for residue testing	Regional Councils under PGS
7	Organic village adoption for manure management and biological nitrogen harvesting	Rs. 10 lakhs/village for adoption of integrated manure management, planting of fertilizer trees on bunds and promotion of legume intercropping through groups/SHGs *etc.* (Maximum 10 village per annum/State will be supported)	State government

Sl. No.	*Component*	*Pattern of assistance*	*Implementation Agency*
8	Training and demonstration on organic farming	Capacity building of stakeholders. Rs. 20,000/- per demonstration for a group of 50 participants or more	NCOF/State Government-not clear
9	Support to research for development of organic package of practices specific to state and cropping system	Against specific proposal	ICAR/SAUs/ other research institutions/state government agencies
10	Setting up of separate organic agriculture research and teaching department	Against specific proposal	SAUs

Source: Adapted from DoAC (2014a).

Besides specific support for organic agriculture, the NMSA also encourages the formation of farmer producer companies by farmers for marketing organic products with support for certification and financial assistance as eligible under both Farmer Producer Organisations (FPOs) and NMSA norms. It suggests linkages between FPOs and market federations for strengthening marketing for organic products (DoAC, 2014a).

3.3 Paramparagat Krishi Vikas Yojana

In 2015, the government announced the Paramparagat Krishi Vikas Yojana (PKVY) as part of the soil health component of the NMSA for increasing organic production and certification of organic produce. Intended as a cluster based programe, PKVY envisages the development of 10,000 clusters covering 5 lakh acre with each cluster consisting of 50 or more farmers and 50 acres. Financial assistance of INR 20000 is proposed for the farmer for organic production including harvest and transport of produce to the market besides also not having to bear the any expense for certification. There is emphasis on promoting use of traditional resources besides creating market linkages. A sum of INR 300 crore was allocated for the year 2015-16 for PKVY.

Given the introduction of PKVY in the NMSA the objectives included previously under the INM and organic farming component have been revised and reorganized between this component and the new PKVY component. The INM and organic farming intervention continues to support the establishment of compost production units, biofertilizer and biopesticide production units as well as organic fertilizer quality testing laboratory still remains along with strengthening of laboratory facilities. Furthermore measures related to location and cropping system specific organic package of practices, setting up of research and teaching departments for organic agriculture and promotion of organic inputs (such as manure, vermin-compost, biofertilizers, herbal extracts *etc.*) are also within its purview (DoAC, undated).

On the other hand objectives related to adoption of PGS certified organic farming through cluster approach, the adoption of organic village for manure management and biological nitrogen harvesting in clusters as well as training and demonstration on organic farming have been moved under PKVY (Table 30.3) (DoAC, undated).

Table 30.3: Details of the Components, Proposed Activities and Assistance under PKVY

Component/ Subcomponent	*Proposed Activities and Assistance*
Adoption of PGS certification through cluster approach	
1.1 Mobilization of farmers/local people to form cluster in 50 acre for PGS certification	Conducting of meetings and discussions of farmers in targeted areas to form organic farming cluster @ Rs. 200/farmer Exposure visit to member of cluster to organic farming fields @ Rs. 200/ farmer Formation of cluster, farmer pledge to PGS and identification of LRP from cluster Training of cluster members on organic farming (3 trainings @ Rs. 20000 per training)
1.2 PGS certification and quality control	Training on PGS certification in 2 days @ Rs. 200 per LRP Training of trainers (20) lead resource persons @ Rs. 250/day/cluster for 3 days Online registration of farmer @ Rs.100 per member cluster x 50 Soil sample collection and testing (21 samples/year/cluster) @ Rs. 190 per sample for three years Process documentation of conversion into organic methods, inputs used, cropping pattern followed, organic manures and fertilizer used *etc.*, for PGS certification @ Rs.100 per member x 50 Inspection of fields of cluster member @ Rs. 400/inspection x 3 (3 inspections will be done per cluster per year) Residue analysis of samples in NABL (8 samples per year per cluster) @ Rs. 10, 000/sample Certification charges Administrative expenses for certification
Adoption of organic village for manure management and biological nitrogen harvesting through cluster approach	
2.1 Action plan for organic farming for one cluster	Conversion of land to organic @ Rs.1000/acre x 50 Introduction of cropping system; Organic seed procurement or raising organic nursery @ Rs.500/acre/year x 50 acres Traditional organic input production units like Panchagavya, Beejamruth and Jeevamruth *etc.* @ Rs.1500/unit/acre x 50 acre Biological nitrogen harvest planting (Gliricidia, Sesbania, *etc.*) @ Rs. 2000/ acre x 50 acre Botanical extracts production units (neem cake, neem oil) @ Rs.1000/unit/ acre x 50 acre

Component/ Subcomponent	*Proposed Activities and Assistance*
2.2 Integrated manure management	Liquid biofertilizer consortia (nitrogen fixing/phosphate solubilizing/ potassium mobilizing biofertilizer) @ Rs. 500/acre x 50
	Liquid biopesticides (*Trichoderma viridae*, *Pseudomonas fluorescens*, *Metarhizium*, *Beauveria bassiana*, Pacelomyces, verticillium) @ Rs. 500/ acre x 50
	Neem cake/neem oil @ Rs.500/acre x 50
	Phosphate rich organic manure/Zyme granules @ Rs. 1000/acre x 50
	Vermicompost @ Rs.5000/unit x 50
2.3 Custom hiring centre (CHC) charges	Agricultural implements
	Walk-in tunnels for horticulture
	Cattle shed/poultry/piggery for animal compost
2.4 Packing, Labeling and Branding of organic products of cluster	Packing material with PGS logo + hologram printing @ Rs. 2500/acre x 50
	Transportation of organic produce (Four wheeler, 1.5 tone load capacity) @Rs. 120000 max. assistance for 1 cluster
	Organic Fairs (maximum assistance will be given @ 36330 per cluster)

Source: Adapted from DoAC (undated).

3.4 National Centre of Organic Farming

Under the NMSA, the key activities of the NCOF broadly include promotion of organic agriculture through

- Area expansion
- Technology transfer, promotion and production of bio-inputs
- Overseeing quality control, development of standards and testing protocols of organic fertilizers
- Implementing certification including PGS
- Fostering awareness and publicize organic farming (through publication of newsletters, training manuals and other literature)
- Technical capacity building

Human resource development through various trainings programs is an important objective. These may include international trainer's training and exposure for government officers from national and state agencies, certification personnel and scientists from SAUs and ICAR as well as various certificate courses. The latter include (i) certificate course on organic farming for cultivating organic agriculture extension workers and field workers (ii) training or refresher courses on production and quality control of organic inputs for government officials and personnel form production units (iii) trainers training through customized courses on various aspects for extension officers, KVK and NGO trainers, fertilizer inspectors and as well as production and quality control unit personnel and members associated with PGS Regional Council (iv) training of field functionaries and extension officers on production soil health management and PGS aspects in addition to quality control training for input dealers (DoAC, 2014).

Promoting domestic markets for organic produce is also possibly one of its functions (DoA, 2014a) although this aspects appears to be absent in what may be the more recent operational guidelines for NSMA in which has details of the PKVY scheme have been included (DoAC, undated2).

However recognizing that NCOF has thus far focused on biofertilizer production rather than on the overall coordination of various activities needed to promote organic agriculture, the NMSA under the XII Plan remedy this aspect by through its proposed initiatives (DoA, 2014)

3.5 Network Project on Organic Farming

Initiated in 2004-05 and supported by the ICAR, New Delhi, the Network Project on Organic Farming was undertaken in order to study organic systems comprehensively and develop package of practices under organic farming for various crops and copping systems across multiple sites in the country. Thirteen centers were chosen from amongst research departments of various SAUs and ICAR institutes for conducting the multi-location experiments associated with the study across 9 agro-climatic zones. The Indian Institute of Farming Systems Research, Modipuram was designated as the lead institute. The cropping systems proposed for evaluation included cereals, pulses and oilseeds, vegetables, spices, cotton, medicinal plants and fodder crops (ICAR-IIFSR, 2014).

The study has a threefold objectives (i) investigate the productivity, economic potential, sustainability as well as input use efficiencies of organic crops and cropping systems in various agro-ecological zones (ii) develop under organic farming efficient options for crop and soil management (iii) develop new techniques for farm-waste recycling that are need based as well as cost effective. As part of the experiments, the intention was to assess the performance of cropping systems under organic and conventional farming besides evaluating the efficiencies of various organic inputs including organic manures and other bio-agents. (ICAR-IIFSR, 2014).

3.6 Regulations for Domestic Certification and Labeling

For import and domestic organic market, the NPOP was notified under the Agriculture Produce Grading, Marking and Certification Act as the Organic Agricultural Produce Grading and Marking Rules, 2009. The regulatory authority is the Directorate of Marketing and Inspection headed by the Agricultural Marketing Advisor (AMA) under the Ministry of Agriculture (DoAC, 2010a). Inspection and certifying agencies accredited under NPOP are eligible to seek 'Certificate of Authorization'. Organic produce certified under Agmark have Agmark India Organic label. Nevertheless the use of AGMARK certification still remains a voluntary (Bose, 2014). Recently the regulatory authority for organic domestic market may also have been extended to the Food Safety and Standards Authority of India (FSSAI), which is under Ministry of Health and Family Welfare (Sood, 2013).

3.7 Participatory Guarantee System

Facilitating relatively simpler alternatives to third party certification, for guaranteeing integrity of organic produce is necessary for integrating small and marginal farmers into the organic movement. Participatory Guarantee Systems (PGS) are locally relevant quality assurance initiatives that undertake certification based on stakeholder participation, built on trust, social networks, shared vision and knowledge sharing (DoAC, undated). There is an institutionalized mechanism for certification processes under PGS and PGS certified organic production can be identified by a special logo (NPOF, undated).

The organizational set up for the implementation and administration of the government led PGS program up involves farmers, local groups, regional councils, zonal councils, and an overarching national body, each assuming responsibilities (DoAC, undated). The NCOF functioning as the secretariat and National Advisory Committee are the overarching national functionaries (DoAC, undated). Any farmer willing to engage in organic production, register with the PGS system and follow its norms can join a PGS group. The local group, comprising of a peer group of farmers who are geographically close, is the body that appraises farms and takes decisions on granting certification. Members of the group sign the pledge agreeing to follow the group specific vision and taking collective responsibility. The regional council in the government supported PGS can be either state agencies, certification bodies, NGOs or even assemblage of local groups. Besides undertaking capacity building of local groups, the regional councils facilitate registration of local groups, sampling of local group for farm appraisals and issue of certification numbers to those approved. Zonal councils are responsible for authorizing regional councils and monitoring their activities, coordinating outreach activities, collecting PGS certified samples for analysis and also function as the complaint redressal appellate (DoAC, undated).

4. Support for Organic Agriculture/Organic Practices through Other MoA and FW Led Schemes

In addition to those programs or initiatives that have a sole or major focus on organic agriculture, MoA and FW has other schemes that supports either organic farming or in some cases specific organic practices. These include Mission for Integrated Development of Horticulture, National Food Security Mission and National Mission on Oilseeds and Oil Palm. Previously other schemes like The National Project on Management of Soil Health and Fertility (NPMSH and F), 2008-09 now subsumed under the Soil Health Component of NMSA also promoted INM and therefore judicious chemical use with organic manures, green manures and bio-fertilizers in addition to intending to strengthen soil testing facilities. Another older scheme Balanced and Integrated Use of Fertilizers initiated in 1991-92 and later subsumed under NPMSH and F also had similar objectives.

4.1 Mission for Integrated Development of Horticulture

The Mission for Integrated Development of Horticulture (MIDH) was launched for the holistic development of the horticulture sector with a focus on regionally differentiated strategies and aspects such as research and technology promotion, extension and marketing. This scheme integrates various ongoing schemes including National Horticulture Mission (NHM), Horticulture Mission for North East and Himalayan States which prior to the formulation of MIDH provided financial assistance for adoption of organic practices and certification in relation to horticulture crops (DoAC, 2014b). MIDH continues with the similar objectives. It specifies that organic farming must not be taken up in isolated spaces and should be pursued only in crops which can ensure a premium and where market linkages can be established. It also recognizes that state interventions must not be limited to only procurement and supply of organic inputs (DoAC, 2014b).

4.2 National Food Security Mission

Focusing on the crops rice, wheat, pulses and also coarse cereals, the aim of the National Food Security Mission (NFSM) is to increase production and productivity in these crops, restore soil fertility and enhance farm profits for economic sustenance of farmers. To be implemented in districts displaying low crop productivity but high potential, NFSM provides assistance for demonstrations, seed distribution, plant protection, micronutrients and soil ameliorants and others. The scheme promotes INM and IPM. Under the block demonstration of production and protection technologies in large crop areas support is provided for demonstrations involving use of biofertilizers (*Azotobactor*, phosphate solubilising bacteria (PSB), potash mobilizing and Zinc solubilizing bacteria) in rice, wheat, pulses, maize and millet systems. Additionally green manuring is promoted for rice and wheat under the various interventions for block demonstration. For pulses on the other hand demonstrations of intercropping, use of *Trichoderma* are amongst the supported interventions.

Under INM, financial assistance for the use of biofertilizers (*Rhizobium* and PSB) is only available for pulses at Rs 100/ha or 50 per cent of the cost whichever is lower (DoAC, undated). However for plant protection there is joint assistance for biopesticides along with plant protection chemicals and IPM at Rs 500/ha or 50 per cent of the cost whichever is lower for rice, wheat and pulses. Under IPM for pulses, pheromone traps, mechanical control as well as biological control by use of natural agents and use of *Trichoderma* for seed treatment is promoted according to another policy document. This document also cites higher financial assistance (Rs.750/ha or 50 per cent of the cost whichever is less) for IPM in pulses which includes pesticide, weedicides, bio-pesticides, bio-agents (DoAC, undated).

For pulses, the Accelerated Pulses Production Program was launched in 2010 for the implementation of NFSM pulses to propagate INM and IPM for adoption amongst farmers through demonstration of plant nutrient and plant protection technologies and practices for improving pulse production/productivity in blocks cultivating major pulse crops. Here through provision of a kit containing critical inputs including *Rhizobium*, PSB (besides gypsum and micronutrients, urea) as well as plant protection chemicals (including pheromone traps, *Trichoderma* and biopesticides), herbicides *etc.* was to be provided free of cost to farmers for a maximum area of 2 ha. The approximate cost of the items provided per ha as part of the kit was estimated at 5400 per ha as of 2010 (DoAC, 2010c).

4.3 National Mission on Oilseeds and Oil Palm

The National Mission on Oilseeds and Oil Palm (NMOOP) envisions increasing the production of vegetable based oils from oilseeds and oil palm given the growing consumption of edible oil in India (MoA and FW, 2014-15). NMOOP provides financial assistance for organic inputs. Under the Mini Mission (MM-I) on oilseeds, there is support for biopesticides by way of 50 per cent of the cost limited to Rs 500/ha. Liquid biofertilizers such as *Rhizobium, Azatobactor*, PSB, potash mobilising bacteria, Zinc solubilizing bacteria are recommended for application for which a fixed subsidy is provided. As part of the Mini Mission II on Oil Palm, the cost of construction of vermicompost units is subsidized by 50 per cent limited to Rs15000/unit (MoA and FW, 2014-15).

5. Support for Organic Agriculture and Practices through Ministry of Rural Development

The Ministry of Rural Development (MoRD) supports sustainable practices in the agriculture production including organic practices through some specific programs that are focused on livelihood generation for farming communities such as the Mahatma Gandhi National Rural Employment Guarantee Scheme (MGNREGS) as well as the National Rural Livelihood Mission (NRLM) besides the Saansad Adarsh Gram Yojna (SAGY).

5.1 Mahatma Gandhi National Rural Employment Guarantee Scheme

The Act which underlines the MGNREGS (a rights based program) ensures a minimum of 100 days of wage based employment in a year to each rural household in an effort to promote livelihood security, social protection the poor and fostering democratic empowerment for inclusive growth. It integrates natural resource management and livelihoods generation through the creation of durable assets for water and soil conservation, horticulture plantation and afforestation and for enabling improved productivity of land (DoRD, 2013).

Works allowed as part of MGNREGS under the agriculture domain include NADEP composting, vermicomposting and liquid bio-manures. Specific interventions in the livestock domain are permissible such as building goat shelters which can improve the efficiency of dung collection to be used as organic inputs and promotes integrated farming. Similarly construction of pucca floor and urine tank can be undertaken for cattle which could facilitate better management of cattle urine and cow dung so that it can be used for improving soil fertility. On the other hand works related to water harvesting and water conservation are also allowed (DoRD, 2013).

5.2 National Rural Livelihood Mission

Undertaken as a mission mode program, the NRLM seeks to address poverty in rural communities through enhancing and sustaining livelihood opportunities (self-employment or wage employment) and strengthening the development of grassroots institutions of the poor. Skill and capacity building linked to employment opportunities is one of the main focus areas. In part the Mission aims to lower cultivation costs, improve productivity as well as economic returns to farmers besides also enhancing food and nutritional security in rural communities (MoRD, 2015a).

The NRLM may support organic agriculture/practices as part of its livelihoods enhancement agenda under which is the Sustainable Agriculture for Small Producers (SASP), an intervention to reduce the vulnerability of the poor. SASP is a part of the Community Investment Support Sub component that provides livelihood support to Self Help Groups/federations undertaking productive livelihood activities. The National Rural Livelihood Program envisions the adoption of sustainable organic agriculture in a significant proportion of land owned by the poor. Therefore the SASP seeks to promote community managed sustainable agriculture and implement sustainable agriculture practices in a phased manner. Key aspects include non-pesticide management and ecological cultivation, biodiversity improvement as well as soil management and water use efficiency cultivation methods (MoRD, 2015a).

The NRLM also facilitates the empowerment of women in the agriculture sector by way of provision of sustainable livelihood opportunities through its sub-component *Mahila Kisan Sashaktikaran Pariyojna (MKSP)*. Facilitating sustainable agriculture and improvement in soil fertility is envisioned through the promotion of farming based on local resources where women have greater control over production systems; through establishment of community based organizations (self-help groups) by way of which women themselves manage support systems and also with help from larger community driven institutions such as NGOs, farmer field schools *etc.* The scheme derives its main funding from MoRD (75 per cent), the balance to be borne by state or any other donor agency. Projects may be proposed and implemented by the state government, panchayati raj institutions, civil society and community based organizations, and women based groups or even a consortium of such organizations (MoRD, 2015b).

5.3 Saansad Adarsh Gram Yojana

Based on Mahatma Gandhi's vision of a model village and rural development, the Saansad Adarsh Gram Yojana (SAGY) envisages the integrated development of chosen villages across various socio-economic and human development domains – agriculture, health, education, livelihoods, environment besides gender equality, community participation and cooperation, social justice, local self-governance, accountability and others (DoRD, 2014). The guidelines provide a list of activities that can be undertaken under the various domains. As part of the activities that can be taken up under economic development are those that promote diversified agricultural livelihoods that also includes horticulture and livestock. With the aim to facilitate sustainable agriculture and organic farming the proposed outcomes are to reduce the use of chemical inputs, strengthen organic input production and use of organic pesticides in addition to ensure access to farm machinery at reasonable hiring rates. The strategy is to promote transfer of technology for organic farming to farmers' groups and women farmers and practices such as SRI, preparation of soil health cards, establishing local seed banks and agro service centres as well as encouraging micro/drip irrigation and solar pumps. The schemes that can be leveraged for activities include various schemes of MoA, MKSP under NRLM and also MGNREGS (DoRD, 2014).

6. National Bank for Agriculture and Rural Development (NABARD)

NABARD was established in 1982 with the purpose of providing credit support and related services as well as enabling institutional development for promoting sustainable and inclusive development within the agriculture sector. As described in a previous section the *Capital Investment Subsidy Scheme for Commercial Production Units of Organic Inputs*, one of the components under the National Project on Organic Farming was being implemented since 2004-05 in collaboration with NABARD. As part of the scheme, credit linked back ended subsidy is provided for fruit and vegetable market waste compost and biofertlizers - biopesticides production units. Subsidy for a third component related to vermin-hatcheries was discontinued in 2010 (NABARD, 2011).

NABARD also hosts an Umbrella Programme for Natural Resource Management (UPNRM) as part of its Natural Resource Management policy. With a focus on initiatives that support sustainable natural resource management for enhancing livelihoods of poor and marginalized groups, the aim of UPNRM is to promote projects that use innovative approaches, proffer scope for replicability and can be mainstreamed into the natural resource management domain. Besides focusing on disadvantage groups, the UPNRM is guided by principles for ensuring environmental sustainability, community participation in design and implementation of projects, decentralized governance as well as utilizing an integrated and needs based approach. Core priority areas under UPNRM include Soil and water conservation as well as plantation and horticulture under which projects associated with organic agriculture may be undertaken.

7. Support for Organic Practices through Ministry of New and Renewable Resources

National Biogas and Manure Management Programme (NBMMP) initiated by the Ministry of New and Renewable Resources promotes the setting up of family type biogas plants with the aim to provide clean gaseous fuel for cooking as well as organic manure to rural and semi urban households. Besides fostering improved sanitation and reducing drudgery for women, the scheme also seeks to promote the use of the digested slurry (from the biogas plant) as a bio-manure in order to reduce or supplement the use of chemical fertilizer. This scheme is being implemented through the state nodal departments and is intended to help mitigate climate change. Approximately 47.5 lakh biogas plants have been set up across the country and the target for the year 2014-15 was 1.1 lakh (MNRE, 2014).

8. Conclusion and Ways Forward

Over the last 15 years, the country has laid a sound foundation for the development of organic farming in the nation through the establishment and implementation of the NPOP and NPOF. Along with civil society, state supported programs such as NPOF have helped popularize organic agriculture and its benefits soil health management. A beginning has been made towards understanding the scientific basis of various organic practices and assessing their efficacy with the ICAR network project. With a visible thrust under the NMSA and the recent PKVY and the unveiling of the PGS, there appears to be a clear intention for strengthening organic agriculture in India. However in order to exploit its full potential and facilitate the sustainable expansion of organic farming several aspects must be addressed.

For one, anchoring organic agriculture to a broader national policy on agriculture is necessary in order to reflect its linkages and opportunities for rainfed agriculture and watershed development, food security, natural resource conservation and management, rural development and climate change mitigation. Attention must be paid to policies on land reform and programs that influence availability of labour for agricultural work as they may significantly impact the viability of organic agriculture. Additionally there is a need to identify suitable expansion targets for organic agriculture in various regions based on available local capacities, agroclimatic conditions and socio economic realities. Overall an effective implementation strategy must be formulated based on assessment of needs and feasibility of projects, analyzing the entire value chain and involving various stakeholders at the local level. An in-depth evaluation of the past initiatives and outcomes is vital to inform these efforts (UNEP-UNCTAD, 2008; Deshpande Sarma, 2015).

Harmonization of various schemes supporting organic agriculture is needed to facilitate programmatic efficacy. In relation to financial assistance to farmers, support for inputs and certification could be further strengthened based on a thorough review the utilization pattern. However more importantly there is need to aid farmers financially during the period of conversion from conventional to organic farming –about 3 three years when yields might fall and market opportunities are yet to be identified (Deshpande Sarma, 2015).

In various programs/schemes, there has been excessive focus on commercial input production or use. This needs to be balanced by making agroecological principles the underlying basis for organic agriculture, promoting utilization and management of on farm and local resources for developing input and pest management strategies. In schemes promoting biofertilizer or biopesticide use as part of another larger mandate, emphasis on these aspects is usually diluted and implementation fragmented. There is a need to encourage a wider set of organic practices aligning them with the relevant cropping and agroecosystems. On the other hand, the availability and quality of external inputs- biopesticides and biofertilizers is also problematic and needs to be addressed both through regulatory measures as well as stakeholder deliberations. Mainstreaming organic practices in agriculture policy and practice would promote sustainable agriculture and familiarize farmers with the techniques of organic agriculture (Deshpande Sarma, 2015).

The primary emphasis on inputs production has also resulted in sub-optimal attention to other key areas. Greater attention towards training and capacity building of stakeholders including farmers, research and development, technical and advisory support to farmers, value chain consolidation and development of market linkages, strengthening PGS, building public awareness is essential for the successful development of the sector (DoAC, 2014a).

Widespread adoption of organic agriculture demands emphasis on research and extension support. The proposed thrust towards establishing research centers for organic agriculture and supporting location as well as cropping system specific research for development of appropriate package of practices in institutions is welcome. However there is clear need to mainstream organic agriculture research in agriculture universities and ICAR institutions underlined by interdisciplinary and participatory research with farmers and other stakeholders. Research on the assessing the efficacy of various organic practices including traditional inputs is necessary as is undertaking an evaluation of the suitability of various organic practices in relation to the different cropping and agro-ecologoical systems. Furthermore the already existing research and packages developed under various research projects must be taken forward into the farmers' field in a planned and effective manner in order to validate the same and accommodate it to field realities. However in midst of these initiatives attention must be paid to effective coordination of research across various organizations in addition to collection, management and documentation of data and best practices (Deshpande Sarma, 2015).

A greater thrust in promoting organic practices is urgently required within state extension agencies. For improving the quality and frequency of technical advice provided to farmers, collaboration with local NGOs and training local youth may be explored. Besides consulting with farmers on production issues and encouraging agroecological practices, extension personnel may also provide advice on aspects such as soil health aspects, yield optimization and certification. Training of farmers as well as other stakeholders on production aspects as well as market opportunities, storage and processing, strengthening of value chains – backward and forward linkages is also imperative (Deshpande Sarma, 2015).

Whereas certification standards are comprehensive, third part certification in its present form poses challenges to farmers, especially small and marginal farmers with regards to costs, complex or lengthy procedures and bureaucratic hurdles. Besides developing mechanisms to make third party certification farmer friendly, expansion and strengthening of PGS systems is vital. To reduce problems with quality assurance adequate capacity building for farmers on third party or PGS certification requirements is needed in addition to developing strong internal control systems for organic clusters. Strengthening value chain infrastructure (especially collection centres and storage) could also help. Strict enforcement of regulations related to ensuring integrity of certification process itself may also help in discouraging fake organic produce entering the market (Deshpande Sarma, 2015).

Market development is another domain that has significantly lagged behind in organic programe initiatives. To improve market penetration of organic products measures such as holding mass awareness programs at various institutions (hospitals, hotels, and educational institutions) could help. Facilitating public procurement of organic produce (for example as part of mid-day meals *etc.*) and state retail of organic produce at optimal prices would be useful. To capture markets it is also essential to consolidate fragmented value chains and encourage processing and value addition in addition to establish adequate infrastructure. Support for farmer producer companies and cooperatives could help small farmers is also key (Deshpande Sarma, 2015, IFAD, 2005).

Organic practices and organic farming based on agroecology can help facilitate in sustainable development of the agricultural sector underpinned by restoration of soil health, natural resource preservation and socio-economic development of farming communities and livelihood enhancement. Widespread adoption of organic agriculture is possible if institutional support in relation to policy, production, research and extension, certification, marketing domains strengthened effectively and with adequate financial support. In tandem developing a skilled work force for engaging with organic agriculture will be vital. Here, enhancing the technical capacities of the workforce at NCOF and the regional centers is important. In addition effective liaison and coordination between NCOF and other ministerial departments, state agencies as well as national and international agencies associated with/interested in organic agriculture is needed. Finally prioritizing organic practices and farming in mainstream policies and programmatic initiatives besides sensitizing the public and other stakeholders (even policy makers) of the benefits and the opportunities that it presents will be necessary for successful expansion of organic agriculture (Deshpande Sarma, 2015).

9. Acknowledgement

This paper is based on the learning that emerged from the study "Policy and Institutional Support for Organic Agriculture: Enabling Pathways for Inclusive Sustainable Development" supported by Ministry of Agriculture, GoI. The study was undertaken jointly by TERI and NISTADS.

REFERENCES

Bose S. undated. National Program for Organic Production, Agricultural and Processed Food Products Export Development Authority (APEDA), Ministry of Commerce and Industry, GoI.

Deshpande Sarma S. 2015. Organic Agriculture: An option for fostering sustainable and inclusive agriculture development in India, Discussion paper, June 2015, New Delhi: The Energy and Resources Institute, pp. 28.

DoAC. 2010a. Memorandum for Expenditure Finance Committee (EFC) on National Project on Organic Farming (NPOF), F.No. 3-1/2008-Org. Fmg., 25th February, 2010, Department of Agriculture and Cooperation, GoI. Available at http: // ncof.dacnet.nic.in/Policy_and_EFC/12th_Plan_EFC_Memo.pdf

DoAC. 2010b. Guidelines on National Project on Organic Farming, Department of Agriculture and Cooperation, Ministry of Agriculture, GoI. Available at www. dacnet.nic.in/ncof

DoAC. 2010c. National Food Security Mission Operational Guidelines for Accelerated Pulses Production Programme, Department of Agriculture and Cooperation, Ministry of Agriculture, GoI.

DoAC. 2014. National Program for Organic Production, Department of Commerce, Ministry of Commerce and Industry, GoI.

DoAC. 2014a. National Mission for Sustainable Agriculture (NMSA), Operational Guidelines, Department of Agriculture and Cooperation, Ministry of Agriculture, GoI.

DoAC. 2014b. Mission for Integrated Development of Horticulture, Operational Guidelines, Horticulture Mission, Horticulture Division, Department of Agriculture and Cooperation, Ministry of Agriculture, GoI.

DoAC. 2014c. Rashtriya Krishi Vikas Yojna (RKVY) Operational Guidelines for XII Five Year Plan, Department of Agriculture and Cooperation, Ministry of Agriculture, GoI.

DoRD. 2013. Mahatma Gandhi National Rural Employment Guarantee Act, 2005 (Mahatma Gandhi NREGA) Operational Guidelines, 4th Edition, Department of Rural Development, Ministry of Rural Development, GoI.

DoRD. 2014. Saansad Adarsh Gram Yojna (SAGY) Guidelines, Department of Rural Development, Ministry of Rural Development, GoI.

ICAR-IIFSR. 2014. Annual Report 2013-14, Network Project on Organic Farming, Indian Council of Agricultural Research- Indian Institute of Farming Systems Research, Modipuram, Meerut.

IFAD. 2005. Organic agriculture and poverty reduction in Asia: China and India focus, thematic evaluation, International Fund for Agriculture Development (IFAD) Report No 1664.

MNRE. 2014. Implementation of National Biogas and Manure Management Programme (NBMMP) during 12th Five Year Plan – Administrative Approval – reg, NO. 5-5/2014-BE (NBMMP), Biogas Technology Development Group, Ministry Of New And Renewable Energy, GoI, dated: 30.06. 2014, Available at http: //mnre.gov.in/file-manager/dec-biogas/biogasscheme.pdf

MoA. 2005. Organic Farming Policy, Department of Agriculture and Cooperation, Ministry of Agriculture, GoI. Available at http: //ncof.dacnet.nic.in/Policy_and_EFC/Organic_ Farming _Policy_2005.pdf

MoA and FW. 2014-15. Operational Guidelines for National Mission on Oilseeds and Oil Palm (NMOOP) during XII Plan (Effective from 2014-15), National Mission on Oilseeds and Oil Palm (NMOOP), GoI.

MoRD. 2015a. NRLM- National Rural Livelihoods Mission Programme Implementation Plan, Ministry of Rural Development, GoI

MoRD. 2015b. Mahila Kisan Sashaktikaran Pariyojana (MKSP), MKSP Guidelines, Ministry of Rural Development, GoI.

NABARD. 2011. Capital Investment Subsidy Scheme for Commercial Production Units of Organic Inputs under National Project on Organic Farming, Ref.No.NB. TSD.LD/275 to 303/NPOF-4/2011-12, Circular No. 88/TSD- 1/2011, 09 May 2011, National Bank for Agriculture and Rural Development.

Sood J. 2013. Organic goes online, Down to Earth, Oct 15, 2013. Available at htpp: //www.downtoearth.org.in/content/organic-goes-online

UNCTAD-UNEP. 2008. Best Practices for Organic Policy- What developing countries can do to promote the organic agriculture sector, UNEP-UNCTAD Capacity Building Task Force on Trade Environment and Development, United Nations Conference on Trade and Development and United Nations Environment Program, United Nations.

Index

C

D

E

F

G

H

I

J

K

L

M

N

O

T

U

V

W

www.ingramcontent.com/pod-product-compliance
Ingram Content Group UK Ltd.
Pitfield, Milton Keynes, MK11 3LW, UK
UKHW021528300726
14060UKWH00011B/104

9 789389 569155